Visible Light Communication

Visible Light Communication

Comprehensive Theory and Applications with MATLAB®

Suseela Vappangi
Vakamulla Venkata Mani
Mathini Sellathurai

CRC Press is an imprint of the
Taylor & Francis Group, an **informa** business

First edition published 2022
by CRC Press
6000 Broken Sound Parkway NW, Suite 300, Boca Raton, FL 33487-2742

and by CRC Press
2 Park Square, Milton Park, Abingdon, Oxon, OX14 4RN

CRC Press is an imprint of Taylor & Francis Group, LLC

ISBN: 9780367632175 (hbk)
ISBN: 9781032043401 (pbk)
ISBN: 9781003191537 (ebk)

DOI: 10.1201/9781003191537

Typeset in Nimbus font
by KnowledgeWorks Global Ltd.

Dedication

To the Almighty Lord,
Loving Parents and Family

Contents

Preface

The rapid proliferation in the acquisition of mobile devices like smart phones, palmtops, laptops and a massive number of connected internet of things (IoT) devices leads to an insatiable demand for data access over the wireless networks. During the last 3 decades, tremendous growth in mobile data traffic has been witnessed. Even though the network providers are providing additional services, still there is a requirement to meet this unquenchable demand for data. This explosive growth in mobile data traffic is further fueled by the huge demand of end users access to many multimedia applications like high definition television, live video streaming, streaming audio, video conferences, high-speed internet access and many other such services. Therefore, in order to accomplish the aforementioned services, the quest for high bandwidth is intended to grow at a rapid pace in the near future. Specifically as the global demand for this bandwidth continues to accelerate, it becomes exceedingly difficult for the existing radio frequency (RF) spectrum to accommodate huge amount of services. A day is approaching in the very near future, where the existing RF spectrum emerges as one of the precious resources. This enormous increase in the demand for data manifests to curtail the reliability of services because of huge increase in latency and decrease in network throughput.

Consequently, to alleviate the RF spectrum congestion problem, it is necessitated to explore novel communication-based technologies. One such communication which emerged is optical wireless communication (OWC). When OWC exploits simple, sustainable, cost-effective, solid-state, energy-efficient opto-electronic devices like light emitting diode (LED), it has resulted in the emergence of a novel communication-based technology called visible light communication (VLC). The transmission of data through light is not new to human kind. In the recent times, VLC-based systems are sparking a significant propulsion due to wide-scale deployment of LEDs in a multitude of applications. The most distinguished characteristics of LEDs include: high tolerance to humidity, lower power consumption, longer lighting hours, cooler and smoother operation, and evolution of the future lighting system as LED-based. Additionally, the solid state lighting has revolutionized the indoor illumination where the current incandescent and fluorescent lamps have been replaced by the LEDs at a rapid pace. The most stunning nature of LED is that it facilitates simultaneous ‘illumination’ as well as ‘communication’ with unimaginable implications. This unique trait of LED can be employed to promote a truly green technology.

VLC can be claimed to be an intriguing alternative to RF-based wireless communication because it offers several remarkable benefits like its huge, wide, unlicensed and unregulated bandwidth in the range of 380 nm to 780 nm, corresponding to a 400 THz to 700 THz frequency range. So, this technology offers 1000 times more bandwidth when compared with RF-based wireless communications. This huge availability of spectrum enables VLC to deliver high data rates in the range of several Gigabits per second. More significant works in the literature reveal that ever since the development of VLC systems, they are able to impart high data rates by employing sophisticated multicarrier modulation formats. Hence, it is obvious that the potential of this technology is even greater. Eventually, these characteristics offer VLC the premise to be a part of future 5G technologies. This necessitates to understand as well as to get acquainted with all the research developments which are associated with this emerging area. Having witnessed the developments of VLC in the last decades, we feel that there is a necessity to publish a new textbook which underlines the current state-of-the-art aspects in this field. The existing textbooks in this area are quite a few. So, there is a necessity to publish a new textbook which is of paramount importance to undergraduate, graduate and post-graduate students, research scholars, professional engineers, etc. Subsequently, there is a necessity for this new textbook to cover the theory, applications and simulations in a comprehensive and concise manner. This book is broadly classified into 8 chapters which deal with the history, background, emergence of VLC, modulation formats and multiple access schemes which are compatible with intensity modulation and direct detection (IM/DD) systems, channel modeling for optical domain, applications of VLC for smart cities, amalgamation of VLC with power line communication (PLC), and prominence of VLC in vehicular communication. Finally, we precisely outline the research challenges which are associated with VLC.

The main attraction of this book is the presentation of in-depth review about several current state-of-the-art aspects and MATLAB codes which helps the reader to have a clearcut understanding about the topic in discussion and will enable them to identify several research gaps in their future research activities. As a quick start, Chapter 1 gives a brief review of wireless communication including a thorough explanation about electromagnetic spectrum and then of the background, motivation and emergence of OWC. This chapter also presents the basic VLC system model and gives an elaborate description of the requirements of a VLC system as well as illustrates the differences between RF-based wireless communications and OWC/VLC. Before understanding the pros and cons of VLC, it is imperative to understand the characteristics of light sources and their types. Further, it is very vital to establish a relationship between the radiometric and photometric parameters. Chapter 2 gives a thorough representation of the aforesaid aspects. Furthermore, in order to design any communication system, it is very essential to characterize the channel environment. Moreover, it is to be understood that VLC channel modeling is entirely different than that of wireless channel modeling. Hence, it necessitates to study about VLC channel modeling which includes the line of sight (LOS) propagation model, non-line of sight (NLOS) propagation model and other types of VLC channel models. Chapter 2 presents VLC channel modeling along with MATLAB codes. Further, a review on IEEE 802.15.7 realistic VLC channel models is also the focus of this chapter. With the motive of enhancing high data rate communication as well as to overcome the limited modulation bandwidth of white phosphorescent LEDs, it is vital to exploit suitable sophisticated modulation formats which are compatible with IM/DD systems. Chapter 3 elucidates in detail about the different modulation formats which are reconciling with the requirements of IM/DD systems. In particular, baseband modulation formats like ON-OFF keying (OOK), pulse width modulation (PWM), and pulse position modulation (PPM) were discussed. The non-coherent emission characteristics of LEDs make IM/DD the most viable modulation scheme for VLC. Hence, there is a requisite to explore different multicarrier modulation formats which are in accordance with IM/DD systems.

The important attraction of Chapter 3 is its characterization and comprehensive presentation about different variants of optical orthogonal frequency division multiplexing (OFDM) like DC-biased optical OFDM (DCO-OFDM), asymmetrically clipped optical OFDM (ACO-OFDM), flip OFDM, pulse amplitude modulation discrete multitone modulation (PAM-DMT), layered ACO-OFDM (LACO-OFDM), hybrid ACO-OFDM (HACO-OFDM), reverse polarity optical OFDM (RPO-OFDM), spectrally efficient OFDM (SEE-OFDM), etc. Furthermore, this chapter portrays the exploitation of other real transformation techniques for implementing OFDM modulation and demodulation like discrete Hartley transform (DHT)-based optical OFDM, discrete cosine transform (DCT)-based optical OFDM which is also termed as fast optical OFDM (FOFDM), discrete sine transform-based optical OFDM. This chapter further presents other multicarrier modulation formats like Hadamard coded modulation (HCM) which is based on fast Walsh Hadamard transform (FWHT) and wavelet packet division multiplexing (WPDM) for VLC. Additionally, this chapter gives detailed description of color shift keying (CSK), carrierless amplitude and phase modulation (CAP) along with the research challenges associated with the exploitation of CAP for VLC and multiple-input-multiple-output (MIMO) for VLC.

Chapter 4 illustrates the non-linearities which are associated with the LEDs and gives in detail several peak to average power ratio (PAPR) reduction schemes along with certain MATLAB codes, as well as simulations. In addition, this growing technology called VLC provides a flexibility for the creation of a small scale cellular communication network within an indoor room environment where each installed LED can act as a base station (BS) rendering services to multiple mobile stations/user equipments which are within the vicinity of LEDs. In this regard, there is an urge to investigate the most prominent multiple access schemes like orthogonal frequency division multiple access (OFDMA), single carrier frequency division multiple access (SC-FDMA) and non-orthogonal frequency division multiple access (NOMA) satisfying the requirements of IM/DD systems. Chapter 5 highlights different multiple access schemes and provides an extensive review of different multiple access schemes like optical orthogonal frequency division multiple access (OOFDMA), optical code division multiple access (OCDMA), optical space division multiple access (OSDMA), as well as compares with non-orthogonal multiple access scheme (NOMA). This chapter mainly emphasizes the current state-of-the-art aspects that are associated with NOMA-VLC systems and details the challenges that need to be addressed while exploiting NOMA with VLC. Additionally, this chapter depicts how this technology can be interfaced with smart city. A clear description of smart lighting systems as well the significance of VLC for enabling a city to evolve as a smart city is the major contribution of this chapter. Chapter 6 presents the integration of VLC with power line communication (PLC). It states the research efforts done so far regarding the interface of VLC with PLC. This chapter gives a thorough review of PLC-VLC cascaded channel modeling, design of multicarrier and multiple access schemes-based PLC-VLC systems, applications of PLC-VLC systems, etc. Chapter 7 is dedicated to the application part, i.e., the role of VLC in enhancing road safety. In line with the advancements of VLC, intelligent transportation sector (ITS) has decided to exploit VLC or the purpose of enabling communication between moving vehicles on the road to exchange real time information pertaining to high traffic areas, congested areas, distance between moving vehicles, etc. The major contribution of this chapter is to emphasize the major challenges which arise when VLC is employed for enabling vehicular communication. Chapter 8 presents the research challenges of VLC which include exploitation of organic light emitting diodes (OLEDs) for VLC, synchronization aspects, usage of robust multicarrier modulation formats for VLC, and challenges associated in the field of indoor positioning.

For MATLAB product information, please contact:

The MathWorks, Inc.
3 Apple Hill Drive
Natick, MA 01760-2098 USA
Tel: 508-647-7000
Fax: 508-647-7001
E-mail: info@mathworks.com
Web: https: //www.mathworks.com

Authors

Suseela Vappangi received a Bachelor of Technology degree in Electronics and Communication Engineering and a Master of Technology in the stream of Communications and Signal Processing in 2013 and 2015 from Jawaharlal Nehru Technological University, India. She also earned a PhD degree from National Institute of Technology Warangal, India in 2019. Currently, she is working at Vellore Institute of Technology Amaravati University (VIT-AP) as Assistant Professor Senior Grade 1 in the School of Electronics Engineering (SENSE). She is an IEEE member. Her research interests include signal processing for communications, communications with major emphasis on modulation and broadband wireless communications, and visible light communication. She has published 11 scholarly articles in the area of visible light communication.

Vakamulla Venkata Mani received undergraduate (Engg.) and post-graduate (Engg.) degrees from Andhra University, Andhra Pradesh, India in 1992 and 2003, respectively, and a PhD degree from Indian Institute of Technology (IIT) Delhi, India in 2009. She is currently working as Associate Professor in the Department of Electronics and Communication Engineering at National Institute of Technology Warangal, India. She is an IEEE senior member. Her areas of interest include wireless communications, signal processing for communications, coding for communications, networking, localization positioning and visible light communication. She has supervised 8 PhD students and has published over 30 scholary articles in diverse areas of wireless communications and visible light communication. She is actively involved in different projects at IIT Delhi. While she was pursuing her PhD, she worked on her dissertation, "Smart Antenna Design for UWB Communications." She spent one month at Rice University in 2014. During her visit she worked on WARP (wireless open access research platform) for prototyping wireless technologies. Currently, she is guiding research scholars who are working in diverse areas such as: UWB beamforming, MIMO-OFDM, spatial modulation for 5G, prototyping of GFDM for 5G and several aspects such as the modulation formats that are compatible with intensity modulated and direct detected systems, peak to average power ratio (PAPR), non-linearity aspects of LEDs, and several other aspects of visible light communication. She completed one project funded by Anurag, DRDO Hyderabad on "O-QPSK Modulator and Demodulator Design and Its Performance Analysis for IEEE802.15.4 Personal Area

Networks." She is also involved in ongoing DST and SPARC projects titled, "Visible Light Communication Testbed for Indoor Applications" and "VLC-Based Vehicular Communication for Enhancing Road Safety in Smart Cities." Much of the work with her research group was already done in wireless and visible light communication where real-time implementation of generalized frequency division multiplexing as well as test-beds for VLC was done using hardware like universal software radio peripheral (USRP) platform.

Prof. Mathini Sellathurai is currently Professor of Signal Processing with the Heriot-Watt University, Edinburgh, U.K. and leading researcher in signal processing and artificial intelligence techniques in a range of applications including radar, RF networks, MIMO wireless communications (including 4G and 5G), physical layer secrecy, and satellite communications. She has been active in the area of signal processing research for the past 20 years and has a strong international track record in MIMO signal processing with applications in radar and wireless communications research. Her present research includes passive radar topography, localisation, massive-MIMO, non-orthogonal multiple access (NOMA) and waveform designs, assistive care technologies, IoT, hearing aid, advanced, coded-modulation designs based on auto-encoders, channel prediction and imaging using mm-wave and UAV and physical layer secrecy designs. She has 5 years of industrial research experience. She held positions with Bell Laboratories, New Jersey, USA, as a Visiting Researcher (2000); and with the Canadian (Government) Communications Research Centre, Ottawa, Canada as a Senior Research Scientist (2001-2004). Since 2004, she has been working in academia. She also holds an honorary Adjunct/Associate Professorship at McMaster University, Ontario, Canada and an Associate Editorship for the *IEEE Transactions on Signal Processing* from 2009–2018. She is an IEEE senior member and has published 200 IEEE entries, given invited talks, and has written a book and several book chapters on topics related to this project. She is a Fellow of the Higher Education Academy. She was also member of the IEEE SPCOM Technical Strategy Committee from 2014–2018, and editor for the *IEEE Transactions on Signal Processing* from 2009–2014 and from 2015–2018. She received the IEEE Communication Society Fred W. Ellersick Best Paper Award in 2005; Industry Canada Public Service Awards for contributions in science and technology in 2005; and a Best PhD Thesis medal from NSERC, Canada in 2002. She was also the General Co-Chair of IEEE SPAWC2016 in Edinburgh. Prof. Sellathurai is one of the Foreign Investigators for the SPARC project titled, "VLC-Based Vehicular Communication for Enhancing Road Safety in Smart Cities." She was an organizer for the IEEE International Workshop on Cognitive Wireless Systems, IIT Delhi, India in 2009, 2010 and 2013, and the IEEE WCNC 2013 - WORKSHOP - New Advances for Physical Layer Network Coding in 2013, Shanghai, China. She was a Technical Program Committee member for the IEEE International Conference of Communications from 2004–2012 and was a National Publicity Champion for ICC 2013, Budapest. She is also a peer review college member of the Engineering and Physical Sciences Research Council, UK.

Acknowledgments

The authors would like to extend their sincere thanks to the Ministry of Human Resource Development (MHRD)-Government of India for granting the project under the Scheme for Promotion of Academic and Research Collaboration (SPARC) with grant number (2019/249), and the grant Full Duplex and Cognitive Radio Architectures for Spectrally Efficient Communications (UGC -UKIERI 2016-17-058), for supporting our work.

1 Introduction to OWC-VLC

1.1 CURRENT STATE-OF-THE-ART

The explosive magnification of the wireless communication industry over the last decade revolutionized change in our daily life. The present day lifestyle has become very flexible and almost everyone residing on the globe relishes the advances in wireless services which include live video streaming, web browsing, online gaming, and many other such services. Additionally, the expeditious escalation of smartphone applications and overwhelming ultimatum for efficient and convenient wireless communication paved the way for the origination of a huge amount of business opportunities for both mobile manufacturers and wireless network operators. In this scenario, it becomes very difficult for the network providers to capture the needs of the end users. This is valid because during the past few decades, the exceptional augmentation in traffic carried out by the telecommunication networks has been endorsed around the globe [436]. Consequently, this enormous demand for several high-speed internet-oriented services—including high definition television (HDTV), streaming audio, video, video conferences, video calls, cloud-based computing, machine-to-machine (M2M), and augmented and virtual reality (AR/VR) — has reinforced the need to formulate innovative research problem statements, as well as ushered the emergence of novel communication-based technologies which are proficient in imparting high data rate communication to the end subscribers.

Hence, in this sort of scenario where huge amounts of data are required, the radio frequency (RF) spectrum becomes overcrowded [231]. According to the predictions of CISCO, the traffic that is fostered because of AR/VR was expected to increase twenty-fold between 2016 and 2021. Furthermore, the overall intensification of networked devices was anticipated to reach 27.1 billion by 2021. Also, these devices are expected to generate an overall annual global internet traffic of 3.3 zettabytes. Explicitly by the year 2021, the total traffic induced by several wireless and mobile devices was expected to account for an increase of 73% of the total internet traffic. The drastic increase in the growth of mobile data traffic during 2016 and 2021 is delineated in Fig. 1.1, which clearly shows that in 2021, the mobile data traffic is envisaged to increase exponentially by 49 exabytes per month. In addition, even in the coming years, it is predicted that mobile data traffic will increase at an alarming rate. As a result, because of continuous increased depletion of the RF spectrum, it will be one of the most inadequate resources in the near future. This is, in turn, referred to as a "Spectrum Crunch" which is one of the most vital aspects needed to be urgently addressed. It is because this issue is more pronounced, it manifests deterioration of the reliability of services by enforcing a huge increase in latency and decrease in network throughput.

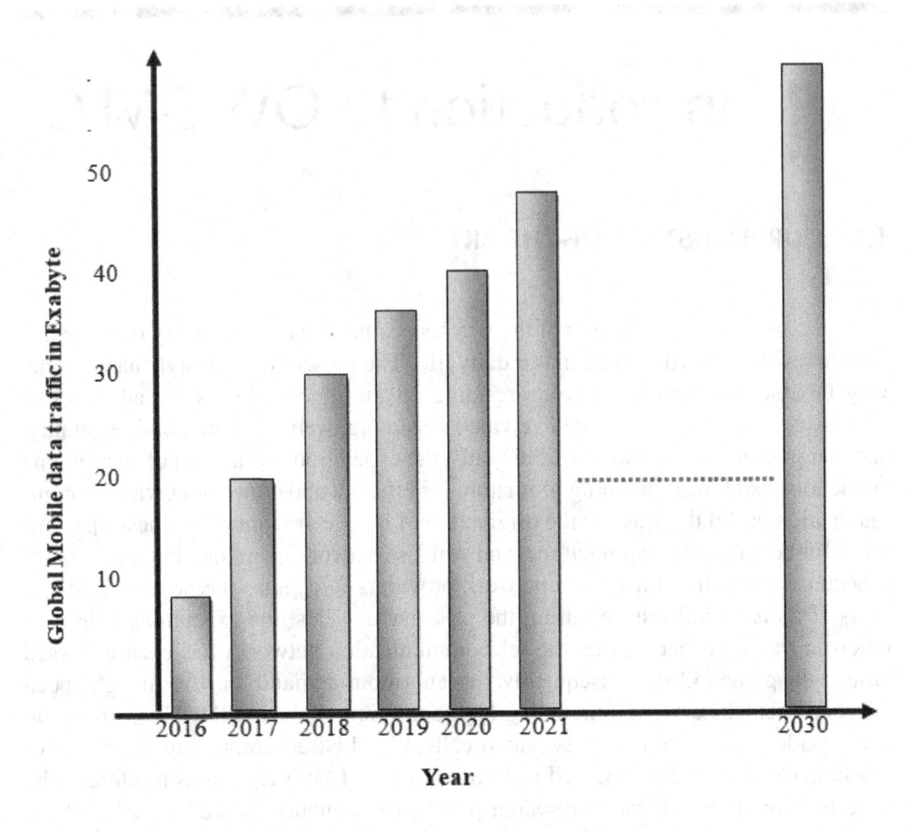

Figure 1.1: Global mobile data traffic between 2016–2030 in the range of exabytes as predicted by CISCO.

So, to alleviate this RF spectrum scarcity problem, it is of paramount importance to explore novel communication-based technologies. One such communication which can be exploited is optical wireless communication (OWC), the most alternative form of communication that utilizes infrared (IR) and visible and ultraviolet (UV) subbands as the propagation medium. Compared with RF-based wireless communication, OWC provides diverse and remarkable advantages that include its huge bandwidth in the order of several terahertz (THz) frequency, which is not being subjected to any sort of electromagnetic interference, as well as is very cost-effective, since it does not impose any licensing charges for the offered bandwidth, and ensures a huge amount of security.

1.2 HISTORY OF THE ORIGIN OF OWC

The most startling peculiarity of OWC is that it can be claimed to be any form of telecommunication that utilizes light as a transmission medium of communication. Maneuvering OWC for rendering communication goes back to the old days

Figure 1.2: Exploitation of smoke signals for communication support.

when smoke signals and beacon fires were employed as shown in Fig. 1.2. It is reported in the literature that semaphore lines, which can also be termed as installation that is employed for transmitting optical signals, can be confirmed to be the preliminary form of technological application of OWC. In 1792, the fundamental optical telegraph network was built by the French engineer, Claude Chappe. Principally, this work witnessed the transmission of 196 information symbols by utilizing the semaphore towers. Heliograph is another example of OWC, a simple and effective instrument used during the late 19th and early 20th centuries. Particularly, this instrument is meant for enabling instantaneous optical communication over long distances. Generally, this device, which is called wireless solar telegraph, is meant for signalling flashes of sunlight by means of pivoting a mirror or interrupting the beam with a shutter. Later, in 1836, the invention of Morse Code facilitated navy ships to communicate with onshore lighthouses by means of a signal lamp. In [58], it is clearly evident that Alexander Graham Bell exemplified the first implementation of a free space optical (FSO) link in the form of the photophone, which is basically a telecommunication device that allows for the transmission of a speech signal upon a beam of light. The illustration of a photophone is depicted in Fig. 1.3. The underlying phenomenon involved in the operation of a photophone is that by employing a vibrating mirror at the transmitting end and crystalline selenium cells at the focal point of a parabolic receiver, a voice message signal is modulated onto a light signal. From early research as stated in [177], it can be portrayed that the advancements of OWC technology gained significant momentum only after the tremendous efforts instigated by Gfeller and Bapst in 1979. This research work illustrates the caliber of OWC for high-capacity in-house networks where the system has the potential to impart 260 Mbps data rates by employing a simple ON-OFF keying (OOK) modulation

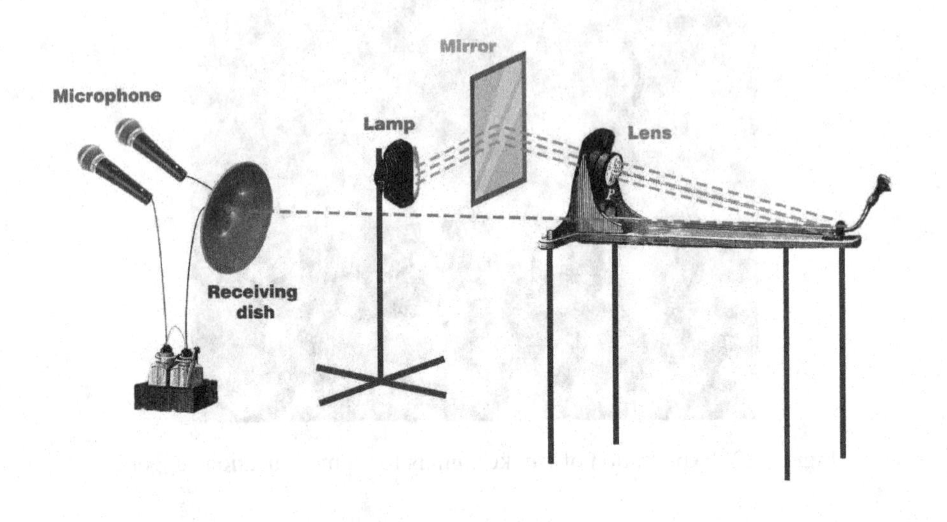

Figure 1.3: Visual of photophone.

technique at a center wavelength of 950 nm in the IR spectrum. This is further fueled by the rapid amelioration in semiconductor-based lighting sources which facilitated the evolvement and success of OWC.

One classification of OWC is FSO, which is intended to support long-distance coverage, and, in fact, can be professed that the fortune of OWC has been changed in the 1960s with the invention of several optical sources, more importantly, the laser. Though the study of laser-based OWC is beyond the scope of this book, it is worthwhile to have an understanding of its usage in OWC. In general, an FSO link is established by the exploitation of highly directional laser diodes at the transmitting end, while photodiode (PD) is employed at the receiving side. The outbreak of FSO demonstrations were witnessed during the early 1960s and late 1970s. The authors in [333] demonstrated the transmission of 50 Mbps data at low bit error rates (BERs) by exploiting diffused infrared radiation. This work employs OOK as the modulation format and employs decision feedback equalization (DFE) to counteract the effects of multipath induced intersymbol interference. In the later works as witnessed in [84], the authors demonstrated the implementation of faster OOK IR systems which achieved data rates of 70 Mbps. The research work in the aforementioned reference [84], portrays the design and analysis of the performance of a prototype angle diversity infrared communication system. In this system, the achievable BER is 10^{-9} over a 4 m range. The other branch of OWC, which is directed toward the successful implementation of indoor mobile wireless networks, is through the exploitation of solid-state opto-electronic illuminating devices like light emitting diodes (LEDs) as optical transmitters. The approach of using fast switching speed LEDs to facilitate data support while fulfilling the illumination aspects can be substantiated in [394] in 1999. In fact, this sort of OWC technology which employs LEDs at the transmitting end and PDs at the receiving end can be termed as visible light communication (VLC). The usage of white LEDs for the purpose of conveying data transfer in addition to the foremost task of facilitating illumination in an indoor environment was illustrated in the renowned work by Tanaka et al. [468]. This work signifies that the usage of high power lighting equipment makes it possible to achieve communication support in the entire room; moreover, it is very easy to install such LED-based illumination system which is aesthetic to the human eye.

In general, Tanaka et al. presented several works illustrating the exploitation of LEDs to render the dual functionality of 'illumination' and 'communication' in an indoor home network [271,276,277,470]. The work in [470] illustrates that the indoor VLC system imparting data rates of 400 Mbps was realized by exploiting fundamental modulation format like OOK. Further, as stated in [277] the fundamental analysis of a VLC system using white LED lights is analyzed by taking into consideration the requirements of optical transmission and optical lighting. This work clearly illustrates that similar to any wireless-based system, the performance of a VLC system is degraded due to the influence of reflection and intersymbol interference (ISI).

1.2.1 ADVANTAGES AND APPLICATIONS OF OWC

In recent times, the tremendous interest shown by modern society towards the wireless communication technologies has resulted in the explosive increase in demand for more robust, reliable and cost-effective wireless communication-based services. Addressing this alarming demand for wireless communication-based services exemplifies as the most challenging aspect in the future. At the earlier stages, these wireless communication devices were capable of providing only voice and some primitive data services. Later, with the passing of time, they are having enough potential to offer sophisticated services which allow for the successful realization of high-speed data networks that facilitate high-speed internet browsing, online gaming, live data and video streaming, etc. Despite offering these services, still there is a need to assure for high data throughputs and low latency services. This is further fueled by the rapid proliferation of smart mobile devices and several other things which are on the internet of things (IoT). Consequently, the tremendous breakthrough in the mobile data traffic at an exponential rate has resulted in depleting the RF spectrum. Since, the RF spectrum is a limited and valuable source, definitely this emphasizes the necessity to depend on new and complementary wireless communication-based technologies to relieve spectrum overcrowding and congestion. Thus, it can be pronounced that OWC is one such propitious and alternative means of wireless communication which provides a significant number of advantages over the RF-based wireless transmission. Specifically, VLC has gained remarkable popularity in recent times to combat the looming RF Spectrum Crunch. VLC is drastically striving forward for rendering high-speed data transmission with rapid deployment of LEDs [340]. This sort of communication is gaining significant momentum because the deployment of VLC access points (APs) is a straightforward approach, as the existing lighting infrastructural units can be reused. Moreover, in terms of energy saving and reducing the carbon footprint, it is vital to replace the current incandescent and fluorescent lamps with much simpler, sustainable and cost-effective devices like LEDs. The energy to be spent for communication purposes can be saved by piggybacking of data on illumination. Several benefits offered by VLC are detailed in the subsequent sections.

OWC can be applied to a diverse number of applications and a few can be proclaimed in several significant studies as stated in [66, 70, 102, 137, 162, 179, 214, 480, 489]. From Fig. 1.4, it can be shown that OWC can be applied to diverse areas like industry, healthcare, transportation sectors, public places like railway-stations, hospitals, auditoriums, homes, workplaces like office environments, schools, multinational corporate units, shopping complexes, museums, underwater and free spaces, etc. According to the desired application in hand, be it device to device (D2D), machine to machine (M2M), device/user to machine or vice-versa, vehicle to

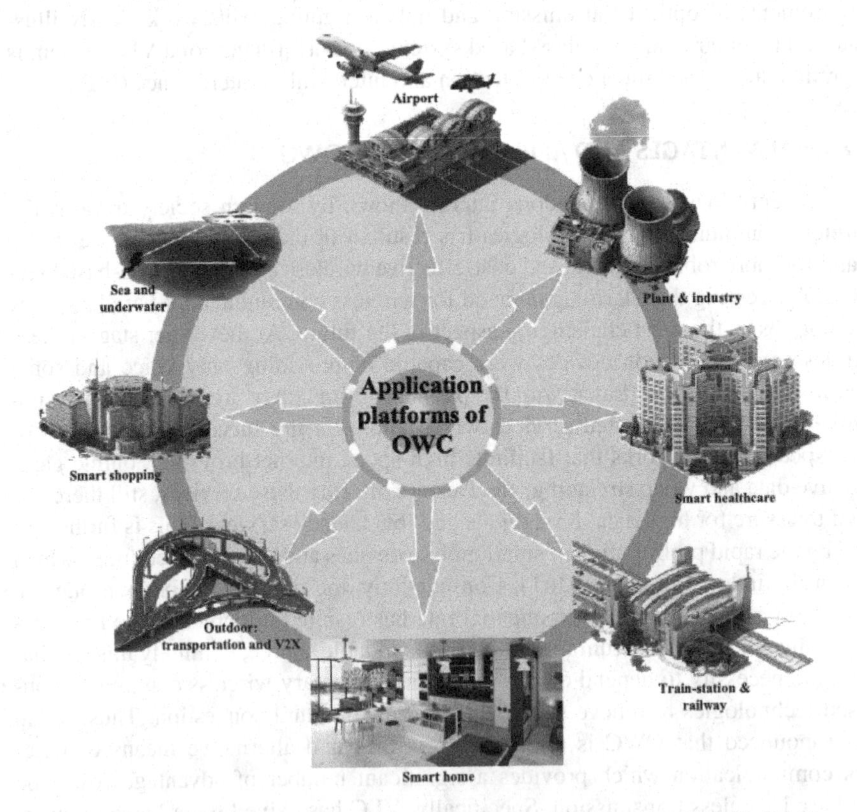

Figure 1.4: Applications of optical wireless communication (OWC) [115].

infrastructure (V2I), infrastructure to vehicle (I2V), vehicle to vehicle (V2V), point to point, point to multipoint, etc. can all be based on OWC technologies. However, based upon the requisites of the application like the desired data speed and platform, several OWC technologies can be exploited.

1.3 FREE SPACE OPTICAL WIRELESS COMMUNICATION

The other type of OWC is free space optical wireless communications (FSOWC). In particular, terrestrial point-to-point FSOWC operates at IR, visible and UV sub-bands. More precisely, FSOWC can be described as a form of optical communication technology that employs light as a propagating medium to enable wireless transmission of data for telecommunications and computer networking. In this sort of technology, there is no requirement of optical fiber cable where the optical beams are sent through free space. In several studies it is reported that a FSOWC system comprises of an optical transceiver at both ends in order to assure for a full duplex communication. In a FSOWC system, the generated signal waveforms are modulated by means of an optical carrier. Then, for this generated optical field to reach the desired destination, it is radiated through the atmosphere. Then at the receiving end, this optical field, which is optically collected, is passed through a photodetector for the purpose of conversion of optical field into suitable electrical current. At the receiving side, a sufficient amount of signal processing techniques are imposed to process this electrical current signal and to recover the original transmitted signal.

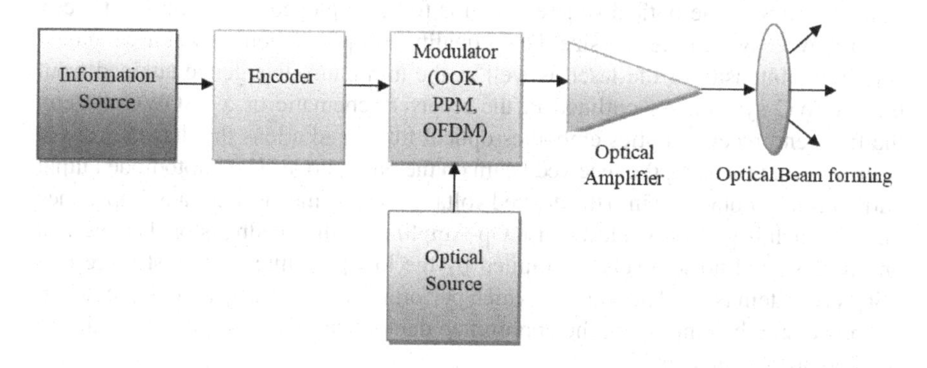

Figure 1.5: The transmitter schematic of FSOWC system.

In general, a FSOWC system looks like the block diagram as delineated in Figs. 1.5 and 1.6. As evident from the transmitter schematic, at the preliminary stage, the incoming data stream which is to be transmitted over a long distance is encoded; the type of modulation format which can be employed is OOK, pulse modulation (PM) or higher multidimensional and multicarrier modulation format like orthogonal frequency division multiplexing (OFDM). As Fig. 1.5, an optical amplifier can be exploited for the purpose of increasing the power intensity of the modulated laser beam. Prior to allowing for the propagation of the beam of light,

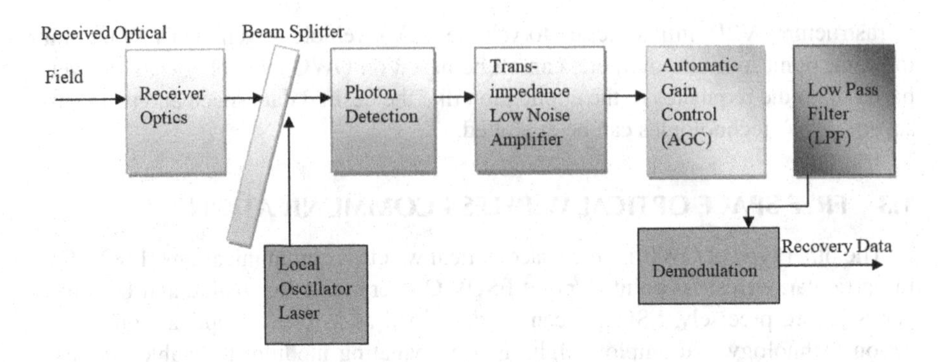

Figure 1.6: The receiver schematic of FSOWC system.

beam forming optics are laid to ensure the collection of the beam of light as well as to refocus it. The exemplary optical source which is used in FSOWC is a laser diode [248]. However, a few manufacturers also employ high-power LEDs with collimators as optical sources in a FSOWC system [4]. Immense care should be taken regarding the choice of exploitation of optical source in a FSOWC system because it should deliver an adequate amount of high optical power over a wide temperature range. Additionally, it should have a long mean time between failures (MTBF), and the corresponding components should have low power consumption. The salient characteristics of the optical source which is to be employed at the transmitter end in a FSOWC system are the size, beam quality and power consumption which ascertain the intensity of the laser as well as the minimum divergence attained from the FSOWC system. As outlined in the receiver schematic of a FSOWC system, the front-end receiver optics embodies optical filters and a lens that has the role of collecting and focusing the received beam on the photodiode. The photodiode output current is then converted into the desired voltage level by means of a trans-impedance circuit which is generally a low-noise Op-Amplifier with a loadresistor. The thermal and background noise levels are limited by the low-pass filter. The last stage in a FSOWC system is the demodulator which performs the necessary demodulation signal processing by employing the appropriate demodulation techniques to obtain the original processing data.

The following are the characteristics possessed by a FSOWC system:

1. Explicitly, as reported in [427], FSOWC is implicitly designated as fiber-free or fiberless photonics, which indicates that the modulated light pulses are transmitted through free space by means of laser beam technology for the purpose of achieving broadband communications. This FSOWC communication is not new and goes back to the early 8th century. Predominantly, a FSOWC system is a line of sight (LOS) communication-oriented technology which involves the transmission of voice, data and video with a maximum data rate of 2.5 Gbps by affirming a bidirectional connectivity.

2. Particularly, free space optics is a flexible network that ensures speeds better than that of the broadband.
3. In particular, a FSOWC system is very easy to install with very low investments, as it takes less than 30 minutes to install at normal locations.
4. This type of system is straightforward to deploy because there is no necessity to worry about the spectrum license charges.
5. A high amount of security can be ensured, as it comprises of a LOS communication path.
6. Since data transmission is done through free space, i.e., air, therefore, the transmission is having the speed of light.
7. It is free from the electromagnetic and radio frequency interference.
8. A FSO system should be capable enough to operate at high power levels for longer distances as well as have the competency to operate over wide temperature ranges.
9. The MTBF must be more than 10 years.
10. It is expedient to rely on high-speed modulation formats for the successful realization of a high-speed FSO system.
11. The low power usage which is associated with the transmission per bit stems out as an important virtue of a FSO system.
12. There is comparably higher bandwidth.

1.3.1 ADVANTAGES AND APPLICATIONS OF FSOWC

Predominantly, FSOWC is generally employed to render high data rate communication between two fixed points spanning over several kilometers. Upon comparison with RF wireless links, FSOWC links offer a huge optical bandwidth and consequently provide a scope for ensuring high data rate communication. A FSOWC system can be realized by using narrow beam lasers where this spatial confinement endeavors a high amount of security, high reuse factor and inherent robustness to electromagnetic interference. Together with this, the amount of bandwidth occupied by this sort of communication is above 300 GHz which is generally categorized as unregulated globally. Therefore, it can be claimed that the systems developed using this sort of technology like FSOWC do not charge any license fee. Thus, it can be anticipated that FSOWC systems can be easily deployed reducing the burden of installation costs of dedicated fiber optics connections. From the literature, it can be evidenced that the penetration of FSOWC into the market is progressed by guaranteeing transmissions rates of 10 Gbps over long distances. The work in [478] illustrates that over a communication distance of 20 m, a 40 Gbps FSOWC link was implemented. The research efforts in [181, 259, 345, 399, 567] clearly state that the data rates achieved by FSOWC systems can compete with the ones achieved by fiber optics. The ease of deployment of FSOWC systems is appealing for a wide range of applications. As portrayed in Fig. 1.7, FSOWC can be employed for facilitating nm-range optical interconnection, satellite-satellite/earth connectivity, seamless connection between vehicles, i.e., car-to-car connectivity, underwater communication,

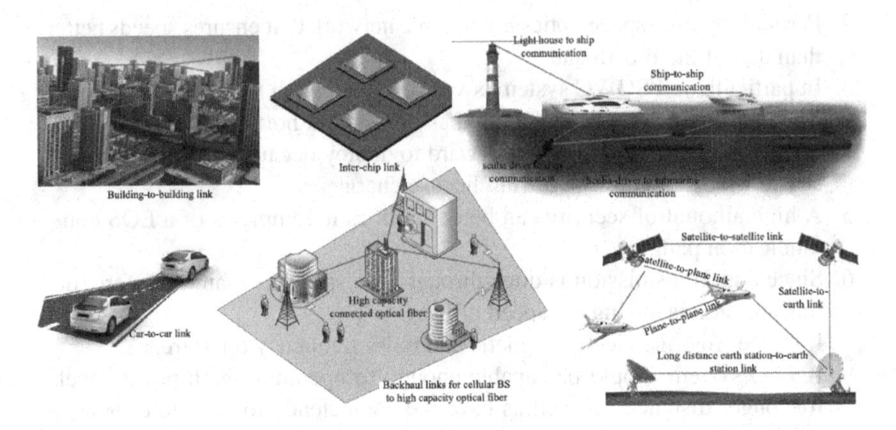

Figure 1.7: Employment of FSOWC in a diverse number of applications [115].

ship-ship/ship-infrastructure communication, cellular backhaul connectivity, etc. The following are some application areas of FSOWC:

- **Connectedness:** FSOWC plays a major role in bridge connectivity among several infrastructural units like campus buildings in schools, multinational corporate offices, etc. and ensures ultra-high-speed data transmission without the cost of dedicated fiber optic cables.
- **Video Vigilance/Supervision:** Particularly in recent times, the installation of surveillance cameras for the purpose of security issues is quite common in several sectors like commercial, law enforcement, public safety and military applications. However, the available wireless communication technologies fail to replenish high throughput for seamless video streaming. Thus, in this regard, FSOWC acts as reinforcement to impart high quality video streaming.
- **Backhaul for Cellular Networks:** The explosive growth of bandwidth-intensive mobile services and the rapid proliferation of a diverse number of devices which are connected over the internet of things (IoT) necessitates the deployment of FSOWC to ensure higher reliable services in terms of reduced latency and high throughputs.
- **Catastrophe Response Rehabilitation:** It is important to gather quick acknowledgment from natural disaster prone areas, emergency areas like terrorist attack areas, etc. Therefore, in such situations where there is a rapid breakdown of infrastructural units, FSOWC links can be easily established within hours.
- **Enhancement of Security:** With the availability of quantum cryptography system, unconditional security can be ensured. Particularly, quantum cryptography systems work reciprocally with fiber optic infrastructure. Hence, in areas where the deployment of fiber optic cables is expensive, FSOWC can be used to work in conjunction with quantum cryptography.

- **Point to Point/Point to Multipoint Connectivity:** FSOWC can be used to build point to point connectivity between two buildings, ship to ship, vehicle to vehicle, satellite to ground, aircraft to ground, point to multipoint links, etc. In this manner communication can be imparted.
- **Broadcasting:** Pertaining to live telecasting of information from sports and live television streaming from remote areas or war zones, signals from a camera should be first conveyed to a broadcasting vehicle that is further connected to a central office via satellite uplink. Therefore, in the process of establishment of connection between a vehicle and a broadcasting unit, FSOWC can be employed to impart high quality data transmission. Moreover, FSOWC can be used to accomplish higher throughputs which are satisfying the requirements of today's high definition television (HDTV).

1.3.2 DRAWBACKS ASSOCIATED WITH FSO

Despite being superior across a wide variety of applications as mentioned above, FSOWC fails to ensure link reliability more specifically in case of long-distance communication because of its high level of susceptibility to several factors, the most prominent being atmospheric effects. The influence of it is apparent from Fig. 1.8 where air is used as a medium of transmission; hence, the influence of the environment is inevitable. A number of such impediments are listed below:

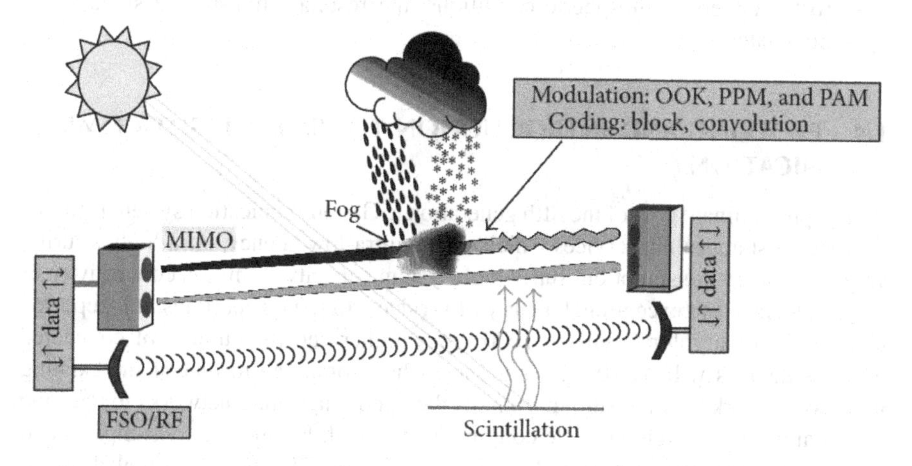

Figure 1.8: Illustration of atmospheric effects on FSOWC system [330].

- **Obstacles Arising due to Different Phenomena:** It is a well-known fact that channel is primarily responsible for deciding the performance of any communication system. In case of a wireless channel, there are several impediments present which include: tall trees, skyscrapers and flying birds can obstruct the line of sight (LOS) path thus hindering the communication in case of FSOWC.

- **Scintillation:** The adverse activities carried out by people on earth lead to the emanation of heat from the earth. Hence, as a result, such drastic variation of temperature has a direct impact on the signal which is received in the FSOWC system. Due to unexpected amplitude variations, the received signal is a distorted version of the transmitted one and it is popularly known as 'image dancing'.
- **Influence of Geometric Loss:** Geometric loss is ably called 'optical beam attenuation.' Generally, this arises due to the spreading of the beam which in turn results in reduction in the power levels of the received signal when the signal is navigated from the transmitting end to the receiving end.
- **Absorption:** In the terrestrial atmosphere, the water molecules are suspended and imbibe the photon power which successively decreases the power density of the optical beam. Therefore, it can be asserted that the availability of transmission is directly affected by the absorption.
- **Prevalence of Atmospheric Turbulence:** Due to atmospheric turbulence, the fluctuations in the density of air arise and lead to a change in refractive index of air. Therefore, the disturbances in the weather and environment impose an effect on the optical beam where the beam is spread and also scintillation is unavoidable.
- **Atmospheric Attenuation:** Heavy rain, fog and haze are the primary weather conditions which have a direct impact on the FSOWC link. Due to the adverse atmospheric conditions, the resultant transmitted signal is attenuated.

1.4 EVOLVING AND INTRODUCTION TO VISIBLE LIGHT COMMUNICATION

The predominant goal of the fifth generation (5G) communication system is to furnish the most appealing services which ensure ultra-low latency, ultra-high security, very low energy consumption, super-high system capacity, seamless connectivity of several devices and exceptional quality of services. In the recent times, it is reported that mobile data volume will be 1000 times higher and the number of connected wireless devices will be 100 times greater when compared to that of the existing wireless networks. As a result, this entails the upcoming future networks to have the potential to impart high data rate communication with low-power consumption and to guarantee minimal or negligible end-to-end delays. This is further fueled by the rapid evolution of IoT era, in which a distinct number of physical devices are connected to the internet. Even though the RF spectrum is extensively employed across various wireless applications, it turns out to become destitute to meet the augmenting demand for 5G wireless capacity as well as to serve the IoT paradigm. In the consequence of this ever-increasing thirst for mobile users all over the globe enforces the wireless communication based on RF will face a severe spectrum shortage problem; moreover, the entire world has witnessed this dramatic increase in the traffic carried out by telecommunication networks. In spite of the adoption of several wireless technologies, i.e., 3*G*, 4*G*, 5*G* and beyond for proliferating the capacity of wireless radio systems, it turns out to rely on alternative communications.

Consequently, when OWC made use of simple, sustainable, cost-effective, energy-efficient and opto-electronic illuminating devices like light emitting diodes (LEDs), it has led to the emergence of novel communication oriented technology like visible light communication (VLC). The exploitation of LEDs to impart simultaneous 'illumination' and 'communication' dates back to the early 2000s where researchers in Keio University in Japan envisaged the usage of white LEDs in homes for developing an access network. This gained a rapid pace driven by the extensive progress in research in Japan, where researchers framed a high-speed communication network through the aid of visible light. Furthermore, the significant research efforts to enable VLC support to many hand-held devices as well as to transport vehicles led to the formation of a Visible Light Communication Consortium (VLCC) in Japan in November 2003. After the formation of VLCC, two standards called Visible Light Communication System Standard and Visible Light ID System Standard were professed. With the advent of time, Japan Electronics and Information Technology Industries Association (JEITA) approved the aforesaid standards as JEITA CP-1221 and JEITA CP-1222. Later, in 2009, the infrared communication physical layer which was propounded by the international Infrared Data Association (IrDA) was assimilated into VLCC. Besides the development of the above-mentioned standards, the hOME Gigabit Access project (OMEGA), which was patronized by the European Union, also strived to adapt the usage of OWC in rendering both 'illumination' and 'communication' support in a manner that will evolve as an intriguing complementary technology to RF. In order to ease the standardization of VLC, Visible Light Communications Associations (VLCA) were confirmed to be a descendant of VLCC in Japan in 2014. With an effort to formulate the standardization of VLC, the IEEE standard for VLC was proposed in 2011 as IEEE 802.15.7, which gives a brief summary regarding the overview of the design specifications pertaining to the physical and link layer aspects.

Furthermore, IEEE 802.15.7 formed a Task Group to revise the current IEEE 802.15.7-2011 standard and announced the development of IEEE 802.15.7 r1 [238]. The technical specification of IEEE 802.15.7 r1 standard provides the optical camera communication (OCC), i.e., VLC exploiting a camera and LED-identification system. Until today, the research efforts are striving forward to revise this IEEE 802.15.7 standard. The IEEE 802.15.7-2018, which is said to be a revised version of IEEE 802.15.7-2011, elucidates physical and medium access control (MAC) sublayer for short range OWC in optically transparent media which is exploiting light wavelengths ranging from 10000 to 190 nm [6]. Moreover, this standard exhibits its potential to impart data rates that are sufficient enough to reinforce both audio and video multimedia services and broadens the scope to include more OWC technologies. This standard takes into consideration several aspects like the mobility of the optical link, the compatibility with several other illumination infrastructures, the impairments due to noise, as well as the interference arising from other sources of ambient light and artificial light sources. It even defines a MAC sublayer that furnishes the exclusive needs of the visible links as well as other targeted light wavelengths. The most vital aspect of this standard is that it complies with eye safety regulations. Additionally, the applicability of OWC in optical wireless personal area networks (OWPANS) is also outlined in this standard and addresses several topics like the usage of OWC in network topologies addressing collision avoidance, acknowledgment, performance quality indication, visibility and dimming support, etc.

When compared with RF-based wireless communication, VLC makes use of higher frequency bands in the visible light spectrum as its transmission medium. The most renowned trait of VLC is that it imparts 'illumination' and 'communication' simultaneously by utilizing a simple device like LED. Because of its numerous advantages, which include its wide, unregulated and license-free bandwidth, VLC is able to relinquish high data transmission rate. Thus, it has quickly become a hot research topic and has sparked a significant amount of attraction in several distinguished and diverse areas of academics, industrial units, research communities, etc. The most fascinating feature of VLC is that it is strongly resistant to electromagnetic interference. Besides, VLC exhibits the potential to deliver high data rate communication by using general solid-state lighting sources like LEDs, and in this sort of communication, data is transmitted by intensity modulating LEDs at a rapid speed which is much faster than the perseverance of the human eye. Additionally, VLC has gained a renewed interest due to the widespread deployment of LEDs. Also, the indoor illumination has been revolutionized by the solid-state lighting devices where the current incandescent and fluorescent lamps have been replaced by more compact and energy-efficient lighting sources like LEDs. On top of this, the recent advancements in LED technology, such as its fast nanosecond switching speeds, has drawn the attention of industrial manufacturers by allowing their interface in many illumination appliances like smart phones, televisions, dashboards, sign boards, advertising panels, street lighting units, traffic lights, vehicular infrastructure, etc.

In the current day scenario, every illumination appliance is becoming LED-based due to their eminent merits of LEDs such as their longer operational lifetime, i.e., lighting hours, high tolerance to humidity, high amount of brightness, smoother operation, smaller size, energy-efficient nature, sustainability, cost-effectiveness and easier maintenance, among others. Taking into consideration these reasons, studies in [400] illustrate that LED-based lighting is expected to have a market share of 84% by 2030. Thus, by making use of a simple device like LED, it is possible to achieve both 'illumination' and 'wireless data communication'. This prominent virtue has enabled VLC to become an intriguing alternative to existing RF-based wireless communication. Despite the existence of several variants of RF-based wireless communication like millimeter wave (mm), infrared, and ultraviolet communication, there are certain drawbacks which are associated with them like having a high pathloss, relying on expensive transmitter and receiver components, etc. Thus, it is worthwhile to mention that VLC-based transmission systems draw their advantage due to the maturity of the transmission technology where comparatively economical and high-performance components are used. Consequently, VLC is emerging as a competitive technology when compared with RF-based wireless communications.

Moreover, communication by using visible light has gained stupendous prevalence in recent times because of its multitude of advantages, which are described as follows:

1. **Shrinking RF Spectrum and Capacity Crunch:**
From the earlier studies in the literature, it can be witnessed that RF spectrum is considered as the dedicated resource of the state. In fact, the RF spectrum can be contemplated as a valuable natural resource much similar to water, land, gas and minerals. Hence, its utilization is carefully regulated for the purpose of removal of signal interference, pollution, as well as to ensure efficient use of spectrum [426]. Consequently, spectrum management turns out to be one of the vital aspects for promoting efficient usage and gaining a net social benefit. The studies in [47] reveal that for the purpose of enhancing capacity, it is much easier for the mobile operators to acquire the spectrum rather than rely on burdensome activities like developing more base stations, which involves a huge amount of installation costs. The continuous inflation of mobile data during the past decade has enabled mobile data operators to concentrate on several wireless technologies as well as to deploy wireless fidelity (Wi-Fi) hotspots. The statics as depicted in [174,525] clearly emphasize the dramatic growth in the wireless network usage.
To alleviate this dwindling RF spectrum, extensive research has been done to tap into the terahertz (THz) frequency range of the spectrum between RF and microwave. However, this would result in the creation of an exclusive class of infrastructure which is in accordance with the wavelength band. Contrarily, visible light offers $\sim 10,000$ times greater bandwidth than that of the RF which lies in the range of approximately ~ 3 KHz to ~ 300 GHz. Hence, it guarantees a huge, wide, unlicensed and unregulated bandwidth in the range of several THz corresponding to a wavelength of 380 to 780 nm. The electromagnetic spectrum as delineated in Fig. 1.9 highlights the visible band of frequencies. Eventually, this technology is envisioned to coexist with the existing RF counterparts as well as have enough potential to meet many bandwidth-hungry applications like streaming audio, video

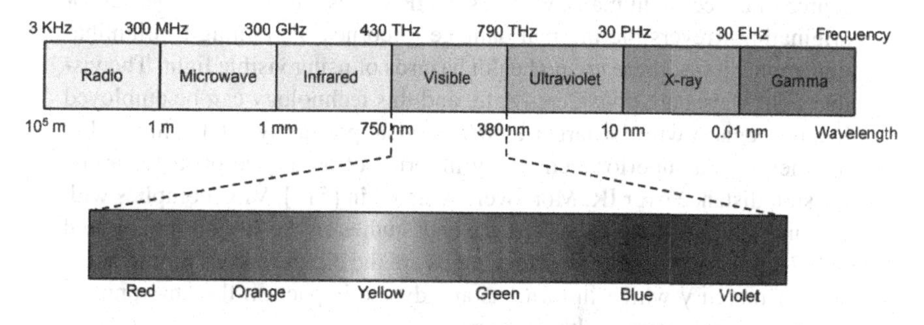

Figure 1.9: Electromagnetic spectrum [398].

and many multimedia applications. Moreover, because of the presence of a huge amount of spectrum, this technology ensures high data rates up to a few tens of Gbps. Furthermore, high data rates can be attained by relying on diversity techniques like multiple input multiple output (MIMO). It is evident from the literature that high data rates are accomplished by using VLC in less than a decade since the development of VLC systems. This kind of striking characteristic feature offers VLC the opportunity to be part of future 5*G* technologies.

2. **Interference-Free Technology:**
   VLC is considered to be fundamentally free from interference with RF signals. Moreover, the remarkable trait of this technology is that it does not have any effect on the fundamental mechanism of the highly sensitive electronics systems. Hence, this technology can be easily interfaced with RF-restrained places like airplanes, hospitals, museums, chemical plants, general street lighting systems, nuclear plants, etc.
3. **Assurance of Security:**
   The important difference between RF-based wireless communication and VLC stems out from the implicit fact of exploitation of electromagnetic waves in the scenario of RF; whereas visible light is used as carrier for the data signals. The radio waves are competent enough to penetrate any non-metallic material, while visible light signals cannot penetrate walls, specifically, they can pervade through transparent materials. Nonetheless, this limited penetration capability has certain drawbacks like limited mobility and coverage area, but the major benefit is that it limits the interferences invading from non-line of sight (NLOS) systems and hence, a high degree of security can be ensured against eavesdropping. Therefore, in contrast to the RF signals which are much susceptible to signal snooping, a VLC system persuades a well-defined coverage zone with enhanced security and prevents the intruder from hacking the data.
4. **Affirmation of Safety:**
   One of the most important features of VLC is that, unlike other wireless communication technologies, it is considered to be intrinsically safe for human health. Upon comparison, radio waves are classified to be a potential source of cancer in humans, whereas the IR waves are direct root causes for originating irreversible thermal damage to cornea. Pertaining to illumination applications, there are no health hazards of using visible light. The visible light waves are considered safe, and this technology can be employed in any scenario with far larger emitted optical power, by establishing VLC to emerge as a superior technology in terms of assurance of larger transmission distance over IR. Moreover, as stated in [515], VLC complies with eye and skin safety regulations. Several studies in the literature as stated in [67,152] reveal that fluorescent lamps are the direct sources for the emission of mercury which in turn has an adverse impact on the environment thereby raising many health concerns.

5. **Energy-Efficient Nature:**
With the primary motive to reduce the emission of greenhouse gases, significant efforts have been made by researchers to obtain impressive results by replacing the existing infrastructure with solid state, cost-effective, energy-efficient, sustainable, illuminating devices like LEDs. The studies in [260] emphasize that LEDs are considered to be highly controllable light sources and a part of green communication technology [406, 484]. Additionally, if LEDs replace all the existing conventional light sources, then the global electricity consumption, carbon dioxide emissions, and crude oil consumption reduce significantly by 50%, 10 gigatons and 962 million barrels, respectively. Substantially, this made VLC evolve as an eco-friendly, green communication technology because there is no requirement for extra energy to impart data transmission [484]. In this scenario, the primary advantage is that the same light can be used to promote illumination as well as act as a carrier signal for the transmission of data signals. These reductions will account for financial savings in excess of one trillion dollars, as well as encounter for total energy savings of 1.9×10^{20} joules over a decade [266]. In the recent report generated from the United States Department of Energy [51], solid state lighting (SSL) will assuredly penetrate the market and by 2025, there is a certainty of saving 217 terawatt-hours of energy.
6. **Co-existence with Existing Infrastructural Units:**
Besides the above benefits, another asset of VLC is its ubiquitous peculiarity. In addition to illumination, VLC guarantees a data transmission capability by making use of the rapid switching speeds of the LEDs at a rate that is not within the perseverance of the human eye. By incorporation of relatively simple and cheap front-end components which are operating in the base-band, VLC can be conveniently interfaced with the existing lighting framework [245]. The synergistic relationship with the indoor energy-efficient lighting units enabled the rapid deployment of VLC transmitters which paved the way for the creation of short-range indoor VLC. Since, VLC is purely based on existing lighting infrastructural units, it has the caliber to deliver high data rate wireless communication wherever there is a possibility for the existence of artificial lighting sources in both indoors as well as outdoors. Thus by leveraging the existing lighting units, an indoor link can be established to render high data rate communication for fast internet connections [382].
7. **Potential in the Field of Vehicular Communication:**
Even in the area of vehicular communication, VLC exhibits its prominence because LEDs are already interfaced with the headlights and brake lights of cars. So, definitely this technology can be used to enable the vehicle drivers travelling on roads to exchange real time information pertaining to traffic updates which include, most congested traffic areas, distance between moving vehicles, etc. The usage of VLC technology is delineated in Fig. 1.10 where information regarding traffic safety is collected from the

Figure 1.10: Use of VLC in vehicular communication [332].

geographical area and is broadcasted toward nearby approaching vehicles. This information includes the distance between two moving vehicles, the velocity of the heavy-load carrying vehicles, the distance between the vehicles and the road-intersection points, speedlimits, etc. In particular, this data is disseminated among the neighboring vehicles by making use of their headlights and brake lights which are fitted with LEDs. As evident from the earlier studies, it can be proclaimed that the adoption of VLC in automotive applications was a breakthrough [25, 394, 395]. As evident from Fig. 1.11,

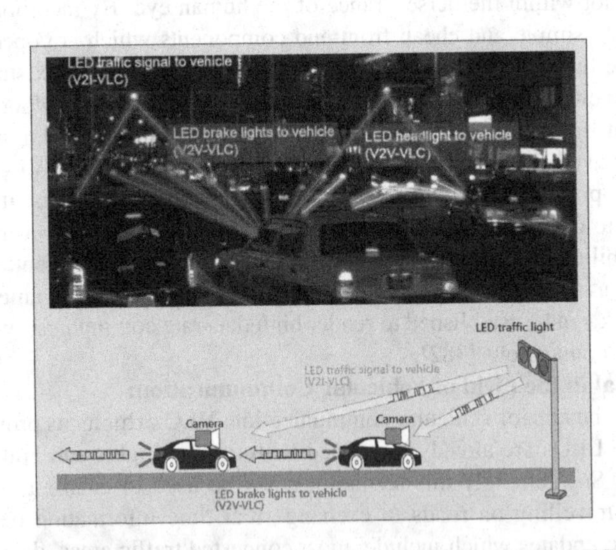

Figure 1.11: Schematic representation of vehicle to infrastructure (V2I) VLC by using the LED-based traffic light, and vehicle to vehicle (V2V) VLC by using the LED headlights and brake lights [542].

a typical representation of a V2I-VLC and V2V-VLC network has been established by making use of LED traffic light, LED headlight and brake lights. In particular, the underlying concept involved behind V2I-VLC is that there is a wide scope for the seamless exchange of critical safety and operational data between moving vehicles and roadway infrastructure, and this concept is familiar to that of the V2I using radio wave technology, except the transmission involves the usage of light waves instead of radio wave technology. By creating a networked environment between vehicles and infrastructure, V2I-VLC facilitates safe driving by adaptive traffic signal control, intersection movement assistance, speed management, and so on. This is also true for V2V-VLC, which involves wireless exchanges of data between moving vehicles travelling in the same area. Potential applications of V2V systems include emergency brake light warnings, forward collision warnings, and control loss warnings.

8. **Cost-effectiveness:**
   VLC offers an unregulated spectrum which subsequently reduces the cost of this technology. Moreover, by installing LEDs and by using the photodiode-based VLC receiver this provides the flexibility for the creation of a small-scale cellular communication network where each installed LED can act as a base station (BS) rendering services to several mobile stations or roaming subscribers which are within the vicinity of the installed lighting fixture. Thus, VLC reduces the burden of high installation costs of RF-based BSs. In contrast to other wireless communication-based technologies, the implementation cost of VLC is considerably less since only a few upgrades of lighting infrastructure are required comparative to the huge amount of installation set-up cost of an entire communication system as desired in RF-based wireless communication. The omnipresence of LEDs have led to their interface with almost every road side unit present in the city; thus VLC even plays a vital role in enhancing the tourism opportunities in several cities.
9. **Amalgamation of VLC with Smart Power Grid:**
   Earlier studies in the literature state that power line communication (PLC) has been used in different applications like automation, control and monitoring. Previously, they were not designed for the purpose of communication. PLC is like any other communication technology whereby a sender modulates the data to be sent, injects it onto a medium, and the receiver de-modulates the data to read it. The major difference is that PLC does not need extra cabling, it reuses existing wiring. Considering the pervasiveness of power lines, this means that with PLC, virtually all line-powered devices can be controlled or monitored. There seems to be an instinctive advantage by leveraging the existing power-line framework for the purpose of ensuring connectivity among the devices while exploiting energy-efficient and sustainable LEDs for wireless downlink. PLC turns out to be an attractive solution as a backbone network for an indoor VLC system, much like the way the Wi-Fi supports the broadband ethernet connections [325].

1.5 BASIC ARCHITECTURE OF VLC SYSTEM MODEL

This section gives an elaborate description of the basic system model of VLC with a thorough summary on VLC transmitter and receiver block diagrams. Particularly, a VLC transceiver is realized by using cost-effective, solid-state, energy-efficient and sustainable illuminating devices like LEDs at the transmitting end, where the data at the VLC transmitter side is modulated by the intensity of an optical source like LED. A VLC receiver comprises of a photosensitive element, usually a positive intrinsic negative (PIN) photodiode or an image sensor which is primarily used for fetching out the data signal from the modulated beam of light. Even though the VLC transmitter and receiver are physically separated, they are connected by means of a VLC channel. The underlying prerequisite for VLC system is that the transceiver depends on LOS conditions. The detailed schematic of VLC transceiver is elucidated in Figs. 1.12 and 1.13. As depicted from the figure, a typical VLC system is realized by employing opto-electronic devices like LEDs and photodiodes. Moreover, an important fact to be considered is that, unlike RF-based wireless communication where data signal is modulated by using an RF-based carrier, the modulation of data in case of VLC is quite different. In contrast to modulation of amplitude, frequency and phase of the data signal in RF-based wireless communication, the data signal is modulated by using the intensity of the light waves in the case of VLC. Therefore, this type of modulation is referred to as intensity modulation (IM), and at the receiving terminal, a simple photodiode is employed and the type of detection is referred to as direct detection (DD). Thus, it can be inferred that IM/DD is the most feasible modulation technique for VLC.

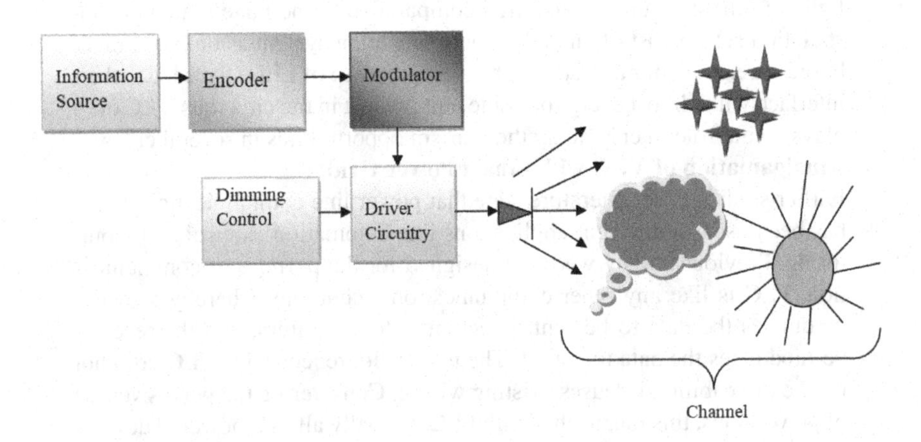

Figure 1.12: Schematic representation of VLC transmitter.

As represented in the transmitter and receiver schematic, the incoming huge stream of data is first encoded in order to allow for the digital representation of the incoming data stream, and then this stream of data is fed into the modulation block. Particularly, the type of modulation techniques to be incorporated must

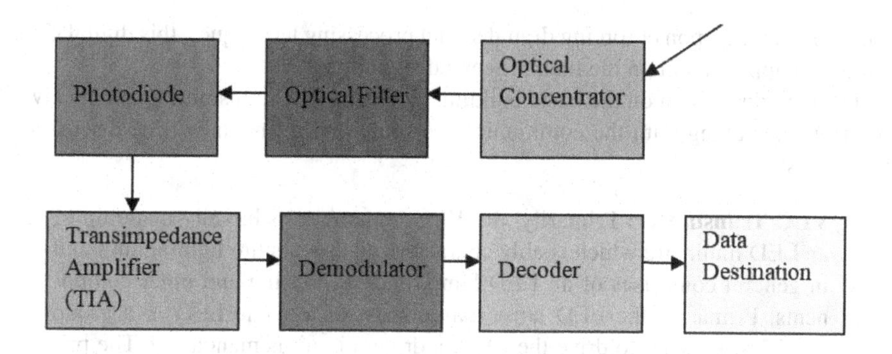

Figure 1.13: Schematic representation of VLC receiver.

comply with the requirements of IM/DD systems. Since light is used for data modulation and as light signal cannot be negative or a complex-valued signal, this enforces a constraint on the transmitted signal where it should be only real and positive-valued (i.e., unipolar). In addition to assurance of high data rate transfer, a VLC system should also provide adequate dimming levels. In case of incandescent and fluorescent lamps, accurate dimming emerges to be a quite difficult task, whereas pertaining to LEDs it is absolutely easy to precisely control the dimming levels because the response time of LEDs during on and off switch operation is very small, most probably in the range of a few tens of nanoseconds. Thereupon, by modulating the driver current at a relatively high frequency, it is conceivable to switch the LEDs on and off at a rapid pace without the change being perceived by the human eye. Therefore, in order to facilitate the LED to be used for communication, the primary task of the driver circuitry is to combine the mapped data signal and the dimming control signal in a manner to drive the LED. Based upon the needs of the application at hand, for example in case of office environments, residential places, etc, lower levels of illumination are required and for public places, stadiums, auditoriums, etc. higher levels of illumination are mandatory.

Hence, for the purpose of saving energy, driver circuitry plays a major role by arbitrarily dimming the light source. Finally, this modulated data signal is intensity modulated through the LED. Thereupon, this transmitted intensity modulated signal is propagated through the optical wireless VLC channel. Further, in order to shape this transmitted signal, this signal is passed through the optical amplifier lens, a collimator, or a diffuser in order to broaden this beam. This intensity modulated signal invades the receiving end upon undergoing several reflections, refractions and diffractions. Hence, in addition to LOS signals, there are NLOS signal components. Therefore, the role of the optical filter is to filter out the unwanted signal components and to accept the desired signal. Moreover, it greatly reduces the effect of noise which is arising due to artificial light sources as well as sunlight. After filtering out the unwanted signal components, the intensity modulated VLC signal is passed through the photodiode in order to convert this light intensity signal into an electrical signal. This electrical signal is further preamplified by means of a transimpedance amplifier (TIA). Upon enforcing digital signal processing techniques, this digital signal is demapped to obtain the final processed signal.

Detailed discussion on the basic building blocks of VLC transmitting and receiving frontends along with the communication-setup arrangements will be discussed as follows:

- **VLC Transmitter:** Primarily, the VLC transmitter is broadly made up of an LED luminaire which is ably characterized as an entire lighting unit and in general comprises of an LED lamp, ballast, housing and other components. Primarily, the LED lamp may consist of a single LED or a group of LEDs. In order to drive the LED, a driver circuit is mandatory. The primary role of the driving/driver circuitry is to control the amount of current flowing through the LED, i.e., it is responsible for adjusting the brightness levels of the LEDs. Moreover, in addition to imparting illumination, whenever, an LED luminaire is meant to be used for rendering communication support as well, and then the driver circuitry is modified in accordance to modulate the data signal by using the intensity of the LED. This concept can be better illustrated by employing a simple example: if a simple modulation format is considered where information symbol 1 and 0 has to be transmitted, then these symbols of information can be transmitted by fixing different intensities of light, i.e., 1 can be assigned with higher intensity of light and 0 with lower intensity of light. An important prerequisite for the VLC transmitter is to provide the dual functionality of illumination and data transfer. Consequently, this ascertains that the wireless data transmission of information should not have any adverse effects on the illumination as well as signalling operations of LEDs. Therefore, it is certain that illumination which is a foremost task of LED not be hindered because of communication purposes. Thus, the performance of VLC system is directly dependent on the choice of the LED luminaire to be employed. Furthermore, it is vital to consider the important aspects that VLC transmitters should not introduce any amount of flickering which can be perceived by the human eye as well as in the prospect to save a considerable amount of energy, it is imperative for the light source to be arbitrarily dimmed as per the specifications of the application at hand. In general, it is a known fact that white light is a universally accepted source of light which is quite commonly employed in both indoor and outdoor applications. Primarily, SSL industry produces the white light by adopting two approaches:
    1. **Manipulating Blue LED with Phosphor:** In this method, white light is generated by utilizing blue LED which is coated with yellowish phosphor. Generally, white light is obtained upon allowing blue light to traverse through the yellowish phosphor coating. Thus, this combination yields white light. Particularly, upon modifying the thickness of the yellowish phosphor layer, different variations of white light can be attained. This sort of LED is also called white phosphorescent LED (WLED).

2. **Merging Red, Green and Blue (RGB) LEDs:** It is apparent that upon mixing three primary colors like red, green and blue, white light can be attained. Thus, by using the amalgamation of three RGB LEDs, white light can be obtained. However, since this sort of approach requires three separate LEDs, the cost of the overall LED luminaire increases when compared to the first approach, i.e., blue LED with yellowish phosphor coating. This sort of realization of white light gives to a special modulation format called color shift keying (CSK) where the data is modulated by using three color intensities.

Upon viewing the above two methodologies of generating white light, using the first approach where white light is acquired by using the blue LED with yellowish phosphor coating is the most widely used approach because of its lower cost and ease of implementation. Nonetheless, in the perspective of rendering communication support, this sort of WLED limits high-speed communication because the slow transient response of yellowish phosphor coating limits the fast switching speeds of LEDs to a few MHz. However, to alleviate this drawback, i.e., to circumvent the limited modulation bandwidth of LEDs and to achieve high data rate communication, extensive research efforts brought forth the proposal of several techniques to render high data rate communication such as:

1. In order to filter out the slow response yellowish phosphor components, it is vital to employ blue filter at the receiving end.
2. Employing pre-equalization circuits at the LED driving module and post-equalization circuits at the receiving end.
3. Extensive improvements in data rates can be achieved by using blue filters to filter out the slow transient response of yellowish phosphor components, pre-equalization and post-equalization circuits can be employed at the transmitting and receiving end.
4. Lastly, in order to accomplish high data rate communication, it is of paramount importance to rely on sophisticated multicarrier modulation techniques like discrete multitone modulation (DMT).

- **VLC Receiver:** In general, there exist two types of VLC receivers to receive the transmitted intensity modulated signal:
  1. **Photodetector-Based VLC Receiver:** As delineated in the block diagram, a typical VLC system can be realized by employing LED at the transmitting end and a photodiode/photodetector at the receiving end. Principally, it is a well-known fact that photodetector is a semiconductor device which is generally used for converting light signal into electrical signal. This type of receiver comprises photosensitive element with a higher bandwidth which ensures the prospect of high data rate communication. Besides, ensuring high-speed communication, the incident parasitic light which is emanating from both artificial and natural light sources eventually leads

to the occurrence of interferences, which in turn results in degradation of the overall system performance. Consequently, in order to eradicate the effects of noise arising due to artificial light sources as well as background solar radiation, it is vital to use optical filters at the receiving end. These optical filters are mainly employed to reject the unwanted spectrum components and to pass the desired signal of interest. Moreover, in order to render high data rate communication, the choice of employing appropriate optical filters for filtering out the slow transient response of yellowish phosphor coating has been proven to be an effective solution. The receiver's field of view (FOV) is a vital aspect which needs to be taken into consideration. The intensity modulated VLC signal can be intercepted by the photodiode only if that signal is within its FOV. The rest of the signal which falls outside the FOV cannot be accepted. Therefore, in order to eradicate the effects of noise, the best way is to narrow the receiver's FOV.

2. **Imaging Sensor or Camera Sensor-Based VLC receiver:** The rapid availability of sophisticated equipment in a number of devices like smart phones, tablets, palmtops, laptops and automobiles enables the easy implementation of modern wireless communication-based technologies. Most of these devices are already fitted with cameras for the purpose of capturing images and videos, so image sensors can also be employed in VLC applications for enabling reception of intensity modulated VLC signals. Primarily, an image sensor or a camera sensor comprises of a huge number of photodetectors which are arranged either as an integrated circuit or as an array of matrix. Principally, such types of devices impart certain limitations when they are employed for VLC applications. For the sake of facilitating high-resolution photography, the numbers of photodiodes are to be increased, but this significantly reduces the number of frames per second (fps) which can be captured by the camera sensor or image sensor. This in turn impairs high data rate communication. Generally, the number of fps which are to be used in low-cost cameras embedded in smart phones should not exceed 50 fps; this implies that such devices achieve low data rates. Therefore, it can be affirmed that the usage of camera sensors for VLC applications cannot ensure high data rate transfer. However, the data rates can be enhanced to a few Kbps by exploiting the 'rolling shutter' property of a camera sensor. Due to the availability of a huge number of photodiodes, it is highly impossible to read the output of each pixel in parallel. So, to combat this effect, most of the modern camera sensors adopt row scanning procedures where the output of the photodectors of one row is read at a time. Accordingly, this phenomenon of reading the output of the photodetector either row-by-row or column-by-

column is referred to as 'rolling shutter'. Despite incorporating the 'rolling shutter' procedure, the results analysis confirms the fact that only static low data rate applications can be accomplished. Eventually, in order to attain higher data rates in the order of several Mbps as well as to achieve long-distance communication, it is desirable to employ high-speed cameras which have a high number of fps, most probably in the range of 1000. The remarkable benefit of image sensor or camera sensor is its enhanced FOV which allows for the wide area data reception. Thus, by taking into consideration all these aspects, camera sensor-based VLC receivers can be potential candidates suitable for mobile long distance communications.

A conclusion can be drawn by giving the differences between the photodiode-based VLC receiver and image sensor-based VLC receiver. Considering the scenario of photodiode-based VLC receivers, where they process the background noise along with the received data signal, the image sensor-based VLC receiver has the capability to spatially isolate the noise sources. The second major difference stems out from the fact that a photodiode-based VLC receiver allows for data recovery by using several digital signal processing techniques, whereas image sensor-based VLC receiver incorporates complex image processing techniques. In most of the cases, the usage of such complex image processing algorithms does not allow for the efficient data decoding, thus making it rely on expensive data processing unit.

1.6 SIGNIFICANT CHALLENGING ASPECTS OF VLC

This section elaborates the following most important challenging aspects which need to be addressed while dealing with the visible light spectrum:

1. **Flickering:** The intrinsic meaning of 'flicker' is that it represents the unwanted fluctuations in the brightness of light. Momentous efforts were carried out by researchers to handle this vital aspect. Because several studies as proclaimed in [61] clearly emphasize that flickering induces serious health hazards like adverse/catastrophic physiological changes in humans when it is perceived by the human eye. Consequently, there is an urge to mitigate the amount of flicker which is resulting from modulating the light sources when they are meant to be used for communication purposes. Therefore, it is of utmost importance to design a suitable modulation format to combat the detrimental aspect of flickering. It should be noted that flickers can be avoided when the brightness of the light falls within the maximum flickering time period (MFTP) where MFTP is delineated as the maximum time period over which the light intensity can change without the human eye perceiving it. Even though there is no universally tolerable flicker frequency number, [61] states that a frequency greater than 200 Hz is contemplated

to be safe [417]. Hence, this implies that the modulation format in VLC should not introduce any levels of flicker during the transmission of data frame and simultaneously should ensure a high data rate transmission.

2. **Dimming:** Predominantly, a divergent number of applications require different levels of luminance to perform certain types of activities. Certain day-time applications require lesser levels of brightness when compared with night-time activities. Lesser illumination of 30 to 100 lux is desired to carry simple visual tasks in public places. Meanwhile, an exorbitant level of illuminance is mandatory to perform activities in offices, stadiums, museums, auditoriums, residential places, etc. Moreover, in order to save energy, it is necessary to arbitrarily dim a light source. With the rapid advancements in LED driver circuitry, based upon the needs of the application at hand, the LED lighting source is dimmed with the perspective to save energy. Hence, dimming turns out to be a more crucial task to save power as well as to maintain energy efficiency. Even while arbitrarily dimming the light source, it is worthwhile to fulfill the requirement of communication. While dimming the LED to a certain arbitrary level, it is crucial to understand its impact on the human perceived light. Generally, it is a proved fact that for lower illumination levels, the human eye enlarges the pupil to allow for more light to enter into the eye. Consequently, such kind of response results in a difference between the measured light and perceived light. Therefore, the amount of perceived light from the measured light can be calculated as follows:

$$Perceived\ Light\ (\%) = 100 \times \sqrt{\frac{Measured\ Light\ (\%)}{100}} \tag{1.1}$$

From Fig. 1.14, it can be inferred that for a lamp that is dimmed to 10% of its measured light, the human eye perceives it as if it is being dimmed to only 32%. This seems to be indispensable in terms of VLC because in the perspective of saving the energy, based upon the needs of the application, the user may arbitrarily choose the dimming levels, but immense care should be taken that the communication not be affected at any cost. Therefore, a conclusion can be drawn that data has to be modulated in such a manner that any desired level of dimming is supported.

1.7 IEEE 802.15.7 PHYSICAL LAYER SUMMARY

This section highlights the IEEE 802.15.7 physical layer aspects of VLC. Particularly, this standard endeavors three physical layer types for VLC:

1. PHY I
2. PHY II
3. PHY III

Going into the details about the description of each physical layer as per the IEEE 802.15.7 standard:

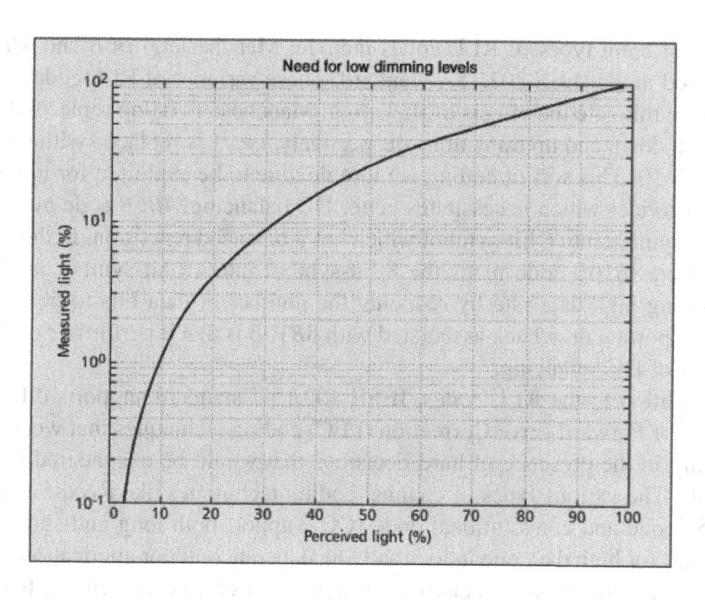

Figure 1.14: Due to the contraction/enlargement of a pupil, the light perceived by the human eye is different [357, 417].

1. PHY I ensures data rates ranging from 11.67 to 266.67 Kbps, while PHY II guarantees data rates from 1.25 to 96 Mbps, and lastly PHY III operates from 12 to 96 Mbps.
2. PHY I and II are explicated for a single type of light source and utilize a modulation format like OOK and variable pulse position modulation (VPPM) technique. Whereas the PHY III employs numerous light sources and uses a special format of modulation source like CSK where the data is modulated by using a different range of frequencies.
3. All the above physical layer techniques along with the modulation schemes which they support are organized in a manner such that they coexist with each other to impart dimming support for the purpose of saving energy as well as mitigating the flickering problem.
4. These physical layer techniques even contain the mechanisms for modulating the light sources in an appropriate manner that sufficient amount of flickering is combated. They employ run length limited (RLL) line coding techniques and channel coding techniques for forward error correction (FCC).
5. In general, the long runs of 1s or 0s reduce the rate of change of light intensity and hence lead to the emanation of flickering as well as potentially lead to clock and data recovery (CDR) problems. RLL codes are mainly used to ensure that the output symbols are having an equal proportion of 1s and 0s. Therefore, the usage of RLL codes mitigates the long runs of 1s and 0s and guarantees DC balance with equal 1s and 0s at each symbol output.

The different types of RLL codes include: Manchester, $4B6B$ and $8B10B$ defined in the IEEE 802.15.7 standard. These varieties of RLL codes work well to mitigate the effects of flickering. Manchester coding replaces 1 and 0 with down and up transitions, respectively, i.e., 1 is replaced with 01 and 0 with 10. This sort of coding is more flexible to be exploited for low data rate services which necessitates better DC balancing. $4B6B$ code maps the 4 bit-symbol into 6 bit-symbol which has a balanced repetition. In the same manner, $8B10B$ code maps the 8 bit-symbol into 10 bit-symbol. Besides ensuring high data rate by reducing the number of data bits to be added, the important drawback associated with $8B10B$ is that it performs poorly in terms of DC balancing.

6. In addition to the RLL codes, IEEE 802.15.7 standard supports different types of Forward Error Correction (FEC) coding techniques that work prudently in the presence of hard decisions that would be engendered by the CDR. The various types of channel coding techniques like Reed-Solomon (RS) code and convolutional code (CC) support both long and short data frames for high data rate indoor and low data rate outdoor applications. Primarily for the outdoor scenarios, stronger codes like concatenated RS and CC codes are widely employed to subdue the path loss problems which generally prevail in longer distances, as well as the interference problems which generally occur due to the existence of artificial lighting sources and ambient lighting sources like sunlight and fluorescent lighting. While, for indoor applications there is a huge amount of requirements for high data rates, RS codes are widely used as FEC coding. In particular, RS codes work in good agreement with RLL codes.
7. The work in [417] says that one of the important parameter determining the performance of each physical layer is "optical clock rate". Predominantly each physical layer modulation mode has an associated optical clock rate which is "divided down" by the various coding schemes to fetch the final data rates as delineated in Tables 1.1, 1.2 and 1.3.
   A lower optical clock rate of $\leq$400 KHz is chosen for PHY I because this layer is designed to be suitable for outdoor applications. Moreover, the second reason for the consideration of such a lower optical clock rate for PHY I is that LEDs can be more appropriate to be used in traffic lights where they switch slowly because of the requirement of high currents driving the LEDs. While a much higher optical clock rate of $\leq$120 MHz is exclusively dedicated for PHY II because it is designed to be applicable for high data rate indoor applications especially in mobile and portable devices. PHY III has an optical clock rate of 24 MHz, which is the maximum clock rate reinforced by the current infrastructure white LEDs.

Table 1.1
Operational specifications of PHY I and achievable data rates [417].

Modulation	RLL Code	Optical Clock Rate	FEC		Data Rate
			Outer Code (RS)	Inner Code (CC)	
OOK	Manchester	200 KHz	(15,7)	1/4	11.67 Kbps
			(15,11)	1/3	24.44 Kbps
			(15,11)	2/3	48.89 Kbps
			(15,11)	None	73.3 Kbps
			None	None	100 Kbps
VPPM	4B6B	400 KHz	(15,2)	None	35.56 Kbps
			(15,4)	None	71.11 Kbps
			(15,7)	None	124.4 Kbps
			None	None	266.6 Kbps

Table 1.2
Operational specifications of PHY II and achievable data rates [417].

Modulation	RLL Code	Optical Clock Rate	FEC	Data Rate
VPPM	4B6B	3.75 MHz	RS (64,32)	1.25 Mbps
			RS (160,128)	2 Mbps
		7.5 MHz	RS (64,32)	2.5 Mbps
			RS (160,128)	4 Mbps
			None	5 Mbps
OOK	8B10B	15 MHz	RS (64,32)	6 Mbps
			RS (160,128)	9 Mbps
		30 MHz	RS (64,32)	12 Mbps
			RS (160,128)	19.2 Mbps
		60 MHz	RS (64,32)	24 Mbps
			RS (160,128)	38.4 Mbps
		120 MHz	RS (64,32)	48 Mbps
			RS (160,128)	76.8 Mbps
			None	96 Mbps

1.8 IEEE 802.11 LIGHT COMMUNICATIONS AMENDMENT-TASK GROUP "BB"

Dating back to the early 1880s, the first illustration of the development of OWC systems was through the invention of the photophone by Alexander Graham Bell. Until the early 1990s, there have not been any significant standardization efforts pertaining to OWC technologies. In an attempt to formulate interoperable infrared

Table 1.3
Operational specifications of PHY III and achievable data rates [417].

Modulation	Optical Clock Rate	FEC	Data Rate
4-CSK	12 MHz	RS (64, 32)	12 Mbps
8-CSK		RS (64,32)	18 Mbps
4-CSK	24 MHz	RS (64,32)	24 Mbps
8-CSK		RS (64,32)	36 Mbps
16-CSK		RS (64,32)	48 Mbps
8- CSK		None	72 Mbps
16-CSK		None	96 Mbps

(IR)-based OWC, infrared data association (IrDA) was formed in 1993, followed by the development of IEEE 802.11 standard in 1997. However, this standard failed to succeed due to the limited data rate it offered and hence, it was overtaken by IEEE 802.11b, which is also known as Wi-Fi 1. Later on in 2000, with the rapid breakthrough of LEDs, research efforts in relevance to OWC shifted toward the visible light portion of the electromagnetic spectrum. Eventually, this paved a platform for the formation of a new standard in VLC domain, i.e., the VLC Consortium, IEEE 802.15.7, IEEE 802.15.13 and the International Telecommunication Union-Telecommunication Standardization Sector (ITU-T) G.9991, which is also known as G.vlc. Concurrently, from a list of earlier findings, it can be affirmed that RF-based Wi-Fi technologies have encountered numerous amendments, and this is confirmed with the release of several Wi-Fi standards, i.e., IEEE 802.11a/g/n/ac, more precisely the Wi-Fi 2 to 5 standards. Moreover, the recent advancement in Wi-Fi is apparent through the release of IEEE 802.11ax which is the Wi-Fi 6 standard [342].

By taking into consideration the prediction as directed by Cisco Visual Networking Index, the mobile data traffic seems to increase seven-fold by 2022 [59]. Eventually, this implies that the mobile data traffic increases substantially by approximately 10^9 gigabytes by 2022. Also, from several significant findings, it can be affirmed that almost 54% of the mobile data traffic in 2017 was offloaded to Wi-Fi networks. However, due to the pragmatic increase in data traffic, this offloading rate of mobile data traffic is expected to increase by 59% by 2022. Moreover, Wi-Fi alone turns out to be the most appropriate candidate to handle more than half of the total internet protocol (IP) traffic in the coming years. Consequently, this strongly confirms the prominence of Wi-Fi for future wireless communication technologies. Eventually, to cope with this enormous increase in the demand for data, researchers shifted toward exploring the higher spectrum more accurately the 60 GHz Wi-Fi which is ably known as WiGig [408]. Besides, the optical spectrum can even be utilized as well. Thus, in July 2018, a task group, namely the IEEE 802.11 Light Communications Amendment- Task Group "bb" (TGbb), was formulated with the major agenda to bring together Light Fidelity (Li-Fi)-based technology with the IEEE 802.11 ecosystem. The pervasive aspiration of this amendment i.e., IEEE 802.11bb is to offer a fully mobile and networked solution for the OWC technologies, more specifically the Li-Fi. It is interesting to note that this stems out as the major difference between this newly proposed standard, i.e., IEEE 802.11bb and the other prevailing OWC standards. The purview of this task group, i.e., TGbb, is to enumerate a new physical (PHY) layer as well as to set forth a few modifications to the IEEE 802.11 MAC layer [408]. In addition, it is worthwhile to specify that at the very inception of this project, it was contemplated to expedite the coexistence of Wi-Fi technology with Li-Fi by adding the Fast Session Transfer protocol in the MAC layer.

In particular, the vision of the task group, i.e., TGbb is shared below as follows:

- All the uplink and downlink operations are facilitated to be carried out in the band ranging from 380 to 5000 nm.
- The second major goal is to enable all the PHY modes of operations are enabled to accomplish minimum single link throughput ranging of up to 10 Mbps and notably one mode of operation should attain single-link throughput of at least 5 Gbps as measured at the MAC data service access point (SAP).
- A strict sense of potential to exchange and exploit information exists among the solid state illuminating sources possessing various modulation bandwidths. More precisely, interoperability is mandated to be maintained between the solid state light sources.

1.9 COMPARISONS BETWEEN RADIO FREQUENCY-BASED WIRELESS COMMUNICATION AND OWC

Even though there are several forms of OWC, since the major emphasis of this book is VLC, and, moreover, taking into consideration the assured data rates as well as the standardization formats, we compare only the two types of OWC, i.e., VLC and FSOWC with RF-based wireless communications. Principally, it is a well-known fact that different wireless communication-based technologies retains peculiar characteristic features. Conventionally, they supervene different types of standardizations and encompass different types of modulation formats. In the case of OWC like VLC, data is modulated by using the intensity of the light wave and taking into consideration the non-coherent emission characteristics of LED; therefore, this constrains the transmitted signal to be real and positive-valued. Hence, the traditional modulation formats which are existing cannot be enforced in a straightforward manner in case of OWC. The modulation techniques which can be enforced like amplitude shift keying (ASK), phase shift keying (PSK), pulse modulation (PM) techniques, code division multiple access (CDMA), multicarrier modulation formats like orthogonal frequency division multiplexing (OFDM), generalized frequency division multiplexing (GFDM) and space diversity techniques like multiple input and multiple output (MIMO) techniques all have to comply with the requirements of IM/DD systems. Moreover, the major difference stems from the type of the transmitted medium, the type of the transceiver elements employed, the principle operation involved, and the levels of security. The comparison differences between RF-based wireless communications and different forms of OWC like visible light and FSOWC are summarized in Table 1.4.

These tables clearly illustrate the differences between the architectures, basic principle methodologies, application scenarios, and limitations involved behind each wireless communication technologies. As stated in the table, VLC can be used to impart simultaneous 'illumination' and 'communication' as well as be used in localization. VLC is more effective in case of short-distance communication especially indoors, where it can render high data rate communications. While, in the outdoor environments, the high data rate communication is perturbed because of the presence of noise like ambient sunlight and interference which is arising due to the presence of artificial light sources. Even FSOWC systems can ensure high data rate communication and ultra-long-range communications but their performance is drastically affected because of adverse environmental effects. An important aspect which can be inferred from the table is that the assurance of data rates by exploiting OWC is higher than that of the RF-based wireless communications. Furthermore, the range of communication which is guaranteed by the OWC is also quite higher than that of the RF-based wireless communication. OWC-based VLC systems are much more flexible to be realized because of the rapid deployment of cost-effective solid-state illuminating devices like LEDs, and, moreover, the level of interference in case of OWC is less when compared with RF communication, but OWC systems suffer from blocking and shadowing. The amount of security which is ensured in case of OWC is greater since light signals cannot penetrate walls and the data can be protected from the intruder.

By considering the aforesaid aspects, it can be envisaged that OWC-based wireless communications and RF-based wireless systems will coexist rendering seamless services to the endusers.

Table 1.4

Differences between different types of wireless communication-based technologies.

Key Aspect	RF	VLC	FSOWC
Transceiver Components	Antenna	LEDs (Transmitter), Photodiodes, Camera or Imaging Sensor (Receiver)	Laser (Transmitter), Photodiodes (Receiver)
Standardization	Matured	Matured (IEEE 802.11bb)	Well Developed
Types of Modulation Formats	ASK, PSK, PM, CDMA, OFDM, GFDM, MIMO	Since data is modulated by using the intensity of light, IM/DD is the most viable modulation format for VLC and amplitude and phase modulation techniques cannot be applied. OOK, PWM, PPM, OFDM, MIMO, GFDM, CSK, CAP modulation	Similar to VLC, holds good OOK, PM, OFDM, etc.
Range of Communication	More than 100 Km can be ensured by using a microwave link	In the range of several meters	More than 10000 Km
Levels of Interference	Very High	Low	Low
Impact of Noise	Noise induced due to the presence of all electronics and electrical components	Background solar radiation, artificial light sources	Sunlight and artificial light sources
Effect of Environment	Major Effect	No Effect	Drastically Affected
Security	Assurance of secured communication is quite low	Since light signals cannot penetrate through walls, a high degree of security can be assured against eavesdropping	High Amount of Security

–continued

Table 1.4
Differences between different types of wireless communication-based technologies (continued).

Key Aspect	RF	VLC	FSOWC
Generated Data Rate	6 Gbps for IEEE 802.11 & frequencies around 60 GHz	In the range of several 10 Gbps	40 Gbps
Band of Spectrum	Radio Waves	Visible Light	Infrared/ Visible Light
Regulation of Spectrum	Regulated	Unregulated	Unregulated
Path Loss	High	Very High for NLOS	High
Major Motive	Communication & Positioning	Communication & Illumination	Communication
Limitations	Affected by Interference	Mainly short distance communication	Drastically impacted by environment

2 VLC CHANNEL MODELS

2.1 INTRODUCTION

It is a widely known fact that the design of any communication system should effectively comprehend the limitations imposed by the channel environment in which it operates and the same implies to VLC. Predominantly, the VLC channel environment plays a dominant role in ultimately deciding the performance of the overall communication system. Particularly, it is vital to accurately characterize the parameters of the VLC channel for the purpose to establish a high-quality communication link. Furthermore, effective channel modeling is of significant importance for ensuring a reliable, secure and efficient design of the VLC system with a tolerable amount of performance. This chapter focuses on VLC channel modeling exclusively for indoor applications. This chapter presents an overview about the different propagation modes that are associated with optical wireless channels. Furthermore, the major emphasis of this chapter is to portray a thorough analysis along with simulation results for the indoor channel environment using both single and multiple illumination devices. Additionally, this chapter outlines the current state-of-the-art research aspects in the area of VLC channel modeling. Moreover, a comprehensive review on the mathematical modeling pertaining to different VLC channel models that have been proposed in the literature were summarized. Also, this chapter discusses briefly the realistic VLC channel model that was acquired by the IEEE 802.15.7*r*1 Task Group.

2.1.1 REVIEW ON DIFFERENT PROPAGATION MODES

Particularly, for indoor links, based upon the existence or non-existence of LOS paths between the transmitter and receiver, six different configurations have been defined by the authors in [248]. Accordingly, the configurations which are associated with the optical wireless channel are listed below:

1. Directed LOS
2. Non-directed LOS
3. NLOS configurations/diffuse systems

Principally, in case of the establishment of point-to-point links, it is mandatory for the existence of LOS between the transmitter and receiver, while ignoring the effect of reflections. Predominantly, this sort of link, i.e., point-to-point, circumvents the problem of interference that emerges from ambient stray light, while furnishing high data rate communication with minimal signal attenuation. In spite of their significant nature, these links suffer from severe blocking and shadowing effects. Consequently, the path loss is calculated from the knowledge of transmitter, beam divergence, receiver size and separation distance. Primarily, the LOS channel is characterized by the receiver FOV and the transmitter LED's (which is also called emitter) beam angle. Pertaining to the scenario of directive links, the transmitter and receiver comprise of small divergence angle and FOV. A typical directive link is delineated in the Fig. 2.1. The stringent requirements of the transmitter divergence angle and FOV of

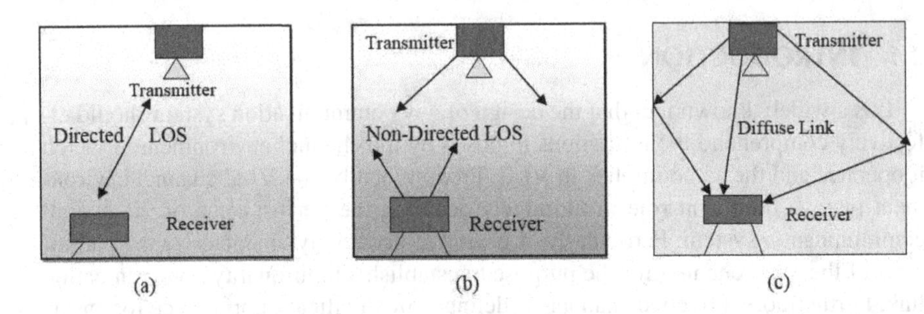

Figure 2.1: Illustration of different link configurations: (a) Directed LOS. (b) Non-directed LOS. (c) Diffuse.

the receiver necessitate these links to possess accurate alignment. Moreover, mobile scenarios like the movement of people within the room as well as the presence of objects in the room enable these links to suffer from blocking effects. Meanwhile, in case of non-directive links, both the transmitter and receiver comprise of wide angles as illustrated in the figure. On the other hand, the diffuse/NLOS configuration is one of the widely employed configurations and, as emphasized in the figure, the location of the transmitter plays a vital role to determine the levels of power at various points within the indoor room. The transmitter and receiver in the diffuse/NLOS link generally have a wide beam angle and FOV, respectively. Particularly, the wide FOV of the receiver is employed to collect the diffused light emanating from the walls, ceiling, floor and other objects present in the room. More specifically, the NLOS/diffuse systems in an indoor environment take into account the reflections originating from almost every object present in the room surface which includes furniture, walls, ceiling, objects, etc. Furthermore, NLOS configuration counteracts the effect of blocking and shadowing by taking into consideration the multiple diffuse reflections that originate from the walls, ceiling, floor, objects, etc. Inevitably, NLOS configuration facilitates the receiver to receive signals from a number of paths thereby ensuring the link availability at all times. However, the multipath induced ISI results in the reduction of the offered data rate. Note that in case of OWC, the ISI is dependent on the data rate and the FOV on the transmitter and the receiver.

Essentially, for the assurance of high data rate communication, it is imperative to rely on LOS path due to the fact that the amount of achievable data rate is limited by non-directed LOS or diffused configurations. Specifically, LOS path is significant enough to ensure a higher received light intensity, more particularly, the amount of attained SNR is high and hence, the chances of occurrence of ISI will be drastically minimized. However, this remarkable nature is only at the expense of limited mobility because when mobility within the indoor room environment is taken into account then, definitely, shadowing and beam blockages emanate. Therefore, when considering mobility, it is vital to rely on the tracking mechanism in order to ensure the perfect alignment between the transmitter and receiver. Dealing with the scenarios for a fixed transmitter on the ceiling there exist three different tracking mechanisms as listed below:

1. Full Tracking
2. Half-Tracking
3. Non-Tracking

These three different tracking mechanisms are outlined in the Fig. 2.2. As elucidated

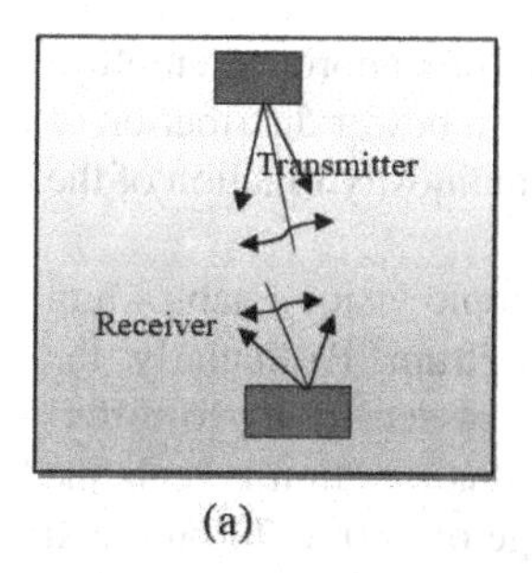

(a)

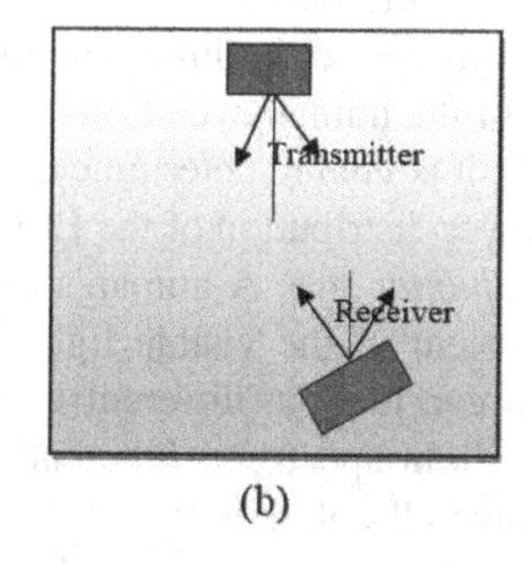

(b)

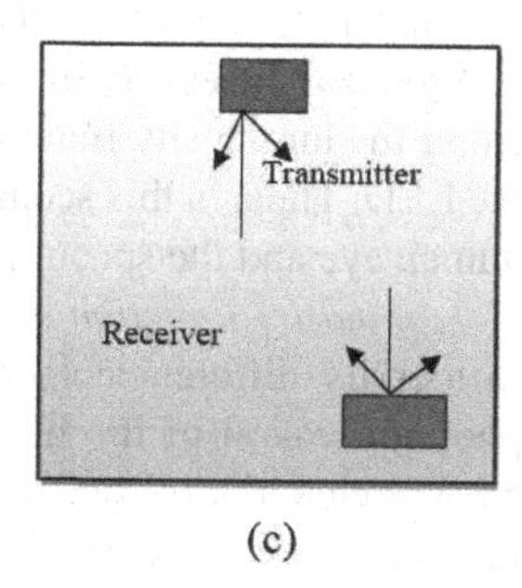

(c)

Figure 2.2: Illustration of the classification of LOS links with respect to mobility. (a) Full Tracking. (b) Half-Tracking. (c) Non-Tracking.

in the figure, the full tracking mechanism at the transmitter and receiver facilitates the alignment of transmitter and receiver with small apertures, thereby ensuring high SNRs at the receiver. While, in the half-tracking mechanism, the transmitter tracks the receiver enabling the exploitation of lower complexity tracking mechanism more suitable for multiuser systems. As for the case of non-tracking mechanism, the orientations of the transmitter and receiver are fixed and vertical to each other, thereby making it a cost-effective system.

2.2 REVIEW ON PHOTOMETRY

Before going into details about VLC channel modeling, it is of utmost importance to know the basics of photometry. Unlike a conventional wireless channel, modeling of VLC channel is not straightforward; this is due to the fact that VLC employs opto-electronic illumination devices like LEDs to render simultaneous illumination and communication. Consequently, it is imperative for the researcher working in this emerging area like VLC to thoroughly understand the significance of photometric and radiometric parameters and then formulate the relationship between the two parameters. This section focuses on basics of photometry where much of the literature pertaining to it is adopted from [398]. The photometric parameters evaluate the characteristics of light such as the brightness, color, etc., as anticipated by the human eye. Particularly, they aid in determining the illumination aspects of the LEDs. On the other hand, the radiometric parameters help in quantifying the communication-related aspects of LEDs. The useful parameters which we need to determine prior to the calculation of transmitted power, path loss and the received power of the LOS and the NLOS links include the luminous flux, luminous intensity, illuminance, and the Lambert radiator.

2.2.1 LUMINOUS FLUX

Principally, the luminous flux can be calculated by using either of the two approaches: spectral integral and the spatial integral, based upon the availability of the parameters for a given LED transmitter.

Spectral Integral: In order to derive the luminous flux, this approach employs either the luminosity function of the human eye or the spectral power distribution of the LED. Thus, in this scenario, it is vital to determine the luminosity function of the human eye and the spectral power distribution of the LED.

Luminosity Function of the human eye: A human's photopic vision enables him to identify different colors present in the visible light spectrum. Particularly, the photopic vision of the human eye has got different levels of sensitivity to different wavelengths of the visible light spectrum. Inherently, this function reveals the fact that human eye can perceive all colors within the range of 380 to 780 nm and exhibits maximum levels of sensitivity to the yellowish-green region, corresponding to a wavelength of 555 nm.

Spectral Power Distribution of the LED: This parameter portrays the power of the LED at all the wavelengths of the visible light spectrum. Primarily, this is a radiometric parameter which is measured in Watt/nm. Figure 2.3 delineates the spectral power distribution of three different colored LEDs. It can be emphasized that all the three LEDs exhibit higher radiant power particularly at yellow and blue regions. It is apparent that white light is obtained by making use of blue LED with yellowish phosphor coating. In general, in accordance with the type of white color, i.e., whether warm, natural or cool, the levels of blue and yellow emissions were controlled with the aid of phosphor coating. From the figure, it is evident that when compared to cool white LED, both warm and natural white facilitates more yellow light. Consequently, by superimposing the luminosity function of the human eye with the spectral power distribution of the LED, the luminous flux, i.e., the amount of perceived power is calculated as follows

$$\phi = 683(lumens/Watt)\int_{380\ nm}^{780\ nm} S_T(\lambda)V(\lambda)d\lambda \tag{2.1}$$

The constant value $683(lumens/Watt)$ as represented by equation (2.1) indicates the maximum luminous efficiency. Substantially, the luminous efficiency depicts substantially the ratio of the luminous flux to the radiant flux. Precisely, the luminous efficiency gives a measure of how well the radiated electromagnetic energy and the required electricity of the LED were transformed to furnish visible light illumination. As evident from the Fig. 2.3, the human eye is more sensitive toward the yellowish region, i.e., corresponding to a wavelength of 555 nm. Eventually, this implies that

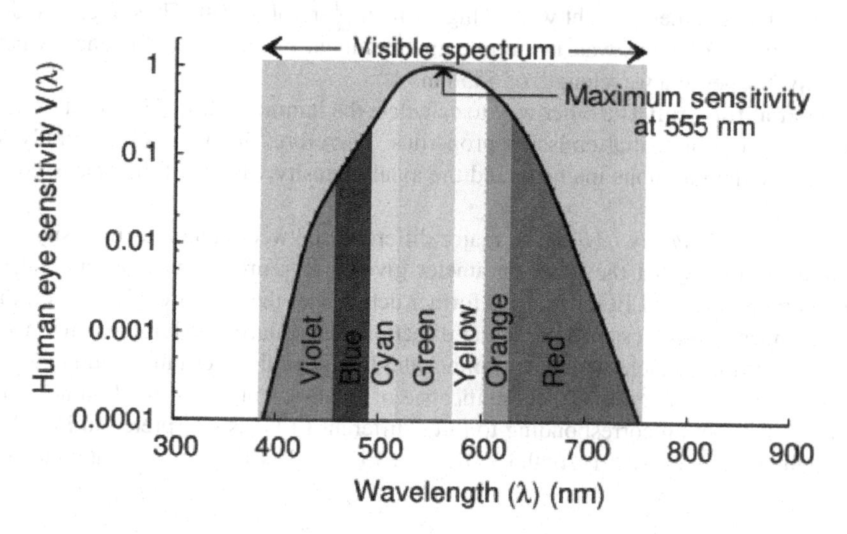

Figure 2.3: Luminosity function $V(\lambda)$ depicting the sensitivity of the human eye to different wavelengths of the visible light spectrum [398].

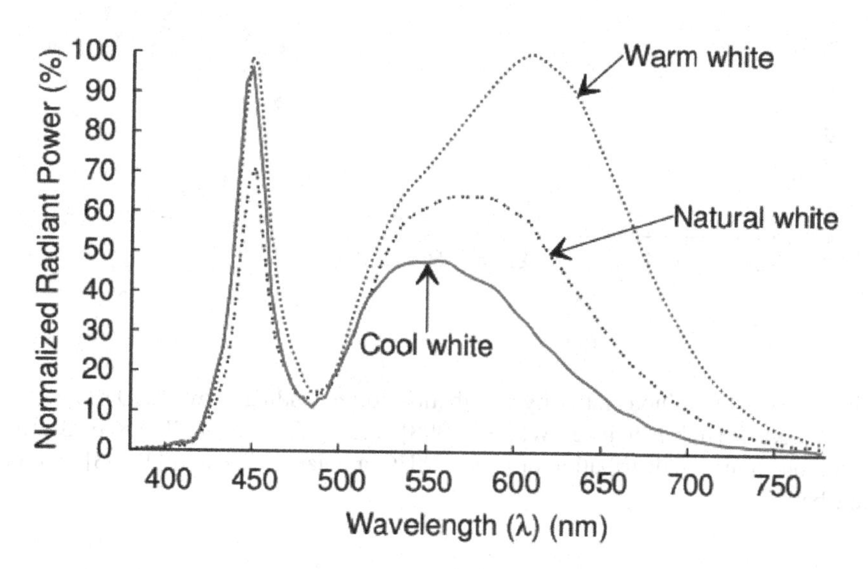

Figure 2.4: Power spectral distribution for LED comprising of three different colors: warm white, natural white and cool white [2, 398].

the amount of electrical power that is vital to obtain one lumen of light corresponding to the wavelength of 555 nm is determined to be $\frac{1}{683}$rd of a watt [532]. Consequently, this inherent fact implies that for any other light source, the amount of power required to furnish one lumen of light will be higher than $\frac{1}{683}$rd of a Watt. Thus, it can be deduced that 683 lumens/watt is the maximum amount of luminous efficiency which usually invades at a wavelength of 555 nm.

Spatial Integral: The other way to calculate the luminous flux of LED is through the utilization of spatial emission properties. Therefore, this necessitates firstly to determine the luminous intensity and the axial intensity which will be discussed in detail below:

Luminous Intensity $I_t(\theta)$*:* The major difference between luminous intensity and luminous flux is that the latter parameter gives a measure of the total amount of light emitted by the LED, while the former determines the measure of how well the LED imparts brightness in a particular direction. Particularly, the luminous intensity is measured in candela which determines the luminous flux per unit solid angle. In order to clearly understand the significance of luminous intensity, the luminous intensity distribution corresponding to three different LEDs is interpreted in Fig. 2.5. As delineated in Fig. 2.5(a), both LEDs, i.e., LED 1 and LED 2 emit light at wider an-

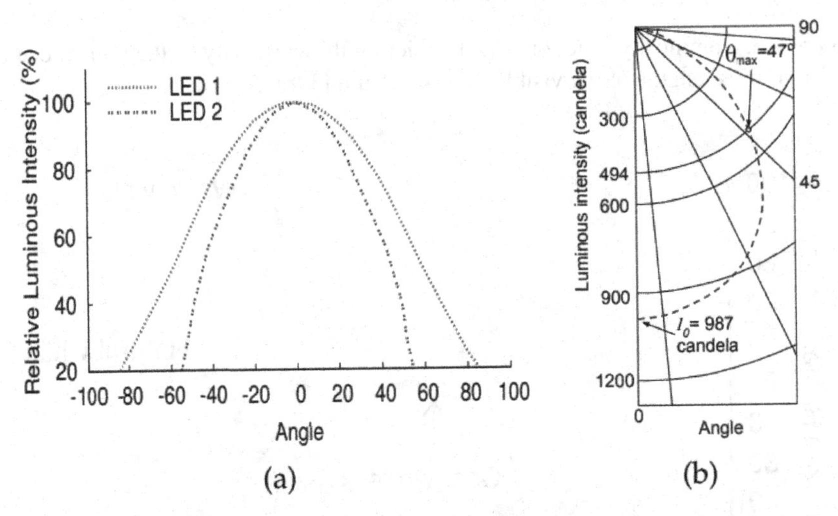

Figure 2.5: (a) Luminous intensity distribution corresponding to two LEDs: (1) Cree XLamp XP-E High-Efficiency White [2, 398], and (2) Cree XLAMP XR-E [3]. (b) Luminous intensity distribution of Cree LMH6 in polar coordinates [1, 398] and its half-beam angle.

gles offering better levels of illumination in many directions. On the other hand, the third LED as illustrated in Fig. 2.5(b) exhibits a narrower beam of emission. Furthermore, in the literature [274] it is revealed that most of the LEDs possess Lambertian beam distribution which indicates that the LED glows uniformly in all directions. The Lambertian emission implies that light intensity drops as the cosine of the incident angle. The other two important parameters which need to be considered in deriving the intensity distribution are: axial intensity and the half-beam angle.

Axial Intensity: It is the luminous intensity at 0°. As shown in the Fig. 2.5, the axial intensity is 987 candela.

Half-Beam Angle: It determines the angle at which the light intensity drops to half of the axial intensity. Specifically, the half-beam angle is determined from the entire beam angle Ω_{max} as follows

$$\Omega_{max} = 2\pi\left(1 - cos\left(\theta_{max}\right)\right) \tag{2.2}$$

From (2.2), the parameter θ_{max} corresponds to the half-beam angle. Thus, by exploiting the spatial integral approach, the luminous flux is determined by integrating the luminous intensity function over the entire beam solid angle which is formulated as

$$\phi = \int_{o}^{\Omega_{max}} I_0 I_t\left(\theta\right) d\Omega \tag{2.3}$$

From (2.3), I_0 specifies the axial intensity and $I_t\left(\theta\right)$ is the normalized luminous intensity. Accordingly, by differentiating equation (2.3), the following expression is attained

$$d\Omega_{max} = 2\pi sin\left(\theta_{max}\right) d\theta_{max} \tag{2.4}$$

Thereupon, substituting the expression as specified by equation (2.4) into the mathematical expression as stated by (2.3) as well as changing the order of integral, yields the following expression

$$\phi = I_0 \int_{0}^{\theta_{max}} 2\pi I_t\left(\theta\right) sin\left(\theta\right) d\theta \tag{2.5}$$

2.2.2 LUMINOUS INTENSITY

As discussed above, the luminous intensity is defined as the luminous flux per unit solid angle which is given by

$$I = \frac{d\phi}{d\Omega} \tag{2.6}$$

Ω in (2.6) denotes the solid angle and g represents the luminous intensity.

2.2.3 ILLUMINANCE

Illuminance is defined as the luminous flux received in a unit area which is given as

$$E = \frac{d\phi}{dS} \tag{2.7}$$

Where from equation (2.7), the area of the lighting is specified by dS and the luminous flux received from the surface dS is delineated by $d\phi$. For the scenarios where uniform lighting source is employed, the illumination imparted is very large, in such cases, the illumination is determined as the total luminous flux ϕ divided by the total area S.

2.2.4 LAMBERT RADIATOR

Typically, the Lambert radiator is the radiation model, and the LEDs which are widely deployed in VLC can be approximated as Lambertian types. Notably, the illuminance level in case of Lambertian radiator is uniform.

2.3 INDOOR VLC CHANNEL MODELING

Intensity modulation (IM) with direct detection (DD) is the most widely deployed modulation format for a VLC system due to the inherent nature of reduced cost and implementation complexity. Particularly, the intensity of the input signal in VLC is modulated, rather than the amplitude and phase. Primarily, as delineated in Fig. 2.6, an indoor VLC system employs LED as the source where its light intensity is modulated by a message signal, and a large-area PD is employed for reception at the receiving end. Particularly, in case of VLC, as stated above, the intensity of

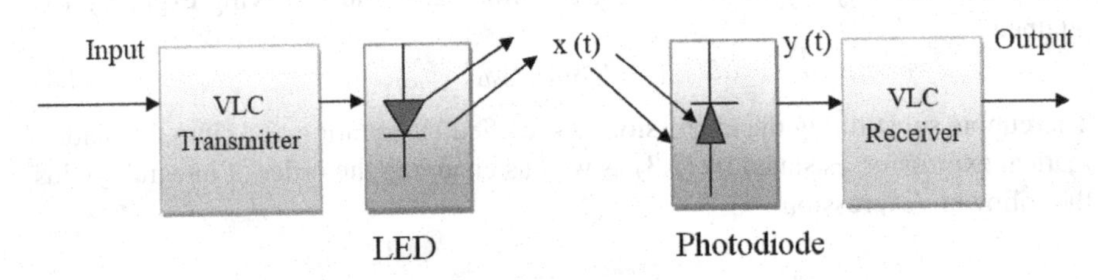

Figure 2.6: Schematic representation of intensity modulation and direct detection communication channel.

the input signal is modulated, rather than the amplitude or phase information. As delineated in the figure, the intensity modulated signal $x(t)$ is passed through the channel environment and is received by the photodiode to attain the signal $y(t)$. Mathematically, this can be put up as follows

$$y(t) = Rx(t) \circledast h(t) + n(t) \tag{2.8}$$

R in equation (2.8) represents the responsivity of the photodiode, $x(t)$ and $y(t)$ depict the transmitted and received signal, the channel impulse response is denoted by $h(t)$, and $n(t)$ corresponds to the additive white Gaussian noise (AWGN). Upon solving the convolution of two signals in equation (2.8), the following expression is attained

$$y(t) = \int_{-\infty}^{\infty} Rx(\tau)h(t-\tau)\,d\tau + n(t) \tag{2.9}$$

The mathematical expression delineated by equation (2.9) represents a linear filter channel with AWGN, and it differs from the conventional RF wireless channel due to the fact that the instantaneous optical power is proportional to the generated electrical current. Thus, the signal $x(t)$ depicts the power rather than the amplitude. Consequently, this imposes two important constraints on the transmitted signal. The first constraint is that the transmitted signal should be non-negative valued which implies

$$x(t) \geq 0 \tag{2.10}$$

The second constraint is that the maximum amount of transmitted optical power should adhere to the eye safety requirements. Thus, this imposes that the average value of the transmitted signal should not exceed a specified maximum power value P_{max}, i.e.,

$$P_{max} = \underset{T \to \infty}{Lim} \frac{1}{2T} \int_{-T}^{T} x(t)\, dt \tag{2.11}$$

Eventually, these differences impose a profound effect on the overall VLC system design. The major focus of this chapter is to determine the impulse response of the

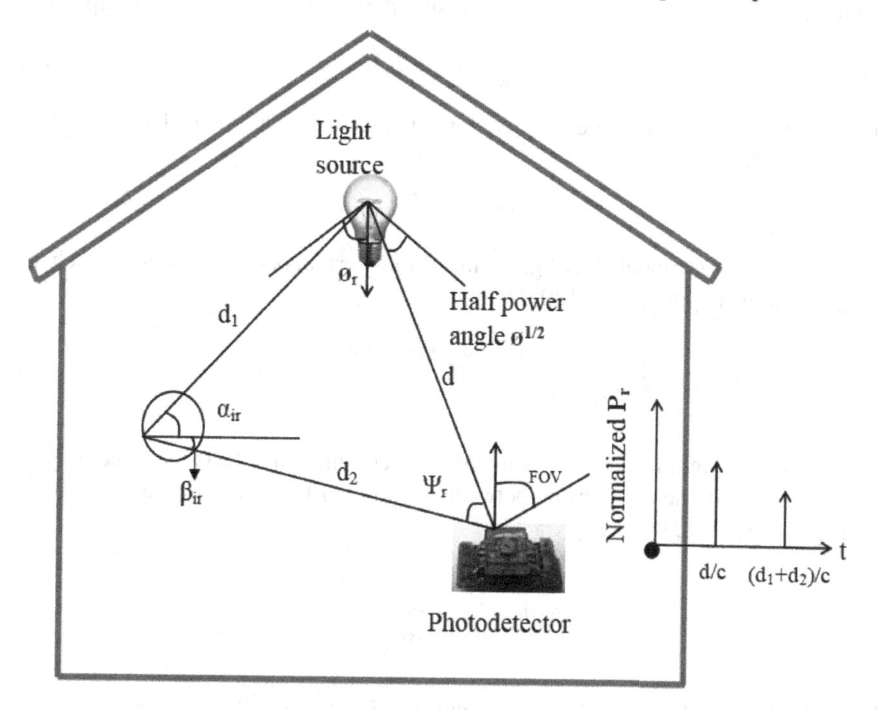

Figure 2.7: Typical indoor room environment comprising of optical transmitter installed on the ceiling and the photodiode-based receiver terminal.

VLC channel for both single source and multiple opto-electronic illuminating source scenarios.

2.3.1 VLC CHANNEL MODELING FOR THE SINGLE SOURCE SCENARIO

In general, due to the multipath propagation environment, the transmitted signal from the LED transmitter undergoes several reflections due to the presence of walls, floor, ceiling, furniture, etc. Hence, in addition to the LOS component, there are several NLOS/diffuse components present and this is delineated in the Fig. 2.7. Thus, it is vital to determine the impulse response of the channel pertaining to both LOS and NLOS components. For the NLOS scenario, it becomes difficult to anticipate the optical path loss because it relies on a number of multiple factors such as the dimensions of the room, reflections due to walls, ceiling and several other objects within the room environment. Moreover, the other parameters like the relative position and orientation of the transmitter and receiver, window size, place and other physical matters within the room also play a vital role. Firstly, for the sake of simplicity if we ignore the diffuse component and take into account only LOS paths, then the intensity of the received signal is dependent only on the transmitter radiation pattern, the photodiode active area and the receiver optics. If we assume the transmitter's optical intensity as P_t, then the optical power corresponding to the receiver can be expressed as

$$P_R = H(0)P_t \tag{2.12}$$

In equation (2.12), the parameter $H(0)$ corresponds to the DC channel gain which is represented as

$$H(0) = \int_{-\infty}^{\infty} h(t)\,dt \tag{2.13}$$

If the transmitter is modeled as a generalized Lambertian pattern, then the DC channel gain can be expressed as [180, 248, 274]

$$H(0) = \begin{cases} \frac{(m+1)A_{PD}cos^m(\phi)T_s(\psi)g(\psi)cos(\psi)}{2\pi d^2}, & 0 \leq \psi \leq \psi_c \\ 0, & 0 \geq \psi_c \end{cases} \tag{2.14}$$

From (2.14), m signifies Lambert's mode number which manifests the directivity of the source beam. The relationship between m and the LED semi-angle at half-power $\phi_{\frac{1}{2}}$ is given by [158, 180, 398]

$$m = \frac{-ln2}{ln\left(cos\phi_{\frac{1}{2}}\right)} \tag{2.15}$$

Further, from the mathematical expression as emphasized by equation (2.14), the parameters A_{PD} denote the photodiode's active area, while ψ_c depicts the FOV of the photodiode. It is mandatory for the received signal to be within the range of photodiode's FOV, in order to ensure successful detection; otherwise, the received signal cannot be detected if it falls outside the range of the receiver's FOV. Thus, the detector can be modeled as an active area A_{PD} collecting the radiation incident at angles ψ smaller than the detector's FOV. Moreover, utilizing a detector with large photoactive area will definitely facilitate collecting as much power as desired, but in practicality this limits the available bandwidth due to increased junction capacitance. Consequently, it entails employing a non-imaging concentrator for attaining an enhancement in the overall efficacious area. Therefore, the optical gain of an ideal non-imaging concentrator having internal refractive index n is given by [180]

$$g(\psi) = \begin{cases} \frac{n^2}{sin^2(\psi_c)}, & 0 \leq \psi \leq \psi_c \\ 0, & \psi \geq \psi_c \end{cases} \tag{2.16}$$

Also, the optical filter which is having an optical gain of $T_s(\psi)$ is employed to filter out the noise emanating from the ambient light sources and other artificial light sources. Furthermore, Fig. 2.8 portrays a typical indoor VLC system comprising

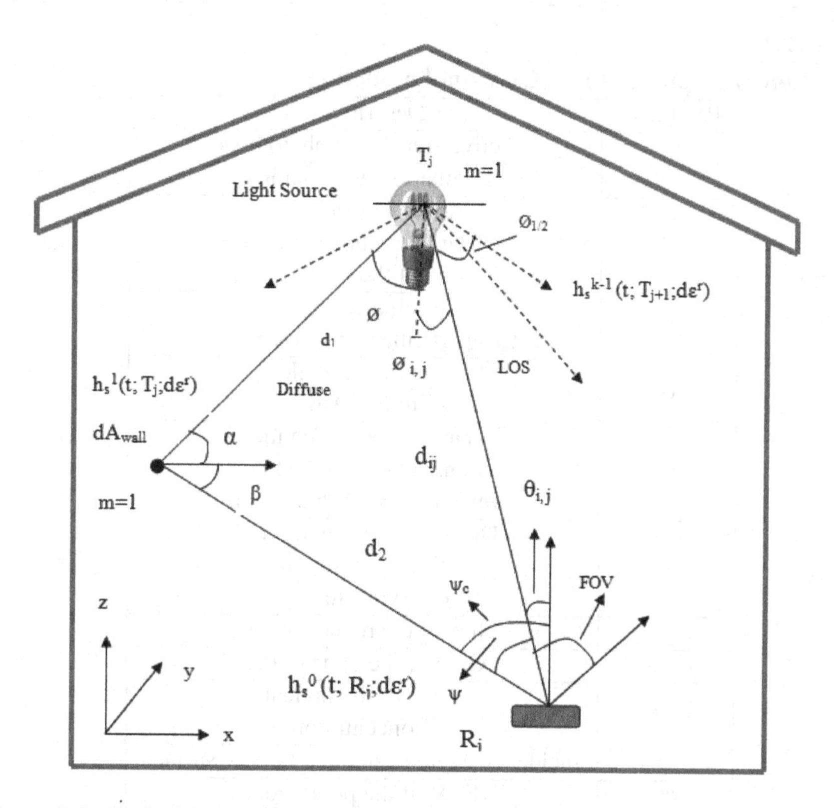

Figure 2.8: Typical indoor room environment comprising of optical transmitter installed on the ceiling and the photodiode-based receiver terminal emphasizing that the transmitted signal undergoes k number of reflections.

of optical transmitter like LED and photodiode-based receiver. The signal from the transmitting LED to the receiver photodiode reaches through the direct LOS as well as through reflected paths. Therefore, by taking into account the power emanated due to the NLOS paths, the DC channel gain corresponding to the reflected path can be formulated as [177]

$$\underset{Refl}{dH(0)} = \begin{cases} \frac{(m+1)A_{PD}}{2\pi d_1^2 d_2^2} \rho dA_{Wall} cos^m(\phi) cos(\alpha) cos(\beta) T_s(\psi) g(\psi) cos(\psi), & 0 \leq \psi \leq \psi_c \\ 0, & \psi \geq \psi_c \end{cases} \tag{2.17}$$

The parameters as represented by equation (2.17) and as specified in the figure are represented in Table 2.1. Thereupon, considering both multipath propagation envi-

Table 2.1
Parameters employed for VLC channel modeling.

Parameter	Description
A_{PD}	Active area of the photodiode
m	Lambert's mode number
β	Angle of irradiance from the reflective area of the wall
ϕ	Irradiance angle to a particular reflective point
α	Irradiance angle to the wall
d_1	Distance between the transmitter and the first reflective point to the wall
d_2	Distance between the wall and the point to the receiving surface
dA_{Wall}	Size of the reflective area
$T_s(\psi)$	Gain of the optical filter
$g(\psi)$	Gain of the optical concentrator
ψ	Incidence angle from the reflective surface
ψ_c	FOV of the photodiode

ronment and LOS component, the total received power can be expressed as [178,180]

$$P_R = \left(H_{LOS}(0) + \int_{reflections} dH_{Refl}(0) \right) P_t \tag{2.18}$$

where H_{Refl} represents the reflected path. There are several factors on which the reflection characteristics of object surfaces within a room environment depend such as, the transmission wavelength, surface material, the angle of incidence and roughness of the surface relative to the wavelength. Further, the electrical SNR in terms of the received power is given by

$$SNR = \frac{(RH(0)P_R)^2}{\sigma_{Total}^2} \tag{2.19}$$

From equation (2.19), the total noise variance is depicted as σ_{Total}^2. It is to be noted that the transmitting light which arrives at the receiver has undergone a definite number of reflections or bounces. Thus, the impulse response of the channel emphasizing that the light signal exactly undergoes k bounces of reflections from the transmitter to the receiver can be put up as

$$h(t;T_j,R_j) = \sum_{k=0}^{\infty} h_S^{(k)}(t;T_j,R_i) \tag{2.20}$$

In equation (2.20), the impulse response pertaining to the kth reflection is depicted by $h_S^{(k)}(t;T_j,R_i)$, and the contribution of the LOS channel impulse response is given as [178]

$$h_S^{(0)}(t;T_j,R_i) = VI(\phi_{ij})\left(\frac{A_{Ri}g(\psi)}{d_{ij}^2}\right) \times \delta\left(t - \frac{d_{ij}}{c}\right) \tag{2.21}$$

As already stated, d in equation (2.21) represents the distance between the source and the receiver. V delineates the visibility function which is taken to be unity provided the LOS path between the transmitter and the receiver remains unobstructed. On the other hand, if the LOS path is obstructed, then the visibility factor is taken to be 0. Therefore, the range for the visibility factor is $0 < V \leq 1$. The other parameters in equation (2.21), i.e., $I(\phi_{ij})$, A_{Ri}, c and $g(\psi)$ denote the luminous intensity, optical collection area, speed of light and the optical gain function of the receiver, respectively. Meanwhile, the optical gain function of the receiver can be expressed as

$$g(\psi) = \begin{cases} cos(\psi), & if\ 0 \leq \psi \leq \frac{\pi}{2} \\ 0, & Otherwise \end{cases} \tag{2.22}$$

So far, we discussed the impulse response of the channel pertaining to the LOS component. In the same manner, the impulse response of the channel corresponding to the kth bounce can be determined by employing the response of $(k-1)$ bounce as follows [86]

$$h_S^{(k)}(t;T_j,R_i) = \int_S \rho d\varepsilon^r . h_S^{(k-1)}(t;T_j,d\varepsilon^r) \otimes h_S^{(0)}(t;d\varepsilon^t,R_i) \tag{2.23}$$

In (2.23), the integral is performed over the surface S, $d\varepsilon^t$ and $d\varepsilon^r$ represent the differential surface of area where the parameter $d\varepsilon^t$ acts as the receiver with respect to the transmitter T_j and the other parameter $d\varepsilon^r$ acts as the source with respect to the receiver R_i. Also, the mathematical expression as portrayed in (2.23) draws an important inference that as the number of reflections tends to reach infinity, i.e., $k \to \infty$, then the value $\left\| h_S^{(k)}(t;T_j,d\varepsilon^r) \right\|$ will become zero. Thus, this entails estimating the overall channel impulse response by taking into consideration the first M

bounces/reflections, respectively. Consequently, the impulse response of the channel for M bounces is given as [86]

$$h_S(t;T_j,R_i) \approx \sum_{k=0}^{M} h_S^k(t;T_j,R_i) \tag{2.24}$$

The earlier significant works as portrayed in [86] illustrate that exact approximations of the channel impulse response can be obtained for the value of M ranging from 3 to 10.

2.3.2 CHANNEL MODELS FOR MULTIPLE SOURCES

A typical indoor room environment comprising of M illuminating devices is delineated in Fig. 2.9 where it is clear that due to the multipath propagation environment, N NLOS components are also inherently present between the transmitter and the receiver. From the link geometry which is shown in the figure, it is clear that the receiver terminal R_j receives the radiation emitted from the multiple illuminating sources including the transmitter T_i via the LOS components as well as k number of reflections from walls, ceiling and floor within the room. From this setup, the channel impulse response pertaining to a particular illuminating device T_i and the receiver terminal R_j can be expressed as [87]

$$h_S^{(0)}(t;T_i,R_j) = \frac{I(\phi_{ij}A_{Ri})}{d_{ij}^2} T_S(\psi_{ij})g(\psi_{ij})cos(\psi_{ij})\,rect\left(\frac{\psi_{ij}}{\psi_c}\right)\delta\left(t-\frac{d_{ij}}{c}\right) \tag{2.25}$$

From the mathematical expression as shown in equation (2.25), the distance between the transmitter T_i and the receiver R_j is given by d_{ij}, meanwhile, the function $\delta(.)$ implies the Dirac Delta function and the rectangular function $rect(y)$ is defined as

$$rect(y) = \begin{cases} 1, & for\ |y| \le 1 \\ 0, & for\ |y| > 1 \end{cases} \tag{2.26}$$

The channel impulse response for the kth bounce is expressed as [86]

$$h_S^{(k)}(t;T_i,R_j) = \sum_{n=1}^{M} \rho d\varepsilon_n^r h^{(k-1)}(t;T_i,d\varepsilon_n^r) \otimes h_S^{(0)}(t;d\varepsilon_n^t,R_j) \tag{2.27}$$

Therefore, the overall impulse response of the channel which considers the multiple transmitters and multiple reflections can be expressed as [86]

$$h(t;T_i,R_j) = \sum_{i=1}^{M}\sum_{k=0}^{\infty} h_i^{(k)}(t;T_i,R_j) \tag{2.28}$$

2.3.3 SIGNAL TO NOISE RATIO ANALYSIS

The major sources of noise that hinder the performance of high data rate communication in indoor visible light link are listed below:

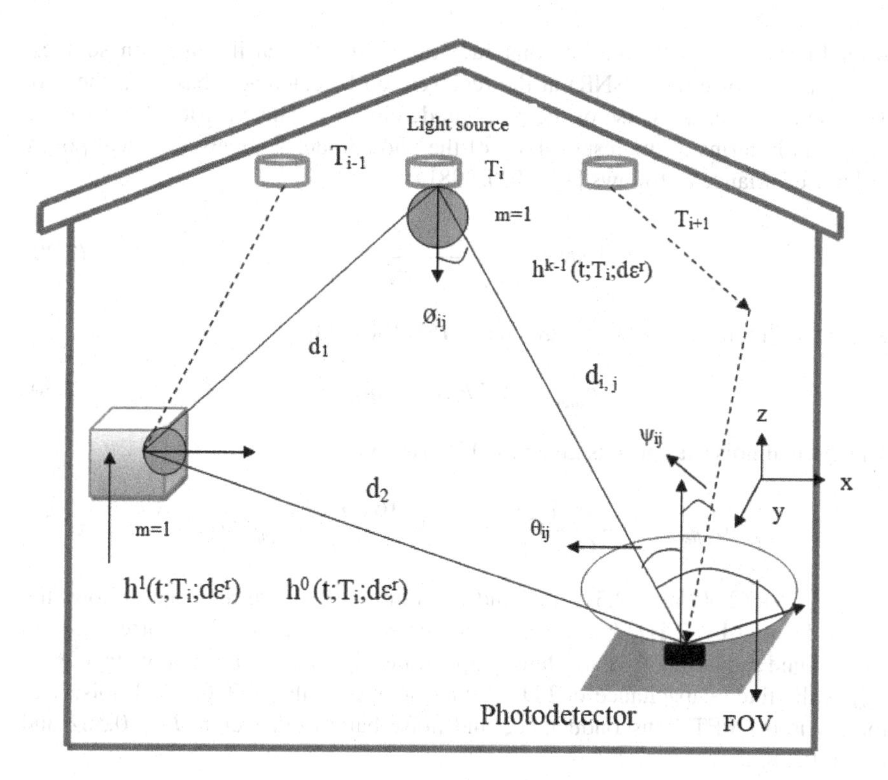

Figure 2.9: Typical indoor room environment comprising of multiple optical transmitters installed on the ceiling and the photodiode-based receiver terminal emphasizing that the transmitted signal undergoes k number of reflections.

- *Ambient light noise:* Predominantly, the ambient noise can be identified as the noise emanating due to the sunlight and other artificial light sources. In particular, the noise due to the solar radiation penetrates through windows, door, etc. Additionally, even the manifestation of other illumination sources like incandescent and fluorescent lamps also imparts significant amount of noise. Primarily, the noise imparted by such sources; i.e., the solar radiation and the artificial light sources result in ambient noise floor which is a DC interference. Therefore, the effects of such noise can be mitigated with the aid of high pass filter at the receiving end. Recent research efforts suggest the usage of sophisticated optical filters to filter out such noise.
- *Shot noise*: The major source of noise that usually hinders the assurance of high data rate communication is the shot noise that is induced due to solar radiation.
- *Electrical pre-amplifier noise* is also known as thermal noise of the photodiode.

After filtering the noise due to solar radiation and artificial illumination sources, the signal to noise ratio (SNR) at the receiver can be calculated based on the shot noise and the thermal noise of the photodiode circuitry. The electrical SNR can be expressed in terms of the responsivity of the photodiode, R, received optical power and noise variance as follows [158, 180, 398]

$$SNR = \frac{(RP_r)^2}{\sigma^2_{Shot} + \sigma^2_{Thermal}} \tag{2.29}$$

where the shot noise variance is given by [158, 180, 398]

$$\sigma^2_{Shot} = 2qRP_rB + 2qI_BI_2B \tag{2.30}$$

The thermal noise variance is given as [158, 180, 398]

$$\sigma^2_{Thermal} = \frac{8\pi kT_k}{G_{ol}} C_{pd}AI_2B^2 + \frac{16\pi^2 kT_k\Gamma}{g_m} C^2_{pd}A^2I_3B^3 \tag{2.31}$$

From (2.29), (2.30) and (2.31) the bandwidth of the electrical filter that follows the PD is denoted by B Hz, k is Boltzmann's constant, I_B is the photocurrent due to background radiation, T_k is absolute temperature, G_{ol} is the open-loop voltage gain, C_{pd} is the fixed capacitance of PD per unit area, Γ is the FET channel noise factor, g_m is the FET transconductance and noise-bandwidth factors $I_2 = 0.562$ and $I_3 = 0.0868$.

2.3.4 OTHER PARAMETERS ASSOCIATED WITH VLC

This subsection highlights the other important parameters that are associated with VLC systems that aids in evaluating the entire VLC system performance. These parameters are listed below:

- **Root Mean Square (RMS) Delay Spread**: It is apparent that due to the presence of multipath propagation environment, the received optical signal is subjected to delay spread. This is due to the fact that due to the prevalence of several obstacles including the presence of people, furniture, floor, ceiling, walls, etc., the transmitted signal from the transmitting LED undergoes several reflections, refractions, diffractions and scattering in order to reach the receiver terminal. As a result, several copies of the same transmitted optical signal reach the receiver at different instants of time. Moreover, with the changes in the orientation of the receiver, a considerable amount of change in the propagation path can be witnessed. Thus, the received optical signal at the receiving end can be seen as a combination of pulses with different time delays. Consequently, the difference between the arrival of the first signal component and the last reflected component can be termed as delay spread. In particular, the RMS delay spread is the most appropriate parameter to measure the channel delay spread as well as the intersymbol interference (ISI) that is induced due to the multipath propagation environment. Essentially, the channel RMS delay spread is an important parameter that accurately determines the SNR by removing the effect of ISI. Pertaining to the channel impulse response $h(t)$, the RMS delay spread can be calculated as [178, 545]

$$\tau_{RMS} = \sqrt{\frac{\int_{-\infty}^{\infty} (t - \tau_0)^2 h^2(t)\, dt}{\int_{-\infty}^{\infty} h^2(t)\, dt}} \tag{2.32}$$

From equation (2.32), the propagation delay time and the average delay time are indicated by t and τ_0, respectively. The average delay time is defined as

$$\tau_0 = \frac{\int_{-\infty}^{\infty} t h^2(t)\, dt}{\int_{-\infty}^{\infty} h^2(t)\, dt} \tag{2.33}$$

- **Path Loss**: This is the important parameter that is introduced by the propagation environment between the transmitter and the receiver. This parameter is closely associated with the channel link design, where the diffuse link is the simplest and most stable link; however, the path loss is higher in it.
- **LED Modulation Bandwidth**: As a matter of fact, the major limitation that impedes the assurance of high data rate communication in VLC systems emerges due to the limited modulation bandwidth of LEDs. Essentially, the bandwidth of white phosphorescent LEDs (i.e., the LEDs manufactured by using blue LED with yellowish phosphor coating) is limited due to the slow response of yellowish phosphor coating. Hence, in order to enhance high data rate transmission, a number of solutions have been proposed in the literature. Exploiting pre-equalization of the driver circuitry seems to increase the bandwidth up to 50 MHz, but at the expense of increase in SNR penalty of 20 dB. The other solution is to make use of discrete multitone modulation to enhance the limited modulation bandwidth of LEDs, but this necessitates to rely on complex driving circuitry at the transmitter or to incorporate frequency domain equalization at the receiving end, which in turn results in an increase in the overall computational complexity.

2.3.5 ILLUSTRATION OF THE DISTRIBUTION OF POWER AND SNR IN INDOOR VLC SYSTEM

For the derived mathematical expressions for the channel impulse responses pertaining to both LOS and NLOS paths, the following section presents the simulation results of the distribution of received power related to both LOS and NLOS components for the indoor VLC system exploiting single illuminating and multiple illuminating sources. Some of the steps of the below MATLAB codes are taken from [180].

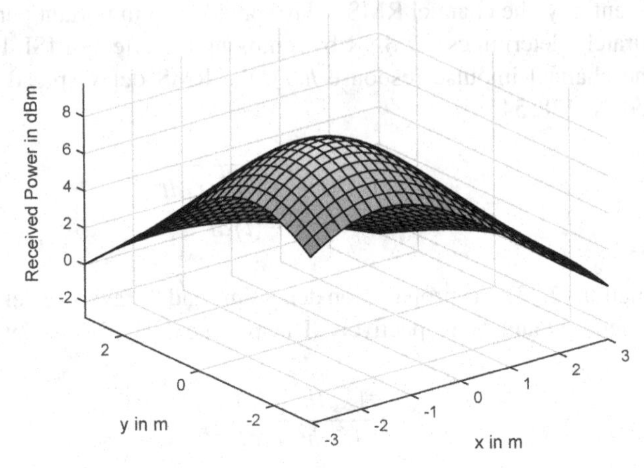

Figure 2.10: Illustration of the received power corresponding to the LOS path in indoor VLC system exploiting single illuminating source.

Listing 2.1: MATLAB code for the computation of total received power corresponding to the LOS component of indoor-VLC system employing a single illuminating source as illustrated in the simulation result shown in Fig. 2.10.

```
1  clc;
2  clear all;
3  close all;
4  theta=70; %Semi angle of the LED at half power
      illumination
5  m=-log10(2)/log10(cosd(theta)); % Lamberts Mode
      Number
6  Power_LED=20; %Transmitted Power of the LED
7  No_LED=60; %Total nummber of LEDs
8  Total_Power=No_LED*No_LED*Power_LED; %Total amount of
      Power contributed by all the LEDs
9  APD=0.001; %Area of the Photodiode
10 rho=0.8; %Reflectance Parameter
11 Ts=1; %Gain of the Optical Filter
12 Refractive_Index=1.5; %Refractive Index of the Lens
13 FOV_PD=70; %Field of View of the Photodiode
14 Concentrator_Gain=(Refractive_Index^2)/(sind(FOV_PD)
      .^2); %Gain of the optical concentrator
15 lx=6; ly=6; lz=3; %Dimensions of the Room Environment
      6 X 6 X 3 m^{3}
16 h=2.25; %Distance between the transmitter and the
```

```
    receiver plane
17 [XT,YT]=meshgrid([-lx/4 lx/4],[-ly/4 ly/4]); %
    Position of LED
18 Nx=lx*5; Ny=ly*5; %Total number of grids present in
    the receiver plane
19 x=linspace(-lx/2,lx/2,Nx);
20 y=linspace(-ly/2,ly/2,Ny);
21 [XR,YR]=meshgrid(x,y); %Illustrates the receiver
    plane grid
22 D=sqrt((XR-XT(1,1)).^2+(YR-YT(1,1)).^2+h^2); %
    Computation of Distance from the transmitter LED
    to the receiver photodiode
23 cosphi_A1=h./D; % angle vector
24 receiver_angle=acosd(cosphi_A1);
25 H_A1=(m+1)*APD.*cosphi_A1.^(m)./(2*pi.*D.^2); %
    Channel DC gain corresponding to the LOS path
26 P_received=Total_Power.*H_A1.*Ts.*Concentrator_Gain;
    % Total amount of Received power
27 P_received_dBm=10*log10(P_received);
28 %% Figure
29 surfc(x,y,P_received_dBm);
30 title('Received Power in Indoor-VLC System
    corresponding to the LOS path')
31 xlabel('x in m');
32 ylabel('y in m');
33 zlabel('Received Power in dBm ')
34 axis([-lx/2 lx/2 -ly/2 ly/2 min(min(P_received_dBm))
    max(max(P_received_dBm))]);
```

Listing 2.2: MATLAB code for the computation of total received power and distribution of signal to noise ratio (SNR) in indoor-VLC system employing multiple illuminating sources as depicted in the simulation results shown in Figs. 2.11, 2.12 and 2.13.

```
1 clear all;
2 close all;
3 theta=70; %Semi angle of the LED at half power
    illumination
4 m=-log10(2)/log10(cosd(theta)); % Lamberts Mode
    Number
5 Power_LED=20; %Transmitted Power of the LED
6 No_LED=60; %Total nummber of LEDs
7 Total_Power=No_LED*No_LED*Power_LED; %Total amount of
    Power contributed by all the LEDs
8 APD=0.001; %Area of the Photodiode
```

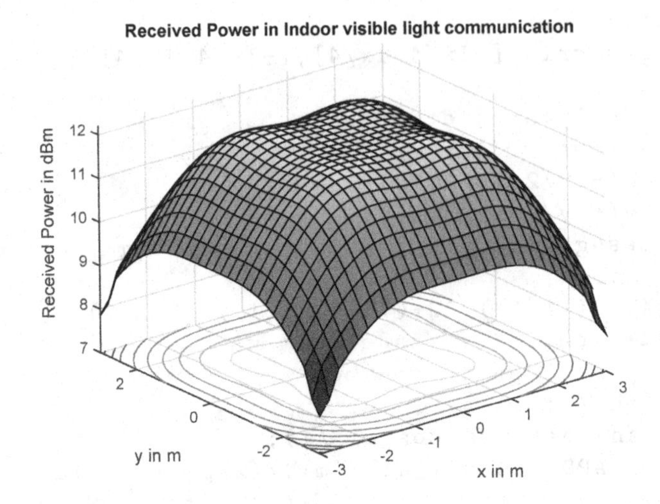

Figure 2.11: Illustration of the received power in indoor VLC system using multiple illuminating sources for the LED semi-angle as 60°.

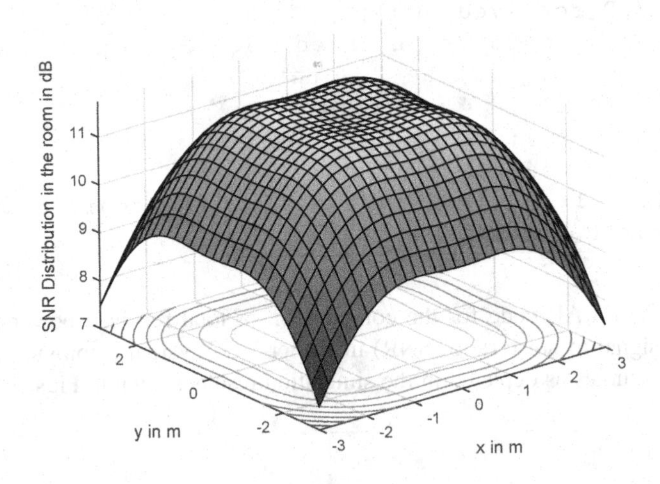

Figure 2.12: Illustration of the distribution of signal to noise ratio (SNR) in indoor VLC system using multiple illuminating sources for the LED semi-angle as 60°.

```
9  rho=0.8; %Reflectance Parameter
10 Ts=1; %Gain of the Optical Filter
11 Refractive_Index=1.5; %Refractive Index of the Lens
12 FOV_PD=70; %Field of View of the Photodiode
13 Concentrator_Gain=(Refractive_Index^2)/(sind(FOV_PD)
```

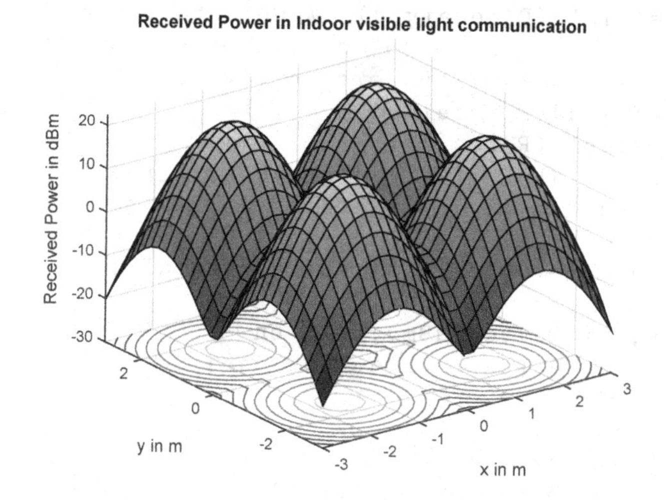

Figure 2.13: Illustration of the received power in indoor VLC system exploiting multiple illuminating sources for the semi-angle of the LED as 12.5°.

```
        .^2); %Gain of the optical concentrator
14  lx=6; ly=6; lz=3; %Dimensions of the Room Environment
        6 X 6 X 3 m^{3}
15  h=2.25; %Distance between the transmitter and the
        receiver plane
16  [XT,YT]=meshgrid([-lx/4 lx/4],[-ly/4 ly/4]); %
        Position of LED
17  Nx=lx*5; Ny=ly*5; %Total number of grids present in
        the receiver plane
18  x=linspace(-lx/2,lx/2,Nx);
19  y=linspace(-ly/2,ly/2,Ny);
20  [XR,YR]=meshgrid(x,y); %Illustrates the receiver
        plane grid
21  D=sqrt((XR-XT(1,1)).^2+(YR-YT(1,1)).^2+h^2); %
        Computation of Distance from the transmitter LED
        to the receiver photodiode
22  cosphi_A1=h./D; % angle vector
23  receiver_angle=acosd(cosphi_A1);
24  H_A1=(m+1)*APD.*cosphi_A1.^(m)./(2*pi.*D.^2); %
        Channel DC gain corresponding to the LOS path
25  P_received=Total_Power.*H_A1.*Ts.*Concentrator_Gain;
        % Total amount of Received power corresponding to
        single LED source 1
26  P_received(find(abs(receiver_angle)>FOV_PD))=0;
```

```
27 P_rec_A2=fliplr(P_received); % Computation of
      Received power from source 2, due to symmetry
      separate calculations are not required
28 P_rec_A3=flipud(P_received);
29 P_rec_A4=fliplr(P_rec_A3);
30 P_received_total=P_received+P_rec_A2+P_rec_A3+
      P_rec_A4; %Total received power from both LED
      sources
31 P_received_dBm=10*log10(P_received_total);
32 %% Fig.
33 surfc(x,y,P_received_dBm);hold on
34 title('Received Power in Indoor visible light
      communication')
35 xlabel('x in m')
36 ylabel('y in m')
37 zlabel('Received Power in dBm ')
38 %Computation of SNR = Signal Power/Noise Power
39 noisebandwidth_factor=0.562; %Noise Bandwidth Factor
40 dataRate=512000; %Data Rate
41  Bn=noisebandwidth_factor*dataRate; %Total Bandwidth
42  q=1.6*10e-19; %Charge of the electron
43  R=1; %Responsivity of the Photodiode
44 Sigma_shot=2*q*R*(P_received_total)*Bn; %Computation
      of Shot Noise
45  Ba=4.5E6;
46  Amplifier_current=0.01;
47  Sigma_Amplifier=Amplifier_current^2*Ba; %Computation
       of receiver circuitry noise
48  Sigma_Total=Sigma_shot+Sigma_Amplifier; %Total Noise
49  SNR=(R*P_received_total)^2./Sigma_Total; %
      Determination of Signal to Noise Ratio (SNR)
50  SNRdb=10*log10(SNR); %SNR in dB
51  figure(2)
52  surfc(x,y,SNRdb)
53  xlabel('x in m');
54  ylabel('y in m');
55  zlabel('SNR Distribution in the room in dB');
```

2.3.6 REVIEW ON REALISTIC CHANNEL MODEL FOR VLC SYSTEMS

Much similar to any other communication systems, realistic channel models play a significant role for the design, analysis of performance and testing of VLC systems. Based upon these grounds, the reference channel models that were endorsed by the IEEE 802.15.7r1 Task Group for the evaluation of the VLC systems will be

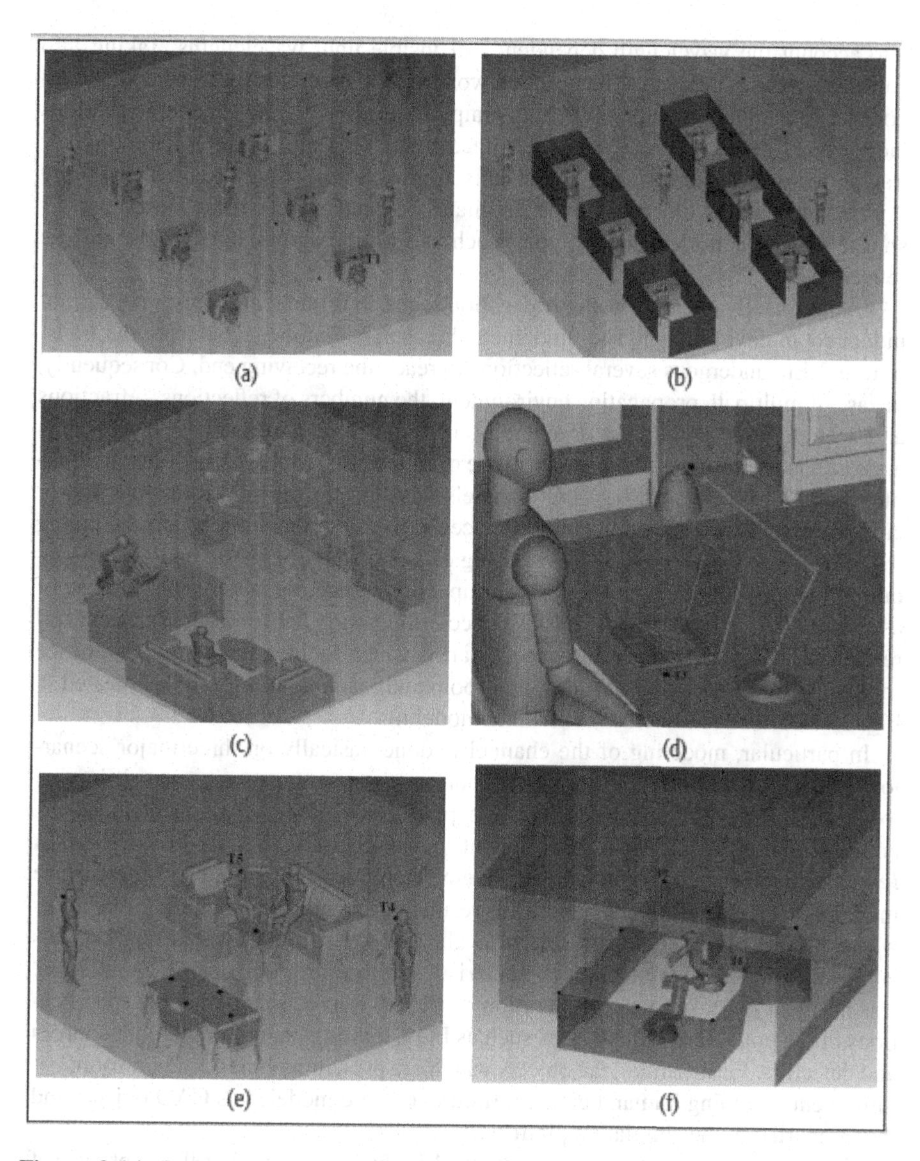

Figure 2.14: Reference scenarios for channel modeling [487]:(a) workplace with open office; (b) workplace with cubicles; (c) office room with secondary light; (d) enlarged version showing secondary light, i.e., desklight; (e) living room; (f) manufacturing cell.

discussed in this section. As discussed in the previous subsections, many significant works pertaining to indoor VLC channel modeling were developed by exploiting idealistic assumptions such as deploying the illumination sources which have a Lambertian radiation pattern, etc. Moreover, much of the research works do not take into account the wavelength dependency of visible light wavelengths. Taking into consideration these aspects, the research work in [347] presents the most realistic indoor VLC channel modeling, where the impulse response of the channel for various indoor environment scenarios is determined by making use of the accelerating ray tracing features of Zemax. In general, this approach of making use of accelerating ray tracing features of Zemax even facilitates to deduce the channel's impulse response for any non-ideal source types, such as specular and mixed specular-diffuse reflections.

Predominantly, the multipath propagation scenario is inherently present in case of indoor room environment. The fundamental reason is that the signal from the transmitting LED undergoes several reflections to reach the receiving end. Consequently, in case of multipath propagation environment, the numbers of reflections, refractions and scatterers are significant. Eventually, in order to acquire better accuracy, this approach is used for handling a wide range of reflections. Making use of the methodology as proposed in [347], VLC channels were originated for distinctive indoor environments which include common places such as home, public places including the office environments and manufacturing cells, and later on these models were acquired by the IEEE 802.15.7r1 Task Group as reference models for the purpose of assessing VLC system proposals [487]. According to the IEEE 802.15.7r1 technical reference documents, four reference scenarios which include the workplace, office room employing secondary light, living room and manufacturing cells delineated in Fig. 2.14 have been selected for channel modeling.

In particular, modeling of the channel is done basically on three major scenarios. If we take into account an indoor room environment scenario, first, it is vital to create a three-dimensional simulation platform in Zemax, which includes the entire geometry of the room environment, more precisely, the size of the room in $length \times breadth \times height$ in m^3, shape, etc. Moreover, the presence of the surface materials such as floor, ceiling and walls results in a significant amount of reflections which in turn play a dominant role in evaluating the path loss. Consequently, this entails to compute the reflection characteristics of the surface materials. In addition, it is necessary to take into consideration as well as to precisely define the specifications of the opto-electronic devices such as FOV, lighting pattern of the light sources and detectors. Furthermore, the objects which are present within the indoor room environment including human beings, furniture, etc. are modeled as CAD objects and then imported to the simulation platform.

The second major phenomenon involved is that by employing the approach of the non-sequential ray tracing features of Zemax, the detected optical power and path lengths from the source to detector is calculated. Further, the fundamental aspect which is usually involved behind ray tracing is that the rays emitted by the source follow a given statistical distribution, whereupon these sets of emitted rays are traced along a physically realizable path until they intercept an object. By taking into consideration the multipath propagation scenario, the transmitted signal from the transmitter undergoes several reflections, refractions and diffractions before reaching the receiving end. Thus, the same phenomenon holds well even for the indoor room environment where the transmitted signal from the transmitting illuminating source undergoes several reflections due to the presence of walls, floor, ceiling and several other objects which are present. Thus, along with the presence of LOS components, NLOS components also emanate. Finally, the major objective behind the exploitation of the Zemax ray tracing tool is to originate an output file which encompasses the detected power and path length for each ray. Thereupon, the generated data is imported to MATLAB and by availing this information, the impulse response of the channel can be determined as [487]

$$h(t) = \sum_{i=1}^{N_r} P_i \delta(t - \tau_i) \tag{2.34}$$

From equation (2.34), P_i and τ_i signify the optical power and the propagation time corresponding to the ith ray, respectively, while $\delta(t)$ characterizes the Dirac delta function. Whereas, N_r specifies the total number of rays received by the detector. According to the IEEE 802.15.7r1 technical requirements document [488, 490] four scenarios which include workplace with open office floor and cubicles, office room with secondary light, living room and manufacturing cells were selected for channel modeling. In all these scenarios a major inference can be drawn that if the separation between the source LED and the receiving PD is very large, then it can be asserted that a huge number of scattering components can be observed due to reflections from walls, ceiling and floor. Consequently, these enormous reflections lead to substantial enhancement in the delay spread.

2.4 REVIEW ON VLC CHANNEL MODELS

This section presents in detail a comprehensive review on the existing research efforts that strived to implement theoretical analysis on VLC channel characteristics by making use of the impulse response of the channel $h(t)$ in order to determine the properties of the VLC channels. Particularly, when compared with indoor infrared communication channels, VLC channel modeling exhibits several similarities and differs only in the type of media exploited. Therefore, it can be anticipated that since VLC channel modeling shares several similar features with infrared communication channel models, so much of the VLC channel modeling is largely based on infrared channel modeling, which in turn helps to characterize the multipath propagation environment.

There are many significant works pertaining to VLC channel modeling in the literature. We confine our discussion on the most widely used channel models. The prior works on indoor infrared communication channel models dates back to 1979, when Gfeller and Bapst [177] proposed a reflectance model suitable for indoor infrared communication. However, the work presented by Gfeller and Bapst accounts for only single reflection. Eventually, a considerable number of research works emerged to compute the impulse response of the VLC channel by taking into account multiple reflections. Later on, in 1993, the work in [52] presents a recursive method to determine the impulse response of an indoor free-space optical channel comprising of Lambertian reflectors. The stunning notable feature of this method is that it can compute the impulse response of the channel for any number of reflections. Substantially, this method facilitates to accurately evaluate the effects of multipath dispersion on high-speed indoor OWC system. Furthermore, from this work, it can be witnessed that the proposed recursive method for determining the impulse response of the channel was validated by means of computer simulations for both LOS and diffuse transmitter configurations. From the simulation results, it can be inferred that multiple order reflections are the major source for the emergence of ISI. Pertaining to VLC, the first break through dates back to 2000 when Tanaka et al. proposed an indoor wireless optical communication system that exploits white-colored LEDs dedicated for wireless home links [468]. The authors presented basic analysis and conducted simulations on VLC channel models.

The same authors extended the work as stated in [468] by determining the optical path difference as well in [469]. This work portrays that the influence of the optical path difference on the wireless link has been investigated, and the authors employed two approaches to reduce the amount of delay in order to ensure high data rate communication. The first approach is through the usage of ON-OFF keying (OOK) return-to-zero (RZ) coding, and the second approach employs optical OFDM modulation format. From the simulation results, it can be witnessed that the former approach mitigates the induced delay when the data rate was 100 Mbps, but a remarkable improvement in the overall performance cannot be witnessed for higher data rates of 400 Mbps. On the other hand, the latter approach, i.e., OFDM dominates the former by reducing the amount of induced delay. The performance of indoor diffuse optical channel environment was analyzed by enforcing an equalization technique using artificial neural network (ANN) in a PPM-based VLC system [419]. Generally, this channel model is based on the infrared (IR) channel model which was established by Carruthers and Kahn et al. [85]. A wireless IR communication channel in an indoor environment is presented in [246]. The IR channel in this scenario is modeled as a parallel combination of both LOS (direct) and diffuse (scattered) components. While considering the diffuse component alone, an integrating sphere is employed to determine the properties of the channel. Whereas, in the case when LOS component is present, the transfer function is dependent upon the Rician factor K. In order to ensure a distortion-free data transmission in the range of 100 Mbps, a K factor of 13 dB is required.

The work in [139] signifies that the VLC channel model utilizes a very fast method to simulate the channel based on ray tracing algorithm. Essentially, this algorithm is based on the principality of Lambert-Phong reflection model where it can work well even in complex environmental conditions, i.e., in the presence of furniture, curved surfaces, etc. The indoor VLC channel modeling as depicted in [288] and [289] employs the wavelength characteristics of white LEDs as well as the spectral reflectance of indoor reflectors. Besides, upon comparison of VLC's power delay profile with IR communication it can be inferred that the received power from the reflected paths as well as the root mean square (RMS) delay spread of VLC is smaller than that of IR scenario, which implies that VLC has large optical transmission bandwidth. The channel model as illustrated in [579] considers the multiple reflections of an indoor room environment where the authors have evaluated the bit error rate (BER) performance of the system when affected with multiple reflections. A survey on VLC channel modeling which details the earlier works is reported in [545].

NLOS-OWC link with IM and DD is more appropriate for high-speed indoor communication systems. This is because, upon comparison with LOS links, NLOS links offer a huge coverage area, and high mobility without any accurate alignment between the transmitter and receiver. In addition, when compared with diffuse configurations, NLOS-OWC links provide lower path loss, lower ISI caused by multipath reflections and ensures higher transmission bandwidth. Therefore, taking into consideration the aforementioned advantages of NLOS-OWC links, the authors in [526] propose an optimised Lambertian order (OLO) of LEDs for an indoor NLOS-OWC system. Using the conventional Lambertian LED model, an expression for OLO, was derived and then the channel characteristics in terms of optical power distribution and multipath time dispersion for an indoor multicell NLOS-OWC system were analyzed for the two scenarios of one-cell and four-cell configurations. The result analysis infers that there is a significant improvement in transmission bandwidth of NLOS-OWC system which is based on OLO of LEDs.

A realistic mobile VLC channel based on a non-sequential ray tracing phenomenon was developed in [346]. This work employs the non-sequential ray tracing algorithm to determine the channel impulse response for each point over the user movement trajectories followed by the calculation of path loss and RMS delay spread as a function of distance. However, the obtained results depict large variations in received power. Consequently, while designing the system it necessitates incorporating adaptive system which is based on luminaries selection for the purpose of choosing the transmission parameters according to the channel conditions. Such kind of adaptive system shows improvement in terms of spectral efficiency when compared with non-adaptive systems.

In [402], the channel properties of NLOS ceiling-to-device and device-to-ceiling are evaluated by taking into account the receiver's orientation and variable FOVs. Furthermore, through experimental validations, it can be deduced that shadowing, which arises due to the movement of people, leads to a significant reduction in the received power level. The work in [114] exploits the combined approach of deterministic and modified Monte Carlo methods to determine the channel impulse response of an indoor optical wireless room environment with sufficient number of reflections from walls, floor and ceiling. This approach of calculation of channel impulse response incorporates the advantages of both these methods namely deterministic and Monte Carlo methods and further allows to efficiently utilize the full power of modern multicore computer processors.

The performance of the VLC system which is analyzed in [118] depicts that the impulse response of the channel is determined by taking into consideration the dynamic indoor room environment. This work reports that the performance of the system, which is affected with the movement of the people, is evaluated both analytically as well as experimentally for three different indoor frameworks which include corridor, furnished room and emptyhall. Additionally, this work reports that the channel characteristics were investigated by using the cumulative distribution function of the received power distribution and the RMS delay spread. The authors in [190] considered a complex indoor room environment with walls, floor and ceiling and investigated the impact of multipath reflections on a two-dimensional VLC positioning system. This work employs the combined deterministic and modified Monte Carlo method to compute the impulse response of the channel, and calibration approaches were proposed to reduce the effect of these multipath reflections in a VLC positioning system.

Much recent literature pertaining to VLC channel modeling can be reported in the below works; Principally, most of the earlier works as stated above assume that the indoor VLC system is static. However, pertaining to the real-world scenario, indoor VLC channel is affected by the density of people present within the indoor room environment, shadowing effects, dimming, background lights and interiors create a dynamic effect. Thus, this infers that the time-varying channel effect cannot be ignored while modeling the VLC system. Moreover, the effect of dynamic channel cannot be mitigated just by enhancing the LED transmission power. Possibly, the best way to model the channel under dynamic conditions is to have a clear estimate of the channel under dynamic VLC channel environments. Consequently, much relevant work which works in this direction, i.e., which takes into account the time-varying channel effect can be reported in [34], where the authors consider a dynamic VLC channel environment which clearly suggests that a decrease in the normalized received power follows Rayleigh distribution. By employing the variants of least mean square (LMS) algorithm which includes the normalized LMS (NLMS), zero attracting LMS (ZA-LMS), block LMS (BLMS) and fast block LMS (FBLMS), the channel coefficients were estimated. Furthermore, the applicability of various adaptive algorithms were validated in terms of mean square error (MSE) and tap-weights convergence, computational complexity and the number of pilot symbols required in VLC dynamic channel. Additionally, from the simulated results, it is clear that FBLMS and NLMS algorithms exhibit superior performance when compared with the other algorithms. A much similar work which takes into account the impact of position, size and shape of the obstacle while modeling the VLC channel can be found in [361]. Particularly, the illumination area of the LED was bifurcated into independent solid angles, where by means of beam streaming the ray was transmitted into every solid angle. Thereupon, the point cloud of the obstacle was identified by employing the rays that were received by the photodiode. The authors in this work use the convex hull model to realize the point cloud, where a complex VLC channel model was established and then followed by derivation of the impulse response of the channel. The simulation results of this work depict that the position, shape and size of the obstacle have a significant impact on the characteristics of VLC channel. Also, beam steering enhances the uniformity of the distribution of SNR and the available bandwidth.

A VLC channel is modeled as a multipath propagation environment due to the presence of specular and non-specular signal components on the photodiode. Particularly, in such environments, the Rician K-factor is employed to determine the superiority of LOS signal power over that of the NLOS link. Toward this end, the work in [366] exploits the Rician K-factor to evaluate the VLC link, where several metrics of the link include the SNR, channel capacity and outage probability. In several studies, it is revealed that the performance of VLC-based systems is limited in certain scenarios especially when the users are mobile with respect to the transmitting luminaire. Predominantly, the relative motion of the mobile users with respect to the transmitting LED luminaire enables the overall VLC channel to be time-varying. In order to tackle this aspect, it is vital to rely on appropriate mechanism at the receiver end. In addition to the user mobility, ISI, the non-linear characteristics of LEDs, turns out to be a major limiting factor that impairs the throughput of VLC systems. Consequently, to tackle these impairments, some of the techniques like Volterra/Hammerstein-based receivers seems to be the better choice. However, such receivers suffer from modeling error due to the truncation of the polynomial kernel till second order terms. In the recent times, sparse reproducing kernel Hilbert space (RKHS)-based methods are appropriate to ensure universal approximation with reasonable computational simplicity. However, this method is also not devoid of drawbacks, the choice of a single hyper-parameter restricts its ability to model time-varying channels. Thus, taking into consideration these aspects, the work in [349] proposes a novel RKHS-based post-distorter method that adaptively learns a sparse dictionary based on the incoming observations, and monitors validity of the dictionary based on a proposed metric in RKHS.

The work in [36] analyzes the secrecy performance of MIMO-VLC wiretap channel. The basic system model of this work consists of three nodes, namely, the transmitter equipped with multiple LED lighting fixtures, one receiver and one eavesdropper equipped with multiple photodiodes. Further, the VLC channel is modeled as a real-valued amplitude-constrained Gaussian channel, while the eavesdropper is randomly located in the coverage area. The authors in this work propose a low-complexity precoding scheme that is based on generalized singular value decomposition (GSVD) with the major motive to augment the secrecy performance of the overall system. Further, the achievable average secrecy rate was evaluated by using truncated discrete generalized normal (TDGN) distribution and the proposed scheme was validated by means of extensive simulations. A much similar work in [37] investigates the secrecy performance of the multiple input single output (MISO) VLC wiretap channel. Here, this system model consists of three nodes, namely, the transmitter (Alice) which is equipped with multiple fixtures of LEDs, an appropriate receiver which is called Bob and an eavesdropper called Eve which is comprising of one photodiode. Here, in this work, the VLC channel is modeled as a real-valued amplitude-constrained Gaussian channel and Eve is assumed to be located randomly in the same area as that of Bob. Hence, as a result, artificial noise (AN)-based beamforming is employed as a transmission strategy with the motive to degrade Eve's SNR. Furthermore, this work derives the closed-form expression for the achievable secrecy rate as a function of beamforming vectors and input distribution. To account for the random location of Eve, the average secrecy performance of the system is evaluated by using stochastic geometry. In order to evaluate the performance of the proposed scheme, this work adopts the TDGN as a discrete input distribution. The analytical results were validated by means of Monte Carlo simulations and the proposed scheme exhibits significant improvement when compared with the existing ones.

Even though, VLC exploiting LEDs has drawn a significant amount of attention due to its energy-efficient nature and high data rate communication. However, the performance of indoor VLC systems can be affected due to the shadowing induced by the presence of mobile obstacles as shown in Fig. 2.15. Thus, the presence of these obstacles block the LOS link, thereby deteriorating the performance of the overall system. To address this problem, the authors in [148] model the link blockages as the incomplete channel matrix with sparse missing elements and reconstruct the signal which is transmitted by means of 0-minimization enforcing additional constraint on shadowing loss. Indoor VLC channel is highly correlated; therefore, in order to further improve the accuracy of the overall system performance, a transmit beamforming has been proposed to reduce the total coherence. Researchers also worked

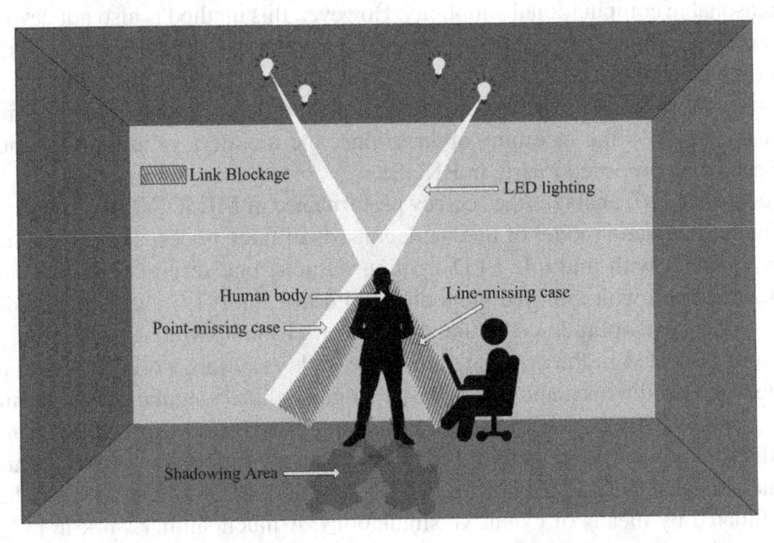

Figure 2.15: Impact of shadowing induced by the obstacles in MIMO-VLC systems.

in the direction to overcome the computational complexity that is usually involved in VLC channel modeling. In particular, a ray-tracing-based channel modeling technique is very powerful in evaluating the characteristics of VLC channel. However, in the practical realizations of VLC channel, it is essential for the ray-tracing-based channel modeling approaches to consume a less amount of time. Toward this end, the work in [302] proposes a channel model that exploits optical simulations which are based upon non-sequential ray-tracing that works well with the self-developed algorithms in MATLAB. Furthermore, the authors in this work propose an approach that is capable of effectively managing time to determine the dependence of a receiver position on the VLC characteristics along the diagonal of the floor of a room.

The ray-tracing information pertaining to each ray which is invading on the diagonal stripe is exported to MATLAB to facilitate further analysis. In this manner, the proposed approach is able to determine 26 number of receiver positions along with the diagonal axis of a room. Furthermore, it is also interesting to note that upon utilizing only one ray-tracing simulation, 26 receiver positions were determined without enforcing much repetition in the ray-tracing simulations. The work in [195] presents an architecture for MIMO simulator that works well for radio propagation channels. The sophisticated nature of this architecture is that it can work well with time-varying 802.15.7 VLC signals, and this hardware channel simulator is designed on a FPGA Virtex-VII FPGA from Xilinx. Moreover, this architecture is employed to evaluate the "on-table" VLC indoor propagation channels. Further, this simulator utilizes the photodiode power (PDP) model by generalizing Barry's model to attain the impulse response of the channel with a frequency range of 650 THz. This work analyzes the accuracy of the output signals, the occupation of FPGA and the architecture latency.

In addition to the indoor VLC channel modeling, research is rapidly progressing to model an outdoor VLC channel as well. In this regard, much relevant work pertaining to this context can be witnessed in [27], where the authors propose an outdoor vehicle to vehicle VLC (VVLC) channel model. Due to the absence of a realistic channel model that takes into consideration the mobility of the vehicles and the propagation nature of light, this work proposes a regular-shaped geometry-based stochastic channel model (RSGBSM). More accurately, an ellipse shape was utilized to determine the LOS and the reflection or the NLOS components of the proposed VVLC channel. The important characteristic feature offered by this channel model is that it offers the viability to determine the locations of the vehicles, the speed and direction of the approaching vehicles along with stationary and non-stationary reflectors that are rapidly and continuously changing. From the simulated results, it can be deduced that the direction of motion of the vehicles mostly affects the LOS component; whereas, the relative speed of the vehicles imposes a huge impact on the reflection components.

A much similar recent finding which is related to modeling of VLC channel pertaining to the outdoor vehicular applications is portrayed from [250] where the authors evaluate the performance limits of vehicular VLC systems. First, in order to yield the impulse response of the channel corresponding to the vehicle to vehicle link over several weather conditions, the authors employ the non-sequential ray-tracing algorithm. By making use of the obtained channel impulse responses, this work presents a closed form path loss expression whose summation depends upon the geometrical loss and attenuation loss, as well as takes into account the asymmetrical patterns of the vehicle light sources and the geometry of the vehicle to vehicle transmission. Furthermore, the proposed expression explicitly depends on the link distance, lateral shift between two moving vehicles, the type of the weather conditions, the aperture diameter of the receiver and the beam divergence angle of the transmitter. In general, this proposed path loss expression was used to arbitrate the maximum achievable link distance of vehicular applications while maintaining a desired BER under several weather conditions such as clear, rainy, foggy, etc.

The work in [425] characterizes the availability of LOS link in indoor VLC networks, where the impact of human behavior on the availability of VLC link is quantified. In general, humans affect the availability of the VLC link in three distinct ways: More particularly, in the first way, users turn on and off lights in each room, secondly, the users carry mobile devices, and thirdly, the channel between the transmitter and the receiver is blocked or shadowed since users act as mobile obstacles. Consequently, to address these aspects, this work develops a mathematical framework that characterizes the VLC link availability and as well as develop a probabilistic model for the VLC network that takes into consideration the behavior of humans in indoor environments. Furthermore, this work designs a simulation set-up for a realistic multiuser indoor VLC system that takes into account both static and mobile VLC devices which usually prevails in smart home environments. Finally, this work presents four sets of inferences from the simulation results: firstly, the authors employ the statistics of the blockage durations of VLC links to categorize the VLC links. Secondly, this work demonstrates the performance of selection diversity versus maximal ratio combining for mobile VLC devices that were carried by humans in the smart home environment. Thirdly, the authors show that the VLC link blockages which are usually caused by humans have an impact on the optimal LED resource allocation policies among multiple users. Finally, the transition of humans between rooms will facilitate the VLC link blockages in different rooms to become dependent.

The work in [362] exploits the non-sequential ray-tracing algorithm to model the MIMO VLC channels under various practical wiring and cabling topologies. In order to improve the data rates, MIMO communication emanates as the most appealing solution for VLC. In such scenarios, multiple luminaires are exploited for the purpose of illumination. Predominantly, the LED luminaire comprises of several LED chips, and the wiring topology is meant to determine how well the LED chips are connected within the luminaire. On the other hand, the cabling topology assesses whether the luminaires are properly connected to the communication access points. Thus, based upon the type and length of the cabling and wiring, there are chances for the introduction of significant amount of delays; therefore, it is vital to take into consideration such delays while modeling the VLC channel. Toward this end, the work in [362] develops MIMO VLC channel models that take into account both wiring and cabling topologies and compare the performance analysis of different MIMO techniques which include repetition coding (RC), spatial multiplexing (SMUX), and spatial modulation (SMOD). The simulation results of this work clearly illustrate the effect of both wiring and cabling delays and offer deep insights into the optimized design of lighting infrastructure and luminaires to facilitate VLC to emerge as an add-on service.

2.4.1 REVIEW ON INDOOR VLC CHANNEL CHARACTERISTICS MODELED BY LEE ET AL.

Gfeller and Bapst [177] report the first study on channel characteristics pertaining to infrared communications. In [248], a computer simulation for the channel modeling was presented by exploiting a recursive method. These studies take into account a narrowband near-monochromatic infrared light source, where the reflectance parameter of all the reflectors within the indoor room environment is modeled as a constant by ignoring its dependence on wavelength. The research contribution in [275] and [189] present in detail several results about the VLC channel modeling by using a recursive algorithm. It is to be noted that a constant value of reflectance was assumed for all the wavelengths, much similar to infrared communication. Similarly, the work in [125] also reports some experimental validations of the VLC channel model where the reflectance parameter was not considered. Thus, the authors in [290] extend the recursive algorithm as portrayed in [248] in order to formulate a VLC channel model exclusively for indoor applications by taking into consideration the wavelength-dependent nature of the reflectors. Very similar to a conventional wireless channel, the received optical signal in case of an indoor room environment undergoes the effect of time dispersion which generally arises due to the reflection emanating from walls and other objects which are present within the room environment. In general, the reflections arising from several interior materials within the room environment are purely diffusive in nature. Here, we focus on VLC channel model that exploits the recursive algorithm and takes into account the reflectance parameter of the reflectors [290].

A VLC channel model illustrating the communication from the nth LED to the photodiode is illustrated in Fig. 2.16 which is adopted from [290]. This channel model takes into consideration the reflections emanating from the walls, floor and other objects present within the room. Primarily, this work compares the simulations of the characteristics of VLC channel model with that of the infrared communications. Particularly, as evident from the figure, the phosphor-based white LEDs radiate wideband visible light, and the radiant power emanated by such LEDs is widely distributed over the entire visible light spectrum ranging from 380 to 780 nm. The two important parameters which should be considered while modeling the indoor optical channel environment are the power spectral distribution (PSD) and the spectral reflectance. First, the PSD of an illuminator in general determines the radiant power per unit wavelength. On the other hand, the second important parameter is the spectral reflectance which determines the reflectivity that is varying in accordance with the wavelength. Essentially, due to the effect of multipath propagation environment, i.e., the occurrence of reflections from the walls, floor, ceiling, etc., leads to time spreading of the received signal. Therefore, the power delay profile (PDP) can be employed to evaluate the effect of multipath dispersion between the VLC transceivers. Consequently, the PDP for recursive algorithm comprises of the distance dependent path-loss, time-delay and the reflected power. In order to determine the reflected light power, a method that takes into consideration both PSD of optical illuminating source and the spectral reflectance of the reflector has been proposed in [290].

When multiple LEDs are installed on top of the ceiling, after k reflections, the impulse response of the channel can be expressed as [290]

$$h(t) = \sum_{n=1}^{NLED} \sum_{k=0}^{\infty} h^{(k)}(t; \Phi_n) \tag{2.35}$$

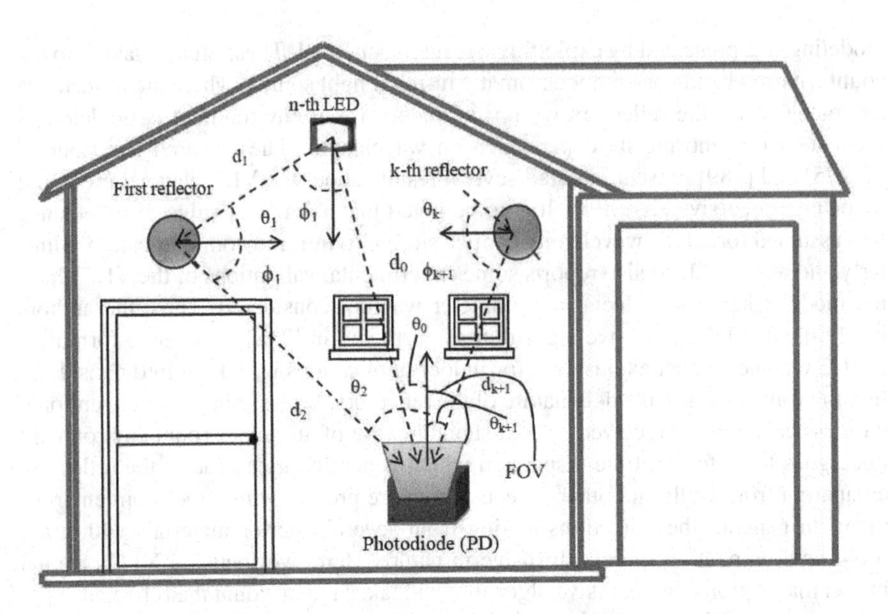

Figure 2.16: VLC channel model proposed by Lee et al.

From (2.35), the parameters $NLED$, Φ_n represent the total number of LEDs that were installed on the ceiling and PSD, respectively. Furthermore, the transmitted power of these LEDs is equal. A generalized expression for the channel impulse response indicating the response of the nth LED experiencing the kth reflection can be formulated as [290]

$$h^{(k)}(t;\Phi_n) = \int_S \left[L_1 L_2 L_3 \cdots L_{k+1} \Gamma_n^{(k)} rect\left(\frac{\theta_{k+1}}{FOV}\right) \delta\left(t - \frac{d_1 + d_2 + d_3 \cdots + d_{k+1}}{c}\right)\right] dA_{ref}, k \geq 1 \tag{2.36}$$

Where the path loss terms $L_1, L_2, L_3, \cdots L_{k+1}$ as depicted by equation (2.36) can be determined as [290]

$$L_1 = \frac{A_{ref}(m+1)\cos^m(\phi_1)\cos(\theta_1)}{2\pi d_1^2}$$

$$L_2 = \frac{A_{ref}(m+1)\cos(\phi_2)\cos(\theta_2)}{\pi d_2^2}$$

$$\vdots \quad \vdots$$

$$L_{k+1} = \frac{A_{PD}\cos(\phi_{k+1})\cos(\theta_{k+1})}{\pi d_{k+1}^2} \tag{2.37}$$

Furthermore, from equation (2.36), the parameter S corresponds to the area of all reflection surfaces, while the area of the reflecting element is denoted by A_{ref}. The irradiance angle and the incident angle are indicated by ϕ_k and θ_k, respectively. The distance between the LED transmitter and the receiver is specified by d_k, the photodiode's FOV is indicated as FOV and, in general, the photodiode detects the light signal provided if its incidence angle is less than the FOV of the photodiode. The speed of light is denoted by c and the function $rect\,(y)$ corresponds to the rectangular function which can be put up as [290]

$$rect\,(y) = \begin{cases} 1 & |y| \leq 1 \\ 0 & |y| > 1 \end{cases} \tag{2.38}$$

The parameter $\Gamma_n^{(k)}$ in equation (2.36) characterizes the power of the reflected ray after k bounces from the nth LED and is given by [290]

$$\Gamma_n^{(k)} = \int_\lambda \Phi(\lambda)\rho_1(\lambda)\rho_2(\lambda)\cdots\rho_k(\lambda)\,d\lambda \tag{2.39}$$

In (2.39), $\rho_1(\lambda)\rho_2(\lambda)\cdots\rho_k(\lambda)$ represents the reflectance and Φ_λ denotes the radiant power. Further, from (2.39), both parameters are dependent on the wavelength. Meanwhile, the PDP corresponding to the LOS component is given by [290]

$$h^0\,(t;\Phi_n) = L_0 P_n rect\left(\frac{\theta}{FOV}\right)\delta\left(t - \frac{d_0}{c}\right) \tag{2.40}$$

where L_0 in (2.40) can be expressed as [290]

$$L_0 = \frac{A_{PD}\,(m+1)\,cos^m\,(\phi_0)\,cos\,(\theta_0)}{2\pi d_0^2} \tag{2.41}$$

In order to evaluate the performance of the proposed VLC channel model, simulations were carried out and, further, the received powers of both VLC and infrared communication were compared explicitly. The simulated results infer that the total received power emanated from the reflected paths, and the RMS delay spread pertaining to VLC is quite small when compared to infrared communication. Consequently, this implies that VLC has a larger optical transmission bandwidth than that of the infrared communication due to the inherent fact that reflectivity of visible light is lower than that of the infrared communications.

2.4.2 REVIEW ON DIFFUSE INDOOR OPTICAL WIRELESS CHANNEL MODELED IN ACCORDANCE TO RAJBHANDARI ET AL.

The research work in [419] evaluates the BER performance of a pulse position modulation (PPM) scheme over NLOS indoor optical links which utilizes channel equalization technique based on artificial neural network (ANN). The simulated results of this work affirm that the neural network equalization is an appropriate tool for mitigating the effects of ISI. The channel model that was exploited in this work was assumed to be time invariant and the channel model pertaining to the diffuse indoor optical wireless link was based upon the widely deployed ceiling bounce model comprising of impulse response [85] which can be expressed as

$$h(t) = H(0)\frac{6a^6}{(t+a)^7}u(t) \tag{2.42}$$

From equation (2.42), the height of the ceiling from the transmitter and receiver is represented by H, and $u(t)$ corresponds to the unit step function. Further, the parameter a as shown in equation (2.42) depicts the value given as $a = \frac{2H}{c}$. Moreover, it is associated to the root mean square (rms) delay spread which is determined as

$$\tau_{rms}(h(t)) = \frac{a}{12}\sqrt{\frac{13}{11}} \tag{2.43}$$

Where τ_{rms} in (2.43) specifies the delay spread. It is vital for the parameter a to be modified in accordance to the most realistic channel model comprising of multiple reflections. Therefore, for the scenarios of unshadowed and shadowed channels, Carruthers and Kahn provide the mathematical expressions for the parameter a as delineated by the expressions (2.44) and (2.45).

$$\underset{Unshadowed}{a} = 12\sqrt{\frac{11}{13}}\left(2.1 - 5.0s + 20.8s^2\right)\tau_{rms}(h_1(t)) \tag{2.44}$$

$$\underset{Shadow}{a} = 12\sqrt{\frac{11}{13}}(2.0 + 9.4s)\,\tau_{rms}(h_1(t)) \tag{2.45}$$

From (2.44) and (2.45), the parameter s represents the ratio of the horizontal transmitter-receiver separation to the diagonal distance between the two [545].

2.4.3 REVIEW ON INDOOR VLC CHANNEL MODEL PROPOSED BY DING DE-QIANG ET AL.

The work in [139] analyzes the indoor VLC multipath channel model. This work presents a reflection model that is based on Lambert-Phong pattern. Furthermore, by means of Monte Carlo ray-tracing simulation which is based on Lambert-Phong pattern, this work establishes the VLC optical channel models for both single source and multisource systems. Therefore, the impulse response of VLC channel is determined by exploiting the ray-tracing algorithm that is based on Lambert-Phong pattern. Accordingly, the impulse response of the channel can be expressed as the sum of LOS and diffuse components as shown by equation (2.46)

$$\underset{VLC}{h(t)} = \underset{LOS}{h(t)} + \underset{\substack{NLOS\\(Diffuse)}}{h(t)} \tag{2.46}$$

From (2.46), the impulse response corresponding to the LOS component can be expressed as [139]

$$\underset{LOS}{h(t)} = \sum_{i=1}^{l} P_i \delta\left(t - \frac{|r_R - r_S|}{c}\right) \tag{2.47}$$

P_i in (2.47) indicates the emitted optical power relevant to the ith ray, c is the speed of light and $|r_R - r_S|$ determines the distance between the light source and the receiver. Meanwhile, the impulse response of the channel pertaining to the diffuse component is given by [139]

$$\underset{\substack{NLOS\\(Diffuse)}}{h(t)} = \sum_{j=1}^{J}\left[\prod_{k=1}^{K} \alpha_{j,k} P_j\right] \delta\left(t - \frac{W_j}{c}\right) \tag{2.48}$$

In equation (2.48), the total number of reflections is denoted by K, the path loss of the jth ray undergoing k reflections is signified by $\alpha_{j,k}$, the optical power transmission corresponding to the jth light ray is represented by P_j and the propagation delay pertaining to the jth light source is given by W_j. Thus, by making use of (2.47) and (2.48), the impulse response of the VLC channel exploiting single source can be formulated as [139]

$$\underset{VLC}{h(t)} = \sum_{i=1}^{l} P_i \delta\left(t - \frac{|r_R - r_S|}{c}\right) + \sum_{j=1}^{J}\left[\prod_{k=1}^{K} \alpha_{j,k} P_j\right] \delta\left(t - \frac{W_j}{c}\right) \tag{2.49}$$

Now, for the scenario of multiple illuminating sources, the VLC channel impulse response can be represented as

$$\underset{VLC}{h(t)} = \sum_{l=1}^{NLED}\left[\sum_{i=1}^{l} P_{l,i} \delta\left(t - \frac{|r_R - r_{S,l}|}{c}\right) + \sum_{j=1}^{J}\left[\prod_{k=1}^{K} \alpha_{l,j,k} P_{l,j}\right] \delta\left(t - \frac{W_{l,j}}{c}\right)\right] \tag{2.50}$$

From the expression (2.50), the total number of LEDs used is indicated by $NLED$.

Finally, this model which was proposed by Ding De-qiang et al. was based upon the classic Lambert-Phong reflection model that works in association with the ray-tracing algorithm. Also, as seen earlier, it is interesting to note that the channel impulse response is expressed as the combination of both LOS and diffuse components; thus, this sort of VLC channel modeling draws more insights about the multipath propagation environment in an indoor room environment.

2.4.4 REVIEW ON THE EFFECT OF HIGHER-ORDER LIGHT REFLECTIONS ON VLC CHANNEL MODELING

Even though the VLC channel models proposed in the aforementioned research works seems to attain an enhancement in simulation efficiency, the computation consumes a significant amount of time. Moreover, with the increase in the order of reflections, the computation complexity increases significantly. Thus, in order to make sure whether the proposed channel models can be simulated within the stipulated time period, researchers focused to use the channel models by taking into consideration only the first few bounces/reflections; most probably three reflections were employed in order to represent the complete model. This assumption holds well only for low-speed transmissions. However, with the rapid shift toward high data rate transmissions in the range of Gbps, the aforesaid assumptions no longer hold well, producing invalid results. Eventually, there is a need to consider higher order reflections while modeling VLC channels. In this context, the work in [579] evaluates the effect of higher order reflections on VLC channel models. Upon employing the attained results, the authors in this work formulate general rules to model the error distributions for the increase in the order of reflections. Moreover, a calibration approach has been proposed to reduce the errors caused by RMS delay spread as well as to maintain an enhancement in computational efficiency. A typical indoor optical

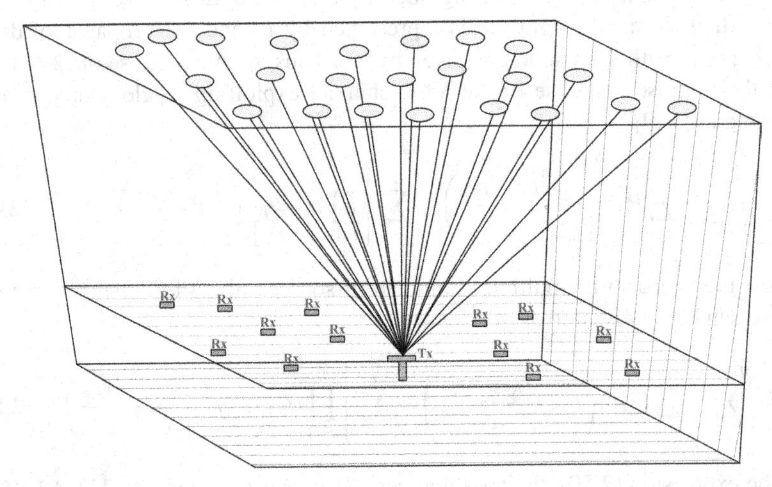

Figure 2.17: Illustration of the VLC channel model proposed by Zhou et al.

wireless communication system as delineated in Fig. 2.17 comprises of a transmitter placed in the center of the room. This transmitter generates multiple scattering diffuse spots on the ceiling, and the receivers are generally located in the room which are meant to receive the data from these diffuse spots. Indoor optical wireless channels are described by means of impulse responses. Thus, these different multiple diffuse spots as elucidated in the figure bounce the data off the ceiling and act as spatially diverse transmitters. Particularly, the impulse responses are attained only when the receiver is located at different locations in the room. Therefore, if the total number of spots that were located in the room were termed to be M and the number of receiver locations were assumed to be N, then such an indoor room environment is characterized by an impulse response matrix which is designated as $H_{M\times N}$ and is shown by equation (2.51) [579].

$$H_{M\times N} = \begin{bmatrix} h_{11}(t) & h_{12}(t) & \cdots & h_{1N}(t) \\ h_{21}(t) & h_{22}(t) & \cdots & h_{2N}(t) \\ \vdots & \vdots & \ddots & \vdots \\ h_{M1}(t) & h_{M2}(t) & \cdots & h_{MN}(t) \end{bmatrix} \tag{2.51}$$

From the mathematical expression as represented by (2.51), the element in the matrix which is designated by $h_{ij}(t)$ indicates the impulse response from the ith diffuse spot to the jth location of the receiver. Since multiple diffuse spots were employed to implement the phenomenon of spatial diversity, equal gain combining can be employed to determine the overall impluse response of the jth location which is given by

$$h_j(t) = \sum_{i=1}^{M} h_{ij}(t) \tag{2.52}$$

By means of diffused multipath propagations, the total impulse response of the channel can be determined. In particular, a scattering channel is characterized by the emanation of a large number of time extensions and delays in LOS signals. Accordingly, the impulse response of the VLC channel pertaining to the ith transmitter and the jth receiver can be formulated as follows

$$h_{ij}^{(0)}(t) = \frac{cos(\phi_{ij})\, cos(\theta_{ij}) A_{Rj}}{R_{ij}^2} \delta\left(t - \frac{R_{ij}}{c}\right) rect\left(\frac{\theta_{ij}}{FOV_j}\right) \tag{2.53}$$

From (2.53), the subscript in the term $h_{ij}^{(0)}(t)$ signifies the LOS component indicated that from ith transmitter to jth receiver, the number of reflections experienced are zero. Meanwhile, the angles ϕ_{ij} and θ_{ij} specifies the irradiance and incidence angles, respectively. The effective area of the receiver is denoted by A_{Rj}, while the distance between the transmitter illuminating source and the receiver photodiode is represented by R_{ij} and the photodiode's FOV, i.e., the jth receiver is emphasized by FOV_j. While, the impulse response from the ith transmitter to the jth receiver undergoing exactly k bounces can be determined from the channel impulse responses experiencing $k-1$ bounces and LOS impulse response, respectively.

$$\begin{aligned} h_{ij}^{(k)}(t) &= \sum_{i=0}^{Q} h_{il}^{(0)}(t)\, h_{lj}^{(k-1)}(t) \\ &= \frac{1}{\pi} \sum_{l=0}^{Q} \frac{\rho_l cos(\phi_{il})\, cos(\theta_{il}) A_{Rj}}{R_{il}^2} rect\left(\frac{\theta_{il}}{FOV_l}\right) h_{lj}^{(k-1)} \times \left(t - \frac{R_{il}}{c}\right) \end{aligned} \tag{2.54}$$

From equation (2.54), the reflectivity of the reflector l is indicated by ρ_l, and Q delineates the total number of diffuse reflectors present in the channel model. Thus, by recursively exploiting the mathematical expression for the channel impulse response in accordance to the diffuse components, the impulse response pertaining to any arbitrary number of reflections can be generated. Accordingly, the impulse response of the first k reflections corresponds to the sum of the impulse responses that experience exactly l orders, where $l = \{1,2,3\cdots k\}$ and this is given by

$$h_{ij}^{k}(t) = \sum_{l=0}^{k} g_{ij}^{l}(t) \tag{2.55}$$

2.4.5 REVIEW ON INTEGRATED SPHERE MODEL

The application of integrated sphere as a diffuser in a wireless infrared communication system was investigated in [403], and its use can be reported in [112,140,246]. Particularly, the integrated sphere model was the first modeling scheme that was employed for infrared communication to estimate the diffuse portion of an indoor room environment. The important characteristic feature of this modeling scheme is that the diffuse signal gain is assumed to be constant everywhere inside the room. When this model is applied to VLC, the frequency response of the channel can be put up as follows

$$H(f) = \sum_{l=1}^{N} H^{(0)}(0;S_l,R_j)\, exp\left(-j2\pi f \Delta\tau_{LOS,l}\right) + H_{DIFF}\frac{exp\left(-j2\pi f \Delta\tau_{DIFF}\right)}{1+j\frac{f}{f_0}} \tag{2.56}$$

In equation (2.56), the parameter $H^{(0)}(0;S_l,R_j)$ corresponds to the DC channel gain pertaining to the LOS component which is given by the following expression

$$H^{(0)}(0;S_l,R_j) = \begin{cases} \frac{(m+1)A_{PD}}{2\pi D^2} cos^m(\phi)\, T_s(\psi)\, g(\psi)\, cos(\psi), & 0 \leq \psi \leq \psi_c \\ 0, & \psi > \psi_c \end{cases} \tag{2.57}$$

From equations (2.56) and (2.57), the parameters S_l denote the lth LED source, R_j represents the jth receiver, the Lambertian mode number of the LED is signified by m, A_{PD} indicates the physical area of the photodetector, ϕ and ψ correspond to the irradiance and the incident angle, respectively, while the distance between the transmitting LED and the receiver is given by D. Also the gain of the optical filter and the optical concentrator is specified by $T_s(\psi)$ and $g(\psi)$, respectively. Usually, in most of the earlier works, the gains of the optical filter and the optical concentrator are assumed to be unity. Furthermore, in equation (2.56), the channel DC gain in accordance to the diffuse signal is given by H_{DIFF}, the signal delays corresponding to both LOS and diffuse components are denoted by $\Delta\tau_{LOS}$ and Δ_{DIFF}, respectively, and f_0 represents the cut-off frequency of the purely diffuse channel. While, the diffuse channel gain is given by

$$H_{DIFF} = \frac{A_{PD}}{A_{ROOM}}\frac{\rho}{1-\rho} \tag{2.58}$$

From equation (2.58), A_{ROOM} denotes the effective area of the room and ρ represents the reflectivity. Further, the impulse response of the diffuse channel pertaining to the integrated sphere model is represented by

$$H_{DIFF}(t) = \begin{cases} \frac{H_{DIFF}}{\tau} exp\left(-\frac{t}{\tau}\right), & t \geq 0 \\ 0, & t < 0 \end{cases} \tag{2.59}$$

The decay time in equation (2.59) can be determined as

$$\tau = \frac{1}{2\pi f_0} \tag{2.60}$$

Finally, the channel impulse response of the integrated sphere model can be manifested as

$$h(t,R_j) = \sum_{l=1}^{N} H^{(0)}(t;S_l,R_j)\,\delta(t-\Delta\tau_{LOS}) + h_{DIFF}(t-\Delta\tau_{DIFF}) \tag{2.61}$$

2.5 CONCLUSION

This chapter presents a comprehensive review on VLC channel modeling. Since VLC exploits opto-electronic illuminating devices like LEDs to render simultaneous illumination and communication, illumination which is the first and foremost task of VLC should not be hindered for the purpose of communication. Thus, this entails focusing on illumination related properties of LEDs. Consequently, in this chapter we stressed upon several photometric parameters such as luminous flux, luminous intensity, etc. Unlike RF wireless channel modeling, modeling VLC channels is entirely different. This is due to the inherent fact that the transmitted signal should adhere to the requirements of IM and DD, where the transmitted signal should be both real and unipolar. Therefore, this chapter focuses on detailed analysis of the distribution of power within the indoor room environment, illuminance characteristics of the transmitter, in-depth mathematical analysis of the impulse responses pertaining to VLC channel modeling. Furthermore, much earlier significant works elucidate the fact that VLC channel modeling shares certain similarities with infrared communication, and the two differ only in the type of media exploited.

Moreover, the major emphasis of this chapter is to provide an in-depth analysis of the current state-of-the-art research aspects relevant to VLC channel modeling. In this context, we provide a detailed comprehensive review of different VLC channel modeling scenarios carried out in the literature. Besides, much emphasis is laid on realistic indoor channel modeling that was endorsed by the IEEE 802.15.7r1 Task Group as well. From a list of findings, it is clear that even till data research is progressing to model the VLC channel in both indoor as well as outdoor vehicular communications under different weather conditions like clear, rainy and foggy. Also, from many significant works, it is ascertained that one of the most vital aspects that need to be taken into consideration while modeling the VLC channel environment is the mobility of people within the indoor room environment. Much of the

VLC channel modeling is carried out either by using ray-tracing/Monte Carlo simulations which consume much time. Therefore, future research efforts should work in the direction of developing sophisticated and accurate methods with minimal time consumption to model the VLC channel environment.

3 MODULATION FORMATS FOR VLC

Predominantly, the non-coherent emission characteristics of LEDs facilitate intensity modulation (IM) and direct detection (DD) as the most viable modulation formats for VLC. Principally, the optical intensity in IM/DD-based VLC systems is necessarily real and positive valued. Therefore, this constraint of IM/DD systems mandates that the type of modulation scheme to be used for VLC must assure for a real and unipolar, i.e, positive valued signal. Thus, the modulation schemes which exhibited remarkable benefits in RF-based wireless communication cannot be directly exploited in the optical domain as they might not offer the same advantages when employed in VLC. Generally, VLC has gained a significant amount of propulsion in recent times because of the rapid deployment of the off-shelf opto-electronic LEDs which enabled VLC to evolve as a cost-effective and high-speed solution to meet the looming spectrum crisis. The massive production of these inexpensive devices garnered the attraction of several research communities belonging to government, research and industrial agencies to promote their interface with almost every road side unit present in the city, thereby making it evolve as a smart city. Principally, LEDs exhibits a unique characteristic feature of offering a high switching speed which enabled it to become one of the most suitable light sources for VLC. Moreover, with the blooming of energy-efficient resources for retailing, residential and commercial units, it is foreseen that LEDs will definitely replace traditional lighting fixtures. Consequently, this emphasizes stressing on the modulation formats which are appropriate to be employed for IM/DD systems to assure for high data rate communication as well as to maintain a sufficient amount of illumination which is pleasing to the human eye.

In case of IM/DD systems, data is modulated on subtle changes of the light intensity; as a result, it is vital to take into consideration the effect of non-linearity of the voltage-luminance characteristics. Therefore, to alleviate this non-linearity issue of LED, pre-distortion techniques have been proposed in the literature to linearize the non-linear region of LED. Primarily, the majority of the OWC systems that are being deployed for both indoor and outdoor applications adopt the IM/DD scheme as the most appropriate modulation format. Pertaining to the outdoor applications, adverse weather conditions, especially, heavy fog and mist, pose a major problem for the realization of a highly, reliable communication link because under such conditions, there is a significant reduction in the intensity of the light which is propagating through thick fog. Eventually, the best feasible solution to overcome this sort of attenuation is to concentrate and focus more optical power into smaller areas. Nonetheless, the eye safety regulation should be taken into account because, for indoor applications, the eye safety limit on the amount of transmitted optical power is quite stringent. There is a significant amount of differentiation between the RF-based wireless channel and the VLC channel. In contrast to RF-based wireless communication, where the amplitude, phase and frequency of the carrier wave are modulated, in the context of OWC, i.e., VLC, the intensity of the optical carrier is modulated in most of the systems which are operating below 2.5 Gbps data rates. Thereupon, for data rates generally greater than 2.5 Gbps, it is necessitated to adopt external modulation formats.

This chapter is directed toward the elaborate discussion of a variety of modulation formats that are more flexible to be implemented for IM/DD-based VLC systems. Furthermore, major emphasis is laid on the remarkable benefits offered by such techniques in terms of bandwidth and power efficiency. While designing the modulation formats which are to be used for IM/DD-based VLC systems, it is of utmost importance to consider the illumination aspects of the LEDs. Particularly, illumination which is the foremost task of VLC should not be disrupted when the LED is meant to be used for the purpose of communication. Thus, the modulation formats to be employed for IM/DD systems should reconcile with the requirements of illumination as well. Predominantly, the modulation techniques that are widely employed for VLC can be elucidated from Fig. 3.1 where it is clear that the modulation techniques which are employed for VLC can be broadly classified into baseband modulation formats, multicarrier modulation formats and special modulation formats. The details of each modulation format will be discussed in the subsequent sections below.

3.1 BASEBAND MODULATION FORMATS

Baseband modulation formats like ON-OFF keying (OOK), pulse amplitude modulation (PAM), pulse width modulation (PWM) and pulse position modulation (PPM) provide the flexibility to be applicable for IM/DD systems because of their ease of implementation. Particularly, these modulation formats are suitable when low-to-moderate data rates are desired. However, when there is an increment in the spectral efficiency, the performance of the OWC systems which are employing the aforementioned modulation formats results in deterioration of the overall system performance due to the emanation of ISI under dispersive conditions of the optical channel environment. Thereupon, in order to attain an improved system performance in terms of reduced probability of error, it obliges these schemes to rely on complex equalization techniques like decision feedback equalization (DFE), maximum likelihood sequence detection (MLSD), linear feed forward equalization (LFFE), etc. The overview of each baseband modulation format will be presented in sequel as follows:

3.1.1 ON-OFF-KEYING (OOK)-BASED VLC SYSTEMS

The simplest form of baseband modulation format to be employed for IM/DD system is ON-OFF-keying (OOK). This has gained tremendous popularity due to its ease of implementation and power efficiency. The basic phenomenon associated with OOK is that a binary bit "1" is interpreted by an optical pulse which occupies

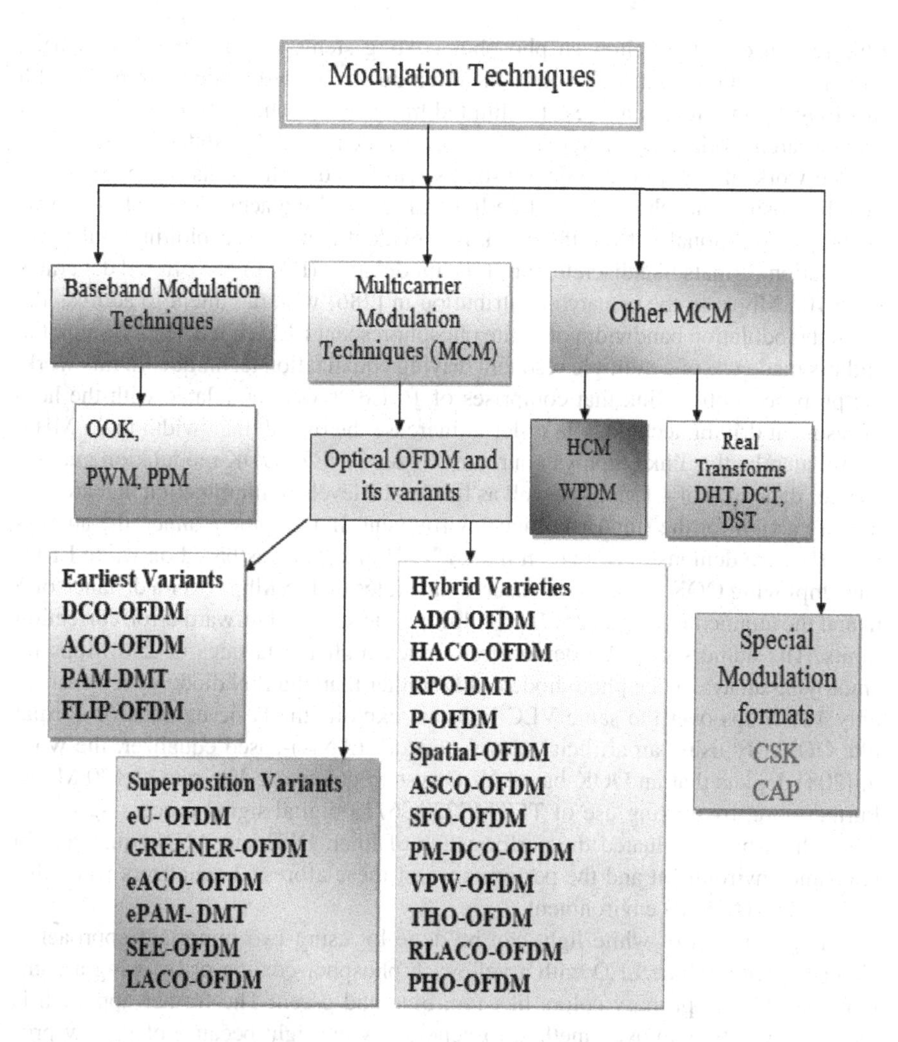

Figure 3.1: Modulation techniques compatible with IM/DD systems for VLC.

the total or a portion of bit duration; whereas, the absence of an optical pulse represents the binary bit "0". Precisely, the incoming bit stream, i.e., the binary bits "1" and "0" are modulated by switching the LED between "ON" and "OFF" states. In the earlier times, OOK was generally employed for wireline communication. Much of the early works of exploiting OOK modulation format for VLC exploits white phosphorescent LEDs. As already discussed in the previous chapter, in the process of generation of white light, such LEDs utilize the combination of blue LED with yellowish phosphor coating. However, in spite of its low production cost, the slow time response of the yellowish phosphor coating stems out as a major drawback which impedes high data rate communication due to limited modulation bandwidth of LEDs. In an effort to address the limited modulation bandwidth of LEDs, extensive research works have been published and can be reported as stated below.

The work in [188] clearly demonstrates wireless data transmission over an optical link using phosphorescent white light LEDs, and the achievable data rate was 40 Mbps. Additionally, from this work it is evident that upon exploiting multilevel modulation formats like discrete multitone modulation (DMT), the offered data rates were 100 Mbps. In the research contribution in [286] with the intent to address the limited modulation bandwidth of white phosphorescent LED-based VLC system, the authors made use of a multiple resonant driving equalization technique. In this work, the proposed optical link that comprises of 16 LEDs was modulated with the help of resonant driving technique in order to increase the overall bandwidth to 25 MHz. Consequently, this link employs nonreturn-to-zero (NRZ) OOK modulation guarantees for data rates of 40 Mbps as well as fulfills the levels of illumination that are sufficient enough for the standard office environment. In the same manner, the authors in [501] have demonstrated that an indoor VLC link that was based on white LEDs and employing OOK modulation ensured data rates of 125 Mbps over a distance of 5 m and the attained BER was 2×10^{-3} fulfilling the standard forward error-correction limits. The authors in [503] demonstrate much higher data rates of 230 Mbps by employing an avalanche photodiode (APD) rather than the PIN diode which ensures only 125 Mbps over the same VLC link that exploits the basic modulation format like OOK. By using an artificial neural network (ANN)-based equalizer, the work in [208] depicts that an OOK-based VLC system delivers a data rate of 170 Mbps. Furthermore, by making use of TI TMS320C6713 digital signal processing board, the authors have evaluated the performance of liner, DFE and ANN equalizers in real-time environment and the performance of these aforesaid equalizers were also verified in MATLAB environment.

The generation of white light can be done by using two important approaches either by using a blue LED with a yellowish phosphor coating or by using a combination of three primary colors like red, blue and green. The former approach is the most widely employed method to generate white light because of its low production cost compared to the latter approach, because this sort of way of generation of white light incurs high production costs due to the requirement of three separate driver circuits. However, the authors in [172] proposed to modulate the data by using only red LED while feeding constant current for the green and blue LEDs for illumination purposes. Thus, in this manner both illumination and communication were accomplished. Consequently, this VLC link made up of red LED with a low-cost PIN photodiode demonstrates a maximum data transmission speeds of 477 Mbps by using the simple modulation format like OOK. Furthermore, from the earlier findings it can be deduced that tremendous research efforts strived to tackle the limited modulation bandwidth of WLEDs-based VLC systems. The contribution in [295] portrays that a post-equalization circuit that comprises of two passive and one active equalizer has been exploited to meet the limited bandwidth of VLC systems. In this work, by making use of blue filtering as well as by employing a post-equalization circuit, the VLC system was capable of attaining a bandwidth of 151 MHz, while allowing an OOK-NRZ data transmission of up to 340 Mbps and demonstrating a BER of 2×10^{-3} which is within the limit of forward error correction. The research efforts in [297] demonstrate a real-time VLC system, where the VLC link exploits white light LED and NRZ-OOK modulation scheme. This link exhibits its potential to deliver a data rate of 550 Mbps at a BER of 2.6×10^{-9} when the VLC link operated at a distance of 60 cm. Upon further increasing the VLC link distance by 160 cm, this sort of VLC system was capable of rendering 480 Mbps data rate and the achievable BER was 2.3×10^{-7}. Moreover, by making use of pre-emphasis and post-equalization circuits, a significant improvement in the 3 dB bandwidth from 3 to 233 MHz can be witnessed from this work. A similar approach of employing pre-emphasis and post-equalizing circuits can be reported in [171] where the VLC link employs a duo-binary technique to impart a data rate of 614 Mbps.

Generally, any type of modulation format that is to be employed for VLC should adhere to the requirements of opto-electronic devices, i.e., it should as ensure high data rate transmission while fulfilling the requirements of opto-electronic devices, i.e., firstly, even while dimming the light source, the data transmission should not be hindered, and secondly, flickering must be avoided while transmitting data. Therefore, in order to save sufficient amount of energy, it is required for the light sources to be dimmed based upon the needs of the application at hand. Thus, the modulation formats should support adequate levels of dimming as well. In this regard, as suggested by the IEEE 802.15.7, there exist two ways to ensure the support of dimming when OOK modulation scheme was exploited for VLC. The first way is to redefine the ON and the OFF intervals, and the second approach is to add compensation periods. In the first approach, by assigning the ON and the OFF levels to different light intensities, a sufficient amount of dimming can be achieved. The noteworthy feature of this scheme is that without relying on additional communication overhead, an adequate amount of dimming support can be attained. From the literature it can revealed that by incorporating this sort of approach, the achievable data rate of NRZ-OOK modulation technique can be maintained at the expense of decrement in the range of communication at lower dimming levels. The principal drawback that is associated with this technique is that employing lower levels of intensities and ON/OFF levels enables the LEDs to be operated at lower dimming currents that eventually lead to changes in the levels of emitted colors of LEDs [359].

While, in the latter approach, the ON and OFF levels of the modulation scheme prevail as it is, but only the additional compensation periods are added when the LEDs are fully turned ON (ON periods) or OFF (OFF periods). Generally, in this approach, by taking into consideration the desired levels of dimming, the compensation period duration is defined. More particularly, if the desired level of dimming is more than 50%, the ON periods are added; and if the desired level of dimming is less than 50%, then OFF periods are added. The research contribution in [258] determines a methodology to fetch the desired level of dimming and to enumerate the percentage time of active data transmission X within the transmission interval as represented below

$$X = \begin{cases} 2(1-D), & D > 50\% \\ 2D, & D \leq 50\% \end{cases} \tag{3.1}$$

It is to be noted that when the desired level of dimming is associated with OOK, then the maximum communication efficiency E_p can be determined by using the concept of information theoretic entropy as follows

$$E_p = -Dlog_2 D - (1-D)\, log_2\, (1-D) \tag{3.2}$$

From (3.2) it can interpreted that communication efficiency E_p is a triangular function of the dimming level with maximum efficiency at 50% dimming level. Furthermore, it can be inferred that as the dimming level decreases to 0% or when it increases to 100%, there is a significant drop in the communication efficiency [398]. Finally, it can be concluded that the addition of compensation periods results in reduction in data rate even though the range of communication can be increased by maintaining the similar intensities of ON/OFF signals. Substantially, this reduction in data rate can be addressed by exploiting inverse source coding phenomena while accomplishing the required levels of dimming [283].

3.1.2 PULSE WIDTH MODULATION (PWM)-BASED VLC SYSTEMS

Despite offering a computationally efficient strategy, OOK modulation hinders high data rate communication when supporting different levels of dimming. Accordingly to address this drawback, research strived to use complementary modulation formats that are based on pulse width and pulse position. This sub-section gives a thorough investigation on the performance of VLC systems exploiting PWM. From the previous works, it can be surmised that both data transmission as well as dimming support can be effectuated through the use of PWM. The stunning feature of PWM is that based upon the requirement of dimming levels, the width of the pulses is adjusted while the pulses themselves are meant for transporting the data. An important aspect to be taken into consideration is that the PWM when used for VLC systems employing LEDs allows for the transmission of modulated signal during the period of pulse while the LED illuminates by exhibiting full levels of brightness during the pulse. Dimming is an indispensable functionality of modern lighting systems. In the perspective of LED-based VLC applications, PWM emerges as the most expedient modulation format that is more flexible to control the illumination levels of LED without inducing any color rendering of the emitted light. Primarily, PWM provides the ease of controlling the illumination or brightness levels of LEDs by appropriate adjustment of the duty cycle of the pulse as well as by square pulse modulation of the driving current. Most of the earliest works in the literature proposed several significant research aspects to tackle the dimming levels of LEDs. Much relevant work that seems to be closely associated can be delineated in [461] where the authors have proposed dimming control methods in the physical layer by using two approaches, i.e., either by relying on PWM or by appropriately adjusting the modulation depth in a manner to fulfil the seamless transfer of data along with the control of brightness levels of LEDs. Even though this method accomplishes simultaneous transmission of data adhering to the dimming requirements, the assured data rate is limited to 4.8 Kbps. Successively, to overcome such limited data rate of this approach, the authors in [379] proposed to integrate PWM with DMT for facilitating concurrent transfer of data along with the assurance of adjustment of dimming levels. The schematic rep-

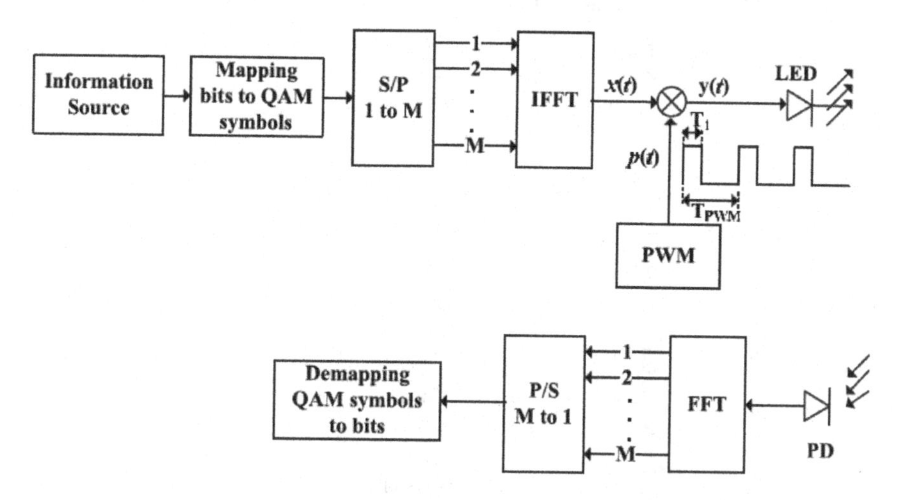

Figure 3.2: PWM-DMT-based VLC system [379].

resentation of PWM-DMT-based VLC system is outlined in Fig 3.2. As depicted in the figure, at the transmitting end, the incoming bit stream of information that needs to be transmitted is to be converted into a stream of symbols by using M-QAM modulation format. Thereupon, these sets of complex symbols are fed to the serial to parallel convertor to enable parallel transmission of data, and, further, this stream of frequency domain data is transformed into time domain format by making use of inverse fast Fourier transform (IFFT) to fetch the multicarrier signal $x(t)$. Upon that, this multicarrier signal is then multiplied by means of a periodic PWM pulse train which is symbolized by $p(t)$. It is to be taken into account that this PWM pulse train comprises of duty cycle $D = \frac{T_1}{T_{PWM}}$, where the notations T_1 and T_{PWM} represent the duration of the PWM pulse and period of the PWM signal, respectively. The amount of dimming level is depicted as $\delta = 1 - D$ which clearly confirms the fact that the amount of dimming is directly dependent on the duty cycle. Finally, the received signal can be expressed as the product of $x(t)$ and the PWM signal $p(t)$. The pictorial representation of PWM-waveform indicating the ON duration of the LED as well as the period of the PWM signal both for 80 and 20% dimming is shown in Fig. 3.3.a, while the PWM-dimming control signal that was superimposed with a DMT signal is elucidated in Fig. 3.3.b. At the receiving side, this signal is intercepted by means of a photodiode and then undergoes a frequency domain translation by means of FFT operation. Finally, after getting applied to parallel to serial convertor, the resulting data symbols are estimated by using the corresponding demapping scheme to obtain

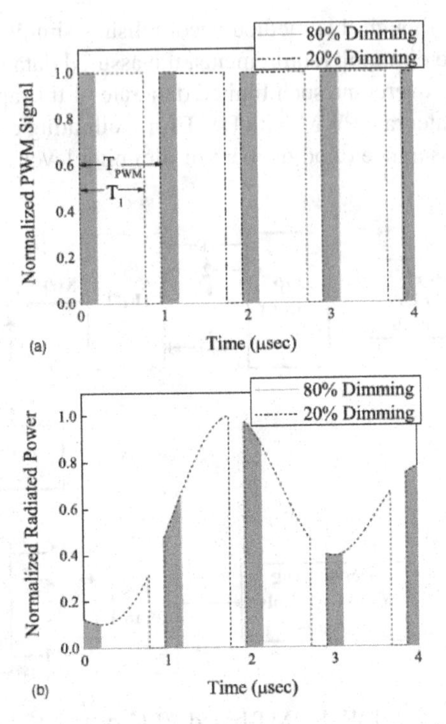

Figure 3.3: (a) Normalized PWM waveform. (b) Combination of PWM-controlled dimming signal with DMT signal [379].

the final output. Furthermore, taking into consideration the importance of dimming in the perspective to save energy, extensive research efforts were carried out to jointly combine PWM with the robust multicarrier modulation format like OFDM leading to several variants of optical OFDM, where the detailed methodologies of such variants will be discussed in subsequent sections.

3.1.3 PULSE POSITION MODULATION (PPM)-BASED VLC SYSTEMS

Similar to PWM, the other pulsed-based modulation method which can be exploited for VLC systems is PPM. The underlying concept involved behind the realization of PPM for VLC systems is that the duration of the symbol is divided equally into t slots, while the pulse being transmitted in one of the t slots. Unlike PWM, where the width of the pulse is varied, in case of PPM, the transmitted symbol is determined by the position of the pulse. PPM has attracted an appreciable amount of significance in many of the earliest designs of OWC systems due to its implementation simplicity. However, since only one pulse can be transmitted during the symbol duration, this modulation technique limits the achievable data rate and it turned out to be a spectrally inefficient scheme. Hence, to address such drawbacks, with the

Table 3.1
Different variants of PPM.

S No.	Variant of PPM
1	*Overlapping PPM (OPPM)* [44]
2	*Multipulse PPM (MPPM)* [444,462]
3	*Overlapping MPPM (OMPPM)* [385,387–390]
4	*Differential PPM (DPPM)* [585]
5	*Differential Overlapping PPM (DOPPM)* [386]
6	*Expurgated PPM (EPPM)* [373]
7	*Multilevel EPPM (MEPPM)* [374]
8	*Variable PPM (VPPM)* [5]

advent of time, diverse variants of PPM have been proposed in the literature and few such varieties are summarized in Table 3.1. Among the aforementioned varieties, overlapping PPM (OPPM) can be pronounced as the generalized version of PPM which allows for the transmission of more than one pulse during the symbol duration. The contribution in [44] investigates the performance of OPPM in terms of dimming range, flicker severity index, cut-off data rate, bandwidth and power efficiencies. Furthermore, OPPM was compared with other modulation schemes like OOK and PPM. The result analysis of this work emphasizes the fact that OPPM exhibited the constant and smallest flicker severity index over all the dimming levels; additionally, OPPM demonstrates higher spectral efficiency. In terms of both illumination and communication perspective, OPPM manifests superior performance than that of the OOK and PPM. The other variant of PPM much similar to OPPM is multipulse PPM (MPPM) which allows for the transmission of multiple pulses during the symbol duration where the data information is transmitted by availing different combinations of the positions of these pulses [462]. MPPM has been proposed to improve the band-utilization efficiency of PPM, and the work in [444] clearly reports the fact that when compared with OPPM, MPPM guarantees high spectral efficiency. The work in [385] signifies that OPPM was combined with MPPM to generate the other variant of PPM which can be called overlapping MPPM (OMPPM). Similar to the aforementioned variants of PPM, OMPPM allows for overlap of more than one pulse position for each optical pulse. This work reveals that OMPPM achieves a significant improvement in information rate as well spectral efficiency when compared with PPM and MPP without allowing for any bandwidth expansion. The authors in this work enforce Reed-Solomon (RS) codes for OMPPM in order to obtain high information efficiency. Moreover, OMPPM exhibits a dominant performance in terms of the achieved capacity when compared with the other modulation formats like OOK, PPM and MPPM.

In [387], the performance of OMPPM was analyzed over noisy photon channel. The work in [389] infers that OMPPM exploiting fewer pulse slots as well as more pulses per symbol duration exhibits better cut-off rate performance. Additionally, investigations on Trellis-coded OMPPM were carried out in direct detection channels with background noise [388, 390]. The work in [585] proposes another variant of PPM ably called differential PPM (DPPM). The basic concept of DPPM is similar to that of the PPM where the OFF symbols after the pulse in a PPM symbol were deleted and the next symbol starts immediately after the pulse of the previous symbol. The work in [444] clearly stipulates the fact that the requirement of average power for DPPM is substantially less than that of the PPM for a given bandwidth over an optical channel environment. Much significant work in [386] proposes differential overlapping PPM (DOPPM) which applies the concept of differential deletion of OFF symbols to OPPM and, furthermore, illustrates the fact that the performance of DOPPM in terms of achievable spectral efficiency is much better than that of the PPM, DPPM and OPPM. The work in [373] proposed expurgated PPM (EPPM) by using the concept of symmetric balanced incomplete block designs (BIBD). In this variant of PPM, by expurgating the symbols of MPPM, the symbols in this modulation technique were obtained. The research work in [378] signifies that EPPM achieves the same amount of spectral efficiency as that of the PPM, and by varying the number of pulses per symbol, this modulation format can accomplish an arbitrary level of illumination support for VLC systems.

By making use of many pulses per symbol, the detrimental aspects of flickering can be mitigated by using EPPM when compared with the traditional PPM modulation format. The research efforts in [374] propose multilevel EPPM (MEPPM) that extends the concept of EPPM by providing the flexibility to support multiple amplitude levels with the intent to increase the spectral efficiency and constellation size. This work presents the calculations of analytical symbol error probabilities and spectral efficiencies along with numerical results, which depicts the performance comparisons of MEPPM with other modulation techniques like OOK and PAM. This technique is shown to outperform the other relevant counterparts over dispersive channels. The other attractive feature of MEPPM is that it offers the dual functionality of extending dimming support as well as guarantees a flicker-free communication over the VLC link. The IEEE 802.15.7 standard proposed variable PPM (VPPM) which can be termed as the combination of PPM and PWM. Generally, in this variant of PPM, the data bits were encoded by appropriately choosing different positions of the pulse and even with the width of the pulse can also be altered based upon the needs of the application. VPPM allows for different levels of dimming by varying the width of the pulse. With the intent to facilitate joint dimming support and seamless high-speed data transfer, the authors in [448] proposed a special variant of PPM called variable rate multi-PPM (VRPPM). Furthermore, the work in [547] employs the combination of MPPM with reverse polarity optical OFDM (RPO-OFDM) to ensure concurrent dimming support and enhancement in spectral efficiency.

3.2 STATE-OF-THE-ART MULTICARRIER MODULATION FORMATS COMPATIBLE FOR IM/DD SYSTEMS

Even though baseband modulation formats exhibit their ease of simplicity for implementation in VLC, their usage in the scenarios of high data rate communication obliges these schemes to employ complex equalization techniques. Furthermore, the deleterious effects such as DC wandering and flickering due to the other fluorescent lamps have a huge impact on the performance of the system especially at lower frequency bands of the used bandwidth. Thus, with the intent to ensure high data rate transfer as well as to reduce the implementation complexity, it necessitated VLC to borrow a robust multi dimensional modulation format like orthogonal frequency division multiplexing (OFDM), one of the radical multicarrier modulation schemes of RF that has been widely used in many standards of wireless communication and has also been extensively used in digital audio broadcasting (DAB), digital video broadcasting (DVB), third generation partnership project long term evolution (3GPP-LTE), wireless interoperability for microwave access (WiMAX), digital subscriber lines (DSL) internet access, power-line networks, and many more applications like high-definition television, video conferencing, online gaming, Voice over Internet Protocol (VoIP), etc. The stunning feature of OFDM is that it converts the frequency selective fading of the communication channel into flat fading channel just by making use of computationally efficient, simple and one-dimensional frequency domain equalization technique. Ever since its introduction, this multicarrier modulation format like OFDM has gained enormous popularity because of exhibiting several remarkable benefits like its assurance of high data rate communication, resilience against narrowband interference, combating the effects of intersymbol interference (ISI), efficient utilization of the spectrum, flexibility and scalability. An additional intrinsic prevalence of OFDM is that it can support multiuser communications in a manner where each user is appropriated with certain subcarriers. Furthermore, OFDM has the potential to adapt the channel utilization to the frequency response of the channel by making use of adaptive bit and power loading algorithms, thus maximizing the overall system performance. The most noteworthy feature of OFDM is that at the medium access control (MAC) level, OFDM ensures a straightforward multiple access scheme which is not easy to achieve with other modulation formats like OOK, PPM and M-PAM.

Typically, in practice, at the transmitting end of OFDM, the incoming huge stream of data is modulated into data symbols by the choice of an appropriate modulation technique like M-ary quadrature amplitude modulation (M-QAM), M-ary phase shift keying (M-PSK) and real mapping technique like M-ary pulse amplitude modulation (M-PAM). Thereupon, these mapped symbols, are then loaded into orthogonal subcarriers where the spacing between each subcarrier is equal to the multiple of symbol duration. The set of these parallel symbols, which are in the frequency domain, can then be multiplexed into a single time domain format by making use of the inverse fast Fourier transform (IFFT) module. It is a well-known fact that OFDM modulation and demodulation can be realized by exploiting IFFT and FFT transformation modules. For complex mapping of the input to the IFFT module, its output will be complex-valued. In particular, the optical link can be realized by exploiting off-the-shelf cost-effective opto-electronic devices like LEDs and the photodiodes, and the fact that these light sources generate incoherent light enables the OFDM time domain signal formats to modulate the intensity of the light source. Specifically, the modulation format in case of RF-based wireless communication modulates the amplitude and phase information of the carrier signal; meanwhile, due to non-coherent emissions of the light sources in case of VLC, the data is modulated by varying the intensities of LED at a rapid pace that is imperceptible by the human eye. Thus, the special modulation and demodulation formats like IM and DD have been exploited for VLC. Thus, this compels the OFDM time domain signal formats to be both real and unipolar (i.e., positive) in nature. Eventually, in order to accomplish a real and positive valued signal, the conventional OFDM as used in RF-based wireless communication cannot be implemented in VLC in a straightforward manner without any necessary modifications to its input and output frame structure. Therefore, in order to attain a real-valued signal, the input frame structure of the IFFT module is enforced with the Hermitian Symmetry constraint, so that the time domain signal at the output of IFFT will be assured of its real nature rather than a complex valued signal transmission. Consequently, in case of an optical OFDM system which is based on the Fourier signal processing mandates the exploitation of Hermitian Symmetry criteria for the purpose of guaranteeing a real valued signal.

However, in addition to real valued signal transmission, it is also of immense prerequisite to ensure a positive, i.e., unipolar signal transmission. In this regard, a myriad variants of optical OFDM have been proposed to achieve unipolar signal transmission. Each variant of OFDM has certain trade-offs in terms of spectral efficiency, power efficiency, computational complexity, BER performance, etc. This subsection focuses on presenting a comprehensive review of the performance analysis of each variant of OFDM. The different variants of optical OFDM as described in the literature for the generation of non-negative valued signal generally can be grouped into three categories: earliest unipolar optical OFDM variants, superposition optical OFDM variants and hybrid variants of optical OFDM. The up-to-date research contributions on the performance analysis of different categories of optical OFDM will be detailed below.

3.2.1 PERFORMANCE ANALYSIS OF EARLIEST UNIPOLAR OPTICAL OFDM VARIANTS

The earliest variants of optical OFDM that have been proposed to attain a unipolar signal are as follows:

1. DC-biased optical OFDM (DCO-OFDM)
2. Asymmetrically clipped optical OFDM (ACO-OFDM)
3. Pulse amplitude modulated discrete multitone modulation (PAM-DMT)
4. Flip-OFDM
5. Unipolar OFDM (U-OFDM)

The above optical OFDM variants can be claimed to be the earliest variants that have been proposed to generate a non-negative valued signal. Even though several optical OFDM variants have been proposed in the literature, the majority of them are combination of these basic variants. Hence, it is vital to have a clear understanding about the system implementation along with the performance analysis of the above-mentioned class of optical OFDM.

DCO-OFDM: From the literature, DCO-OFDM can be reported as the earliest variants of optical OFDM, where this technique comprises of addition of sufficient amount of DC bias to the bipolar signal in order to generate a unipolar or non-negative valued signal. A conventional optical OFDM system is delineated in Fig. 3.4. As depicted from the figure, the incoming high-speed data stream is being encoded into M-PSK, M-QAM and M-PAM symbols where M signifies the constellation order which varies from 2, 4, 8, 16, 32, 64, 256 and 1024. For ensuring perfect reception of the transmitted data, estimation and tracking of the channel is of major requisite. Hence, pilot symbols are inserted based upon comb-type pilot arrangement. Further, this stream of symbols are split into a large number of low-speed data sets with the help of a serial to parallel (S/P) converter.
Accordingly, the resultant signal can be expressed as

$$X[k] = X\left[mN_f + l\right] = \begin{cases} x_P[m], l = 0 \\ Data,\ l = 1, 2, 3, \cdots, N_f - 1 \end{cases} \tag{3.3}$$

From (3.3), N_f is the pilot symbol insertion frequency, N denotes the total number of subcarriers, $m = 0, 1, 2, \cdots, N_P - 1$, N_P specifies the number of pilot subcarriers which can be obtained as $N_P = \frac{N}{N_f}$ and $X[k]$ is the frequency domain representation of the data symbols.
Thereupon, these frequency symbols are fed as input to the IFFT transformation block to yield the time-domain signal. It is apparent that, for complex mapping of the input data to the IFFT, its output will be definitely a complex-valued signal. Therefore, this necessitates constraining the input to the IFFT module to satisfy the Hermitian Symmetry criteria. According to Hermitian Symmetry criteria, if there are a total of N number of subcarriers, only $\frac{N}{2}$ will be utilized for data transmission and the remaining half are flipped complex conjugate versions of the previous ones. The pictorial representation of the phenomenon involved behind the usage of Hermitian Symmetry criteria is illustrated in Fig. 3.5. Mathematically, this can be represented as:

$$X[N-k] = X^*[k],\ k = 1, 2, \cdots \frac{N}{2} \tag{3.4}$$

To avoid any imaginary part at the output, the first and the middle subcarriers are set to 0, i.e.,

$$X[0] = X\left[\frac{N}{2}\right] = 0 \tag{3.5}$$

Hence, the input to the IFFT module has the following representation as shown by (3.6)

$$X = \left[0, X_1, X_2, X_3, \cdots X_{N/2-1}, 0, X^*_{N/2-1}, \cdots X^*_2, X^*_1\right]^T \tag{3.6}$$

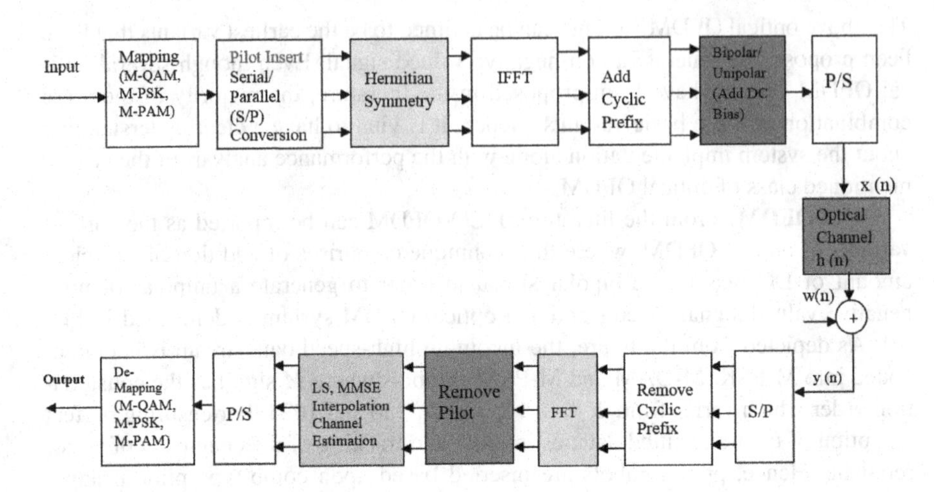

Figure 3.4: Schematic block of channel estimation in Hermitian Symmetry imposed IFFT-based DCO-OFDM.

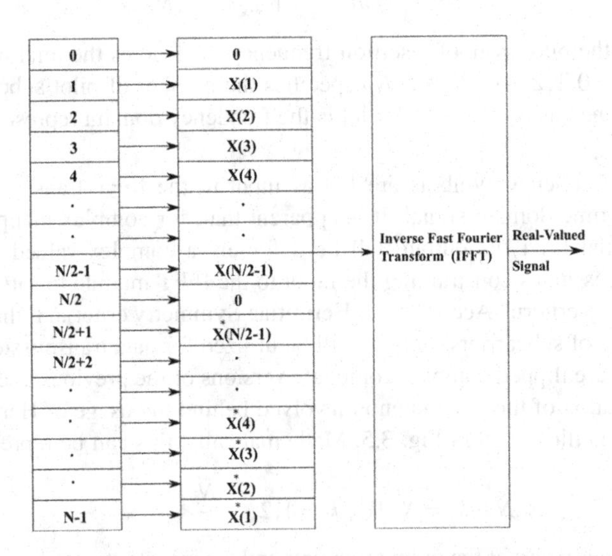

Figure 3.5: Hermitian Symmetry imposed IFFT.

From (3.6), it is evident that due to Hermitian Symmetry criteria, only the first half of the subcarriers carry data, while the second half are flipped complex conjugates of the first half, i.e, out of N subcarriers only $\frac{N}{2}$ are utilized to carry data. The discretized signal at the output of the IFFT can be expressed as

$$x[n] = \frac{1}{N}\sum_{k=0}^{N-1} X[k]\, e^{\frac{j2\pi nk}{N}} \tag{3.7}$$

Therefore, (3.7) can be solved by incorporating the aforementioned constraints to obtain the time domain signal as

$$x[n] = \frac{1}{N}\left[X[0]\, e^{\frac{j2\pi n(0)}{N}} + \sum_{k=1}^{\frac{N}{2}-1} X[k]\, e^{\frac{j2\pi nk}{N}} + X\left[\frac{N}{2}\right] e^{\frac{j2\pi n\left(\frac{N}{2}\right)}{N}} + \sum_{k=\frac{N}{2}+1}^{N-1} X[k]\, e^{\frac{j2\pi nk}{N}}\right] \tag{3.8}$$

Incorporating (3.5) in (3.8), then (3.8) can be reduced to

$$x[n] = \frac{1}{N}\left[\sum_{k=1}^{\frac{N}{2}-1} X[k]\, e^{\frac{j2\pi nk}{N}} + \sum_{k=\frac{N}{2}+1}^{N-1} X[k]\, e^{\frac{j2\pi nk}{N}}\right] \tag{3.9}$$

Letting $N - k' = k$ and changing the order of limits in (3.9), and rearranging yields

$$x[n] = \frac{1}{N}\left[\sum_{k=1}^{\frac{N}{2}-1} X[k]\, e^{\frac{j2\pi nk}{N}} + \sum_{k'=1}^{\frac{N}{2}-1} X\left[N-k'\right] e^{-\frac{j2\pi n\left[N-k'\right]}{N}}\right] \tag{3.10}$$

Making use of (3.4) in (3.10) and reordering

$$x[n] = \frac{1}{N}\left[\sum_{k=1}^{\frac{N}{2}-1} X[k]\, e^{\frac{j2\pi nk}{N}} + \sum_{k=1}^{\frac{N}{2}-1} X^*[k]\, e^{-\frac{j2\pi nk}{N}}\right] \tag{3.11}$$

Further, by making use of Euler's Identities $e^{j\theta} = cos\theta + jsin\theta$ and $e^{-j\theta} = cos\theta - jsin\theta$, (3.11) can be solved as

$$x(n) = \frac{1}{N}\left[\sum_{k=1}^{\frac{N}{4}-1} (X[k] + X^*[k])\, cos\left(\frac{2\pi nk}{N}\right)\right] + j\frac{1}{N}\left[\sum_{k=1}^{\frac{N}{4}-1} (X[k] - X^*[k])\, sin\left(\frac{2\pi nk}{N}\right)\right] \tag{3.12}$$

By utilizing the real and imaginary parts of the signal which are given as $x[t]_{RC} = \frac{x[t]+x^*[t]}{2}$

and the imaginary component is $\underset{IC}{x[t]} = \frac{x[t]-x^*[t]}{2j}$. Upon substitution of the above-mentioned signal processing, (3.12) reduces to

$$x[n] = \frac{2}{N}\left[\sum_{k=1}^{\frac{N}{2}-1} \underset{RC}{X[k]} cos\left(\frac{2\pi nk}{N}\right) - \sum_{k=1}^{\frac{N}{2}-1} \underset{IC}{X[k]} sin\left(\frac{2\pi nk}{N}\right)\right] \tag{3.13}$$

where $\underset{RC}{X[k]}$ and $\underset{IC}{X[k]}$ in (3.13) denote the real and imaginary components of $X[k]$. Furthermore, from (3.13), it is clear that upon enforcing Hermitian Symmetry criteria, a real valued signal is attained. To surpass the effects of ISI, a certain amount of cyclic prefix/guard interval is inserted. The cyclic prefix to be inserted is $\frac{1}{4}^{th}$ of the subcarrier's size used and the cyclic prefix added discrete time-domain signal can be expressed mathematically as

$$x_{cp}[n] = x_g[n] = \begin{cases} x[N+n], & n = N_g, N_g - 1, \cdots, 1 \\ x[n], & n = 0, 1, 2, \cdots, N-1 \end{cases} \tag{3.14}$$

In (3.14), N_g is the number of samples in the guard interval. This parallelized signal is then converted into a serial signal for transmission through the channel. Generally, this cyclic prefix added signal (i.e., $x_{cp}[n]$) is real but not necessarily unipolar. Apparently, for the conversion of this bipolar signal into a positive (unipolar) signal, it entails relying on bipolar to unipolar conversion strategies. One such straightforward approach for converting the bipolar signal into unipolar signal is DCO-OFDM methodology.

This variant of optical OFDM consists of addition of a certain amount of DC bias to the bipolar signal in order to convert it into a unipolar signal. To guarantee a non-negative signal, the required amount of DC bias which is to be added equals the absolute value of the maximum negative amplitude of the bipolar optical OFDM signal. In spite of offering significant advantages, the high peaks which arise due to superimposition of huge number of subcarriers make OFDM prone to high peak to average power ratio (PAPR). Consequently, this high PAPR results in the increase of addition of DC bias for the purpose of ensuring non-negativity in the signal which is to be transmitted. In particular, for large values of the number of subcarriers, the optical OFDM signal amplitude can be approximated as a Gaussian distribution. Therefore, in order to circumvent the addition of surplus DC bias and at the same time to obtain a reduction in the required optical power, it is vital to utilize the DC bias B_{DC} proportional to the power of $x[n]$. The DC bias B_{DC} is usually relative to the electrical power of the signal $x_{cp}[n]$ and is represented as

$$B_{DC} = k\sqrt{E\left\{x_{cp}^2[n]\right\}} \tag{3.15}$$

where, k is the proportionality constant in general, it is the clipping factor, and $E\{\}$ denotes the expectation operator.

However, the amount of DC bias which is to be added is given in the literature as [38]

$$B_{DC} = 10log_{10}\left(k^2+1\right)\ dB \tag{3.16}$$

The resultant discretized DC biased added signal can be expressed as

$$x_{DC}\left[n\right] = x_{cp}\left[n\right] + B_{DC} \tag{3.17}$$

In order to prevent the added DC bias from being excessive, the peaks of the negative signal must be clipped. Hence, this introduces a noise called clipping noise which is denoted as η_{clip} [38]. Therefore, the resultant signal after the inclusion of clipping noise is represented as

$$x_{DC}\left[n\right] = x_{cp}\left[n\right] + B_{DC} + \eta_{clip} \tag{3.18}$$

Thus, it can be inferred from the above analysis that a real and unipolar signal is attained and this can be represented as

$$\underset{unipolar}{x\left[n\right]} = x_{DC}\left[n\right] = x_{cp}\left[n\right] + B_{DC} + \eta_{clip} \tag{3.19}$$

Primarily, PAPR of OFDM increases with the accumulation of subcarriers. Consequently, clipping noise can be minimized upon increasing B_{DC}. Aiding this, in order to let the clipping noise be small for higher orders of constellation like 64, 256 and 1024 PSK/PAM/QAM, the amount of DC Bias B_{DC} to be added must be large. One interesting fact to note is, in DCO-OFDM due to imposition of Hermitian Symmetry constraint on IFFT, out of N subcarriers only $\frac{N}{2}$ subcarriers are utilized for conveying the data and the remaining half are flipped complex conjugates of the previous ones. Eventually, (3.15) clearly illustrates that a proportionality constraint exists between the optical power and the OFDM signal amplitude. In addition, the electric signal modulates the intensity of the optical transmitter; therefore, this confirms that the required optical power is proportional to the OFDM signal amplitude. Hence, as a matter of fact, it can be affirmed that DCO-OFDM is afflicted due to addition of DC bias, as there is a huge power inefficiency. But in the literature, it is revealed that high amount of powers are desired to fulfil the illumination requirements [28], [237]. Thereupon, this DCO-OFDM signal is transmitted over the optical/VLC channel environment. While, at the receiver end, as depicted in the schematic block diagram, inverse operations were performed at the receiver which includes removal of cyclic prefix, enforcing FFT operation in order to convert the time-domain signal into frequency domain signal followed by extraction of pilot tones. Further, the channel estimated employing least square (LS) and minimum mean square error (MMSE) channel estimation algorithms and then appropriate demodulation techniques are enforced to successfully recover the transmitted data. Figure. 3.6 illustrates the BER versus SNR performance of Hermitian Symmetry imposed FFT-based DCO-OFDM system exploiting 7 dB and 13 dB of DC bias. From the simulated result, it is clear that an increase in the amount of DC bias will result in more amount of SNRs.

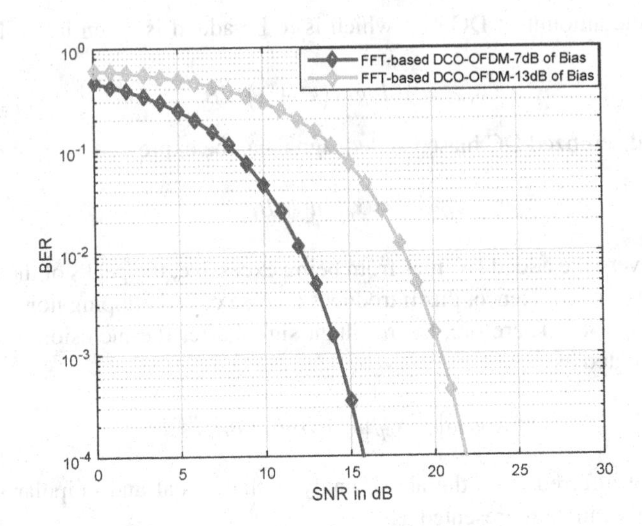

Figure 3.6: Performance analysis of FFT-based DCO-OFDM system with 7 dB and 13 dB of bias.

Listing 3.1: MATLAB code to depict the performance of DCO-OFDM system exploiting 7 and 13 dB of DC bias for the simulated result as illustrated in Fig. 3.6. (Note: This MATLAB code works with R2014b Version).

```
1  clc;
2  clear all;
3  close all;
4  m=512; %Total number of OFDM symbols
5  N=1024; %Length of each OFDM symbol
6  M=4; %Size of the Constellation (M can be 4, 8, 16,
     32, 64, 128, 256)
7  Ncp=256 ; %Length of the Cyclic Prefix
8  Tx=modem.qammod('M',M); %Choosing the modulation
     format as M-ary Quadrature Amplitude Modulation (M
     -QAM)
9  Rx=modem.qamdemod ('M',M); %Fixing up the
     demodulation format at the receiving end as M-ary
     QAM (Note: In order to simulate the code in MATLAB
      R2020a version, please remove Tx and Rx
     instantiation and assign DataMod as DataMod=qammod
     (Data, M);
10 %DCO-OFDM Transmitter
11  Data=randi ([0 M-1],m,N); %Generation of Random bits
      Matrix of size m by N
12 DataMod=modulate(Tx,Data); %Performing Data
     Modulation, for MATLAB R2020a version, assign
```

```
    DataMod=qammod(Data, M);
13  DataMod_serialtoparallel=DataMod.'; %Performing
        Serial to Parallel Conversion
14  datamat=DataMod_serialtoparallel; %Assigning the
        total data to a variable called datamat
15  % Computation of Hermitian Symmetry Criteria
16  datamat(1,:)=0; %Assigning the First subcarrier to
        Zero
17  datamat(513,:)=0; %Assigning the Middle Subcarrier to
        Zero
18  datamat(514:1024,:)=flipud(conj(datamat(2:512,:))); %
        Illustrating that only half of the subcarriers are
        exploited for data transmission as the remaining
        half are flipped complex conjugate versions of the
        previous ones.
19  d_ifft=ifft((datamat)); %Computation of IFFT
        operation
20  d_ifft_paralleltoserial=d_ifft.'; %Parallel to Serial
        Conversion
21  CP_part=d_ifft_paralleltoserial(:,end-Ncp+1:end); %
        Addition of Cyclic Prefix
22  DCOOFDM_CP=[CP_part d_ifft_paralleltoserial]; %
        Transmissin of DCO-OFDM signal
23  bdc=7; %DC BIAS
24  clip=sqrt((10.^(bdc/10))-1); %clipping factor k
25          bdcc=clip*sqrt(DCOOFDM_CP.*DCOOFDM_CP);%
                Computation of DC bias
26          DCOOFDM_BIAS=DCOOFDM_CP+bdcc; %Addition of DC
                bias to the cyclic prefix added signal
27  count=0;
28  snr_vector=0:1:80; %size of signal to noise ratio (
        SNR) vector
29  for snr=snr_vector
30      SNR = snr + 10*log10(log2(M));
31      count=count+1 ;
32      DCOOFDM_with_chann=awgn(DCOOFDM_BIAS,SNR,'
            measured' ) ; %Addition of AWGN
33      %Receiver of DCO-OFDM
34      DCOOFDM_with_chann1=DCOOFDM_with_chann-bdcc;%
            Removal of DC bias
35      DCOOFDM_removal_CP=DCOOFDM_with_chann1(:,Ncp+1:N+
            Ncp); %Removal of Cyclic Prefix
36      DCOOFDM_serialtoparallel=DCOOFDM_removal_CP.'; %
            Serial to Parallel Conversion
```

```
37      DCOOFDM_parallel_fft=fft(DCOOFDM_serialtoparallel
           ) ; %Computation of FFT operation
38
39      DCOOFDM_Demodulation=qamdemod(
            DCOOFDM_parallel_fft.',M);
40    [~,s_e1(count)]=symerr(Data(:,2:512),
           DCOOFDM_Demodulation (:,2:512)) ;
41  end
42  %%%%%%With 13 dB Bias
43  m=512; %Total number of OFDM symbols
44  N=1024; %Length of each OFDM symbol
45  M=4; %Size of the Constellation (M can be 4, 8, 16,
       32, 64, 128, 256)
46  Ncp=256 ; %Length of the Cyclic Prefix
47  Tx=modem.qammod('M',M); %Choosing the modulation
       format as M-ary Quadrature Amplitude Modulation (M
       -QAM)
48  Rx=modem.qamdemod ('M',M); %Fixing up the
       demodulation format at the receiving end as M-ary
       QAM
49  %DCO-OFDM Transmitter
50   Data1=randi ([0 M-1],m,N); %Generation of Random
        bits Matrix of size m by N
51  DataMod1=modulate(Tx,Data1); %Performing Data
       Modulation, for MATLAB R2020a version, assign
       DataMod1=qammod(Data1, M);
52  DataMod_serialtoparallel1=DataMod1.'; %Performing
       Serial to Parallel Conversion
53  datamat1=DataMod_serialtoparallel1; %Assigning the
       total data to a variable called datamat
54  % Computation of Hermitian Symmetry Criteria
55  datamat1(1,:)=0; %Assigning the First subcarrier to
       Zero
56  datamat1(513,:)=0; %Assigning the Middle Subcarrier
       to Zero
57  datamat1(514:1024,:)=flipud(conj(datamat1(2:512,:)));
        %Illustrating that only half of the subcarriers
       are exploited for data transmission as the
       remaining half are flipped complex conjugate
       versions of the previous ones.
58  d_ifft1=ifft((datamat1)); %Computation of IFFT
       operation
59  d_ifft_paralleltoserial1=d_ifft1.'; %Parallel to
       Serial Conversion
```

```
60 CP_part1=d_ifft_paralleltoserial1(:,end-Ncp+1:end); %
      Addition of Cyclic Prefix
61 DCOOFDM_CP1=[CP_part1 d_ifft_paralleltoserial1]; %
      Transmissin of DCO-OFDM signal
62 bdc1=13; %DC BIAS
63 clip1=sqrt((10.^(bdc1/10))-1); %clipping factor k
64         bdcc1=clip1*sqrt(DCOOFDM_CP1.*DCOOFDM_CP1);%
              Computation of DC bias
65         DCOOFDM_BIAS1=DCOOFDM_CP1+bdcc1; %Addition of
               DC bias to the cyclic prefix added signal
66 count=0;
67 snr_vector=0:1:80; %size of signal to noise ratio (
      SNR) vector
68 for snr=snr_vector
69     SNR = snr + 10*log10(log2(M));
70     count=count+1 ;
71     DCOOFDM_with_channel=awgn(DCOOFDM_BIAS1,SNR,'
          measured') ; %Addition of AWGN
72     %Receiver of DCO-OFDM
73     DCOOFDM_with_chann2=DCOOFDM_with_channel-bdcc1;%
          Removal of DC bias
74     DCOOFDM_removal_CP2=DCOOFDM_with_chann2(:,Ncp+1:N
          +Ncp); %Removal of Cyclic Prefix
75     DCOOFDM_serialtoparallel2=DCOOFDM_removal_CP2.';
          %Serial to Parallel Conversion
76     DCOOFDM_parallel_fft2=fft(
          DCOOFDM_serialtoparallel2) ; %Computation of
          FFT operation
77
78     DCOOFDM_Demodulation2=qamdemod(
          DCOOFDM_parallel_fft2.',M);
79    [~,s_e2(count)]=symerr(Data1(:,2:512),
         DCOOFDM_Demodulation2 (:,2:512)) ;
80 end
81   % Plotting the BER curves
82 semilogy(snr_vector,s_e1,'rd-','LineWidth',2);hold on
      ;
83 semilogy(snr_vector,s_e2,'gd-','LineWidth',2);
84 legend(3, 'FFT-based DCO-OFDM-7dB of Bias', 'FFT-
      based DCO-OFDM-13dB of Bias');
85 axis([0 30 10^-4 1]);
86 xlabel('SNR in dB');
87 ylabel('BER');
88 grid on;
```

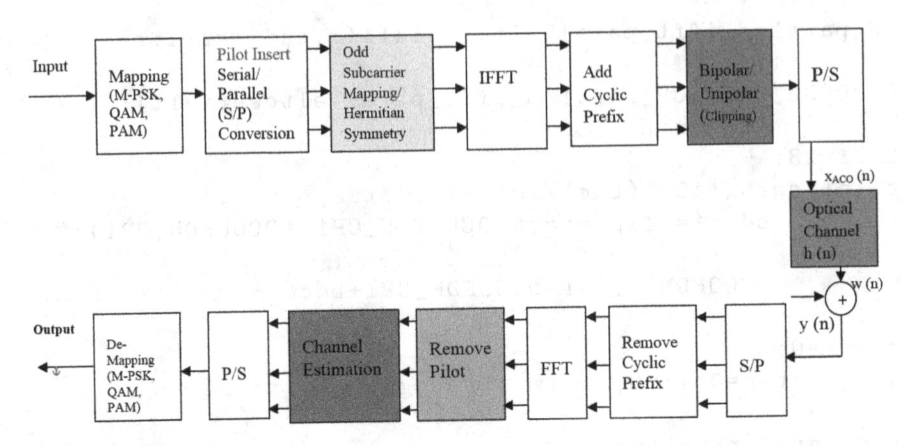

Figure 3.7: Schematic representation of channel estimation in Hermitian Symmetry imposed-IFFT-based ACO-OFDM system [498].

ACO-OFDM: Substantially, to overcome the drawback of power inefficiency in DCO-OFDM, ACO-OFDM methodology can be employed to yield a positive-valued signal. Predominantly, a unipolar signal can be generated without the requirement of addition of DC bias. In contrast to DCO-OFDM methodology, a positive-valued signal is obtained by reliably choosing the subcarriers for data modulation. As evidenced from Fig. 3.7, the frequency domain representation of the signal follows equation (3.3). Here, only odd subcarriers are modulated and are assigned with Hermitian Symmetry, while the even subcarriers are assigned zero. Thus, the resultant signal can be put up as

$$X[N-k] = \begin{cases} X^*[k] & k \text{ is odd} \\ 0 & k \text{ is even} \end{cases} \tag{3.20}$$

Hence, the input to the IFFT resembles the following

$$X = \left[0, X_1, 0, X_3, \cdots X_{N/4-1}, 0, X^*_{N/4-1}, 0, \cdots X^*_1\right]^T \tag{3.21}$$

It is apparent that from (3.21) only odd subcarriers are modulated and the even subcarriers are set to zero. Hence, out of N subcarriers only $\frac{N}{2}$ are utilized and due to Hermitian Symmetry constraint only $\frac{N}{4}$ are meant for carrying data because the rest $\frac{N}{4}$ are flipped complex conjugate versions of the previous ones. Therefore, (3.4) can be modified as

$$X\left[\frac{N}{2}-k\right] = X^*[k],\ k = 1, 2, \cdots \frac{N}{4} \tag{3.22}$$

By making use of (3.7), the discretized time-domain signal at the output of the IFFT can be solved as

$$\begin{aligned} x[n] = \frac{1}{N}\Bigg[& X[0]\, e^{\frac{j2\pi n(0)}{N}} + \sum_{k=1}^{\frac{N}{4}-1} X[k]\, e^{\frac{j2\pi nk}{N}} + \\ & X\left[\frac{N}{4}\right] e^{\frac{j2\pi n\left(\frac{N}{4}\right)}{N}} + \sum_{k=\frac{N}{4}+1}^{\frac{N}{2}-1} X[k]\, e^{\frac{j2\pi nk}{N}} \Bigg] \end{aligned} \tag{3.23}$$

Upon further solving, (3.23) can be reduced as

$$x[n] = \frac{1}{N}\left[\sum_{k=1}^{\frac{N}{4}-1} X[k]\, e^{\frac{j2\pi nk}{N}} + \sum_{k=\frac{N}{4}+1}^{\frac{N}{2}-1} X[k]\, e^{\frac{j2\pi nk}{N}}\right] \tag{3.24}$$

Now applying change in variable transformation, i.e., letting $\frac{N}{2} - k' = k$ and the resultant expression is

$$x[n] = \frac{1}{N}\left[\sum_{k=1}^{\frac{N}{4}-1} X[k]\, e^{\frac{j2\pi nk}{N}} + \sum_{k'=\frac{N}{4}-1}^{1} X\left[\frac{N}{2} - k'\right] e^{\frac{j2\pi\left(\frac{N}{2}-k'\right)n}{N}}\right] \tag{3.25}$$

Changing the limits in (3.25) yields the following expression

$$x[n] = \frac{1}{N}\left[\sum_{k=1}^{\frac{N}{4}-1} X[k]\, e^{\frac{j2\pi nk}{N}} + \sum_{k'=1}^{\frac{N}{4}-1} X\left[\frac{N}{2} - k'\right] e^{\frac{-j2\pi\left(\frac{N}{2}-k'\right)n}{N}}\right] \tag{3.26}$$

By making use of (3.22) in (3.26) and then rearranging, the following expression can be attained

$$x[n] = \frac{1}{N}\left[\sum_{k=1}^{\frac{N}{4}-1} X[k]\, e^{\frac{j2\pi nk}{N}} + \sum_{k=1}^{\frac{N}{4}-1} X^{*}[k]\, e^{\frac{-j2\pi nk}{N}}\right] \tag{3.27}$$

Making use of Euler's inequalities and signal processing identities, (3.27) can be derived as

$$x[n] = \frac{1}{N}\left[\sum_{k=1}^{\frac{N}{4}-1} 2X_{RC}[k] cos\left(\frac{2\pi nk}{N}\right) - \sum_{k=1}^{\frac{N}{4}-1} 2X_{IC}[k] sin\left(\frac{2\pi nk}{N}\right)\right] \tag{3.28}$$

(3.28) unveils that a real-valued signal is attained at the expense of reduced throughput as only $\frac{N}{4}$ subcarriers are involved for data transmission. Further, to this signal, cyclic prefix is added and the resultant signal $x_{cp}[n]$ is assured of its positivity by hard clipping the entire negative excursion and then transmitting only the positive signal. This operation can be represented as

$$\underset{Unipolar}{x[n]} = x_{ACO}[n] = \begin{cases} x_{cp}[n] & if\ x_{cp}[n] \geq 0 \\ 0 & if\ x_{cp}[n] \leq 0 \end{cases} \tag{3.29}$$

It is interesting to note that this time domain signal has an anti symmetric property. The main reason involved behind employing odd subcarrier modulation and clipping the even subcarriers is to ensure that clipping noise falls only on the even subcarriers. The anti symmetric property can be proved mathematically as: Using the expression for discretized time-domain signal as shown in (3.7),

$$x\left[n + \frac{N}{2}\right] = \frac{1}{N}\sum_{k=0}^{N-1} X[k]\, e^{\frac{-j2\pi\left(n+\frac{N}{2}\right)k}{N}} \tag{3.30}$$

On further solving, (3.30) reduces as

$$x\left[n+\frac{N}{2}\right]=\frac{1}{N}\sum_{k=0}^{N-1}X[k]e^{\frac{-j2\pi nk}{N}}e^{j\pi k} \tag{3.31}$$

$e^{j\pi k}=(-1)^k$; therefore (3.31) can be reduced to

$$x\left[n+\frac{N}{2}\right]=\frac{1}{N}\sum_{k=0}^{N-1}X[k]e^{\frac{-j2\pi nk}{N}}(-1)^k \tag{3.32}$$

Since, only odd subcarriers are utilized for data modulation and the even subcarriers are set to zero. Hence, we can consider only odd subcarriers. Therefore, $(-1)^k=-1$ since our point of interest is only odd subcarriers.

$$x\left[n+\frac{N}{2}\right]=-\frac{1}{N}\sum_{k=0}^{N-1}X[k]e^{\frac{-j2\pi nk}{N}} \tag{3.33}$$

Subsequently from (3.33) it can be inferred that

$$x\left[n+\frac{N}{2}\right]=-x[n] \tag{3.34}$$

Therefore, it can be deduced that the time-domain signal has an anti symmetry property. Even though the negative peaks are clipped, for each clipped negative peak a positive peak with the same absolute value will be transmitted. As a result, clipping does not lead to loss of information. However, clipping noise arises in ACO-OFDM.

Figure 3.8 clearly depicts the performance of ACO-OFDM system exploiting M-QAM constellation over VLC channel environment. It is evident from the figure that with the increase in the orders of modulation, due to the domination of clipping noise, it becomes difficult to reach a reduced error floor. Figure 3.9 illustrates the BER vs. SNR performance of LS and MMSE channel estimation algorithms aided ACO-OFDM system over VLC channel environment. From the figure, it is clear that MMSE channel estimation algorithm outperforms the LS channel estimation due to the inherent fact that LS is more susceptible to noise. However, the superior performance of MMSE is achieved only at the expense of increase in the computational complexity of the receiver design.

Listing 3.2: MATLAB code depicting the performance of channel-estimation aided ACO-OFDM system over VLC channel environment for the simulated result as illustrated in Fig. 3.9. It is to be noted that this code works for 4 QAM; the reader can simulate it for other increasing orders of QAM modulation. The M-codes for the functions MMSEesti.m and interpolate.m can be exploited from reference cho2010mimo. (Note: This MATLAB code works with R2014b Version).

```
1 clc;
2 clear all;
```

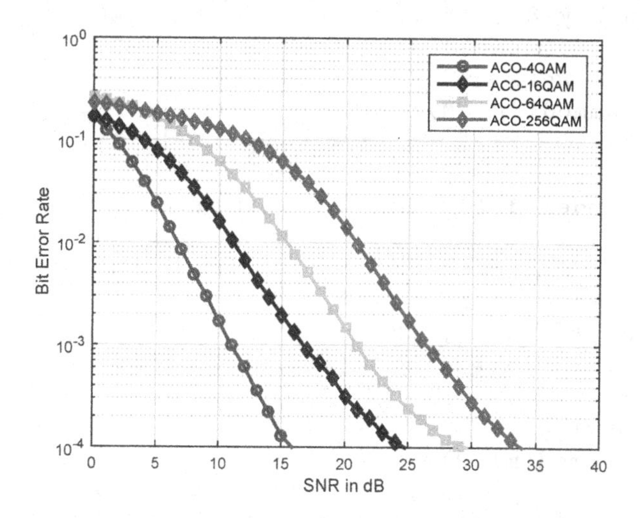

Figure 3.8: BER vs. SNR performance of ACO-OFDM system using M-QAM constellation considering NLOS components over VLC channel [498].

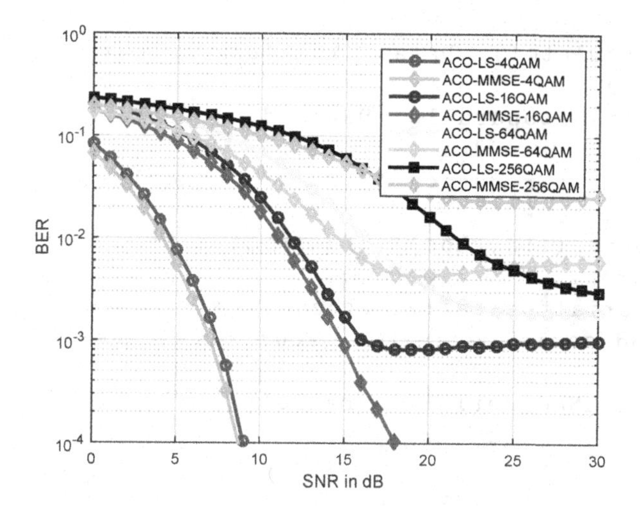

Figure 3.9: BER vs. SNR performance of LS and MMSE aided channel estimation algorithms in ACO-OFDM system over VLC channel [498].

```
3  close all;
4  m=512; %Total number of OFDM symbols
5  N=1024; %Length of each OFDM symbol
6  M=4; %Size of the Constellation (M can be 4, 8, 16,
      32, 64, 128, 256)
```

```
7  pilotFrequency=8 ; %Pilot Symbol insertion frequency
8  E=2; %Energy of each Pilot Symbol
9  Ncp=256 ; %Length of the Cyclic Prefix
10 Tx=modem.qammod('M',M); %Choosing the modulation
      format as M-ary Quadrature Amplitude Modulation (M
      -QAM)
11 Rx=modem.qamdemod ('M',M); %Fixing up the
      demodulation format at the receiving end as M-ary
      QAM
12 %ACO-OFDM Transmitter
13  Data=randi ([0 M-1],m,N); %Generation of Random bits
        Matrix of size m by N
14 for k1=1:m
15      for m1=1:N
16          if mod(m1,2)==0 %Performing Modulo operation
                to extract the even subcarriers
17              Data(k1,m1)=0; %Setting the Even
                    Subcarriers to Zero
18          end
19      end
20  end
21 DataMod=modulate(Tx,Data); %Performing Data
      Modulation
22 DataMod_serialtoparallel=DataMod.'; %Performing
      Serial to Parallel Conversion
23 PLoc = 1:pilotFrequency:N; %Fixing the location of
      Pilot carrires
24 DLoc = setxor(1:N,PLoc); %Fixing the location of Data
       subcarriers
25 DataMod_serialtoparallel(PLoc,:)=E*
      DataMod_serialtoparallel(PLoc,:); %Inserting Pilot
       carriers
26 datamat=DataMod_serialtoparallel; %Assigning the
      total data including the pilots to a variable
      called datamat
27 % Computation of Hermitian Symmetry Criteria
28 datamat(1,:)=0; %Assigning the First subcarrier to
      Zero
29 datamat(513,:)=0; %Assigning the Middle Subcarrier to
       Zero
30 datamat(514:1024,:)=flipud(conj(datamat(2:512,:))); %
      Illustrating that only half of the subcarriers are
       exploited for data transmission as the remaining
      half are flipped complex conjugate versions of the
```

```
      previous ones.
31 d_ifft=ifft((datamat)); %Computation of IFFT
      operation
32 %Ensuring that only the positive portion of the
      signal is transmitted
33  for k2=1:N
34    for m2=1:m
35       if(d_ifft(k2,m2)<0)
36         d_ifft(k2,m2)=0;
37        end
38     end
39  end
40 d_ifft_paralleltoserial=d_ifft.'; %Parallel to Serial
      Conversion
41 CP_part=d_ifft_paralleltoserial(:,end-Ncp+1:end); %
      Addition of Cyclic Prefix
42 ACOOFDM_CP=[CP_part d_ifft_paralleltoserial]; %
      Transmissin of ACO-OFDM signal
43 % VLC Channel Modeling
44 theta = 70; %LED semi-angle
45 ml=-log10(2)/log10(cos(theta)); %Computation of
      Lambertian Mode Number
46 APD=0.01; %Area of the photodiode
47 lx=5; ly=5; lz=3; %Size of the Dimensions of the
      Indoor Room Environment
48 h=2.15;
49 [XT,YT]=meshgrid([-lx/4 lx/4],[-ly/4 ly/4]);
50 Nx=lx*5; Ny=ly*5;
51 x=linspace(-lx/2,lx/2,Nx);
52 y=linspace(-ly/2,ly/2,Ny);
53 [XR,YR]=meshgrid(x,y);
54
55 D1=sqrt((XR-XT(1,1)).^2+(YR-YT(1,1)).^2+h^2);
56 D2=sqrt((XR-XT(2,2)).^2+(YR-YT(2,2)).^2+h^2);
57 cosphi_A1=h./D1;
58 receiver_angle=acosd(cosphi_A1);
59
60 H_A1=3600*((ml+1)*APD.*cosphi_A1.^(ml+1)./(2*pi.*D1
      .^2)+(ml+1)*APD.*cosphi_A1.^(ml+1)./(2*pi.^2*D1
      .^2*D2.^2)); %Computation of VLC channel Impulse
      Response taking into consideration Non Line of
      Sight (NLOS) Environment
61 H_A2=H_A1./norm(H_A1)
62 d_channell = filter(H_A2(1,1:2),1,ACOOFDM_CP.').'; %
```

```
    Illustration of channel effect on the transmitted
    ACO-OFDM signal
63  count=0;
64 snr_vector=0:1:30; %size of signal to noise ratio (
    SNR) vector
65 for snr=snr_vector
66     SNR = snr + 10*log10(log2(M));
67     count=count+1 ;
68     ACOOFDM_with_chann=awgn(d_channell,SNR,'measured'
         ) ; %Addition of AWGN
69     %Receiver of ACO-OFDM
70     ACOOFDM_removal_CP=ACOOFDM_with_chann(:,Ncp+1:N+
        Ncp); %Removal of Cyclic Prefix
71     ACOOFDM_serialtoparallel=ACOOFDM_removal_CP.'; %
        Serial to Parallel Conversion
72     ACOOFDM_parallel_fft=fft(ACOOFDM_serialtoparallel
        ) ; %Computation of FFT operation
73     % Channel Estimation
74     TransmittedPilots = DataMod_serialtoparallel(PLoc
        ,:); %Extracting the transmitted pilots
75     ReceivedPilots = ACOOFDM_parallel_fft(PLoc,:); %
        Extracting the received pilot tones effected
        by channel
76     H_LS= ReceivedPilots./TransmittedPilots; % Least
        Square Channel Estimation
77     for r=1:m
78         H_MMSE(:,r) = MMSEesti(ReceivedPilots(:,r),
            TransmittedPilots(:,r),N,pilotFrequency,
            H_A2(1,1:2),SNR);%Minimum Mean Square
            Error (MMSE)Channel Estimation
79     end
80     for q=1:m
81     HData_LS(:,q) = interpolate(H_LS(:,q).',PLoc,N,'
        spline'); %Interpolation
82     end
83     HData_LS_parallel1=HData_LS.'; %Parallel to
        Serial Conversion
84     HData_MMSE_parallel1=H_MMSE.';
85     ACOOFDM_SERIAL_LS=demodulate(Rx,(
        ACOOFDM_parallel_fft.')./HData_LS_parallel1) ;
         %Demodulation
86     ACOOFDM_SERIAL_MMSE=demodulate(Rx,(
        ACOOFDM_parallel_fft.')./(HData_MMSE_parallel1
        )) ;
```

```
87      %Recovery of Pilots from the Original Transmitted
            signal and Received Signal
88      Data_no_pilots=Data(:,DLoc);
89      Recovered_Pilot_LS=ACOOFDM_SERIAL_LS(:,DLoc);
90      Recovered_Pilot_MMSE=ACOOFDM_SERIAL_MMSE(:,DLoc);
91      %Computation of Bit Error Rate
92      [~,recoveredLS(count)]=biterr(Data_no_pilots
           (:,2:255),Recovered_Pilot_LS(:,2:255));
93      [~,recoveredMMSE(count)]=biterr(Data_no_pilots
           (:,2:255),Recovered_Pilot_MMSE(:,2:255));
94  end
95  %Plotting the BER curves
96  semilogy(snr_vector,recoveredLS,'rd-','LineWidth',2);
       hold on;
97  semilogy(snr_vector,recoveredMMSE,'gs-','LineWidth'
       ,2);
98  axis([0 30 10^-4 1]);
99  grid on;
```

PAM-DMT: In case of ACO-OFDM system, only odd subcarriers are exploited to carry the data, while the even subcarriers are set to zero. As seen in the aforementioned mathematical analysis, the time domain output of the IFFT has an anti-symmetry property which implies that for every clipped negative peak, there will be a positive peak transmitting the same absolute value. Even though ACO-OFDM does not exhibit any energy efficiency loss, it is spectrally in efficient as only $\frac{1}{4}$th of the subcarriers are exploited for transmitting the data. Moreover, the restriction of odd subcarrier mapping imposes a severe degradation in the overall performance of ACO-OFDM system in case of frequency-selective fading channel predominantly when bit allocation algorithm is exploited. Consequently, to overcome this constraint of exploiting only an odd number of subcarriers as well as to enable the system performance to optimally adapt to the frequency response of the channel, an alternative to ACO-OFDM system, namely, pulse amplitude modulation discrete multi tone modulation (PAM-DMT) has been proposed by the authors in [291]. By making use of the Fourier properties of the imaginary signals, a unipolar signal is guaranteed by PAM-DMT. Unlike ACO-OFDM, where only odd subcarriers are modulated, in PAM-DMT, the imaginary components of all of the available subcarriers are modulated while, the real components being off to ensure an anti symmetry in the time domain signal waveform of PAM-DMT. Finally, a positive-valued signal is ensured by enabling asymmetric clipping at zero. Therefore, similar to ACO-OFDM system without the requirement of DC bias, a positive valued signal is attained. Moreover, since all of the subcarriers are exploited for data transmission, it enhances the spectral efficiency, and bit loading algorithm can be exploited more efficiently to adapt the channel frequency response thus facilitating to achieve better performance than that of the ACO-OFDM system.

Similar to DCO-OFDM and ACO-OFDM, PAM-DMT also relies on Hermitian Symmetry criteria to accomplish real signal transmission. Since, the real values of the subcarriers are set to zero, PAM-DMT relies on one-dimensional mapping scheme like M-ary PAM. Like the other IFFT-based optical OFDM variants, PAM-DMT also relies on Hermitian Symmetry criteria for achieving a real-valued signal. Generally, the input to the IFFT module has the following arrangement

$$X[k] = \begin{cases} 0, \quad k = 0, \frac{N}{2} \\ jS[k], \quad k = 1,2,3 \cdots \frac{N}{2} - 1 \\ X^*[N-K] \;\Rightarrow\; -jS^*[N-k], \quad k = \frac{N}{2} + 1 \cdots N-1 \end{cases} \tag{3.35}$$

Referring to equation (3.35), $S[k]$ represents the symbols carrying the appropriate data. Thus, by taking into account this frequency domain representation of the mapped PAM-DMT symbols and substituting in the expression for the time-domain output of the IFFT module, the following resultant signal is attained

$$x[n] = \frac{1}{N}\left\{ \sum_{k=1}^{\frac{N}{2}-1} jS[k]\, e^{\frac{j2\pi nk}{N}} + \sum_{k=\frac{N}{2}+1}^{N-1} -jS^*[N-k]\, e^{\frac{j2\pi nk}{N}} \right\} \tag{3.36}$$

By applying a change in variable transform for the purpose of equalizing the orders of limits in both the summation terms, the resultant expression is obtained as follows

$$x[n] = \frac{1}{N}\left\{ \sum_{k=1}^{\frac{N}{2}-1} jS[k]\, e^{\frac{j2\pi nk}{N}} + \sum_{k=1}^{\frac{N}{2}-1} -jS^*[N-k]\, e^{-\frac{j2\pi nk}{N}} \right\} \tag{3.37}$$

Also, by making use of the fact that according to the Hermitian Symmetry criteria, the second term in the summation implies $S^*[N-k] = S[k]$. Therefore, (3.37) can be reduced to

$$x[n] = \frac{1}{N}\left\{ \sum_{k=1}^{\frac{N}{2}-1} jS[k]\, e^{\frac{j2\pi nk}{N}} + \sum_{k=1}^{\frac{N}{2}-1} -jS[k]\, e^{-\frac{j2\pi nk}{N}} \right\} \tag{3.38}$$

By making use of Euler inequalities, where $cos[\theta] = \frac{e^{j\theta}+e^{-j\theta}}{2}$, equation (3.38) can be rearranged to obtain the final PAM-DMT time domain signal output

$$x[n] = \frac{-2}{N} \sum_{k=1}^{\frac{N}{2}-1} S[k]\, sin\left[\frac{2\pi nk}{N}\right] \tag{3.39}$$

Thus, (3.39) infers that a real-valued signal is attained without any complex value being present. Finally, the negative peak of (3.39) is clipped to attain a positive signal. As stated, similar to ACO-OFDM, PAM-DMT also has anti symmetric property as follows:

$$x[N-n] = \frac{-2}{N} \sum_{k=1}^{\frac{N}{2}-1} S[k]\, sin\left[\frac{2\pi (N-n)k}{N}\right] \tag{3.40}$$

Upon further solving the argument term of sine in equation (3.40)

$$x[N-n] = \frac{-2}{N} \sum_{k=1}^{\frac{N}{2}-1} S[k] \sin\left[\frac{2\pi Nk}{N} - \frac{2\pi nk}{N}\right] \quad (3.41)$$

Upon solving the argument term of sine by using the trigonometric identity $sin(A-B) = sin(A)cos(B) - cos(A)sin(B)$, the expression as denoted by (3.41) reduces to

$$x[N-n] = -\frac{2}{N} \sum_{k=1}^{\frac{N}{2}-1} S[k] \left[sin[2\pi k] cos\left[\frac{2\pi nk}{N}\right] - cos[2\pi k] sin\left[\frac{2\pi nk}{N}\right]\right] \quad (3.42)$$

Upon further using the fact that cosine of even integral multiples of π as 1, equation (3.42) reduces to

$$x[N-n] = \frac{2}{N} \sum_{k=1}^{\frac{N}{2}-1} S[k] \sin\left[\frac{2\pi nk}{N}\right] \quad (3.43)$$

The right-hand side of equation (3.43) represents the time-domain signal $-x[n]$. Thus, clearly, it can be stated that $x[N-n] = -x[n]$. Therefore, this implies that upon clipping the entire negative excursion does not lead to any information losses because the same amount of information is transmitted in the positive part of the frame. At the receiving end, only the imaginary component of the subcarriers is taken into consideration for data detection leaving behind the real components.

Flip-OFDM: This variant of optical OFDM has been proposed as an alternative to ACO-OFDM system. In contrast to ACO-OFDM system, where only odd subcarriers were modulated, in case of Flip-OFDM all of the subcarriers are modulated and exploit the Hermitian Symmetry criteria to fetch the real signal. However, in this optical OFDM, the time-domain signal is subjected to certain signal transformation without relying on biasing requirement to yield a non-negative valued signal. In Flip-OFDM, the resultant bipolar time-domain signal can be regarded as a combination of positive and negative parts which can be expressed as

$$x[n] = x^+[n] + x^-[n] \quad (3.44)$$

meanwhile, the positive and negative parts of (3.44) can be represented as

$$x^+[n] = \begin{cases} x[n], & if x[n] \geq 0 \\ 0, & otherwise \end{cases} \quad (3.45)$$

$$x^-[n] = \begin{cases} x[n], & if x[n] < 0 \\ 0, & otherwise \end{cases} \quad (3.46)$$

Thereupon, in order to convert the bipolar signal in Flip-OFDM into unipolar signal, the positive part of the bipolar signal is transmitted in the first OFDM frame, while the flipped negative part is transmitted in the second frame as follows:

$$x_{2N,Flip}[n] = \begin{cases} x^{+}[n], & n = 0, 1, 2, \cdots N-1 \\ -x^{-}[n-N], & n = N, \cdots 2N-1 \end{cases} \tag{3.47}$$

The entire signal processing mechanism can be pictorially represented as shown in Fig. 3.10. In this manner, the bipolar signal in case of Flip-OFDM is converted into

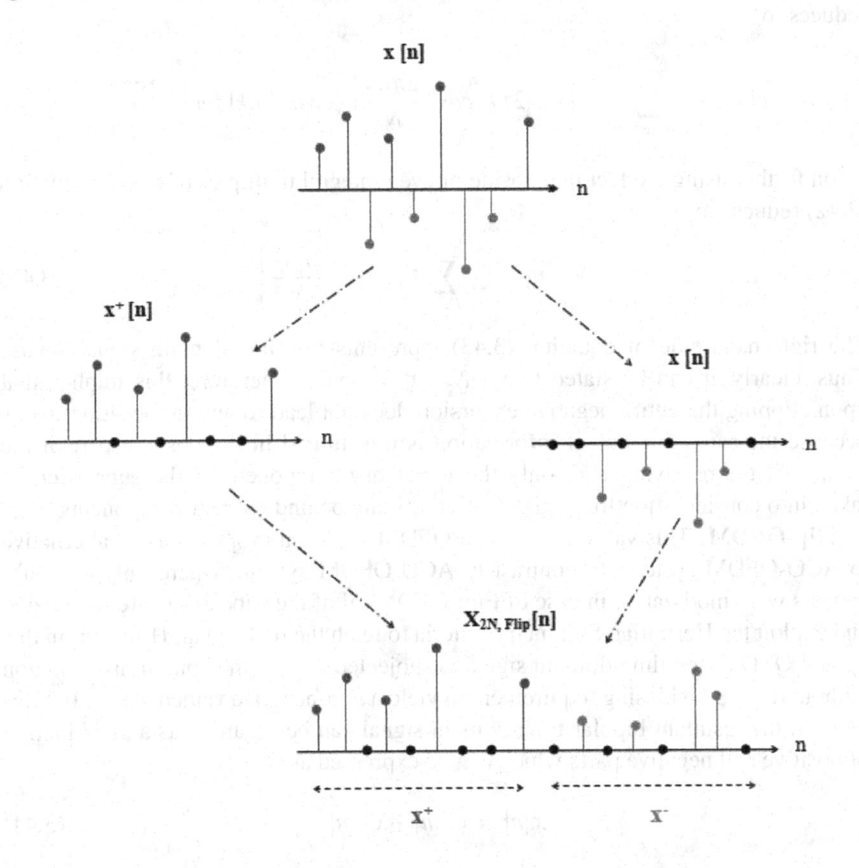

Figure 3.10: The signal processing concept of Flip-OFDM.

unipolar and, at the receiving end, the bipolar signal is recovered by subtracting the samples contained in negative frame from the samples contained in positive frame. The work in [169] signifies the fact that both ACO-OFDM and Flip-OFDM exhibit a similar performance in terms of spectral efficiency, SNR and BER. Furthermore, while employing N-point IFFT and FFT transformation blocks, Flip-OFDM allows for the transformation of twice the data information than that of the ACO-OFDM. Thus, when compared with ACO-OFDM, a Flip-OFDM system considerably reduces the hardware complexity of the receiver by 50%. However, Flip-OFDM is not devoid of its drawbacks: The Hermitian Symmetry criteria is essential in Flip-OFDM to generate a real-valued signal, which not only increases the computational complexity but also increases the power consumption, because in order to modulate N frequency domain symbols it requires double the size of the transmission points. The second major drawback of Flip-OFDM is that it facilitates the generation of only ACO-OFDM signals and fails to be implemented for other power efficient techniques that can be generated by using the methodology of DCO-OFDM.

Unipolar-OFDM (U-OFDM): The methodology of U-OFDM is similar to that of the Flip-OFDM. It is apparent that in case of optical OFDM, IFFT and FFT transformation modules play a vital role in enabling OFDM modulation and demodulation; therefore, in case of U-OFDM, similar to the other variants of optical OFDM, Hermitian Symmetry criteria is enforced to yield a real-valued signal. However, to fetch a positive-valued signal, the time-domain signal is subjected to simple transformation without relying on the addition of DC bias. Here, in this U-OFDM system, the bipolar frame is divided into two frames named as positive and negative frames. The positive frame contains only the positive samples, and the negative samples are replaced with zeros, while the negative frame comprises of the absolute values of all the negative samples and zeros placed in the position of all positive samples. Thus, in this manner, the bipolar frame is divided into two frames and this sort of transformation eventually leads to halving of both the achievable data rate and the spectral efficiency. Thus, the achievable spectral efficiency of the unipolar OFDM can be expressed as:

$$\eta_{U-OFDM} = \frac{log_2 M\,(N_{FFT}-2)}{4\,(N_{FFT}+N_{cp})}\; bits/s/Hz \tag{3.48}$$

Conversely, the spectral efficiency of DCO-OFDM system can be represented as

$$\eta_{DCO-OFDM} = \frac{log_2 M\,(N_{FFT}-2)}{2\,(N_{FFT}+N_{cp})}\; bits/s/Hz \tag{3.49}$$

The spectral efficiency of U-OFDM as illustrated by (3.48) clearly infers the fact that it is reduced to half when compared with DCO-OFDM system. The parameters $log_2 M$ represent the total number of bits that are mapped by using M-QAM constellation, N_{FFT} indicates the size of the FFT/IFFT transform, while $\frac{(N_{FFT}-2)}{2}$ interrupts the fact that the first and the middle subcarriers are set to zero fulfilling the Hermitian Symmetry criteria. At the receiving end, the bipolar optical OFDM signal can be recovered by subtracting the negative samples contained in the negative frame from the positive values of the samples contained within the positive frame. In this process of recovery of the bipolar optical time-domain signal, the AWGN of both frames are added together resulting in doubling the amount of noise at the receiving end, thereby leading to a 3 dB penalty in the SNR when compared with bipolar OFDM. Moreover, power efficiency of U-OFDM slowly dies out even though the positive valued signal is generated without the requirement of addition of DC bias; this is because M^2 QAM-U-OFDM needs to be compared with M QAM-DCO-OFDM system. Thus, as the order of the constellation increases, the amount of requirement in power also increases, thus making U-OFDM a power inefficient scheme.

The aforementioned unipolar variants (ACO-OFDM, PAM-DMT, Flip-OFDM, U-OFDM) have been proposed with the rationale to overcome the power inefficiency of DCO-OFDM system. However, in case of ACO-OFDM system, out of N subcarriers, half of the subcarriers are lost due to the Hermitian Symmetry criteria, and among the remaining subcarriers, only half are exploited due to odd subcarrier mapping, i.e., only $\frac{1}{4}$th subcarriers are exploited for data transmission. Whereas, in case of Flip-OFDM and U-OFDM, the input and output frame structures are modified in such a manner that both these schemes require two time-domain frames to transmit the information. Contrarily, DCO-OFDM system requires just a single frame to transmit the information. Consequently, this encompasses ACO-OFDM, Flip-OFDM and U-OFDM to exploit double the size of the constellation in order to compare their performances with a DCO-OFDM system. Precisely, it is necessitated for ACO-OFDM, Flip-OFDM and U-OFDM exploiting M^2-QAM to be compared with DCO-OFDM system employing M-QAM. Pertaining to the scenario of PAM-DMT, the imaginary parts of the subcarriers are modulated by means of real mapping scheme like M-PAM, while the real parts are set to zero. Therefore, the performance of PAM-DMT exploiting M-PAM constellation is equivalent to the performance exhibited by M^2-QAM. Thus, this implies that the performance of PAM-DMT is similar to that of the ACO-OFDM, Flip-OFDM and U-OFDM. Moreover, when compared to DCO-OFDM system, at the same spectral efficiency, the performances of these unipolar variants of optical OFDM namely ACO-OFDM, PAM-DMT, Flip-OFDM and U-OFDM deteriorates with the increase in the size of the constellation [237]. Consequently, in order to enhance the performance of these aforesaid optical variants of OFDM, improved receivers were designed and the corresponding literature can be found in [129, 130, 228, 229, 533, 539, 552].

The work in [129] proposes a novel receiver for ACO-OFDM system that exploits all the subcarriers in contrast to the conventional ACO-OFDM receiver which discards the even subcarriers that are distorted by means of clipping noise. In this work, a novel receiver structure is designed to facilitate the even subcarriers to gather additional diversity and coding gains resulting in substantial improvement in the overall BER performance. From the simulation results, it can be deduced that the conventional receiver provides a considerable improvement in the SNR gains of up to 10 dB when compared with the conventional receiver in case of NLOS channel conditions. The work in [228] proposes an iterative receiver for Flip-OFDM system for improving the transmission performance of conventional Flip-OFDM system. Principally, the receiver performance of a conventional Flip-OFDM system is improved by enabling the iterative receiver to attain the diversity gain by making use of the signals in both negative as well as positive frames. Moreover, when compared with the conventional Flip-OFDM system, the iterative receiver comprising of only two iterations imparts an appreciable improvement in SNR gain. Figure 3.11 illustrates the BER performance of Flip-OFDM system exploiting different receivers over NLOS channel environment. Among the other receivers, iterative receiver achieves the better BER performance. When compared with the conventional receiver at BER of 10^{-4}, the iterative receiver exhibits SNR gain of around 4 dB. This remarkable

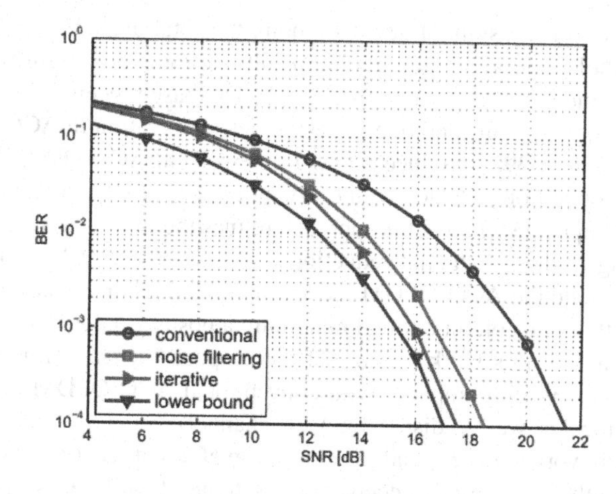

Figure 3.11: BER vs. SNR performance analysis of different receivers in Flip-OFDM system [228].

improvement in SNR gain is due to the fact that the iterative receiver fully exploits the signal structure when compared to the conventional receiver. Additionally, when compared to the noise filtering receiver, iterative receiver gives a better performance demonstrating SNR gain of 1 dB at a BER of 10^{-4} over NLOS channel environment. Iterative receiver in Flip-OFDM in spite of its significant performance, the computational complexity involved behind the implementation of the receiver is very complex. Therefore, to overcome this computational complexity, the work in [552] proposes a novel receiver design for Flip-OFDM systems which exhibits better performance in terms of BER when compared with the conventional receiver of Flip-OFDM. Primarily in case of Flip-OFDM system, it is apparent that two consecutive OFDM frames were employed to transmit both the positive and negative parts of the signal. At the receiving side, the transmitted signal is recovered by exploiting the difference between the two frames of signal. This sort of approach results in in-efficient utilization of the received power. Consequently, in order to overcome this limitation, a new receiver characterizing temporal and spectral diversity combining is proposed in [552]. The simulation results of this work infers that these proposed receivers significantly improve the BER performance by making use of the information attained from both the sum and difference of the two frames. Furthermore, the spectral receiver proposed in this work remarkably reduces the computational complexity of the iterative receiver.

The work in [130] presents a novel receiver design for ACO-OFDM system that exploits the frequency domain diversity combining (FDDC) technique. The FDDC technique when compared to time-domain diversity combining (TDDC) technique exhibits its potential to effectively exploit the frequency selectivity of the channel for enhancing the detection performance. Simultaneously, this work proposes an enhanced version of FDDC receiver which is called eFDDC and enables for the optimal selection of the symbol vectors among the total available multiple sets of candidate symbol vectors. The simulation results of this work confirms the superiority of FDDC and eFDDC receiver over the TDDC-based receiver in ACO-OFDM system. By taking into consideration the asymmetric operation in ACO-OFDM system, the authors in [539] introduce a new framework in ACO-OFDM by adding a reconstruction operation in place of the asymmetric clipping operation before the transmission of optical signals. This approach refines the time-domain signal format, thereby leading to the substantial reduction in the amount of PAPR. Furthermore, in order to facilitate the ACO-OFDM receiver to effectively recover the transmitted signals, the authors exploit an optimal maximum a posteriori detection method. The simulation results of this work reveal that the proposed scheme illustrates a better BER performance when compared with conventional ACO-OFDM system operating over both ideal as well as dispersive VLC channels.

The research works in [229] and [533] propose efficient receivers for PAM-DMT systems. Typically, in case of the receiver of traditional PAM-DMT systems, the clipping noise distorts the real parts of the subcarriers and hence they are ignored in the process of detection. On the contrary, the work in [229] states that the real parts of the subcarriers also contain the information about the transmitted data signal. Therefore, taking this fact into consideration, the authors propose an iterative receiver that exploits both the real and imaginary components of the subcarriers to improve the transmission performance of conventional PAM-DMT systems. In the same manner, the authors in [533] propose a novel receiver design for PAM-DMT systems that exploits the temporal and spectral diversity combining in a manner to extract the effective information in clipping noise for recovering the data. The simulation results validate the superiority of the proposed receiver which exhibits 2.3 dB gain in $\frac{E_b}{N_0}$ at a BER of 10^{-5} over AWGN channel.

3.2.2 ON THE PERFORMANCE OF DIFFERENT SUPERPOSITION OPTICAL OFDM VARIANTS

The different variants of optical OFDM that belong to this category are listed as follows:

1. Enhanced unipolar optical OFDM (eU-OFDM)
2. Generalized enhanced unipolar OFDM (GREENER-OFDM)
3. Enhanced ACO-OFDM (eACO-OFDM)
4. Enhanced PAM-DMT (ePAM-DMT)
5. Spectrally and energy-efficient OFDM (SEE-OFDM)
6. Layered ACO-OFDM (LACO-OFDM)
7. Unipolar orthogonal transmission (UOT)

Predominantly, the aforementioned variants of superposition optical OFDM have been proposed in the literature with a major emphasis to augment the spectral efficiency of the earliest variants of optical OFDM like ACO-OFDM, PAM-DMT and Flip-OFDM. The underlying methodology involved behind the implementation of different superposition optical OFDM variants is that multiple layers of OFDM waveforms have been superimposed with the intent to double the spectral efficiency than that of the earliest variants. Recalling that the implementation of OFDM for OWC is made possible by the introduction of DCO-OFDM, ACO-OFDM, PAM-DMT, Flip-OFDM and U-OFDM. Among these earliest variants of optical OFDM, the concept of DCO-OFDM is very straightforward to implement and its practicality is proved in several applications pertaining to VLC. However, the biasing requirement of LEDs results in considerable energy dissipation and this stems out as the disadvantage of DCO-OFDM. Consequently, the other four variants of optical OFDM have been proposed to relax the requirement of biasing as well as to attain a significant improvement in spectral efficiency. However, these schemes impose several restrictions on their input and output frame structure, thereby resulting in a decrement in spectral efficiency. That is why the earliest enhanced optical OFDM variant which can be termed as enhanced unipolar optical OFDM (eU-OFDM) was proposed to address this reduction in spectral efficiency [479].

The fundamental aspect involved behind the implementation of eU-OFDM is that it increases the spectral efficiency by allowing superimposition of multiple U-OFDM frames to form a single composite time-domain signal. The underlying concept involved behind the implementation of eU-OFDM is pictorially shown in Fig. 3.12. From this sort of arrangement, it is of utmost importance to understand the generation of eU-OFDM frames. As delineated in the figure, *N* indicates the absolute values of the negative valued samples of the bipolar optical OFDM signal with zeros replaced in places of positive values and *P* represents the positive values of the samples of the bipolar optical OFDM signal with zeros added in the places of negative values and *CP* represents the added cyclic prefix. Principally, at depth 1, the time domain signal of eU-OFDM has been generated in a manner exactly like that of the conventional U-OFDM signal as briefed in the previous subsection. Meanwhile, the information stream of U-OFDM at depth 2 is superimposed on the data stream at depth 1 in such a manner that the receiver should have the potential to recover the transmitted data bits in consideration of the added stream follows a certain structure.

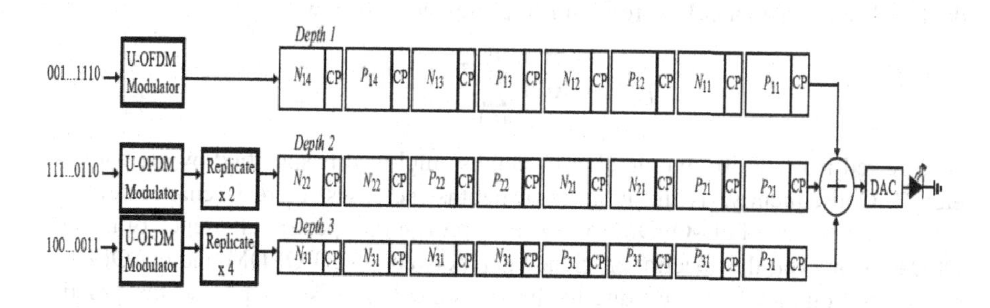

Figure 3.12: Pictorial representation of the underlying phenomena of enhanced U-OFDM [479].

As portrayed in the figure, every individual U-OFDM frame at depth 2 is replicated and transmitted twice in a way that the second frame is an exact copy of the first one, and the fourth frame exactly matches the third one, and so on. In this type of arrangement at depth 2, since two U-OFDM frames contain the same amount of information, with the intent to preserve the overall signal energy at this depth, the amplitude of each frame is scaled by $\frac{1}{\sqrt{2}}$. In a similar manner, even additional data streams can also be added, where at depth 3, the information data stream of U-OFDM is superimposed to the stream contained at depth 2. At depth 3, each U-OFDM is replicated 4 times where the amplitude of each frame instance is scaled by a factor of $\frac{1}{2}$. Thus, in this manner additional data streams can be added provided they are replicated 2^{d-1} times and scaled by a factor of $\frac{1}{\sqrt{2^{d-1}}}$ where d signifies the depth of the respective information stream. At the receiving end, firstly, demodulation proceeds by taking into consideration the data stream at depth 1, where traditional U-OFDM receiver is used as described in the previous subsection.

Meanwhile, the inter stream interference that might be caused due to the superimposition of multiple streams of information can be removed by exploiting the subtraction operation due to the inherent fact that the interference on each positive frame in depth 1 can be considered equivalent to the amount of interference on each negative frame mainly because of the imposed stream structure in the modulation process. Finally, the information at depth 1 is demodulated and the recovered bits are remodulated at the receiving end for the sake of reconstruction of the information signal at depth 1 whereupon this stream is subsequently subtracted from the overall received U-OFDM signal. Thus, the signal contained at depth 1 is completely removed from the received U-OFDM signal. The demodulation process for the information signal belonging to depth 2 proceeds as follows: The two identical frames at depth 2 are summed, and the demodulation process follows a similar approach as that of the conventional U-OFDM receiving algorithm, and finally the recovered bits are remodulated to facilitate for the information signal at depth 2 to be subtracted from the overall U-OFDM received signal. Accordingly, for all the subsequent streams, the demodulation process repeats in a similar manner until and unless the information at all depths is successfully recovered. Thus, the spectral efficiency of U-OFDM can be expressed as the combination of spectral efficiencies of the information signals at all depths. The mathematical expression is represented as follows:

$$\eta_{eU-OFDM} = \sum_{d=1}^{D} \frac{\eta_{U-OFDM}}{2^{d-1}} \tag{3.50}$$

As inferred from (3.50), the number of depths employed is specified by the parameter D. Consequently, as the number of depths increases, there is chance for the spectral efficiency of eU-OFDM to closely approach the spectral efficiency of DCO-OFDM. However, the spectral efficiency gap between DCO-OFDM and eU-OFDM cannot be completely closed due to the implementation issues behind the practical realization of eU-OFDM system. There are several practical limitations on the maximum number of depths to be employed, computational complexity aspects, latency and memory requirements. From the earlier studies it can be inferred that it is impossible to close the spectral efficiency gap between DCO-OFDM and U-OFDM because it mandates relying on the superposition of a large number of information streams. Moreover, the superimposed information streams exploit the same constellation size; therefore, with the major motive to close the spectral efficiency gap between DCO-OFDM and U-OFDM, eU-OFDM was generalized leading to the emergence of generalized enhanced unipolar OFDM (GREENER-OFDM), where the superimposed information streams with arbitrary constellation size and arbitrary power allocation were employed.

The work in [233] reveals the fact that the authors exploited GREENER-OFDM which extends the concept of eU-OFDM, where arbitrary combinations of U-OFDM data streams with varying constellation sizes and power allocations were exploited. Furthermore, a closed form mathematical expression on theoretical bound on the BER performance of GREENER-OFDM was derived and was validated through Monte Carlo simulations. Figure 3.13 depicts the comparison between GREENER-

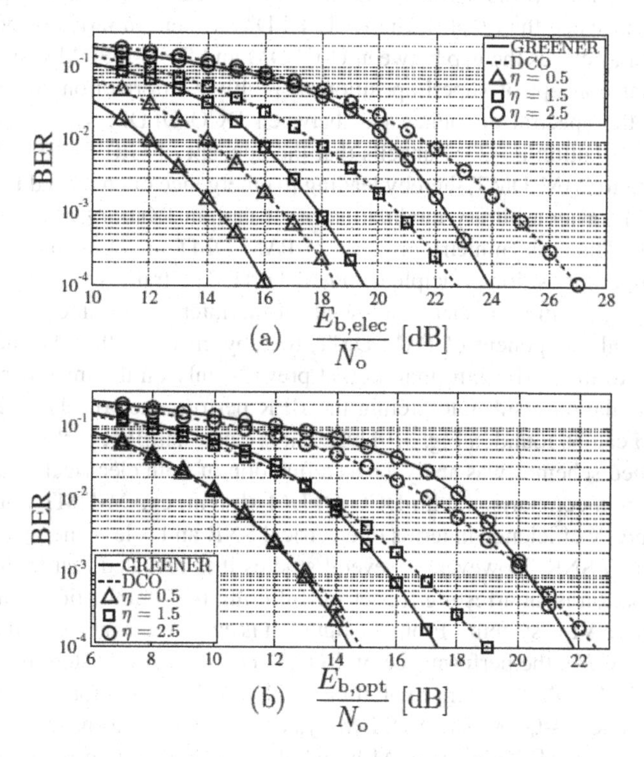

Figure 3.13: Comparison of the performance of GREENER-OFDM and DCO-OFDM in terms of different spectral efficiencies η [233].

OFDM and DCO-OFDM system at different spectral efficiencies which is usually characterized in terms of $bits/s/Hz$. As evident from the figure, at a spectral efficiency η of 0.5 $bits/s/Hz$, $[2-4]$-QAM GREENER-OFDM exhibits a superior performance than that of the DCO-OFDM system employing BPSK modulation format and the amount of energy savings of GREENER-OFDM over DCO-OFDM was 2.6 and 0.4 dB for the electrical and optical energy dissipation, respectively. While at spectral efficiency of 1.5 $bits/s/Hz$, the BER performance of GREENER-OFDM employing $[16-8-4]$-QAM dominates the BER performance of DCO-OFDM system employing 8-QAM modulation. At this spectral efficiency, GREENER-OFDM employing the aforesaid modulation format exhibits a significant amount of energy savings by 3.5 dB for electrical energy and by 1.5 dB for optical energy over DCO-OFDM system. Furthermore, upon increasing the orders of constellation as well as at higher spectral efficiencies of 2.5 $bits/s/Hz$, the energy efficiency of GREENER-OFDM is much superior than that of the DCO-OFDM system. More precisely, at $\eta = 2.5\ bits/s/Hz$, $[64-64-16]$-QAM GREENER OFDM imparts higher amount of energy savings by 3 dB (electrical energy) and 0.75 dB (optical energy) than that of the DCO-OFDM system making use of 32-QAM, respectively. Thus, the simulation results of this works surmises the fact that GREENER-OFDM represents a superior performance than that of the DCO-OFDM system in terms of power efficiency as well as closes the gap between U-OFDM and DCO-OFDM systems.

The similar concept of superimposition of multiple information streams in order to close the spectral efficiency gap between DCO-OFDM system and PAM-DMT, the contribution in [235] unveils that the concept of GREENER-OFDM has been extended to PAM-DMT, thereby leading to emergence of enhanced PAM-DMT (ePAM-DMT) for an arbitrary modulation order and an arbitrary power allocation at each depth. The antisymmetry property of PAM-DMT provides the flexibility to exploit the superimposition principle in ePAM-DMT. The frame hierarchy in ePAM-DMT was designed in a manner to allow all of the inter stream interference to fall only on the real component of PAM-DMT, thereby ensuring that the interference is orthogonal to the useful information that prevails only on the imaginary component. The simulation result interpreting the BER performance of ePAM-DMT and DCO-OFDM can be found in Fig. 3.14 where it is clear that BER performance of the aforementioned schemes was analyzed as a function of both electrical SNR and optical SNR. Upon comparison with DCO-OFDM system, ePAM-DMT imparts high amount of energy efficiency in terms of electrical SNR and is less energy efficient in terms of optical SNR. However, in several studies it is revealed that higher optical energy dissipation is an arresting aspect for fulfilling the illumination requirements of cost-effective VLC systems. From the figure, it is evident that at spectral efficiency of $\eta = 1\ bits/s/Hz$, the performance of 4-QAM DCO-OFDM system is better than that of the ePAM-DMT system exploiting [2, 4]-PAM modulation format. At this spectral efficiency, 4-QAM DCO-OFDM system was shown to have better energy efficiency than that of the [2, 4]-PAM-based ePAM-DMT system with 1.35 dB for the scenario of electrical SNR and 3.27 dB for the case of optical SNR. Thereupon, at spectral efficiency of 2 $bits/s/Hz$, ePAM-DMT system making use of [8, 4]-PAM has a better energy efficiency of 1.1 dB for the scenario of electrical SNR, while it suffers from the energy efficiency loss of 1.65 dB for the optical SNR case. In the same manner, at higher spectral efficiencies, ePAM-DMT system offers higher

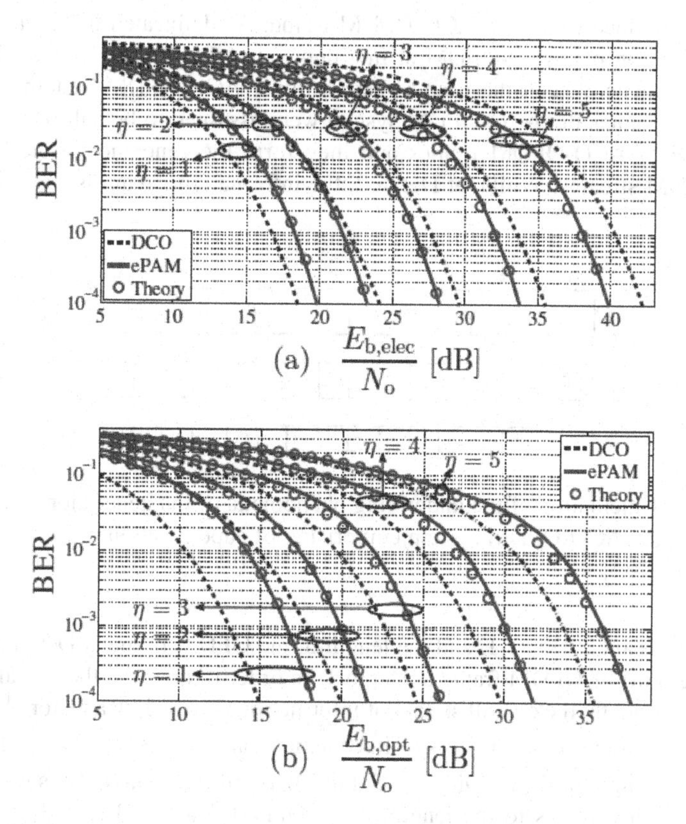

Figure 3.14: Performance analysis of ePAM-OFDM and DCO-OFDM as a function of electrical and optical SNR for different spectral efficiencies η [235].

energy efficiency than that of the DCO-OFDM system for the electrical SNR scenario, but DCO-OFDM system surpasses ePAM-DMT system in terms of energy consumption pertaining to the optical SNR case.

Similar to the superposition variants of optical OFDM as discussed above, the work in [234] proposes enhanced ACO-OFDM (eACO-OFDM) to overcome the spectral efficiency loss of conventional ACO-OFDM as well as to close the spectral efficiency gap between ACO-OFDM and DCO-OFDM. Essentially, the symmetry in U-OFDM lies in its frames, while with ACO-OFDM and PAM-DMT, the symmetry lies in their subframes. Much identical to ePAM-DMT, the frame hierarchy in case of eACO-OFDM system is designed in such a manner that all of the inter stream interference which generally occurs due to the superimposition of several streams to effect only the even-indexed subcarriers at the frequency domain. The possible frame, arrangement of eACO-OFDM system is outlined in Fig. 3.15, where the parameter *CP* specifies the added cyclic prefix, A_{dl}, B_{dl} represents the first and the second half of the unipolar frame, while d and l signify that the frame, at depth-d is confined to the lth ACO-OFDM frame. As delineated in the figure, the signal generation of eACO-OFDM at first depth, i.e., Depth 1 resembles the way the signal is generated in the case of conventional ACO-OFDM system. Particularly, the subframes in eACO-OFDM system were assumed to be half of the length of the original ACO-OFDM frames and they form the fundamental elements of eACO-OFDM streams. While, at Depth 2, the information stream is superimposed

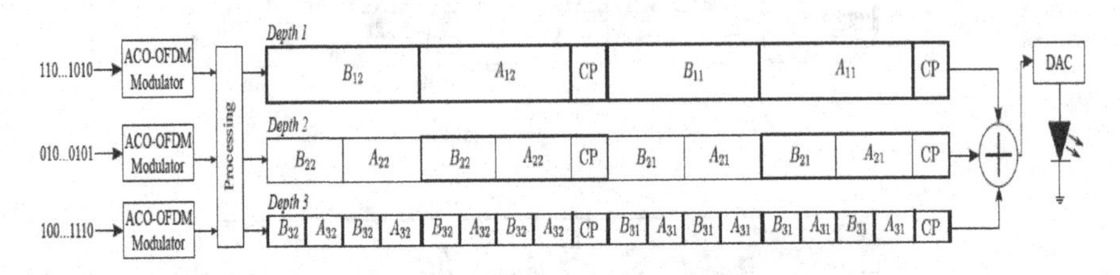

Figure 3.15: Elucidation of the underlying phenomena of the generation of time domain signal in eACO-OFDM system comprising of superposition of three information streams [234].

on the first stream at Depth 1, and the length of the OFDM frame at Depth 2 is half of the length of the OFDM frame at Depth 1. In order to preserve the overall signal energy at this depth, the overall stream of information is scaled by a factor $\frac{1}{\sqrt{2}}$. In the similar manner, additional streams of information can be superimposed at Depth-d provided the length of the OFDM frame at that particular Depth-D obeys $N_d = \frac{N_{fft}}{2^{d-1}}$, where N_{fft} corresponds to the length of the OFDM frame at Depth 1. Additionally, it is necessary for the overall stream of information at Depth-d to be scaled by $\frac{1}{\sqrt{2^{d-1}}}$. At the receiving end, the demodulation process follows a similar approach like that of the eU-OFDM. The transmitted bit streams at Depth 1 are recovered in the same manner as that of the ACO-OFDM system and this is valid because all of the inter stream interference falls only on the even subcarriers. Once the information belonging to the first stream is decoded, then it is again remodulated and then subtracted from the overall received signal. In this manner, the demodulation process repeats for all of the depths. The BER performance of eACO-OFDM as a function of both electrical SNR and optical SNR along with the comparisons with other superimposition variants like ePAM-DMT, GREENER-OFDM over AWGN channel is shown in Fig. 3.16. Also from the figure it is evident that the aforementioned superimposition varieties were compared with the earliest variant of optical OFDM like the DCO-OFDM system. The signal processing steps and the system model statistics of eACO-OFDM and GREENER-OFDM are in good resemblance with each other; thus, the optimal combinations of constellation sizes and their corresponding scaling factors are identical for both eACO-OFDM and GREENER-OFDM systems. Whereas, due to the different design of ePAM-DMT system, the optimal configurations are different in it. The simulation result as emphasized by Fig. 3.16 depicts the performance comparison of optimum configurations in all the superposition optical

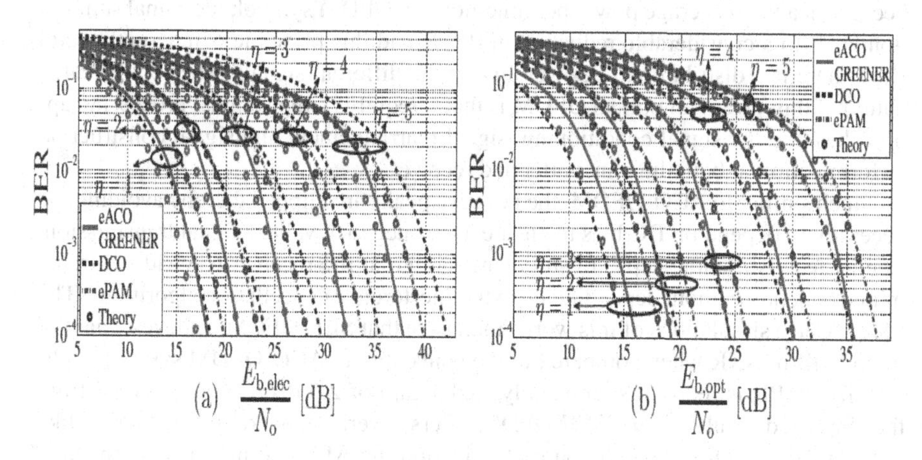

Figure 3.16: Comparison of BER performance of eACO-OFDM, ePAM-DMT, GREENER-OFDM, DCO-OFDM systems over AWGN channel at varying spectral efficiencies $\eta = 1,2,3,4,5$ [234].

OFDM variants with the performance of spectrally efficient DCO-OFDM system. Since, the optimal configurations of eACO-OFDM and GREENER-OFDM are identical to each other, both these variant impart the same performance. Furthermore, the superposition variants illustrate a good amount of energy efficiency in terms of the requirement of electrical SNR than that of the DCO-OFDM system. However, eACO-OFDM and GREENER-OFDM are more energy efficient than ePAM-DMT. While pertaining to the optical SNR scenario, equivalent performance is illustrated by both eACO-OFDM and GREENER-OFDM to DCO-OFDM system. However, the performance of DCO-OFDM system is much efficient than that of the ePAM-DMT system in terms of optical SNR.

The other variant of optical OFDM that has been proposed mainly with the intent to avoid the requirement of DC bias in the process of generation of positive signal is spectral and energy efficient OFDM (SEE-OFDM) [157]. The research efforts of this work describe the fact that without relying on additional forms of interference cancellation and estimation algorithms at the receiving end in the process of recovery of the transmitted information bits, SEE-OFDM doubles the spectral efficiency when compared with other unipolar optical OFDM variants. The underlying phenomenon behind the implementation of SEE-OFDM is that the data or information symbols can be carried by exploiting both the odd and even indexed subcarriers pertaining to the IFFT operation. This is made possible by generating multiple signals by making use of only odd subcarriers and by exploiting different IFFT lengths. Thereupon, the signals arriving from different paths are summed up together to attain a single composite time-domain signal prior to intensity modulation through the LED. This summing of signals is an additional advantage for this variant of optical OFDM, because at a fixed average power per time-domain OFDM symbol, the signal summation leads to a considerable reduction in PAPR and this principle is inherently valid since there is a distribution of power among the different signals. The sophisticated nature of this variant of optical OFDM is that it requires simple signal processing approaches like the signal conditioning, signal rearrangement and signal construction or formulation at both the transmitter and receiver.

Much similar to ACO-OFDM, the negative peaks are clipped thereby avoiding the necessity to depend on DC bias to ensure a non-negative valued signal transmission. Simple reconstruction steps are sufficient enough at the receiver to yield the transmitted data bits without depending on extra interference cancellation algorithms. The Monte Carlo simulations of this work emphasize that SEE-OFDM achieves a gain of up to 6 dB in SNR when compared to the conventional ACO-OFDM system. Additionally, PAPR is reduced significantly, and a gain of 2.5 dB can be observed from the simulated results. Thus, SEE-OFDM offers several remarkable advantages like high data rates, appreciable amount of reduction in PAPR and minimal complexity, thus evolving as an attractive scheme for VLC systems.

The other variant of superposition optical OFDM that has been proposed to overcome the spectral efficiency loss of ACO-OFDM system is layered ACO-OFDM (LACO-OFDM) [508]. In case of ACO-OFDM system, only odd subcarriers are modulated and even subcarriers are set to zero to ensure that the amount of clipping noise falls only on even subcarriers. This sort of arrangement of subcarriers depletes the spectral efficiency of ACO-OFDM system. Thus, in LACO-OFDM system, the subcarriers prior to data modulation are divided into different layers where ACO-OFDM methodology is imposed for each layer and, thereupon, all of the layers are combined to allow for a simultaneous transmission. Since more subcarriers are exploited for data modulation, there is a significant improvement in spectral efficiency in LACO-OFDM when compared to ACO-OFDM. Additionally, this work proposes an iterative receiver with the intent to subtract the clipping distortion of each layer the moment it is detected. This is done mainly to ensure that the signals from different layers could be successfully recovered. The simulated results of this work confirm the fact that the spectral efficiency improvement of LACO-OFDM outperforms the conventional ACO-OFDM system, where a two-time improvement in spectral efficiency is imparted by LACO-OFDM.

The research track of the superposition optical OFDM variants as discussed above focused only on the advantages of the aforementioned techniques. It is also vital to take into consideration the disadvantages associated with the superposition variants. It is to be noted that a considerable amount of system performance can be achieved only at the expense of high complexity involved in the design of transceiver circuitry as well the stringent requirement in the memory, and high computational complexity involved behind the signal processing imposes a major challenging aspect of these superposition OFDM techniques. For instance, if we consider GREENER-OFDM, in the process of generation of one OFDM frame at the transmitting end, there is a substantial increase in the requirement of memory because it requires superimposing of multiple streams of U-OFDM frames. In addition to the huge memory requirement at the transmitting side, it is mandatory to employ an intensive high computational complexity detection mechanism such as the successive interference cancellation (SIC) at the receiving end in order to ensure perfect reception of the transmitted information bits. Thus, it can be stated that all of the superposition variants entail high memory and computational complexity in the realization of hardware resources to achieve a significant improvement in the overall system performance when compared with other classic optical OFDM systems. Ultimately, these drawbacks curtail the practical realization and viability of such techniques for VLC systems. To overcome such issues imposed by the superposition optical OFDM variants, much recent research contribution in [131] follows a new design direction where the authors employed orthogonal waveforms to convert the bipolar signal into a non-negative valued signal.

This work proposes a new low-complexity unipolar transmission scheme which is termed as unipolar orthogonal transmission (UOT) for VLC systems. The sophisticated nature of this work is that without the need of any DC bias or any clipping process, this scheme is shown to be more reliable than that of the ACO-OFDM system. Additionally, in contrast to other unipolar OFDM superimposition schemes, UOT avoids the usage of clipping phenomena and multiple transmission or reception of OFDM frames by exploiting a simple and robust transmission and detection technique. These characteristics, especially the low implementation complexity of UOT, will definitely enable it to become practically viable for IM/DD-based VLC systems. Furthermore, the authors in this work derived the analytical expression for the BER and compared it with the Monte Carlo simulations, where the result analysis described a good agreement. Besides, this contribution presents an exhaustive performance analysis comparison with the widely investigated optical OFDM variants like DCO-OFDM, ACO-OFDM, U-OFDM and Flip-OFDM in terms of the most practical key performance indicators which include the BER, spectral efficiency, PAPR, ICI and computational complexity. The simulated results of this work affirm the indispensable fact that when compared with the other relevant counterparts, the proposed UOT scheme illustrates substantial performance gains in terms of BER without undergoing any compromise in spectral efficiency, PAPR and computational complexity. On top of this, the splendid feature which draws remarkable insights about this novel unipolar transmission scheme like UOT is its robustness against timing jitter error and noise which is exclusively unique when compared with other existing multicarrier VLC systems.

3.2.3 PERFORMANCE ANALYSIS OF DIFFERENT HYBRID OPTICAL OFDM VARIANTS

This subsection provides details about the performance of different hybrid variants of optical OFDM. The different variants of optical OFDM which come under this category are listed below:

1. Reverse polarity optical OFDM (RPO-OFDM)
2. Polar-based OFDM (P-OFDM)
3. Spatial optical OFDM (SO-OFDM)

4. Asymmetrically and symmetrically clipped optical OFDM (ASCO-OFDM)
5. Asymmetrically clipped DC-biased optical OFDM (ADO-OFDM)
6. Hybrid asymmetrically clipped OFDM (HACO-OFDM)
7. Spectrally factorized optical OFDM (SFO-OFDM)
8. Phase modulation DCO-OFDM (PM-DCO-OFDM)
9. Variable pulse width unipolar optical OFDM (VPW-OFDM)
10. Triple-layer hybrid optical OFDM (THO-OFDM)
11. Reconstructed LACO-OFDM (RLACO-OFDM)
12. PAM-DMT-based hybrid optical OFDM (PHO-OFDM)

Many of the significant research works illustrate the fact that conventional OFDM has been modified in a manner to tailor diverse number of aspects pertaining to the reliable realization of VLC link that furnishes the high data rate transfer of information. In this regard, the aforesaid hybrid varieties will address several aspects like dimming issues, assurance of high spectral efficiency, power efficiency and computational efficiency. Several of the previously mentioned modulation formats overlooked the illumination features that are desired for the realization of a realistic illumination and high data rate communication system. More specifically, one of the important vital features of lighting to address both the functional and aesthetic requisites of an indoor or outdoor space as well as to save suitable amount of energy is to expeditiously dim a light source. The paramount benefit offered by a dimmed light source is its less production of heat, thereby spanning its lighting hours. Thus, one of the critical challenging aspects of VLC is that it should ensure efficient dimming functionality for the purpose of conserving the energy while preserving a reliable, computationally less complex and high data rate communication link. Therefore, a much suitable contribution connected with addressing the aforesaid challenging aspect can be found in [160] where the authors have proposed a hybrid optical variant namely RPO-OFDM that offers higher levels of illumination control for VLC systems that are based on off-the-shelf opto-electronic components like LEDs and exploits optical OFDM to render high data rate communication.

The speciality of RPO-OFDM is that, in order to allow for dimming of the light sources to save energy, it combines the unipolar OFDM signal with relatively slow PWM dimming signal such that both of these signals dispense effective brightness of the LED. The important aspect with this hybrid variety is that the dynamic range of the LED is fully utilized to combat the effects of deleterious clipping-induced non-linear distortion. The numerical simulations of this work demonstrate that the proposed RPO-OFDM system attains a higher performance gain than that of the DCO-OFDM system while achieving both dimming and high data rates across a wide range of intensity settings. Even though RPO-OFDM exhibits an inherent advantage of offering both data communication and dimming control, the OFDM signal in RPO-OFDM is based upon unipolar OFDM technique which infers that the spectral efficiency of RPO-OFDM is half that of the earliest variant of optical OFDM, i.e., DCO-OFDM. Thus, as the spectral efficiency increases, the power efficiency advantage of RPO-OFDM begins to decline. Moreover, the duty cycle of PWM was assumed to be known at the receiving end which implies that the side information should be sent before transmission and this mandates the requirement of strict synchronization algorithms between the transmitting and receiving end.

In case of VLC, along with assurance of high levels of illumination, dimming control stems out as an important aspect, as a significant amount of energy can be saved by integrating efficient dimming techniques into VLC. Moreover, it is an immense prerequisite to adjust the brightness levels of LEDs as per the personal preferences. Toward this end, there are several significant research efforts carried out by proposing suitable dimming control approaches for controlling the levels of illumination for optical OFDM variants. Consequently, as illustrated in [518], in order to ensure the dual functionalities of high data rate communication with adequate amount of dimming control, the proposed system model was capable to transmit optical OFDM signal during the ON state of the PWM signal. However, in spite of its ease of implementation, this scheme turned out to be an inefficient scheme due to the fact that the OFF state remained a waste as a result hindering high data rate communication. A much similar work as depicted in [380] clearly emphasizes that the time-domain optical OFDM signal was multiplied by a periodic PWM signal to enable simultaneous transmission, and this scheme failed to impart high data rate communication due to the inherent requirement of the bandwidth of PWM signal to be at least twice as that of the optical OFDM signal.

Consequently, in order to address the dimming issue in case of VLC, a much similar and recent contribution can be found in [292] where the authors have integrated the industry preferred PWM scheme with the spectrally efficient LACO-OFDM system. Thus, this developed system which is ably called reconstructed LACO-OFDM (RLACO-OFDM) can concurrently meet the dimming requirements of cost-effective VLC systems and at the same time ensures a spectral-efficient communication link. However, a direct superimposition of the traditional LACO-OFDM signals with the PWM signals imposes severe clipping induced distortion. Consequently, in order to overcome such clipping induced distortion, the authors in this work proposed a reconstruction procedure of the conventional LACO-OFDM signal by constraining the super imposed signal to operate within the linear dynamic range of the LEDs, thus ensuring that the full dynamic range of the LEDs can be totally exploited. Additionally, the sophisticated nature of RLACO-OFDM system is that the simplicity in its decoding process where the transmitted symbols of RLACO-OFDM can be easily recovered by means of a standard LACO-OFDM receiver without the requirement of any additional detection mechanism for PWM signal, thereby reducing the complexity in the design of receiver. The simulation results of this work extrapolate the fact that upon varying the duty cycle of the PWM signal, the dimming level can be linearly adjusted. In addition, the proposed dimming scheme is having the potential to support a relatively stable and high spectral efficiency.

As discussed in the previous subsection, the techniques to generate non-negative optical OFDM signals intended for IM/DD-based VLC applications generally come in two important categories, i.e., DCO-OFDM and ACO-OFDM. In DCO-OFDM systems, the negative OFDM signal is made positive by adding a suitable amount of DC bias. In particular, the superimposition of a huge number of subcarriers in OFDM system makes it susceptible to very high peak to average power ratio (PAPR) leading to the requirement of high amount of DC bias for the sake of eliminating all the negative peaks. Rather than high DC bias addition, a moderate DC bias value can be added instead where the remaining negative peaks of the optical OFDM signal can be clipped. But, this can be done at the expense of emanation of clipping distortion. Normally, in case of DCO-OFDM system, the data is modulated by exploiting both the even and odd subcarriers, thus the clipping noise affects all of the subcarriers in DCO-OFDM. Moreover, from the earlier findings, the addition of DC bias made this technique a power-inefficient scheme. Meanwhile, in the other earliest variant of optical OFDM, i.e., ACO-OFDM system, the data is carried only on the odd subcarriers and the even subcarriers are set to zero. Precisely, only the odd inputs of the IFFT are mapped, and the even inputs are set to zero. Then, the resulting bipolar or negative valued signal is clipped at zero to ensure a non-negative signal. Even though without the requirement of DC bias, ACO-OFDM guarantees a non-negative valued signal with the cost of reduction in spectral efficiency. Hence, to alleviate the aforementioned drawbacks as well as to individually exploit the advantages of these two variants, the authors in [144] proposed a novel hybrid scheme called asymmetrically clipped DC-biased optical OFDM (ADO-OFDM) where ACO-OFDM methodology has been exploited for odd subcarrier mapping, and DCO-OFDM methodology was utilized for even subcarrier mapping.

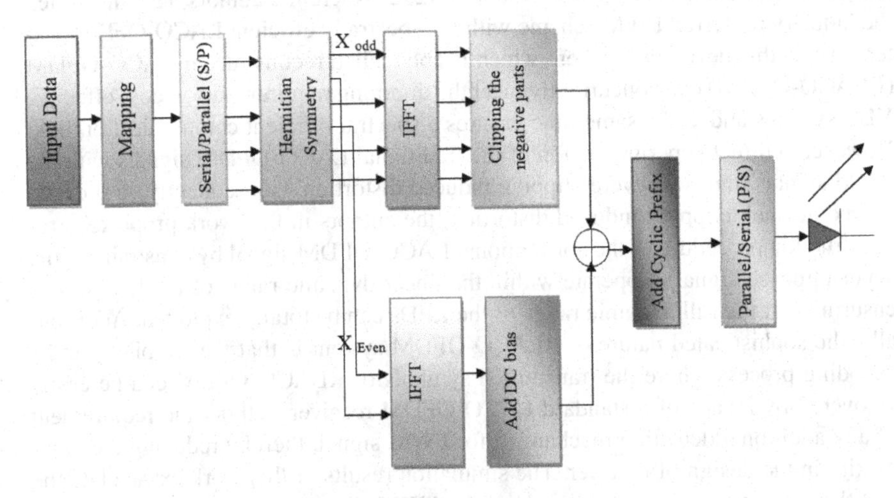

Figure 3.17: Transmitter schematic of asymmetrically clipped DC-biased optical OFDM (ADO-OFDM).

As evident from the transmitter schematic which is shown in Fig. 3.17, both the time domain ACO-OFDM and DCO-OFDM signals that are on the odd and even subcarriers are summed up together, and then the overall signal is intensity modulated through the LED. To understand this mathematically, we proceed further with the elaborate discussion on the generation and detection of time domain and frequency domain signal formats of ADO-OFDM system. Firstly, at the transmitting end, the incoming huge stream of serial and complex-mapped data is transmitted into several low rate parallel data streams with the help of a serial to parallel convertor and since Fourier singal processing implies complex signal processing; therefore, this obliges the huge set of complex symbols denoted by a vector notation X to satisfy the Hermitian Symmetry criteria to ensure a real valued signal transmission. Thereupon, X is further divided into sets of even and odd subcarriers as represented by X_{even} and X_{odd}, respectively. Therefore, the set of even subcarriers can be mathematically represented as

$$X_{even} = [X_0, 0, X_2, 0, X_4, 0, \cdots X_{N-2}, 0] \tag{3.51}$$

Similarly the set of odd subcarriers can also be represented as

$$X_{odd} = [0, X_1, 0, X_3, 0, X_5, 0, \cdots X_{N-1}] \tag{3.52}$$

The frequency domain samples as shown by equations (3.51) and (3.52) are fed as inputs to the IFFT modules to attain the time domain signal formats individually, and these can be denoted as $x_{ACO}[n]$ and $x_{DCO}[n]$, respectively. Generally, in case of ACO-OFDM system, the time-domain signal is made positive by hard clipping it at zero, and the amount of clipping noise which is generated in ACO-OFDM system is given by η_{ACO}. While, in case of DCO-OFDM system, a suitable amount of DC bias β_{DC} is added to ensure a non-negative signal, while the remaining negative peaks are clipped resulting in a clipping noise denoted by η_{DCO}. Thus, at the transmitting end, the individual time domain signals are added to yield the resultant ADO-OFDM signal as follows

$$x_{ADO}[n] = x_{ACO}[n] + \eta_{ACO} + x_{DCO}[n] + \beta_{DC} + \eta_{DCO} \tag{3.53}$$

Therefore, to this signal, a suitable amount of cyclic prefix is added in order to combat the effects of ISI, and then it is transmitted as a serial signal by employing parallel to serial convertor. Thereupon, this intensity modulated signal is then propagated to reach the receiver end. At the receiver side, is is necessary to successfully estimate the data signal corresponding to both ACO and DCO-OFDM systems. For simplicity, the type of channel environment like AWGN can be considered. Thus, the received signal can be expressed as

$$y_{ADO}[n] = x_{ACO}[n] + \eta_{ACO} + z_{ACO}[n] + x_{DCO}[n] + \beta_{DC} + \eta_{DCO} + z_{DCO}[n] \tag{3.54}$$

From the overall time domain expression as stated in (3.54), the terms $z_{ACO}[n]$ and $z_{DCO}[n]$ represent the AWGN of both ACO and DCO-OFDM systems. Now, the task of the receiver is to reliably demodulate the data symbols corresponding to both ACO and DCO-OFDM. Firstly, the data transmitted on the odd subcarriers using ACO-OFDM can be recovered from the corresponding subcarriers at the receiver and then in order to recover the data corresponding to the even subcarriers, i.e., the data transmitted on DCO-OFDM, the ACO-OFDM signal is subtracted from the overall received signal. Even though the DCO-OFDM data on the even subcarriers is recovered, it has huge influence of noise components on it. More precisely, there are three sources of noise components on it, i.e., the clipping noise associated with DCO-OFDM system, the AWGN noise as well as the extra noise component that is incurred due to ACO-OFDM signal in the process of interference cancellation. Hence, from the simulation results it can be inferred that this interference cancellation process leads to a 3 dB noise penalty in the recovery of DCO-OFDM signals. However, since the overall ADO-OFDM system uses more subcarriers for data transmission, the overall performance gain can be achieved.

The authors in [156] have proposed a spectrally efficient unipolar OFDM technique, namely, Polar OFDM, which exhibits twice as much spectral efficiency than that of the ACO-OFDM system. The underlying principle that is associated with Polar OFDM is that the complex constellation symbols that are mapped by using QAM are assigned only to the even indices of the IFFT module, i.e., the odd indexed subcarriers are set to zero $X[2k+1]=0;\ k=1,2,\cdots\frac{N}{2}-1$ and the output of the IFFT exhibits half-wave symmetry property, and this can be mathematically expressed as follows:
The time domain expression for the IFFT output signal is given by

$$x[n]=\frac{1}{N}\sum_{k=0}^{N-1}X[k]\,e^{\frac{j2\pi nk}{N}} \tag{3.55}$$

Since only even subcarriers are utilized for transporting the data, $X[k]$ can be replaced as $X[2k]$ since $X[2k+1]=0;\ k=1,2,\cdots\frac{N}{2}-1$. Therefore, the mathematical equation as represented by (3.55) can be reduced as

$$x\left[n+\frac{N}{2}\right]=\frac{1}{N}\sum_{k=0}^{\frac{N}{2}-1}X[2k]\,e^{\frac{j2\pi\left(n+\frac{N}{2}\right)2k}{N}} \tag{3.56}$$

Finally, upon solving equation (3.56), the resulting expression is

$$x\left[n+\frac{N}{2}\right]=\frac{1}{N}\sum_{k=0}^{\frac{N}{2}-1}X[2k]\,e^{\frac{j2\pi nk}{\frac{N}{2}}}\,e^{j2\pi k} \tag{3.57}$$

Upon using the fact that the exponential term $e^{j2\pi k}$ in (3.57) is 1, the final expression for the time domain P-OFDM signal can be formulated as

$$x\left[n+\frac{N}{2}\right]=\frac{1}{N}\sum_{k=0}^{\frac{N}{2}-1}X[2k]\,e^{\frac{j2\pi nk}{\frac{N}{2}}} \tag{3.58}$$

The right-hand side of equation (3.58) infers that the expression is equivalent to $x[n]$. Thus, $x\left[n+\frac{N}{2}\right]=x[n]$ which confirms the fact that first half of the complex time-domain samples in the OFDM frame are identical to the second half. Therefore, this implies that it is sufficient enough to transmit the first half of the IFFT output. Thus, P-OFDM uses the principle to convert complex valued IFFT samples which are belonging to the Cartesian coordinate system to polar coordinate system where the amplitude and phase information of the samples are transmitted successively in the first and the second half of the frame. At the receiving end, amplitudes and the phases of the transmitted signal are fed as input to the polar to Cartesian convertor in order to represent the transmitted samples in the Cartesian coordinate system. The noteworthy feature of this variant is that it avoids the requirement of Hermitian Symmetry criteria. Thus, when compared with DCO-OFDM system, P-OFDM ensures the same spectral efficiency since only even subcarriers are meant for data transmission. On the other hand, in case of ACO-OFDM system, due to odd subcarrier mapping and the requirement of Hermitian Symmetry criteria, it is a spectrally inefficient scheme when compared with P-OFDM system. Furthermore, the simulated results also emphasize the fact that BER performance exhibits superior performance under the dynamic range constraints of the optical sources.

A much similar work that is based upon the principle methodology of P-OFDM has been proposed in [301], where the research efforts state that a variable pulse width unipolar optical OFDM (VPW-OFDM) has been exploited for indoor VLC systems. Similar to P-OFDM system, without relying on the Hermitian Symmetry criteria, a real-valued signal is generated. Much similar to P-OFDM system, the magnitude and phase information of the complex-valued signals in this proposed system i.e., VPW-OFDM, are transmitted successively. As stated earlier, in case of P-OFDM system, the magnitude and phase components of the complex valued signal are transmitted on two successive time slots with the same duration. In case of optical sources, it is vital to take into consideration the peak radiation power limit, where the signal peaks greater than that of the constraint value are hard clipped, thereby leading to the emanation of clipping distortion resulting in degradation in the overall system performance. Pertaining to P-OFDM system, the phenomenon of clipping happens only for the magnitude component, and the noise that is added on the phase component adequately has a drastic impact on the overall system performance. Therefore, in order to reduce the effects of noise on the phase component, VPW-OFDM system has been proposed where the pulse widths of the magnitude and the phase components can be optimized to manage the received noise power and to improve the overall signal to interference noise ratio (SINR).

From the simulation results, it is evident that by adaptively optimizing the pulse widths, VPW-OFDM system imparts a more remarkable performance than that of the P-OFDM and other state-of-the-art optical OFDM variants like DCO-OFDM, ACO-OFDM and unipolar OFDM systems. Additionally, this work even addresses the effects of illumination requirements on the overall system performance. In order to enhance the system throughput, this contribution exploits simple, single-tap equalizer at the receiving end, and bit loading algorithm is also applied. From the numerical results it can be emphasized that VPW-OFDM system achieves a minimum BER when compared with other optical OFDM variants and bestows with the similar illumination levels like that of the P-OFDM system. Also from the simulated results, it can be inferred that by optimally adjusting the pulse widths, VPW-OFDM system can save more than 2 dB peak radiation power than other optical OFDM variants. However, upon exploiting a severely narrow bandwidth LED, DCO-OFDM system outperforms ACO-OFDM, unipolar OFDM, VPW-OFDM and P-OFDM systems. But, upon exploiting a wider channel bandwidth, VPW-OFDM system remarkably offers a more significant improvement in bit rate than that of the other state-of-the-art optical OFDM variants due to its resistance to the noise on the phase component.

The other hybrid technique which is employed to overcome the problem of power inefficiency as well as to ensure high data rate transmission is hybrid asymmetrically clipped optical OFDM (HACO-OFDM). This sort of optical OFDM system exploits the combination of power-efficient ACO-OFDM and PAM-DMT methodologies that are much suitable for OWC, because one of the major constraining factors of such systems is optical power. Without relying on DC bias, this optical OFDM variant relies on clipping approaches like ACO-OFDM and PAM-DMT to attain a non-negative valued signal. PAM-DMT is one among the clipping-based schemes, where the data signal is modulated on the imaginary part of each subcarrier while setting the real component of each subcarrier to zero; this is done mainly to ensure the clipping noise falls on the real component. A similar scheme is ACO-OFDM where only odd subcarriers are exploited and even subcarriers are set to zero to ensure that the clipping noise falls only on even subcarriers. Thus, both of these clipping-based strategies guarantee that the clipping noise is orthogonal to the transmitted signal, thus facilitating the easy recovery of desired data signal at the receiving end. In this kind of hybrid optical OFDM like HACO-OFDM, both the ACO-OFDM and PAM-DMT signals together are transmitted and since the data is transmitted by using the combination of these optical OFDM schemes, so, only half of the power will be appropriated for ACO-OFDM system. At the receiving end, the clipping noise is first estimated and then cancelled out for the purpose of recovering the PAM-DMT signals. Since, only half of the power is allocated to ACO-OFDM, from the simulation results as depicted in Fig. 3.18, a 3 dB penalty can be observed in the BER performance of HACO-OFDM using 4, 16, 64 and 256 QAM, respectively. In particular, when compared with traditional ACO-OFDM system, the BER performance of HACO-OFDM system incurs a 3 dB degradation in the achievable BER performance. Meanwhile, when conventional PAM-DMT is compared with PAM-HACO-OFDM system as delineated in Fig. 3.19, the degradation in the performance of BER can be observed only for lower electrical energy per bit to noise power ratio $\left(\frac{E_{b(elect)}}{N_0}\right)$; this is due to the effect of the noise incurred while estimating the ACO-OFDM symbols at the receiving end. However, at higher $\left(\frac{E_{b(elect)}}{N_0}\right)$ a good agreement in the BER performance between traditional PAM-DMT and PAM-HACO-OFDM system can be observed because of the gradual reduction in the estimation noise. Finally, this hybrid variant of optical OFDM like HACO-OFDM is power efficient scheme because it does not require the addition of DC bias to attain a positive valued signal and at the same time, the transmitter design is less complex. Moreover, from the simulation results of this contribution, it can be stated that the PAPR of the overall hybrid signal is slightly less when compared with the conventional systems. In spite of its advantages, this scheme

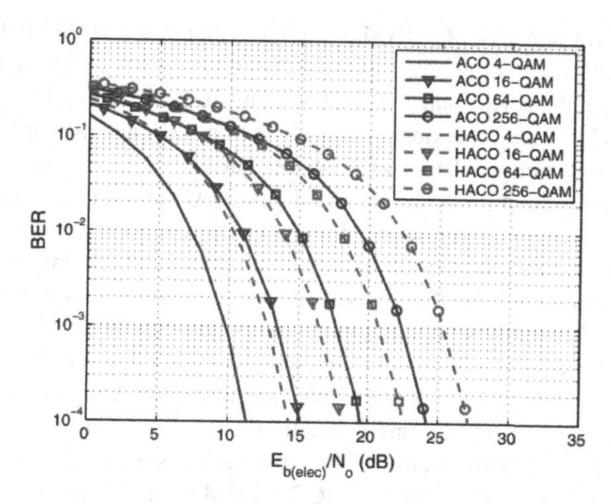

Figure 3.18: Performance comparison of ACO-OFDM and HACO-OFDM for different orders of M-QAM modulation [421].

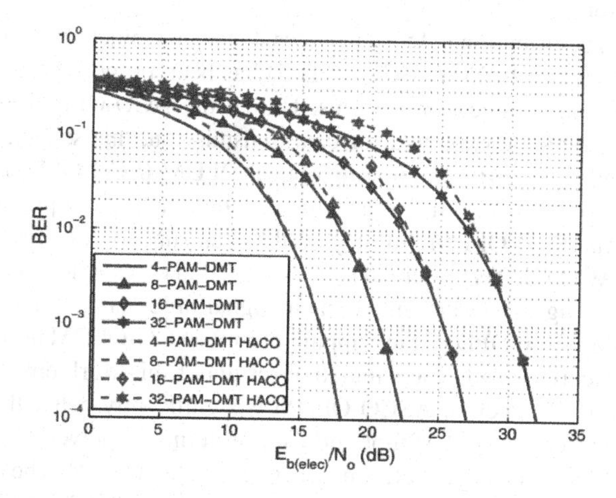

Figure 3.19: Performance comparison of PAM-DMT and PAM-HACO-OFDM for different orders of M-PAM modulation [421].

is not devoid of its drawbacks; even though the transmitter is simple to design, the noise cancellation phenomena involved at the receiving end enable this scheme to rely on extra receiver processing circuitry thus making the design of receiver circuitry more complex. In addition, the receiver design is very straightforward, as it does not exploit the property to fully eliminate the interference between ACO-OFDM and PAM-DMT signals, thereby limiting the overall system performance. Moreover, equal amount of powers were allocated to both ACO-OFDM and PAM-DMT systems. The problem emanates when different orders of modulations were exploited because the performance of M^2-QAM is similar to that of the M-PAM; therefore, this infers that the requirement of power for both these systems is different. Hence, in order to ensure that the performance of both systems in HACO-OFDM are similar, the authors in [507] proposed unequal power allocation mechanisms for both schemes. The research contribution of this work states that an iterative receiver for HACO-OFDM system has been proposed, where the ACO-OFDM and PAM-DMT signals in the frequency domain were detected and then re-generated again in the time domain. After this regeneration in the time domain, the ACO-OFDM and PAM-DMT signals were subtracted from the received signals iteratively. This manner enables to easily distinguish both these signals. Furthermore, by exploiting the signal symmetry properties of both these schemes, namely, ACO-OFDM and PAM-DMT systems in the time domain, pair-wise clipping mechanism has been incorporated to reduce the detrimental aspects of noise and estimation error, thus ensuring an improvement in the overall system performance. The simulation results of this work infer that the proposed iterative receiver-based HACO-OFDM system imparts significant SNR gain over the conventional receiver-based HACO-OFDM system for both equal and unequal power allocations. Nonetheless, the proposed iterative receiver is very complex to implement.

The research contribution in [530] portrays that an asymmetrically and symmetrically clipping optical OFDM (ASCO-OFDM) has been proposed for IM/DD-based optical wireless communication systems. Unlike the ACO-OFDM system, where only odd subcarriers are exploited for data transmission, this scheme uses a novel technique called symmetrically clipping optical OFDM (SCO-OFDM) to modulate the even subcarriers as well. It is interesting to note that the clipping noise of both ACO-OFDM and SCO-OFDM fall only on the even subcarriers. However, the clipping noise of ACO-OFDM system can be estimated and cancelled out at the receiving end. By making use of the time domain signal processing techniques of either unipolar OFDM or flip OFDM, the clipping noise of SCO-OFDM system can be removed at the receiving end. The simulation results of this work emphasize the fact that the spectral efficiency of ASCO-OFDM is significantly better than that of the ADO-OFDM system as well as more efficient in terms of power, since there is no requirement of DC bias to generate a non-negative valued signal. The research work in [569] signifies that a hybrid modulation scheme called PAM-DMT-based hybrid optical OFDM (PHO-OFDM) has been proposed to improve the data capacity of traditional PAM-DMT-based optical OFDM system. The authors in this work replaced one-dimensional mapping schemes like PAM with higher order QAM on the even subcarriers in the frequency domain as shown in Fig. 3.20. The overall time-domain signal of PHO-OFDM system has a larger amplitude range than that of the traditional PAM-DMT system, but with a reduced PAPR. The overall simulation results of this work emphasize the fact that the proposed PHO-OFDM system exhibits remarkable performance in terms of BER, PAPR than that of the conventional PAM-DMT systems, thus manifesting pronounced caliber to be exploited for high-speed VLC systems.

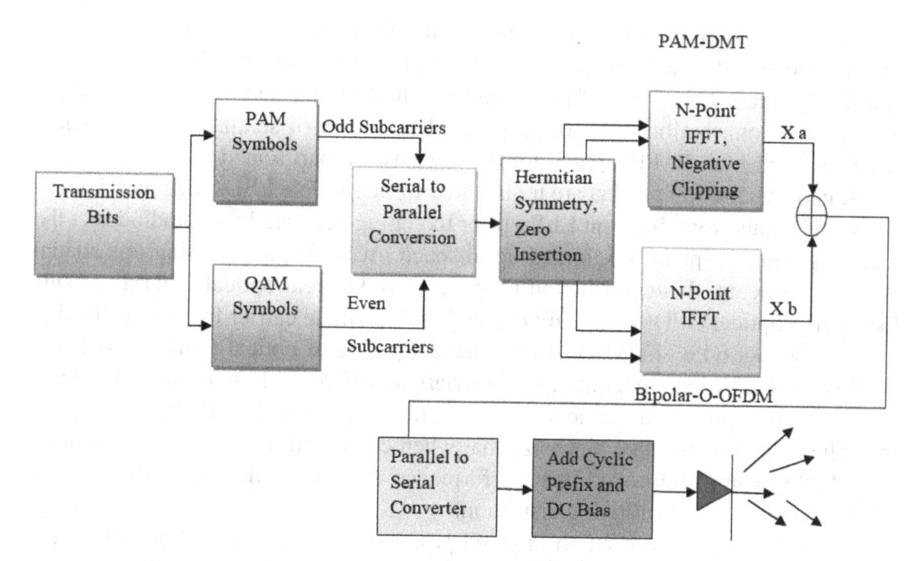

Figure 3.20: Transmitter schematic of PAM-DMT-based hybrid optical OFDM (PHO-OFDM).

The authors in [40] proposed a bandwidth efficient technique to implement the optical OFDM for IM/DD systems, and this sort of optical OFDM technique is called a spectrally factorized optical OFDM (SFO-OFDM). In this work the research contribution highlights the concept that the necessary and sufficient condition for a bandlimited periodic signal to remain non-negative at all the times is that the frequency coefficients form an autocorrelation function. Thus, a unipolar signal satisfying the requirements of IM/DD systems is accomplished by performing the autocorrelation of the complex data sequence prior to data transmission. An interesting aspect is that by employing the phenomenon of spectral factorization, a non-negative optical OFDM signal is obtained without any requirement of additional DC bias, thus assuring an enhanced power efficiency when compared with the earliest variant of optical OFDM like DCO-OFDM system. Additionally, the simulation results report that SFO-OFDM system exhibits a 0.5 dB gain over ACO-OFDM system, and the achievable BER was 10^{-5}. Furthermore, when compared with ACO-OFDM system, there was a significant reduction in the amount of PAPR by almost 30%.

The design of suitable modulation format for VLC should alleviate many practical issues that are concerned with the reliable realization of IM/DD-based VLC systems. It is a well-known fact that VLC leverages on solid-state illuminating opto-electronic devices for creating high-speed data transfer. We also know that IM and DD is the most effective modulation format for VLC. Therefore, for such kinds of systems, the important challenging aspect is imposed by the high PAPR of OFDM signal which is further fueled with the non-linearity of the optical device. Thus, while designing a modulation format in order to be applicable for VLC, immense prerequisites should be directed toward the illumination constraints imposed by the LED luminaries along with the focus on high data rate communication. So, there is a necessity to provide a linkage between the performance of the system in the perspective of communication and the lumination function of the luminary. Thus, taking into consideration these aspects, the contribution in [356] proposes the concept of spatial summing where wideband and high PAPR optical OFDM signals are partitioned into several low-PAPR narrowband signals that are transmitted from multiple LEDs. In general, these different signals from different LEDs prior to getting detected by photodiodes at the receiving end are spatially summed up in space. Accordingly, by using this spatial summing concept, this contribution proposes two different optical OFDM variants like spatial optical OFDM (SO-OFDM) and optical single carrier frequency division multiple access (OSC-FDMA). Both these categories of optical OFDM can either employ the one-to-one mapping of subcarriers to different LED groups or to facilitate for overlapping of subcarriers between different groups of LEDs by using filtering. The simulated results emphasize that when compared to DCO-OFDM system, SO-OFDM exhibits improved BER performance at higher SNRs due to the reduced PAPR and immunity to clipping distortion.

The much recent contribution in [568] proposes a triple-layer hybrid optical orthogonal frequency division multiplexing (THO-OFDM) which ensures higher spectral efficiency meeting the requirements of IM/DD-based VLC systems. Before, starting with a detailed description about the methodology of this scheme, it is vital to have a glimpse of the earliest variants of optical OFDM techniques and the drawbacks associated with them in fetching a unipolar signal. DCO-OFDM system ensures a unipolar signal at the cost of reduced power efficiency. Even though ACO-OFDM and PAM-DMT attain a unipolar signal without the requirement of DC-bias, the lower subcarrier utilization and the one-dimensional PAM mapping scheme evolve these schemes to be spectrally in efficient. Thus, as a result, a multitude of optical OFDM variants have been proposed in the literature to enhance the spectral efficiency. Much significant works which contribute to augment the spectral efficiency can be found in [104, 105, 235, 236, 508]. In general, in order to enhance the spectral efficiency, a layered ACO-OFDM technique has been proposed by combining the L layer ACO-OFDM signals with different effective subcarriers in the time domain for the purpose of improving the spectral efficiency of conventional ACO-OFDM system [508]. Similarly, with the intent to improve the spectral efficiency of conventional PAM-DMT, the research contributions in [235] and [236] propose enhanced PAM-DMT by exploiting superposition modulation streams of multiple PAM-DMT which offer the same spectral efficiency as that of the DCO-OFDM system.

Even though the aforementioned layered optical OFDM schemes are proven to be excellent techniques to enhance the spectral efficiency by removing the gaps between other unipolar OFDM schemes and DCO-OFDM system, pertaining to the real time applications, the efficient realization of such systems is impossible as the superimposition of different layers increases the computational complexity. This is because such systems require extra hardware implementation to carry out distortion cancellation between all the layers in both the time and frequency domains. Thus, in order to overcome the drawbacks of layered optical OFDM techniques, the research efforts in [568] propose a novel spectral-efficient THO-OFDM technique that offers a trade-off between the offered spectral efficiency and the computational complexity in comparison to DCO-OFDM system. This work portrays that by using only three layers, the proposed THO-OFDM reaches the spectral efficiency of LACO-OFDM by combining N-point ACO-OFDM, $\frac{N}{2}$-point ACO-OFDM and $\frac{N}{2}$-point PAM-DMT in just a single time domain frame. Theoretical analysis and simulation results of this work show that the proposed THO-OFDM scheme outperforms the conventional LACO-OFDM in terms of computational complexity and PAPR. When compared with three-layered LACO-OFDM scheme, THO-OFDM offers a 3 dB improvement in PAPR.

3.2.4 OTHER MULTICARRIER MODULATION FORMATS

Having looked at the performance analysis of different variants of optical OFDM employing IFFT and FFT transforms for enabling the OFDM modulation and demodulation, there is a decrement in spectral efficiency as all of the subcarriers are not utilized for data modulation. This is due to the fact of imposition of Hermitian Symmetry criteria to attain a real valued signal. Therefore, taking into consideration this concern, this subsection focuses on highlighting the complexity analysis involved behind the computation of Hermitian Symmetry criteria along with putting major stress on the performance analysis of other real trigonometric transform-based optical OFDM varieties.

3.2.4.1 Complexity involved in the computation of Hermitian symmetry criteria

To be compatible with the characteristics of light sources, it is essential for the OFDM signal to be real and unipolar. One such prevailing approach to yield a real-valued signal is to oblige the frequency domain symbols at the input of the IFFT module to comply with the Hermitian Symmetry criteria. Principally, this kind of modulation format is referred to as discrete multitone modulation (DMT). It is apparent that FFT and IFFT transformation modules are noteworthy in enabling OFDM modulation and demodulation. Consequently, with the increase in the size of IFFT, the power consumption as well as the area on chip increase significantly. On the other hand, pertaining to the optical OFDM system, for the purpose of accomplishing high data rate transmission, it requires double the size of IFFT and FFT blocks due to the enforcement of Hermitian Symmetry criteria on the frequency domain symbols. Assuredly, it requires $2N$-point IFFT and FFT transform sizes for modulating the N frequency domain symbols.

The impact of the sizes of IFFT and FFT transforms on the performance of optical OFDM system is unveiled in [71]. As reported in [71], upon increasing the sizes of IFFT/FFT, the number of data subcarriers increase which in turn increases the dynamic range of the OFDM signal, thereby substantially resulting in the requirement of higher precision in the arithmetic operations. The result analysis as depicted in [71] clearly indicates the effect of varying sizes of FFT/IFFT on the error vector magnitude (EVM) and the size of the FFT/IFFT is chosen from 32 to 1024. Indeed, it can be inferred from the figure that for a given FFT/IFFT bit precision, upon decreasing the sizes of FFT/IFFT, a reduction in the corresponding EVM is observed. As delineated in [122], for a radix-2 algorithm, the number of additions and multiplications which are required for the computation of real-valued FFT is given by

$$\underset{FFT}{Add} = 3Nlog_2N - 3N + 4 \tag{3.59}$$

$$\underset{FFT}{Mul} = Nlog_2N - 3N + 4 \tag{3.60}$$

Therefore, the parameters $\underset{FFT}{Add}$ and $\underset{FFT}{Mul}$ specify the total number of additions and multiplications for FFT transform. In practice, the computational complexity for any transform is defined in terms of the total number of additions and multiplications which are required to compute it. Table 3.2 gives the total number of additions and multiplications required for different sizes of FFT varying from $8, 16, 32, \cdots, 2048$. Figure 3.21 confirms the fact that upon increasing the sizes of FFT, computational complexity increases drastically for a DMT system.

Table 3.2
Illustration of total number of additions and multiplications required for the computation of different sizes of FFT.

Size of FFT	No. of Additions	No. of Multiplications
8	52	4
16	148	20
32	388	68
64	964	196
128	2308	516
256	5380	1284
512	12292	3076
1024	27652	7172
2048	61444	16388

Listing 3.3: MATLAB code to depict the computational complexity of DFT transform for the simulated result as illustrated in Fig. 3.21.

```
1 clc;
2 clear all;
3 close all;
4 N=[8 16 32 64 128 256 512 1024 2048]; %Size of the
      Transform
```

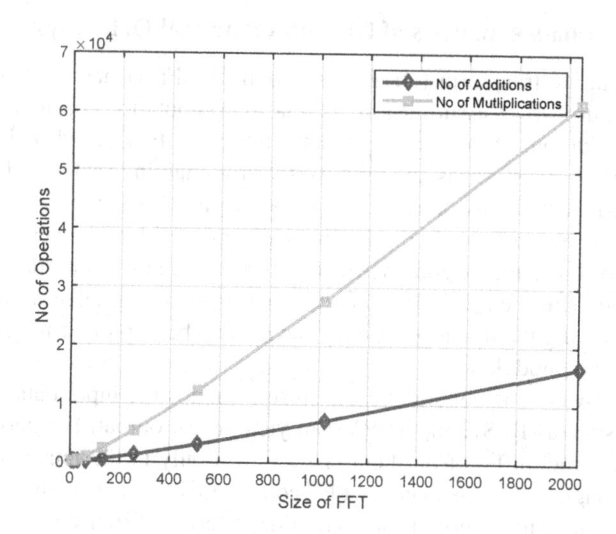

Figure 3.21: Illustration of computational complexity involved behind the split-radix FFT algorithm.

```
5  x1=N.*log2(N)-3*N+4; %Total Number of Additions
      required by FFT transform
6  x=3*N.*log2(N)-3*N+4; %Total Number of
      Multiplications required by FFT transform
7  plot(N,x1,'rd-','LineWidth', 2);hold on;
8  plot(N,x,'gs-','LineWidth', 2);
9  legend('No of Additions', 'No of Multiplications');
10 xlabel('Size of FFT');
11 ylabel('No of Operations')
12 grid on;
13 axis([0 2048 0 7*10^4]);
```

The second important aspect which needs to be taken into account along with the computational complexity is power consumption. Therefore, for maximizing the system performance, Application Specific Integrated Circuits (ASICs) are widely deployed in networking devices [416]. The effects of increasing the size of FFT on the performance as well as the power consumption of real-time ASIC-based DMT transceivers was detailed in [71]. The result analysis of this work confirms the fact that upon increasing the size of FFT and IFFT, the cost and the power consumption increase drastically. This is more pronounced when the size of FFT/IFFT increases from 512 to 1024, where a 40% increase in the area of the chip and 38% increase in the power consumption at the receiver side can be observed.

3.2.4.2 Performance analysis of DHT-based optical OFDM system

A discret Harley transform (DHT)-based optical OFDM has been proposed in [355] for the purpose of assuring a reduction in computational complexity as encountered in traditional Hermitian Symmetry imposed IFFT-based optical OFDM. Typically, Hartley transform is an attractive tool for enabling real signal processing and has gained tremendous popularity because it involves the same digital signal processing in the transmission and reception and works with real algebra. Moreover, Hartley transform is a real trigonometric transform where the forward/direct and inverse transforms are identical. This implies that it possesses self-inverse property. Inevitably, same digital signal processing device can be exploited for enabling the modulation and demodulation.

Being a real-valued trigonometric transform, it maps the input data using a real constellation such as BPSK and M-PAM to yield a real output. Furthermore, Hartley transform kernel differs from Fourier transform only in the imaginary part and the real and imaginary components of FFT coincide with the even and negative odd parts of DHT. Since it has got the same routine in terms of transmission and reception, we need not force its input to satisfy Hermitian Symmetry constraint. Using a simpler implementation scheme, it supports double the input symbols of a standard real-valued FFT. In OFDM, DFT is prominent in implementing OFDM modulation since it can be seen as a bank of modulators whose narrowband channels have mutually orthogonal subcarriers. In the same manner, the mirror-symmetric sub-bands of DHT guarantee for subcarrier orthogonality and the spectral behavior of it enables to carry the data symbols for parallel processing. As a result, DHT can be envisaged to replace the FFT in optical OFDM and to embellish as an attractive modulation scheme to be employed for cost-effective IM/DD systems.

A schematic representation of channel estimation in ACO-OFDM which is making use of DHT over a dispersive VLC channel is presented in Fig. 3.22. As depicted, the incoming input data stream is encoded by utilizing real constellation mapping techniques like BPSK and M-PAM. It is evident from the figure that DHT is used for accomplishing the OFDM modulation and demodulation; therefore, without enforcing the Hermitian Symmetry criteria, a real valued signal is obtained which is a primary requisite of IM/DD system.

The signal processing operation involved behind the transmitter schematic is detailed while dealing with channel estimation in traditional DCO-OFDM and ACO-OFDM systems. Except for the transform technique employed, the rest of the operations are similar. Accordingly, the discrete time-domain signal at the output of IDHT is given by

$$x[n] = \frac{1}{\sqrt{N}} \sum_{k=0}^{N-1} X[k] \left[cos\left(\frac{2\pi nk}{N}\right) + sin\left(\frac{2\pi nk}{N}\right) \right] \tag{3.61}$$

This is equivalent to

$$x[n] = \frac{1}{\sqrt{N}} \sum_{k=0}^{N-1} X(K) \left[cas\left(\frac{2\pi nk}{N}\right) \right] \tag{3.62}$$

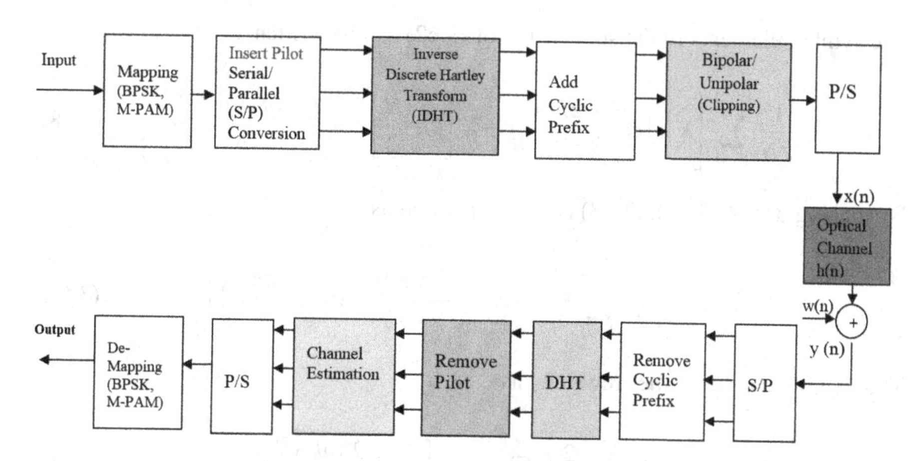

Figure 3.22: Schematic representation of channel estimation in ACO-OFDM using discrete Hartley transform (DHT) over VLC channel [498].

Where $cas\theta = cos\theta + sin\theta$. However, since ACO-OFDM principality is employed, only odd subcarriers are employed for data transmission. Since the forward and inverse transforms are identical, $X[k]$ can be expressed as

$$X[k] = \frac{1}{\sqrt{N}} \sum_{n=0}^{N-1} x[n] \left[cos\left(\frac{2\pi nk}{N}\right) + sin\left(\frac{2\pi nk}{N}\right) \right] \tag{3.63}$$

Similar to DFT-based ACO-OFDM, even while exploiting DHT, it is easy to demonstrate for odd subcarriers. Thus, equation (3.63) can be expressed as

$$X\left[k + \frac{N}{2}\right] = \frac{1}{\sqrt{N}} \sum_{n=0}^{N-1} x[n] \left[cos\left(\frac{2\pi n\left(k+\frac{N}{2}\right)}{N}\right) + sin\left(\frac{2\pi n\left(k+\frac{N}{2}\right)}{N}\right) \right] \tag{3.64}$$

Further, (3.64) can be solved as

$$X\left[k + \frac{N}{2}\right] = \frac{1}{\sqrt{N}} \sum_{n=0}^{N-1} x[n] \left[cos\left(\frac{2\pi nk}{N}\right) cos\pi n - sin\left(\frac{2\pi nk}{N}\right) sin\pi n + \right.$$
$$\left. sin\left(\frac{2\pi nk}{N}\right) cos\pi n + cos\left(\frac{2\pi nk}{N}\right) sin\pi n \right] \tag{3.65}$$

Therefore, (3.65) can be further reduced to

$$X\left[k + \frac{N}{2}\right] = (-1)^n \frac{1}{\sqrt{N}} \sum_{n=0}^{N-1} x[n] \left[cos\left(\frac{2\pi nk}{N}\right) + sin\left(\frac{2\pi nk}{N}\right) \right] \tag{3.66}$$

From (3.66), since we are interested in odd subcarriers, we can establish a relation that

$$X\left[k + \frac{N}{2}\right] = -X[k] \tag{3.67}$$

The symbol elements in the summation of (3.62) can be written as

$$x[n] = \frac{1}{\sqrt{N}} \sum_{k=0}^{\frac{N}{2}-1} \left[X[k]\, cas\left(\frac{2\pi nk}{N}\right) + X\left[k+\frac{N}{2}\right] cas\left(\frac{2\pi n\left(k+\frac{N}{2}\right)}{N}\right)\right] \tag{3.68}$$

By making use of (3.67), (3.68) can be expressed as

$$x[n] = \frac{2}{\sqrt{N}} \sum_{k=0}^{\frac{N}{2}-1} X[k] \left[cos\left(\frac{2\pi nk}{N}\right) + sin\left(\frac{2\pi nk}{N}\right)\right] \tag{3.69}$$

Thus, (3.69) can be further rewritten as

$$x[n] = \frac{2}{\sqrt{N}} \sum_{k=0}^{\frac{N}{2}-1} X(K) \left[cas\left(\frac{2\pi nk}{N}\right)\right] \tag{3.70}$$

Thereupon, the negative peaks in the discretized time-domain signal are clipped and only a positive valued signal is intensity modulated through the LED. The rest of the operations, like pilot tone extraction and channel estimation, are similar to that of the traditional optical OFDM. The BER performance analysis of

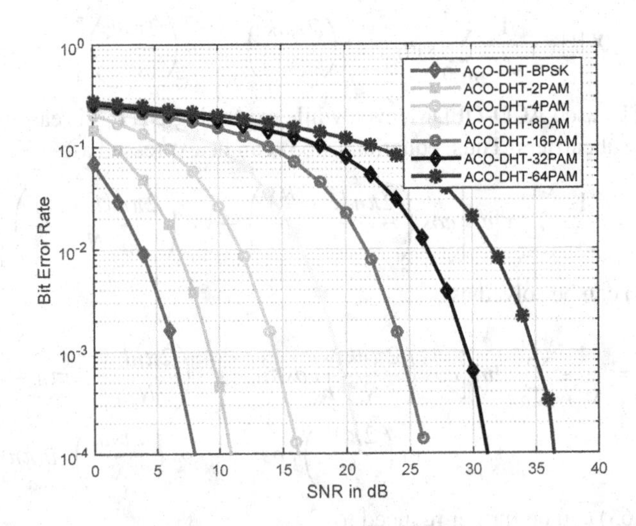

Figure 3.23: BER performance analysis of discrete Hartley transform (DHT)-based ACO-OFDM using BPSK modulation and M-PAM over VLC channel [498].

DHT-based ACO-OFDM system exploiting real mapping techniques like BPSK and M-PAM has been analyzed over VLC channel environment, and this is delineated in Fig. 3.23. From the figure, it can be deduced that when compared to a Hermitian Symmetry-based, DFT-based ACO-OFDM system, in order to achieve the

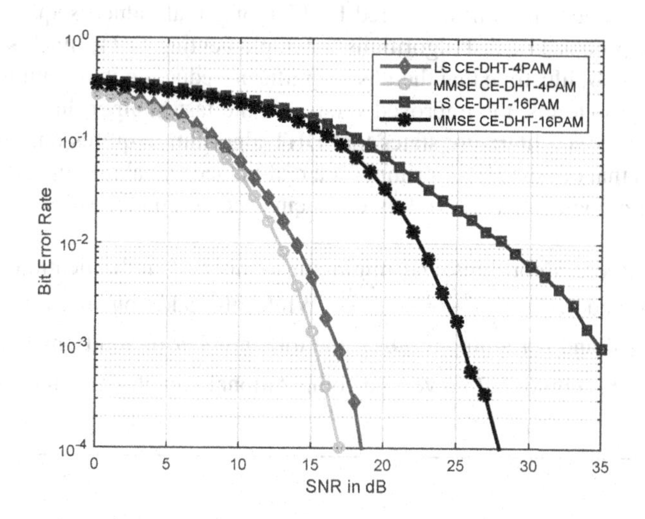

Figure 3.24: Channel estimation in discrete Hartley transform (DHT)-based ACO-OFDM using M-PAM over VLC channel [498].

same amount of performance, a DFT-ACO-OFDM system requires a complex constellation like 4, 16, 64 QAM in contrast to a DHT-ACO-OFDM system which requires simple one-dimensional mapping like 2, 4, 8 PAM, respectively. Therefore, this confirms that DFT-ACO-OFDM system requires double the constellation size when compared with DHT-ACO-OFDM system, thereby exhibiting the superiority of DHT-based optical OFDM system. The performance analysis of channel estimation algorithms like LS and MMSE in DHT-ACO-OFDM system is depicted in Fig. 3.24. Since LS is more prone to noise, MMSE channel estimation algorithm results in a better BER performance than LS. Furthermore, the effect of clipping noise is more pronounced on higher orders of constellation, thereby leading to the emanation of ISI and resulting in the requirement of higher SNRs.

Computational Complexity Analysis of DHT

Several studies as stated in [73,223,456] reveal that FHT requires around the same multiplications as that of the FFT transform when exploiting Hermitian Symmetry criteria. Pertaining to the number of additions, DHT requires more additions than that of the FFT. Nevertheless, for the computation of real-valued FFT, one must rely on additional resources to calculate the complex-conjugate vector. Earlier studies as reported in [456] reveal that in the case of radix-2 algorithm, for both the decimation-in-time and decimation-in-frequency, DHT algorithm has the same multiplications as that of the FFT, but $N-2$ more additions than the corresponding FFT algorithm optimized for a real input vector.

Similarly, even for radix-4, split radix, prime factor and Winograd transform algorithms, as illustrated in [456], the requirement in the number of additions for

DHT slightly exceeds the ones required by FFT of a real valued sequence. However, with the proposal of fast algorithms for implementing DHT, much significant works in [149, 150] illustrate that there is a significant reduction in the number of additions. In precise terms, the improved version of the fastest algorithm as proposed in the aforementioned literature states that DHT algorithm requires only two more additions than that of the FFT algorithm for a real-valued signal thus ensuring a minimal arithmetic complexity as well as enhancing the computational speed of DSP devices.

According to these algorithms, the minimum number of multiplications required for both FFT and DHT is $\frac{(Nlog_2N-3N+4)}{2}$. While, the addition in case of FFT is $\frac{(3Nlog_2N-5N)}{2}+4$ and pertaining to DHT, the total number of additions required are $\frac{(3Nlog_2N-5N)}{2}+6$. Figure 3.25 gives the computational complexity which is defined

Table 3.3
Comparison between DHT-optical OFDM and DFT-optical OFDM.

Type of OFDM	DHT-ACO-OFDM and DHT-DCO-OFDM	DFT-ACO-OFDM and DFT-DCO-OFDM
Hermitian Symmetry	Not required	Required
Constellation	Real (BPSK, M-PAM)	Complex (M-QAM)
Constellation Size	M=L	M=L^2
Self-Inverse Property	Self-Inversive	Not Self-Inversive
Subcarriers Utilized	N/2 for DHT-ACO-OFDM N for DHT-DCO-OFDM	N/2 for DFT-ACO-OFDM N for DFT-DCO-OFDM
No. of Multiplications	$\frac{(3Nlog_2N-3N+4)}{2}$	$\frac{(3Nlog_2N-3N+4)}{2}$
No. of Additions	$\frac{(3Nlog_2N-5N)}{2}+6$	$\frac{(3Nlog_2N-5N)}{2}+4$

in terms of total number of additions and multiplications for both DHT and DFT transforms for the mathematical expressions as portrayed in Table 3.3. Since DHT transform requires only two more additions than that of the DFT, the performance of both transforms as illustrated in the figure is similar. However, the total number of multiplications for both DFT and DHT are similar. Thus, this plot confirms that DHT transform can be exploited for the cost-effective realization of optical OFDM systems.

Listing 3.4: MATLAB code to depict the computational complexity behind the two transforms, namely, DHT and DFT.

```
1  clc;
2  clear all;
3  close all;
4  N=[8 16 32 64 128 256 512 1024 2048]; %Size of the
       Transform
```

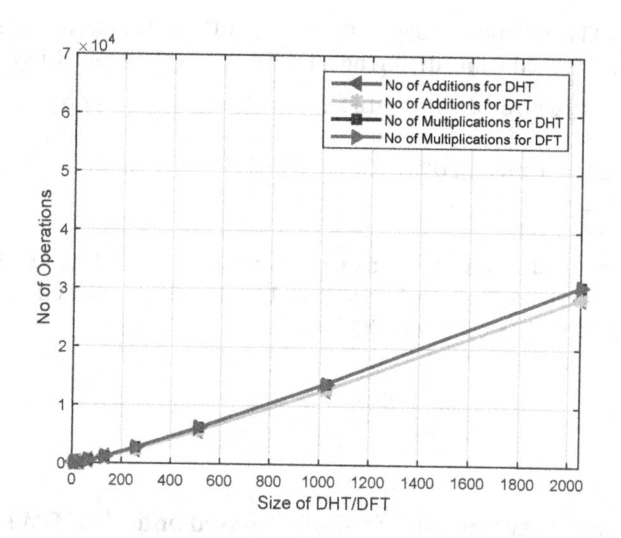

Figure 3.25: Computational complexity in terms of total number of additions and multiplications for both DHT and DFT for the mathematical expressions presented in Table 3.3.

```
5  x=(3*N.*log2(N)-5*N)/2+6; %Total Number of Additions
      required by DHT transform
6  x1=(3*N.*log2(N)-5*N)/2+4; %Total Number of Additions
      required by DFT transform
7  y=(3*N.*log2(N)-3*N+4)/2; %Total Number of
      Multiplications required by DHT transform
8  y1=(3*N.*log2(N)-3*N+4)/2; %Total Number of
      Multiplications required by DFT transform
9  plot(N,x,'r<-','LineWidth', 2);hold on;
10 plot(N,x1,'g*-','LineWidth', 2);
11 plot(N,y,'bs-','LineWidth', 2);
12 plot(N,y1,'m>-','LineWidth', 2);
13 legend('No of Additions for DHT', 'No of Additions
      for DFT', 'No of Multiplications for DHT', 'No of
      Multiplications for DFT');
14 xlabel('Size of DHT/DFT');
15 ylabel('No of Operations')
16 grid on;
17 axis([0 2048 0 7*10^4]);
```

Listing 3.5: MATLAB code to depict the function DHT. It is to be noted that this function can be invoked while desinging a DHT-based optical OFDM system.

```
function inverse_dht=i_dht(DataMod_serialtoparallel,n
    ,k,N)
inverse_dht=zeros(1024,1);
for n=0:N-1
    for k=0:N-1
     inverse_dht(n+1)=inverse_dht(n+1)+(1/sqrt(N))*(
        DataMod_serialtoparallel(k+1)*(cos(2*pi*n*k/N
        )+sin(2*pi*n*k/N)));
    end
end
end
```

3.2.4.3 Performance analysis of DCT/DST-based optical OFDM system

Owing to its remarkable advantages and with the advancements of digital signal processing technology, OFDM has been applied to OWC. With its spectral efficiency feature and resilience against narrow band interference, it sparked a considerable amount of interest in the research community. The underlying physics behind the suitability of OFDM technique for transmission over a dispersive fading channel environment is that it splits a high-speed data stream into several low-speed data streams in such a way that each low-speed data stream can be concurrently transmitted over a number of harmonically related narrowband subcarriers [220]. An improvement in bandwidth efficiency can be achieved by reducing the spacing between the number of subcarriers. With this motive, as well as to emerge as a competitive solution in the market, fast optical OFDM (FOOFDM) has been proposed by setting its target to minimize the transceiver components cost [572].

The architectural design of FOOFDM reduces hardware and software engineering effort, while reducing the computational complexity by employing real signal processing without sacrificing the overall system performance. In contrast to conventional optical OFDM which employs Hermitian Symmetry imposed IFFT for accomplishing a real-valued signal transmission, FOOFDM system exploits real trigonometric transform like discrete cosine transform (DCT) for accomplishing OFDM modulation and demodulation. It is reported in the literature that a DCT-based optical OFDM, which is also referred to as FOOFDM, has been investigated for IM/DD optical transmission systems [184, 185, 575]. In FOOFDM, the spacing of the neighboring subcarrier is half of the symbol rate per subcarrier. It uses simple, real and one-dimensional mapping schemes where only in-phase component is exploited. FOOFDM exhibits an improved system performance in channel estimation and exhibits enhanced robustness to FO and chromatic dispersion [549, 571, 573]. This is due to its excellent energy compaction property.

When compared with traditional Hermitian Symmetry imposed IFFT-based optical OFDM, FOOFDM exhibits lower computational complexity because without the

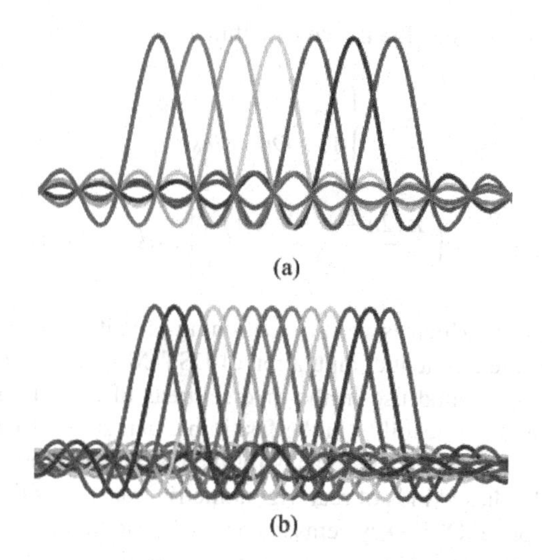

Figure 3.26: Spectrum of DFT-based OFDM and FOOFDM, (a) DFT-based OFDM, (b) DCT-based OFDM [576].

requirement of Hermitian Symmetry criteria, a real-valued signal is attained as well as the operation count in terms of number of additions and multiplications for DCT is less than that of the DFT. As shown in Fig. 3.26, the minimum subcarrier spacing which is required by FOOFDM is just half of that of the DFT-based OFDM. Aiding this, another real trigonometric transform which can be exploited in optical OFDM is discrete sine transform (DST). DST belongs generally to the family of sinusoidal unitary transforms. In particular, a sinusoidal unitary transform is an invertible linear transform whose kernel is described by a set of orthogonal discrete sine basis functions [422].

DSTs, which are ably called discrete trigonometric transforms, consist of eight versions of DST where each transform is classified as even or odd of type I, II, III and IV [338,339,422,517]. In general, even types of DSTs, i.e., DST-I, DST-II, DST-III and DST-IV are widely defined in the literature due to their utilization in several digital signal processing and image processing applications. For a transform of size N, the corresponding four even types of DST as defined in [517] can be expressed as:

$$DST-I = \sqrt{\frac{2}{N}}\left[sin\left(\frac{\pi(n+1)(k+1)}{N}\right)\right], n,k = 0,1,2,\cdots N-2 \tag{3.71}$$

$$DST-II = \sqrt{\frac{2}{N}}\left[\varepsilon_k sin\left(\frac{\pi(2n+1)(k+1)}{2N}\right)\right], n,k = 0,1,2,\cdots N-1 \tag{3.72}$$

$$DST-III = \sqrt{\frac{2}{N}}\left[\varepsilon_n sin\left(\frac{\pi(2k+1)(n+1)}{2N}\right)\right], n,k = 0,1,2,\cdots N-1 \tag{3.73}$$

From (3.72) and (3.73), the value ε_r can be defined as

$$\varepsilon_r = \begin{cases} \frac{1}{\sqrt{2}}, & r = N-1 \\ 1, & Otherwise \end{cases} \tag{3.74}$$

$$DST-IV = \sqrt{\frac{2}{N}}\left[sin\left(\frac{\pi\,(2n+1)\,(2k+1)}{4N}\right)\right], n,k = 0,1,2,\cdots N-1 \tag{3.75}$$

The remarkable aspect which is worth mentioning is that the existence of fast algorithms facilitates the efficient computation of DST. Moreover, these different versions of DST have profound use in the diverse areas of digital signal processing. DST-*IV* has profound applicability in the fast implementation of lapped orthogonal transforms and sine modulated filter banks for the purpose of efficient subband coding [74]. Eventually, these kinds of real trigonometric transforms like DCT and DST can be utilized in optical OFDM systems to enable the modulation and demodulation process. Consequently, such kinds of systems can be envisaged to be cost-effective solutions for IM/DD systems.

Performance Analysis of FOOFDM System

The schematic representation of DCO-OFDM system model which is based on DCT is delineated in Fig. 3.27. The N point DCT and its inverse (IDCT) for a data sequence $m = 0,1,2,\cdots N-1$ as defined in [15, 166] are given as:

$$X_k = \sqrt{\frac{2}{N}}C_k\sum_{m=0}^{N-1} x_m cos\left(\frac{\pi\,(2m+1)\,k}{2N}\right), 0 \le k \le N-1 \tag{3.76}$$

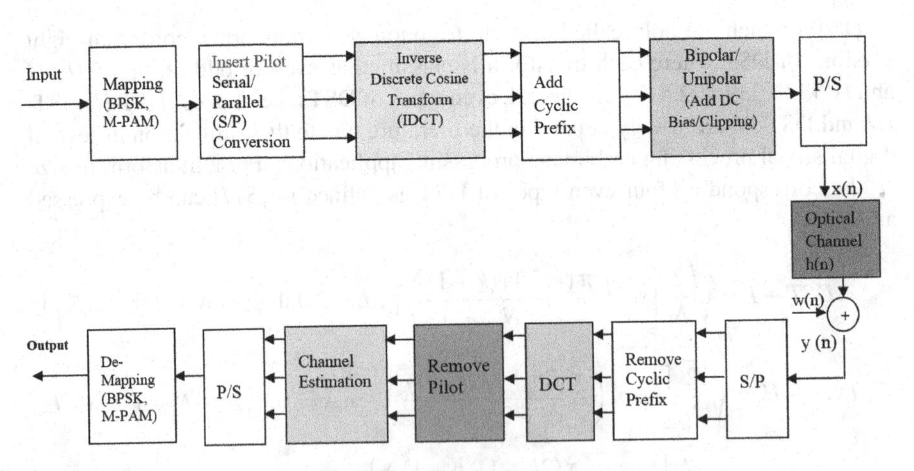

Figure 3.27: Schematic representation for FOOFDM system.

$$x_m = \sqrt{\frac{2}{N}} \sum_{k=0}^{N-1} C_k X_k cos\left(\frac{\pi(2m+1)k}{2N}\right), 0 \leq m \leq N-1 \tag{3.77}$$

Where,

$$C_k = \begin{cases} \frac{1}{\sqrt{2}}, & k = 0 \\ 1, & k = 1, 2, 3, \cdots N-1 \end{cases} \tag{3.78}$$

where X_k and x_m represents the frequency-domain and time-domain samples, respectively. As evident from the block diagram, DCT and IDCT play the role of data modulation and demodulation. In general, the pilot tones are inserted based upon comb-type pilot arrangement pattern into each FOOFDM symbol based upon a specific period. The input vector to the transformation module is mapped by employing real-mapping schemes like BPSK and M-PAM. Without the requirement of Hermitian Symmetry, a real-valued signal is attained. Primarily, while taking into account the optical OFDM system that exploits N-point FFT, for each stage of parallel processing, the input information bit sequence which is mapped by employing M-QAM requires $\frac{N}{2}$ total number of subcarriers for data transmission due to Hermitian Symmetry constraint (i.e., according to Hermitian Symmetry criteria, only half of the inputs carry the data, while the other half are flipped complex conjugate versions of the previous ones). While, the N-point DCT as depicted in equations (3.76) and (3.77) employs all of the transform points, i.e., subcarriers for data transmission therefore ensuring double the data rate. However, in order to ensure non-negativity of the signal, a suitable amount of DC bias is added to bipolar FOOFDM signal, and the remaining negative peaks are clipped to result in a positive DCO-FOOFDM signal. Therefore, the DCO-FOOFDM signal can be expressed as [492]

$$s(n) = \begin{cases} x(n) + \beta_{DC}, & x(n) > \beta_{DC} \\ 0, & x(n) < \beta_{DC} \end{cases} \tag{3.79}$$

The basic principle of DCO-OFDM which includes the addition of DC bias is stated in the previous subsections. Therefore, this real and positive-valued signal is propagated through the optical channel environment. Thus, the received signal can be expressed as

$$y(n) = Rs(n) \circledast h(n) + w(n) \tag{3.80}$$

Where R in (3.80) signifies the responsivity of the PD in $Ampere/Watts$, $\circledast$ specifies the convolution operator and $w(n)$ represents the thermal and shot noise which is modeled as AWGN. For ease of simplicity, the value of R is taken as 1. Equation (3.80) can be solved to attain

$$y(n) = \sum_{l=0}^{L-1} s(n-l)h(l) + w(n) \tag{3.81}$$

where L in represents the total number of paths. At the receiving end, this huge stream of serialized data is converted into parallel by employing S/P converter. Thereupon, pilot tones are extracted and then channel estimation, is enforced, where the transmitted pilots and the received pilots are compared to determine the

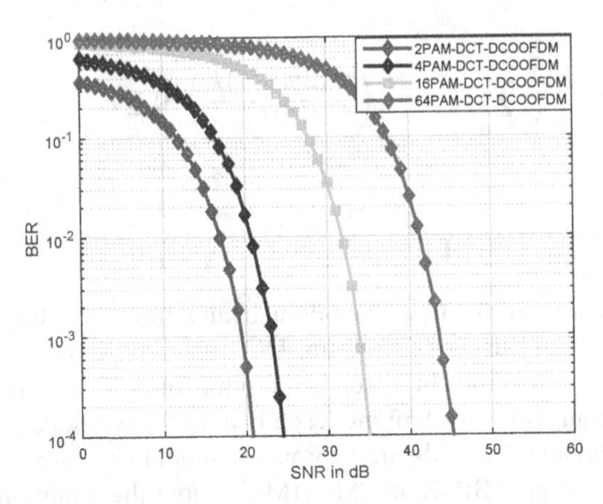

Figure 3.28: BER vs. SNR performance analysis of FOOFDM system using M-PAM.

deterioration of the received signal under the channel effects. Lastly, the signal is processed to obtain the final output. The performance of FOOFDM system which is utilizing M-PAM modulation scheme is shown in Fig. 3.28. This result evidences that FOOFDM system, which is based on the addition of DC bias, requires higher SNRs to achieve the desired BER for 2, 4, 16 and 64 PAM, respectively. This is because, in case of DCO-FOOFDM, the clipping operation results in clipping-induced distortion which reduces the BER performance. Moreover, in order for the BER to reach the forward error correction limit of 10^{-3}, a fixed amount of DC bias addition is not sufficient for all the constellations in DCO-FOOFDM system. Figure 3.29 illustrates the BER versus SNR performance of DCT-based DCO-OFDM system exploiting 7 and 13 dB of DC bias. From the simulated result, it is clear that an increase in the amount of DC bias will result in more SNRs.

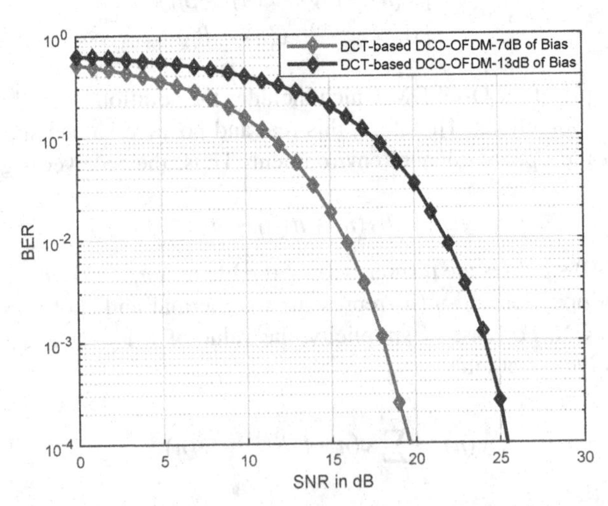

Figure 3.29: Performance analysis of DCT-based DCO-OFDM system with 7 and 13 dB of bias.

Listing 3.6: MATLAB code to depict the performance of DCT-based DCO-OFDM system, exploiting 7 and 13 dB of DC bias, for the simulated result as illustrated in Fig. 3.6. (Note: This MATLAB code works with R2014b Version).

```
1  clc;
2  clear all;
3  close all;
4  m=512; %Total number of OFDM symbols
5  N=1024; %Length of each OFDM symbol
6  M=4; %Size of the Constellation (M can be 4, 8, 16,
      32, 64, 128, 256)
7  Ncp=256 ; %Length of the Cyclic Prefix
8  Tx=modem.pammod('M',M); %Choosing the modulation
      format as M-ary Pulse Amplitude Modulation (M-PAM)
9  Rx=modem.pamdemod ('M',M); %Fixing up the
      demodulation format at the receiving end as M-ary
      PAM
10 %DCO-OFDM Transmitter
11  Data=randi ([0 M-1],m,N); %Generation of Random bits
       Matrix of size m by N
12 DataMod=modulate(Tx,Data); %Performing Data
      Modulation, in order to simulate this code using
      MATLAB R2020a version, please remove the variables
       Tx and Rx. Thereupon, assign DataMod=pammod(Data,
       M);
13 DataMod_serialtoparallel=DataMod.'; %Performing
      Serial to Parallel Conversion
14 datamat=DataMod_serialtoparallel; %Assigning the
      total data to a variable called datamat
15 d_idctt=idct((datamat)); %Computation of IDCT
      operation
16 d_idct_paralleltoserial=d_idctt.'; %Parallel to
      Serial Conversion
17 CP_part=d_idct_paralleltoserial(:,end-Ncp+1:end); %
      Addition of Cyclic Prefix
18 DCOOFDM_CP=[CP_part d_idct_paralleltoserial]; %
      Transmission of DCO-OFDM signal
19 bdc=7; %DC BIAS
20 clip=sqrt((10.^(bdc/10))-1); %clipping factor k
21         bdcc=clip*sqrt(DCOOFDM_CP.*DCOOFDM_CP);%
              Computation of DC bias
22         DCOOFDM_BIAS=DCOOFDM_CP+bdcc; %Addition of DC
               bias to the cyclic prefix added signal
23 count=0;
24 snr_vector=0:1:80; %size of signal to noise ratio (
```

```
      SNR) vector
25  for snr=snr_vector
26      SNR = snr + 10*log10(log2(M));
27      count=count+1 ;
28      DCOOFDM_with_chann=awgn(DCOOFDM_BIAS,SNR,'
          measured' ) ; %Addition of AWGN
29      %Receiver of DCO-OFDM
30      DCOOFDM_with_chann1=DCOOFDM_with_chann-bdcc;%
          Removal of DC bias
31      DCOOFDM_removal_CP=DCOOFDM_with_chann1(:,Ncp+1:N+
          Ncp); %Removal of Cyclic Prefix
32      DCOOFDM_serialtoparallel=DCOOFDM_removal_CP.'; %
          Serial to Parallel Conversion
33      DCOOFDM_parallel_dct=dct(DCOOFDM_serialtoparallel
          ) ; %Computation of DCT operation
34      DCOOFDM_Demodulation=pamdemod(
          DCOOFDM_parallel_dct.',M);
35     [~,s_e1(count)]=symerr(Data(:,2:512),
         DCOOFDM_Demodulation (:,2:512)) ;
36  end
37  %%%%%%With 13 dB Bias
38  m=512; %Total number of OFDM symbols
39  N=1024; %Length of each OFDM symbol
40  M=4; %Size of the Constellation (M can be 4, 8, 16,
      32, 64, 128, 256)
41  Ncp=256 ; %Length of the Cyclic Prefix
42  Tx=modem.pammod('M',M); %Choosing the modulation
      format as M-ary Pulse Amplitude Modulation (M-PAM)
43  Rx=modem.pamdemod ('M',M); %Fixing up the
      demodulation format at the receiving end as M-ary
      PAM
44  %DCO-OFDM Transmitter
45   Data1=randi ([0 M-1],m,N); %Generation of Random
       bits Matrix of size m by N
46  DataMod1=modulate(Tx,Data1); %Performing Data
      Modulation, in order to simulate this code using
      MATLAB R2020a version, please remove the variables
       Tx and Rx. Thereupon, assign DataMod1=pammod(
      Data1, M);
47  DataMod_serialtoparallel1=DataMod1.'; %Performing
      Serial to Parallel Conversion
48  d_idct1=idct((DataMod_serialtoparallel1)); %
      Computation of IDCT operation
49  d_idct_paralleltoserial1=d_idct1.'; %Parallel to
```

```
    Serial Conversion
50 CP_part1=d_idct_paralleltoserial1(:,end-Ncp+1:end); %
    Addition of Cyclic Prefix
51 DCOOFDM_CP1=[CP_part1 d_idct_paralleltoserial1]; %
    Transmission of DCO-OFDM signal
52 bdc1=13; %DC BIAS
53 clip1=sqrt((10.^(bdc1/10))-1); %clipping factor k
54         bdcc1=clip1*sqrt(DCOOFDM_CP1.*DCOOFDM_CP1);%
            Computation of DC bias
55         DCOOFDM_BIAS1=DCOOFDM_CP1+bdcc1; %Addition of
             DC bias to the cyclic prefix added signal
56 count=0;
57 snr_vector=0:1:80; %size of signal to noise ratio (
    SNR) vector
58 for snr=snr_vector
59     SNR = snr + 10*log10(log2(M));
60     count=count+1 ;
61     DCOOFDM_with_channel=awgn(DCOOFDM_BIAS1,SNR,'
        measured' ) ; %Addition of AWGN
62     %Receiver of DCO-OFDM
63     DCOOFDM_with_chann2=DCOOFDM_with_channel-bdcc1;%
        Removal of DC bias
64     DCOOFDM_removal_CP2=DCOOFDM_with_chann2(:,Ncp+1:N
        +Ncp); %Removal of Cyclic Prefix
65     DCOOFDM_serialtoparallel2=DCOOFDM_removal_CP2.';
        %Serial to Parallel Conversion
66     DCOOFDM_parallel_dct2=dct(
        DCOOFDM_serialtoparallel2) ; %Computation of
        DCT operation
67     DCOOFDM_Demodulation2=pamdemod(
        DCOOFDM_parallel_dct2.',M);
68    [~,s_e2(count)]=symerr(Data1(:,2:512),
        DCOOFDM_Demodulation2 (:,2:512)) ;
69 end
70   % Plotting the BER curves
71 semilogy(snr_vector,s_e1,'md-','LineWidth',2);hold on
    ;
72 semilogy(snr_vector,s_e2,'bd-','LineWidth',2);
73 legend(3, 'DCT-based DCO-OFDM-7dB of Bias', 'DCT-
    based DCO-OFDM-13dB of Bias');
74 axis([0 30 10^-4 1]);
75 xlabel('SNR in dB');
76 ylabel('BER');
77 grid on;
```

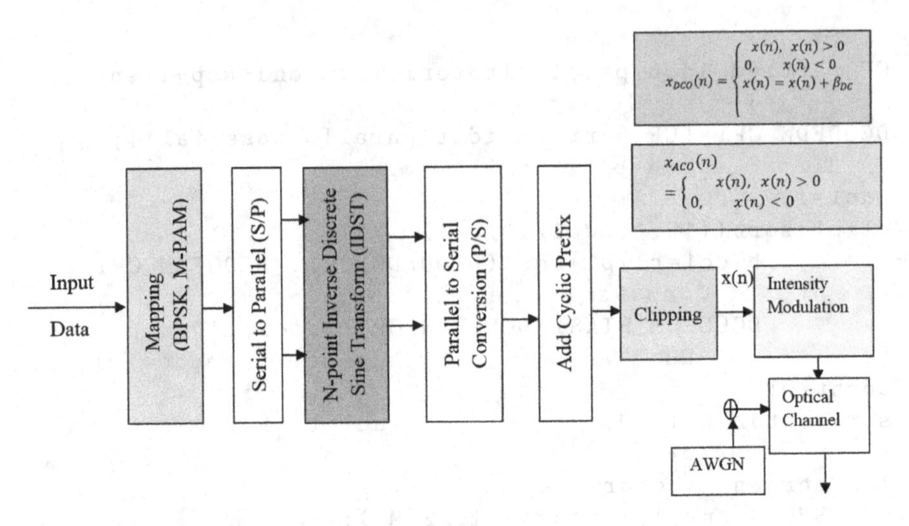

Figure 3.30: Transmitter of DST-DCO/ACO-OFDM system for VLC [493, 495].

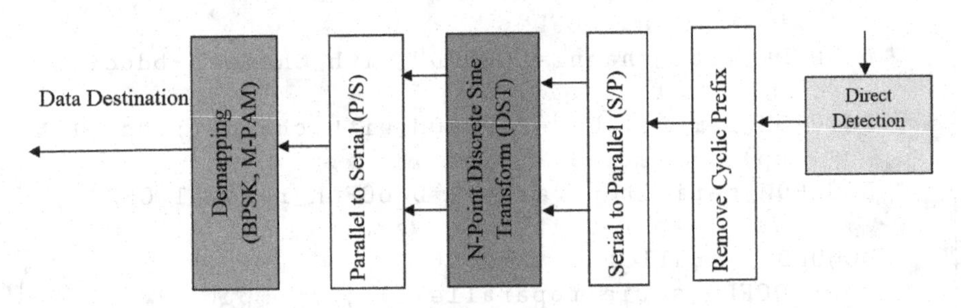

Figure 3.31: Receiver of DST-DCO/ACO-OFDM system for VLC [493, 495].

Performance Analysis of DST-based multicarrier systems

The schematic representations of the transmitter and the receiver of DST-based DCO-OFDM and DST-based ACO-OFDM systems are delineated in Figs. 3.30 and 3.31, respectively. The schematic representations are adopted from the work [495]. Here, in brief, this section portrays the significance of the deployment of real trigonometric transform techniques like DST. From the transmitter and receiver schematics, it is obvious that DST is employed as the suitable transform technique to enable optical OFDM modulation and demodulation rather than the Hermitian Symmetry imposed IFFT-based complex transform technique. Predominantly, the N-point DST and its inverse are defined as [74]

$$X_k = \sqrt{\frac{2}{N}} \sum_{n=0}^{N-1} x_n sin\left[\frac{\pi\,(2n+1)\,(2k+1)}{4N}\right] k = 0, 1, 2, 3, \cdots N-1 \tag{3.82}$$

$$x_n = \sqrt{\frac{2}{N}} \sum_{k=0}^{N-1} X_k sin\left[\frac{\pi(2n+1)(2k+1)}{4N}\right] n = 0,1,2,3,\cdots N-1 \tag{3.83}$$

where X_k and x_n in (3.82) and (3.83) represent the frequency- and time-domain signal, respectively, and furthermore it can be inferred that a real valued signal is attained without enforcing Hermitian Symmetry criteria as required in DFT-based DCO-OFDM/ACO-OFDM system. Similar to the methodology as implemented in case of DCT-based optical OFDM, real and simple one-dimensional mapping like BPSK and M-PAM are sufficient enough to map the huge sets of incoming data stream and, then, this stream of modulated data symbols is transmitted in parallel by employing a S/P. Furthermore, in order to prevent ISI which generally prevails in a multipath environment, cyclic prefix or guard interval is added to this time-domain signal. Prior to intensity modulation of this real valued time-domain signal through the LED, it should be assured of its positivity. Hence, this necessitates to employ DCO-OFDM and ACO-OFDM methodologies.

Pertaining to DST-DCO-OFDM system, in order to ensure the signal positivity, instead of adding a fixed amount of DC bias, it is convenient to add the DC bias per symbol basis as [441, 493]

$$\beta_{DC} = |min\{x_n, n = 0,1,2,3,\cdots N-1\}| + V_{TOV} \tag{3.84}$$

In (3.84), V_{TOV} represents the LED turn on voltage, i.e., the amount of voltage required for the LED to enter into conduction region. Thus, the transmitted signal in case of DST-DCO-OFDM system after the addition of DC bias can be represented as

$$\underset{Unipolar}{x(n)} = \underset{DCO}{x(n)} = x_n + \beta_{DC} \tag{3.85}$$

Where $\underset{DCO}{x(n)}$ in (3.85) denotes the transmitted signal in DST-DCO-OFDM system. In case of DST-ACO-OFDM system, the time-domain signal can be mathematically formulated as: since only odd subcarriers are modulated; therefore, out of N number of subcarriers only $\frac{N}{2}$ are meant for data transmission. Hence, by making use of (3.83), the expression for time-domain signal can be derived to attain [495]

$$\begin{aligned} x_{n+\frac{N}{2}} = &\frac{2}{N} \sum_{k=0}^{\frac{N}{2}-1} X_k sin\left[\frac{\pi}{N}\left(n+\frac{1}{2}\right)\left(k+\frac{1}{2}\right)\right] cos\left[\frac{\pi}{2}\left(k+\frac{1}{2}\right)\right] + \\ &\frac{2}{N} \sum_{k=0}^{\frac{N}{2}-1} X_k cos\left[\frac{\pi}{N}\left(n+\frac{1}{2}\right)\left(k+\frac{1}{2}\right)\right] sin\left[\frac{\pi}{2}\left(k+\frac{1}{2}\right)\right] \\ &\textit{where k is odd} \end{aligned} \tag{3.86}$$

Finally, only the positive part of the time-domain signal is transmitted and can be expressed as

$$\underset{Unipolar}{x(n)} = \underset{ACO}{x(n)} = \begin{cases} x_{n+\frac{N}{2}} & if\ x_{n+\frac{N}{2}} > 0 \\ 0 & if\ x_{n+\frac{N}{2}} \leq 0 \end{cases} \tag{3.87}$$

The obtained time-domain signals pertaining to DST-DCO-OFDM and DST-ACO-OFDM systems as shown by (3.85) and (3.87) are propagated through the channel, and at the receiving side, a PD is employed to convert the light signal to electrical form, and then inverse operations like removal of cyclic prefix, N-point DST, and then demapping are performed to transmit the data to the desired destination. In particular, the obtained real and positive-valued time-domain signals for both the scenarios, i.e., DST-DCO-OFDM and DST-ACO-OFDM as represented by (3.85) and (3.87), are passed through the channel comprising of channel impulse response $h(n)$. Furthermore, the received electrical time-domain signal can be expressed as:

$$y(n) = R \underset{Unipolar}{x(n)} \circledast h(n) + w(n) \tag{3.88}$$

(3.88) can be solved to attain

$$y(n) = \sum_{l=0}^{L-1} \underset{Unipolar}{x(n-l)} h(l) + w(n) \tag{3.89}$$

where L represents the total number of paths, and in general for a VLC system, the channel environment is modeled taking into consideration both the LOS and NLOS scenarios. The details of channel modeling are given in Chapter 2. While dealing with DST-DCO-OFDM, all of the subcarriers are utilized and the corresponding frequency domain signal is passed through the BPSK/M-PAM demapping blocks for the purpose of recovering the data signal. Whereas, in case of DST-ACO-OFDM system, since only odd subcarriers are modulated, the data is extracted corresponding to the odd subcarrier positions.

The performance of DST-based DCO-OFDM and DST-based ACO-OFDM system in terms of BER and SNR is illustrated in Fig. 3.32. The performance of the system is evaluated by taking into consideration the M-ary PAM. While comparing the BER performance of DST-based DCO-OFDM system with that of the DST-ACO-OFDM system, the amount of SNR required to achieve a reduced probability of error in case of DST-DCO-OFDM system is more than that of the requirement in SNR in DST-ACO-OFDM system. This is valid because the addition of DC-bias in DST-DCO-OFDM system results in a significant amount of increase in the levels of requirement of SNR. While, in case of the DST-ACO-OFDM system, even though DC bias is not required to achieve a positive signal, the clipping noise dominates at higher constellation sizes, thereby leading to the difficulty to reach a reduced error floor. It is remarkable to note that DST being a simple real transformation technique employs simple real mapping schemes like BPSK and M-PAM. Upon comparison with Hermitian Symmetry imposed FFT in OOFDM, to achieve the same amount of performance, Hermitian Symmetry imposed FFT-based DCO-OFDM and ACO-OFDM system requires $4, 16, 64, 256$ and 1024 QAM in contrast to $2, 4, 8, 16$ and 32 PAM as required in DST-DCO-OFDM/DST-ACO-OFDM system. Therefore, this confirms that DST supports double the constellation symbols to achieve the same amount of performance upon comparison with Hermitian Symmetry imposed FFT. This is confirmed through the simulated result as shown in Fig. 3.34. Thus, to achieve

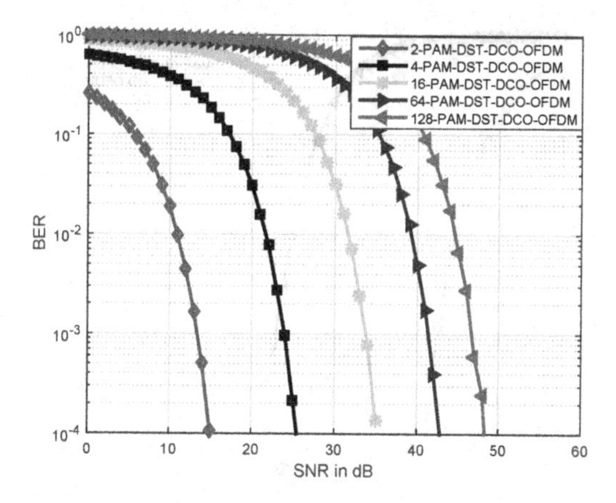

Figure 3.32: Performance analysis of DST-based DCO-OFDM (DST-DCO-OFDM) for VLC [493, 495].

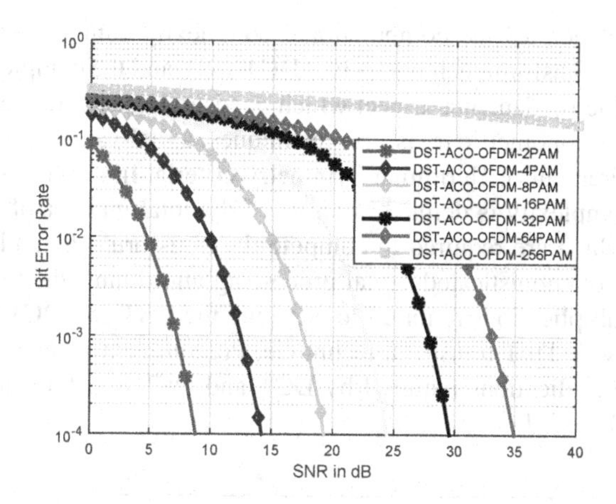

Figure 3.33: Performance analysis of DST-based ACO-OFDM (DST-ACO-OFDM) for VLC [493, 495].

the same error floor, DST-DCO-OFDM system requires lower order modulation format like 2-PAM, whereas DFT-DCO-OFDM system requires 4-QAM. Consequently, this analysis strongly affirms the fact that the spectral efficiency of DST-based optical OFDM system is superior to that of the DFT-based optical OFDM system.

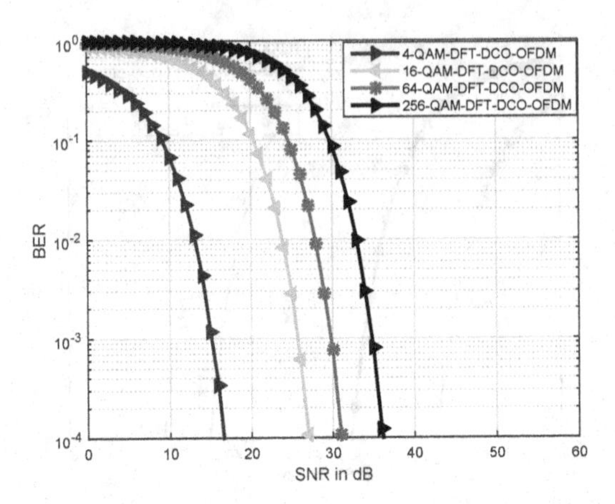

Figure 3.34: Performance analysis of DFT-based DCO-OFDM/conventional optical OFDM (DFT-DCO-OFDM) using M-QAM [495].

Computational complexity analysis of DCT and DST-based optical OFDM systems

This section elaborates the computational complexity comparison between the real trigonometric transform techniques like DCT/DST and the complex Fourier signal processing. The fundamental aspect which needs to be taken into account is that while exploiting trigonometric transform techniques, a real-valued signal is attained without the requirement of Hermitian Symmetry criteria. In general, the amount of computational complexity is defined in terms of the total number of additions and multiplications which are required to compute the transform. Particularly, with the rapid availability of sophisticated signal processing algorithms, the total number of additions and multiplications required to compute the DST and DCT are quite less when compared with DFT transform technique. Predominantly, the total number of additions and multiplications required by DCT and DFT transform techniques are summarized in Table 3.4.

Table 3.4
Comparison of computational complexity [500].

Transform	Additions	Multiplications
DCT	$\frac{(3Nlog_2N-2N+2)}{2}$	$\frac{Nlog_2N}{2}$
DFT	$3Nlog_2N-3N+4$	$Nlog_2N-3N+4$

Upon viewing the table, it is clear that DFT requires $\frac{(3Nlog_2N)}{2}-2N+3$ more additions and $\frac{(Nlog_2N)}{2}-3N+4$ more multiplications than that of the DCT. From the renowned work as reported in [439], it can be deduced that the availability of recursive algorithms for DST-IV transform technique drastically reduced the total number of real additions and multiplications, which can be ably called flop count. The fast algorithms computed in the work as portrayed in [439] make use of the FFT algorithm which reduced the operation count of discrete Fourier transform (DFT) of size N where the flop count was reduced from $4Nlog_2N$ to [243]

$$\frac{34}{9}Nlog_2N - \frac{124}{27}N - 2logN - \frac{2}{9}(-1)^{logN}logN + \frac{16}{27}(-1)^{logN} + 8 \quad (3.90)$$

Using the FFT as computed in [243], the number of flops (i.e., the total number of real additions and multiplications) for DST-IV in [439] were reduced from $2Nlog_2N+N$ to

$$\frac{17}{9}Nlog_2N + \frac{31}{37}N + \frac{2}{9}(-1)^{log_2N}log_2N - \frac{4}{27}(-1)^{log_2N} \quad (3.91)$$

Finally, it can be inferred that by using the improved FFT algorithm where the flop count reduced from $\sim 4Nlog_2N$ to $\sim \frac{34}{9}Nlog_2N$ (i.e., the algorithm where the total number of additions and multiplications are reduced to a significant amount), the earlier works report that new recursive algorithms can be derived for DST-IV to substantially reduce the total number of additions and multiplications for a transform of size $N = 2^m$ from $\sim 2Nlog_2N+N$ to $\sim \frac{17}{9}Nlog_2N$.

Table 3.5

Comparison between DCT/DST and DFT-based optical OFDM systems.

Type of OFDM	DCT/DST-ACO-OFDM and DCT/DST-DCO-OFDM	DFT-ACO-OFDM and DFT-DCO-OFDM
Hermitian Symmetry	Not required	Required
Constellation	Real (BPSK, M-PAM)	Complex (M-QAM)
Constellation Size	M=L	M=L^2
Self-Inverse Property	Self-Inversive	Not Self-Inversive
Subcarriers Utilized	N/2 for DCT/DST-ACO-OFDM N for DCT/DST-DCO-OFDM	N/4 for DCT/DST-ACO-OFDM N/2 for DCT/DST-DCO-OFDM

Comparisons between DFT-based optical OFDM and DCT/DST-based optical OFDM system

In case of DFT-based DCO-OFDM system, in order to generate a real valued signal, the huge stream of input data which is mapped by employing complex constellations like M-QAM is compelled to satisfy Hermitian Symmetry criteria before getting applied to the IFFT block. However, in doing so, only half of the subcarriers, i.e., IFFT points, are utilized for data transmission, while the remaining half are flipped complex conjugate versions of the previous ones. Thus, there is a decrease in throughput. Therefore, the bandwidth efficiency (BE) pertaining to

DFT-DCO-OFDM can be mathematically expressed as [344]

$$BE_{DFT-DCO-OFDM} = \left(\frac{\frac{N}{2}-1}{N+N_G}\right) Blog_2M \ \frac{\left(\frac{bits}{sec}\right)}{Hz} \tag{3.92}$$

Consequently, to address the reduced bandwidth efficiency which is encountered when relying on Hermitian Symmetry imposed IFFT, real transformation techniques like DST can be exploited for DCO-OFDM system. Moreover, the aforementioned mathematical analysis clearly confirms the fact that DST-based DCO-OFDM system involves all of the subcarriers for data transmission. This can be mathematically represented as

$$BE_{DST-DCO-OFDM} = \left(\frac{N-1}{N+N_G}\right) Blog_2M \ \frac{\left(\frac{bits}{sec}\right)}{Hz} \tag{3.93}$$

Meanwhile, in the case of DFT-ACO-OFDM system, due to Hermitian Symmetry criteria, out of N subcarriers, only $\frac{N}{2}$ subcarriers are exploited for the transmission of data and, furthermore, due to the modulation of only odd subcarriers, (since even subcarriers are set to zero) out of $\frac{N}{2}$ subcarriers, only $\frac{N}{4}$ subcarriers are involved for data transmission. Meanwhile, DST-ACO-OFDM system does not rely on Hermitian Symmetry criteria for yielding a real valued signal. Therefore, out of N subcarriers, $\frac{N}{2}$ odd subcarriers are employed for the transmission of data. Mathematically, this can be expressed as [344]

$$BE_{DFT-ACO-OFDM} = \left(\frac{\frac{N}{4}-1}{N+N_G}\right) Blog_2M \ \frac{\left(\frac{bits}{sec}\right)}{Hz} \tag{3.94}$$

$$BE_{DST-ACO-OFDM} = \left(\frac{\frac{N}{2}-1}{N+N_G}\right) Blog_2M \ \frac{\left(\frac{bits}{sec}\right)}{Hz} \tag{3.95}$$

From (3.92), (3.93), (3.94) and (3.95), it can be deduced that, DST-based optical OFDM system achieves double the spectral efficiency when compared with traditional DFT-based optical OFDM system. The parameters N, N_G, B, M in (3.92), (3.93), (3.94) and (3.95) denote the total number of subcarriers, guard interval, the channel bandwidth and the constellation size, respectively. Table 3.5 gives the comparison between DST-based optical OFDM (DST-ACO-OFDM, DST-DCO-OFDM) and DFT-based optical OFDM (DFT-ACO-OFDM, DFT-DCO-OFDM).

The computation complexity involved behind the two transform techniques, namely, DCT and DFT, is exemplified in Fig. 3.35. This figure clearly manifests that the computational complexity which is generally calculated in terms of total number of additions and multiplications of DCT transform is quite less than that of the DFT transform technique and this is mathematically confirmed by the analytical expressions as described in Table 3.4. Therefore, DCT transform technique can be envisaged to be exploited for the cost-effective realization of IM/DD systems for VLC.

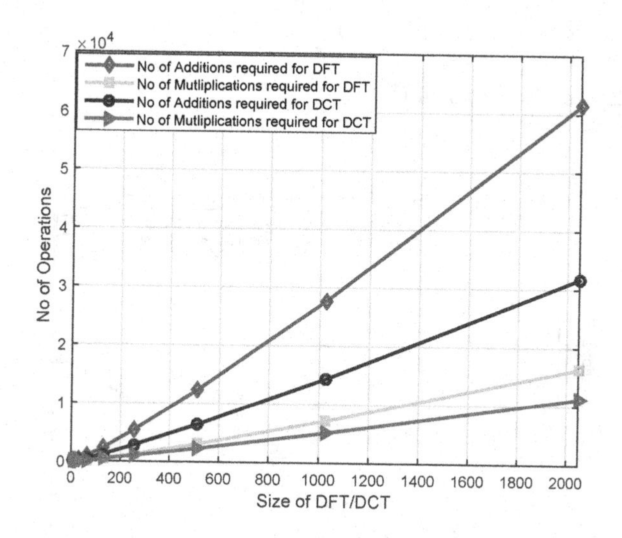

Figure 3.35: Comparison of computational complexity of DCT/DFT transformation techniques.

Listing 3.7: MATLAB code to depict the computational complexity of DCT/DFT transform for the simulated result shown in Fig. 3.35

```
1  clc;
2  clear all;
3  close all;
4   N=[8 16 32 64 128 256 512 1024 2048]; %Size of the
       Transform
5  x=3*N.*log2(N)-3*N+4; %Total Number of Additions
      required by DFT
6  x1=N.*log2(N)-3*N+4; %Total Number of Multiplications
       required by DFT
7  x12=(3*N.*log2(N)-2*N+2)/2; %Total Number of
      Additions required by DCT
8  x11=(N.*log2(N))/2; %Total Number of Multiplications
      required by DCT
9  plot(N,x,'rd-','LineWidth', 2);hold on;
10 plot(N,x1,'gs-','LineWidth', 2);
11 plot(N,x12, 'bo-','LineWidth', 2);
12 plot(N,x11, 'm>-','LineWidth', 2);
13 legend('No of Additions required for DFT', 'No of
      Mutliplications required for DFT', 'No of
      Additions required for DCT', 'No of
      Mutliplications required for DCT');
14 xlabel('Size of DFT/DCT');
```

```
15 ylabel('No of Operations');
16 grid on;
17 axis([0 2048 0 7*10^4])
```

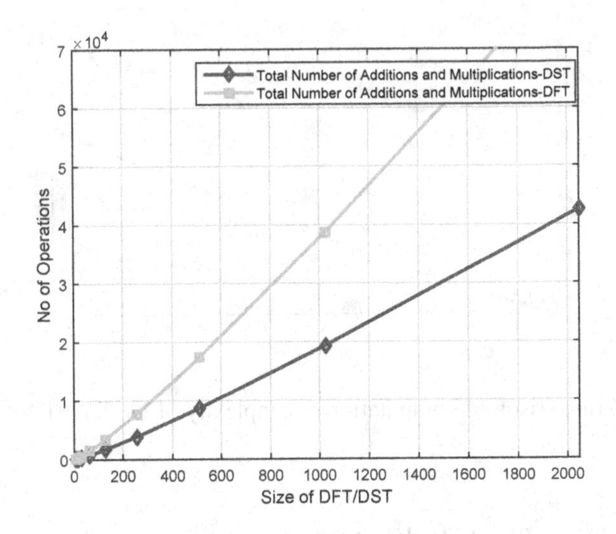

Figure 3.36: Interpretation of computational complexity analysis behind DFT and DST transform techniques.

The computational complexity analysis of DFT and DST transform techniques is clearly elucidated in Fig. 3.36. As stated earlier, due to the availability of sophisticated fast algorithms, the operation count which is also referred to as flop count (i.e., the total number of additions and multiplications together) of DFT algorithm and DST algorithm, is drastically reduced and this is verified through mathematical expressions as enumerated by (3.90) and (3.91). This figure confirms the fact that DST requires less flop count when compared with that of the DFT transform technique.

Listing 3.8: MATLAB code to depict the computational complexity of DST/DFT transform for the simulated result shown in Fig. 3.36.

```
1 clc;
2 clear all;
3 close all;
4  N=[8 16 32 64 128 256 512 1024 2048]; %Size of the
      transform
5  x1=17/9* N.*log2(N); %Total number of Additions and
      Multiplications required by DST
6 x=34/9* N.*log2(N); %Total number of Additions and
     Multiplications required by DFT
7 plot(N,x1,'rd-','LineWidth', 2);hold on;
8 plot(N,x,'gs-','LineWidth', 2);
```

```
legend(3, 'Total Number of Additions and
    Multiplications-DST', 'Total Number of Additions
    and Multiplications-DFT');
xlabel('Size of DFT/DST');
ylabel('No of Operations')
grid on;
axis([0 2048 0 7*10^4])
```

3.2.4.4 Performance analysis of Hadamard Coded Modulation (HCM)-based optical OFDM

The most vital problem encountered with OFDM is that the transmitted signals have high peaks, i.e., high PAPR which imposes a serious signal distortion at the output of the non-linear channels. The authors in [377] proposed a supplementary to OFDM which is based on Hadamard transform that retains several advantages of OFDM and is notably more resilient against non-linearity. Predominantly, Hadamard matrices and Hadamard transform are prominent tools to be employed for communication systems. Particularly, very significant works in [14, 31, 534] report that Hadamard matrices/transform have been proposed as a precoder technique to reduce the amount of PAPR in OFDM systems. Particularly, Hadamard transform is an orthogonal linear transform which is implemented by a butterfly structure in the FFT process and, henceforth, it will not increase the overall computational complexity of the system. Earlier work pertaining to the utilization of Hadamard transform for VLC reports that it has been exploited not as a precoder for the sake of reducing the PAPR, but as a modulation technique for encoding and transmitting the information. In [377], the authors have introduced this multilevel modulation technique called HCM which uses Hadamard matrices as a modulation technique rather than a precoder. This sort of modulation can be realized by using the fast Walsh-Hadamard transform (FWHT), which has the same complexity as that of FFT used in OFDM, i.e., $Nlog_2N$, where N is the size of the Hadamard matrix channel estimation using HCM which is based on FWHT is shown in Fig. 3.37. Similar to DHT, FWHT being a real transform it works with real mapping signals such as BPSK and M-PAM. HCM employs FWHT at the transmitter side to modulate the data, and IFWHT is used at the receiving end to decode the received data.

Followed by pilot insertion, the components of X, i.e., $X_0, X_1, X_2 \cdots X_{N-1}$ are M-PAM modulated and the transmitted signal can be considered as a vector x and is given as [377]

$$x = (XH_N + (1-X)\bar{H}_N) \tag{3.96}$$

Where H_N in (3.96) represents a binary Hadamard matrix of order N which is obtained by replacing -1 by 0 in the original $\{-1,1\}$ Hadamard matrix and $\bar{H}_N$ denotes the complement of H_N. Equation (3.96) can be further expressed as

$$x = X(H_N - \bar{H}_N) + \frac{N}{2}[0,1,1,\cdots 1] \tag{3.97}$$

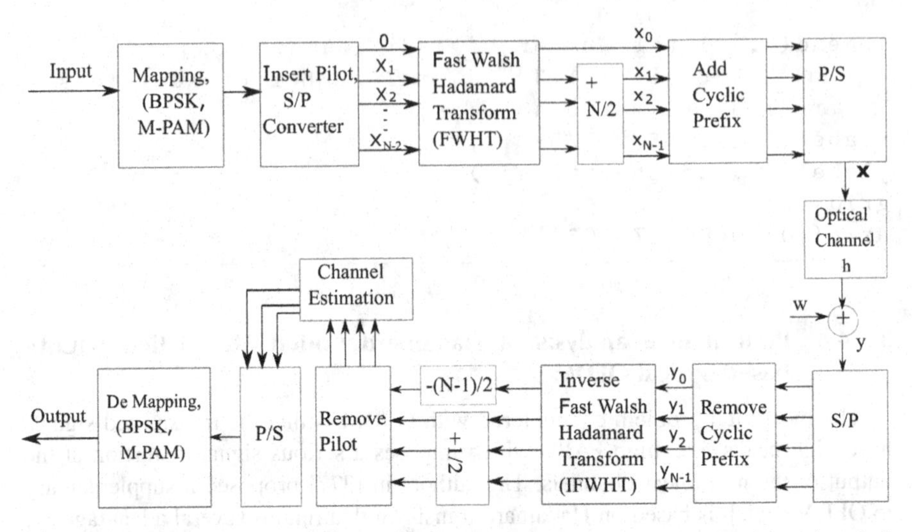

Figure 3.37: Schematic representation of channel estimation using Hadamard coded modulation (HCM) over VLC channel [498].

The matrix $(H_N - \bar{H}_N)$ as evidenced by equation (3.97) denotes the bipolar Hadamard matrix, and hence the first term in (3.97) represents the Walsh-Hadamard transform of the data vector X, while the second term represents the product of a vector of $1 \times N$ all ones with $\bar{H}_N$ [377]. Therefore, the transmitter block can be replaced with fast Walsh Hadamard Transform (FWHT) as illustrated in Fig. 3.37. In brief, at the transmitter side, Hadamard transform using FWHT of size N is applied on the data stream and then a constant of $\frac{N}{2}$ is added to $N-1$ elements to generate the transmitted signal vector x. In this way, a real and unipolar signal is attained fulfilling the requirements of IM/DD systems. At the receiving end, inverse FWHT (IFWHT) is incorporated instead of FFT. The rest of the analysis is similar to the aforesaid system models. Figure 3.38 compares the performance of the two channel estimation algorithms LS and MMSE in HCM system operating over VLC channel environment. From the figure, it is evident that higher orders of modulation require higher amount of SNRs to attain a reduced error floor. Among the channel estimation algorithms MMSE outperforms LS much similar to DHT-based ACO-OFDM system.

3.2.4.5 Wavelet packet division multiplexing (WPDM)-based VLC system

Much similar to other real trigonometric transforms-based optical OFDM system, WPDM exploits inverse discrete wavelet packet transform (IDWPT) instead of the Hermitian Symmetry imposed IFFT transform at the transmitting end; while at the receiving end, discrete wavelet packet transform (DWPT) is exploited. In this manner, OFDM modulation and demodulation is performed by exploiting IDWPT and DWPT at both the transmitting and receiving end. Thus, in this manner, WPDM

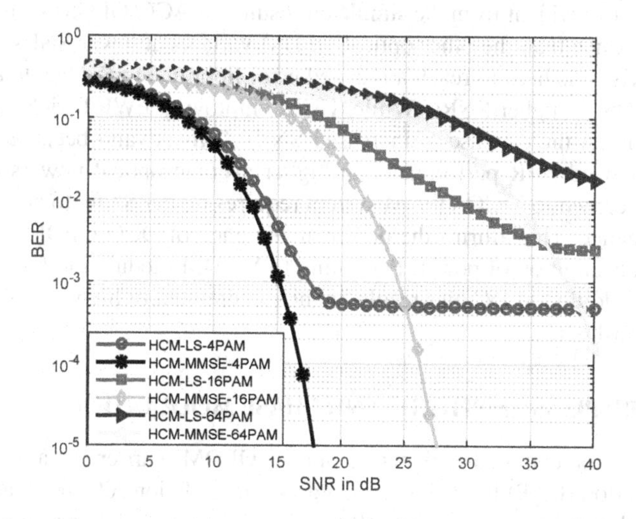

Figure 3.38: Comparison of LS and MMSE channel estimation in HCM system over dispersive VLC channel [498].

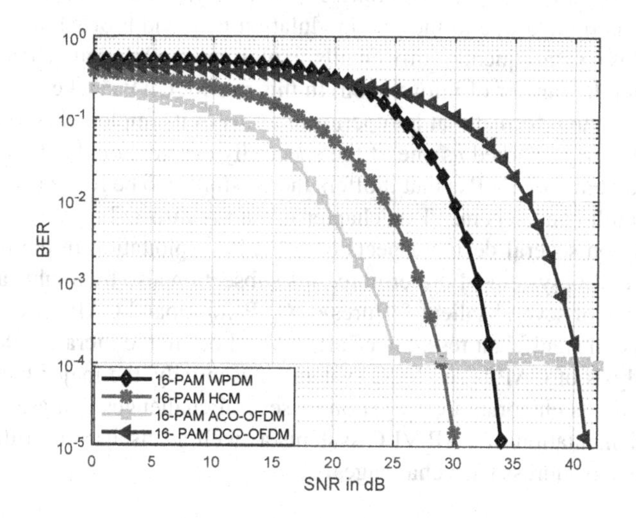

Figure 3.39: BER performance of ACO-OFDM, DCO-OFDM, WPDM and HCM over dispersive VLC channel [498].

relieves the burden of extra signal processing and dependency of extra hardware which is required in Hermitian Symmetry imposed IFFT-based optical OFDM. The Figure 3.39 gives the overall performance comparison of different optical OFDM variants namely ACO-OFDM, DCO-OFDM, WPDM and HCM over VLC channel environment. It is evident from the simulated result that ACO-OFDM system gives a better performance than the other optical OFDM variants at lower SNRs. However, at higher SNRs, it is hard to reach a reduced error floor due to the fact that clipping noise dominates at higher SNRs. While, HCM employing FWHT yields a reduced error floor only at the expense of increased SNRs. This is valid because HCM exhibits an improved BER performance at higher average optical powers due to its low PAPR. Meanwhile, DCO-OFDM system requires higher SNRs to obtain a better BER performance. This affirms the power inefficiency of DCO-OFDM system due to its huge consumption of power. Pertaining to WPDM modulation format, the requirement of addition of DC bias results in the requirement of higher SNRs to attain a reduced error floor.

3.2.5 CARRIERLESS AMPLITUDE AND PHASE MODULATION

Like the aforementioned variants of optical OFDM, carrierless amplitude and phase modulation (CAP) has been proposed as a modulation scheme to address the limited modulation bandwidth of the VLC systems. It is a widely accepted fact that VLC leverages on the illuminating devices like LEDs to render seamless data communication in addition to the illumination support, thus making it a suitable candidate to be deployed in many applications. Prodigiously, the achievable data rate of VLC systems is limited due to the low modulation bandwidth of white fluorescent LEDs (WLEDs). Consequently, this entails exploiting modulation schemes to ensure high data rate transfer of information. In this regard, CAP can be considered as a suitable modulation format that is superior in terms of its implementation simplicity and can also be remarked as one of the spectrally efficient modulation schemes. The renowned nature of CAP is that it offers the feasibility to be realized as a single-band or as a multiband scheme. This chapter is mainly directed toward focusing an in-depth study on several design aspects of CAP, its exploitation in the realization of LED-based VLC systems. Furthermore, this subsection also highlights a comprehensive analysis of several salient features of CAP, multiband CAP (m-CAP) along with the stress on significant research efforts carried out in the literature depicting a thorough analysis of CAP-based VLC systems. Essentially, the major focus of this subsection is on its elaborate investigation of several challenges that are associated with the implementation of CAP-VLC systems as well as discussion of the mitigation techniques to address such challenges.

3.2.5.1 Principle aspects of realization of CAP-VLC systems

In addition to the different variants of optical OFDM, CAP can be regarded as the widely investigated modulation format that is applicable for VLC applications due to the combination of its simplicity in terms of its ease of implementation as well as its spectrally efficient nature. Going into the details of CAP, earlier studies reveal that CAP has been adopted as a standard by asynchronous transfer mode (ATM), ATM forum [176]. CAP was adopted as an early candidate for facilitating signal transmission over the telephone cables broadly meant for asymmetric digital

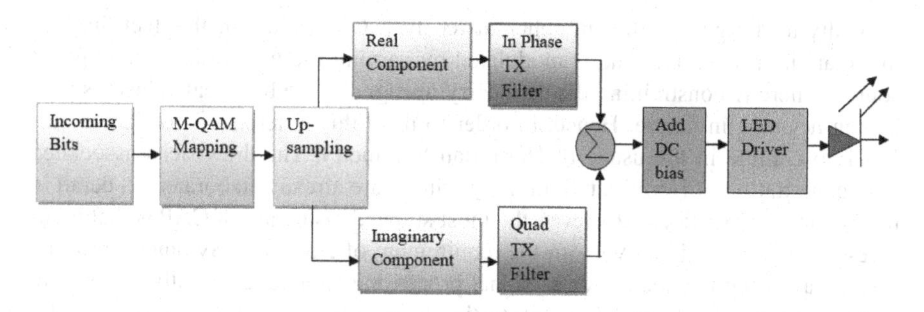

Figure 3.40: Schematic representation of CAP transmitter.

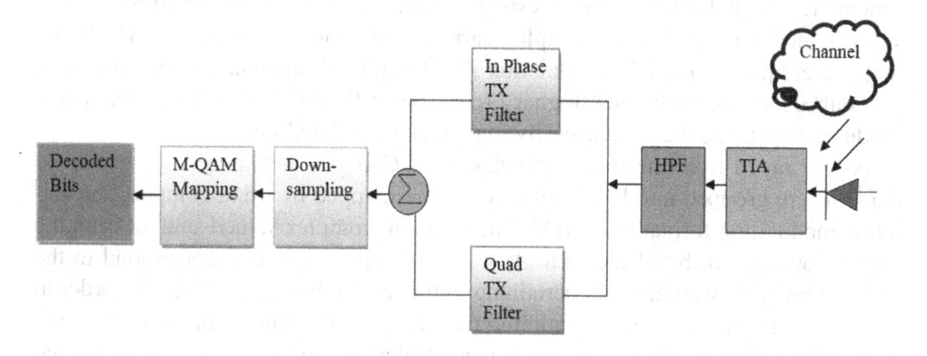

Figure 3.41: Schematic representation of CAP receiver.

subscriber lines (ADSL) [176]. However, the earlier studies also reveal the fact that CAP modulation has been interrupted for a short while since it is imperative to confide on complex equalization resources particularly at high throughputs. Presently, CAP modulation is relishing a resurrection in VLC due to its exceptional characteristic features that paved the way for its utilization in VLC systems. Amidst its several distinctive properties, a remarkable one is its low PAPR. CAP modulation scheme being a single carrier modulation format exhibits a lower PAPR than that of the DMT where high PAPR stems out as one of the major aspects. Typically, this low PAPR is well suited for the cost-effective realization of VLC systems; this is because the transmitter front-end elements in VLC impose a peak power constraint, where the signals with highest peaks are clipped off to ensure that the optical sources operate within their linear regions. However, this process results in a significant amount of clipping distortion, thereby deteriorating the overall system performance.

It is widely known that IM and DD is used as the most viable transceiver approach due to the fact that LEDs being non-coherent light sources enforces a huge difficulty to design an efficient coherent receiver. On account of this fact, in such modulation process, the data is modulated by varying the intensities of the optical sources, thereby constraining the data-carrying signal to be both real valued as well as non-negative in nature. Hence, in order to meet this criteria, it is obligatory for DMT to confide in the usage of Hermitian Symmetry. The drawbacks associated with exploitation of Hermitian Symmetry criteria are already elaborated on detail in the previous subsections. However, the most stupendous nature of CAP is that it ensures a real valued signal without the requirement of Hermitian Symmetry criteria, thereby avoiding the need of extra signal processing techniques usually required in DMT systems. Moreover, it also interesting to note that a CAP transceiver is relatively simple and easy to implement. The CAP transmitter and receiver are delineated in Figs. 3.40 and 3.41, respectively. As exemplified from the transmitter and receiver schematics, digital finite impulse response (FIR) filters were employed to circumvent the necessity to rely on complex carrier modulation and recovery which is a major prerequisite in QAM modulation [7]. Thus, CAP modulation format can be pronounced as the multilevel, higher order, spectrally efficient and uncomplicated modulation scheme that can be easily implemented in VLC systems.

As portrayed from the transceiver design of CAP, the incoming huge stream of data bits are grouped into blocks of bits that are mapped by using complex constellation modulation format like QAM. The resultant complex valued symbol from the mapper output will be of the form $A_i = u_i + v_i$, where u_i and v_i correspond to the real and imaginary component pertaining to the ith symbol, respectively. In order to match with the overall system sampling frequency f_s, the outputs of the mapper are appropriately up-sampled. Thereupon, these real and imaginary components are fed to the in-phase and quadrature pulse shaping filters which are denoted as $p(t)$ and $\tilde{p}(t)$. In general, these digital pulse shaping filters are realized as a product of root raised cosine filter (RRC) comprising of sine and cosine function. Furthermore, it is also interesting to note that these pulse shaping digital filters are orthogonal to each other and form a Hilbert transform pair, where the filters possess same amplitude response but differ in the phase by 90°. The impulse responses of the in-phase and quadrature pulse shaping filter are attained by taking into consideration the product of the sine and cosine with the RRF filter, respectively. For the purpose of illustration, the in-phase and quadrature pulse shaping filters are shown in Fig. 3.42. Then, the outputs of the pulse shaping filters are summed up and a sufficient amount of DC bias is added to ensure that a positive valued signal is attained that is more appropriate for intensity modulation. Thus, the resultant non-negative valued signal is used to intensity modulate the LED to allow for the onward transmission through the VLC channel. Thus, the resultant transmitted signal can be interpreted as

$$s(t) = \zeta(x(t) + \beta_{DC}) \tag{3.98}$$

From (3.98), the parameters ζ, $x(t)$ and β_{DC} specify the electrical-to-optical conversion coefficient, the added DC bias and the transmitted electrical CAP signal,

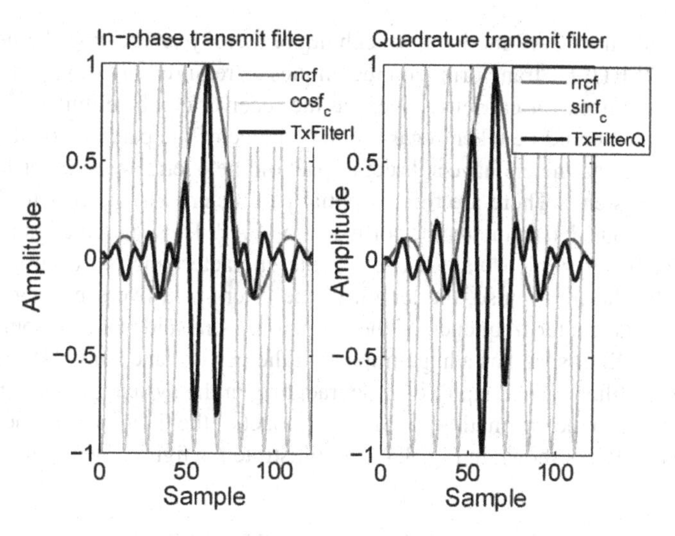

Figure 3.42: Impulse response of in-phase and quadrature pulse shaping filters for CAP modulation [17].

respectively. Also, $x(t)$ can be further expressed as

$$x(t) = \sum_{i=-\infty}^{\infty} \left[u_i p(t-iT) - v_i \tilde{p}(t-iT)\right] \tag{3.99}$$

Where the digital pulse shaping filters are given by the following expressions

$$p(t) = g(t)\, cos(2\pi f_c t) \tag{3.100}$$

and

$$\tilde{p}(t) = g(t)\, sin(2\pi f_c t) \tag{3.101}$$

$g(t)$ in (3.100) and (3.101) corresponds to the RRC, while the duration of the symbol is represented by T, and f_c signifies the orthogonal sinusoids frequency which must satisfy the following condition

$$f_c \geq \frac{(1+\alpha)}{2T} \tag{3.102}$$

It is imperative that the pulse shaping filters that are employed for CAP modulation technique are designed in the digital domain. The underlying fact behind the choice of digital filters is that the problems encountered due to electronic component drift due to the temperature and aging can be overcome, which enables the efficient reproduction of the characteristics of the spectrum without any variations as well as prevents suffering from component tolerance issues. Moreover, these pulse shaping filters are designed as FIR filters which evolve as a prime advantage for the easier implementation of CAP modulation technique, thereby eradicating the necessity to depend on IFFT/FFT transformation techniques as required in QAM and DMT. Additionally, the need for carrier recovery at the receiver can be eliminated. However, immense care should be taken regarding the choice of the parameters of RRC filter because these parameters have a direct impact on the overall system performance of CAP-VLC systems. There are four essential parameters associated with RRC filter whose variations determine the performance of CAP. These parameters include: the roll-off factor (α) which plays a crucial role characterizing the excess bandwidth occupied by the filter pulse, the length of the filter span, oversampling rate of the system and the center frequency of the orthogonal sinusoids (f_c). Among these parameters, the filter span has a huge impact on the performance of CAP; it is because for a limited filter span, a significant degradation in the system performance can be witnessed due to the origination of ISI. The impact of the filter span on the frequency response of RRC is shown in Fig. 3.43. As illustrated in figure, for a span of 50, satis-

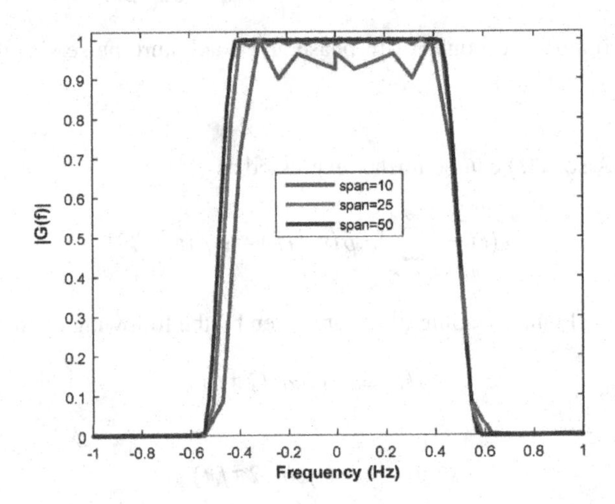

Figure 3.43: Illustration on the impact of filter span on the frequency response of the filter [17].

factory performance of the filter is achieved because, with the increase in the span of the filter, the magnitude response $G(f)$ becomes flat over its spectrum. However, it is desirable to choose the span of the filter by taking into consideration the trade-off between the overall system performance and computational complexity.

At the receiver side as delineated in Fig. 3.41, the transmitted CAP signal from the optical domain is converted into electrical domain by means of a photodiode, whereupon the signal is converted into a voltage signal by means of a transimpedance amplifier (TIA). In general, the high pass filter (HPF) is employed to suppress the DC component of the recovered electrical signal, and then the signal is fed to the matched filters. Therefore, upon traversing through the optical channel, the received

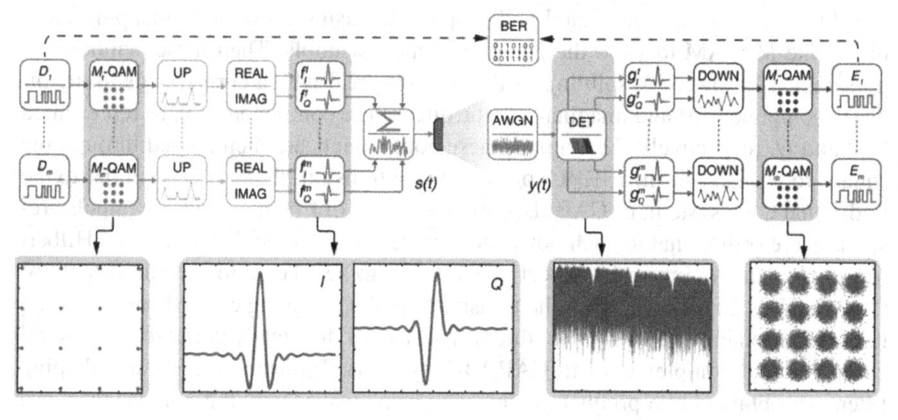

Figure 3.44: Schematic representation of block diagram of m-CAP VLC system [116].

electrical signal can be formulated as

$$y(t) = RP_t h(t) * x(t) + z(t) \tag{3.103}$$

where R represents the responsivity of the photodiode, P_t, $h(t)$ and $z(t)$ imply the total transmitted power, the channel path loss and AWGN with zero mean and double-sided power spectral density $\frac{N_0}{2}$. The matched filters should be designed in such a manner that the overall SNR of the system is maximized, which implies that the matched filters constitute the conjugated and time reversed versions of the transmit pulse shaping filters. As evidenced from the figure, at the final stage, the output of the matched filters is applied as input to the QAM demapper in order to yield the final estimate of the transmitted symbols.

3.2.5.2 Multiband CAP (m-CAP)-based VLC systems

Even though it can be reported from earlier literature findings that a CAP-based VLC system outperforms OFDM-based VLC system in terms of assurance of high transmission speeds, but the necessity of flat frequency response limits the performance of CAP-VLC system over a non-flat frequency response channel. In order to meet this challenging aspect, the concept of multicarrier CAP ably known as m-CAP evolved out. The fundamental principle behind the implementation of m-CAP is that it allows for splitting of the total available bandwidth into m sub-bands, thereby relaxing the requirement of flat-band response as well as facilitating for the exploitation of adaptive bit and power-loading algorithms to allow for the accommodation of a large number of bits per symbol to every individual sub-band or subcarriers. The concept of m-CAP for the first time was implemented for optical fiber links. The schematic representation of m-CAP VLC system is depicted in Fig. 3.44. As delineated in the figure, the independent streams of data as denoted by D_n, where

$n = \{1, 2, 3, \cdots m\}$ are generated and mapped by using a complex mapping technique like M-QAM to yield the complex valued symbols. Then these symbols are up-sampled followed by splitting of the data into real which corresponds to the in-phase component (I) and imaginary or the quadrature component (Q), as represented by I and Q, respectively. Thereupon, these components are then passed through the square-root raised cosine (SRRC) pulse shaping transmit filters. Much similar to the traditional CAP system, m-CAP also employs a pair of FIR filters whose impulse responses are orthogonal to each other, thereby resulting in the formation of a Hilbert transform pair. An important aspect which needs to be taken into account is that, m-CAP requires $2m$ filters at both the transmitting and receiving ends, thereby leading to the requirement of a total of $4m$ filters. This aspect drastically increases the overall computational complexity of m-CAP VLC systems. Typically, these pulse shaping filters are obtained as a product of sine and cosine waves with the impulse response of the SRRC filter pertaining to the in-phase and quadrature filters. In general, the impulse responses corresponding to the in-phase and the quadrature filters can be expressed as [116, 196, 199]:

$$f_I^m[t] = \left[\frac{sin\left[\gamma(1-\beta)\right] + 4\beta \frac{t}{T_s} cos\left[\gamma\delta\right]}{\gamma\left[1-\left(4\beta\frac{t}{T_s}\right)^2\right]} \right] \times cos\left[\gamma(2m-1)\gamma\right] \tag{3.104}$$

$$f_Q^m[t] = \left[\frac{sin\left[\gamma(1-\beta)\right] + 4\beta \frac{t}{T_s} cos\left[\gamma\delta\right]}{\gamma\left[1-\left(4\beta\frac{t}{T_s}\right)^2\right]} \right] \times sin\left[\gamma(2m-1)\gamma\right] \tag{3.105}$$

From the mathematical expressions as emphasized by (3.104) and (3.105), the parameters β, T_s represent the roll-off factor of the SRRC filter and the duration of the symbol, respectively. The roll-off factor determines the bandwidth requirement and β must be within the range $0 \leq \beta \leq 1$, while $\gamma = 1 + \beta$. Particularly, lower values of β are desired to enhance the spectral efficiency of m-CAP-VLC systems. Finally, the transmitted signal is summed up to yield the signal $s(t)$ which can be put up as follows

$$s(t) = \sqrt{2} \sum_{n=1}^{m} \left(s_I^n(t) \otimes f_I^n(t) - s_Q^n(t) \otimes f_Q^n(t) \right) \tag{3.106}$$

The parameters $s_I^n(t)$ and $s_Q^n(t)$ represent the in-phase and the quadrature M-QAM mapped symbols pertaining to the nth subcarrier. The transmitted signal $s(t)$ is then intensity modulated through the LED. At the receiving end, some amount of AWGN is added to the transmitted signal, and then it is passed through the detection block which is denoted as DET in the block diagram. The corresponding received signal can be expressed as

$$y(t) = Rs(t) \otimes h(t) + n(t) \tag{3.107}$$

where R represents the responsivity of the photodiode, $h(t)$ specifies the impulse response of the channel, and $n(t)$ gives the AWGN with zero mean and variance of $\frac{N_0}{2}$, respectively. By exploiting the in-phase and quadrature receiver filters, matched filtering operation is performed followed up by downsampling the data. Finally, the data is demapped by exploiting the M-QAM demapping block. Furthermore, in order to determine the BER performance, the received bit sequences were passed through the estimation block, which is represented as E in the schematic block.

3.2.5.3 Related work on m-CAP-VLC systems

The work in [392] clearly illustrates that m-CAP was formerly designed to be exploited for fiber communication systems. The result analysis of this work infers that when compared with the traditional CAP modulation format, m-CAP scheme accomplished an enhancement in spectral efficiency as well as minimized tolerance toward dispersion. Furthermore, in order to verify the superiority of m-CAP over the traditional 1-CAP system, this work reports an experimental demonstration illustrating the performance comparisons of m-CAP and 1-CAP systems operating over a 15 Km standard single mode fiber (SSMF) and exploiting a bandwidth of 25 GHz. In this work, the m-CAP system was split into six bands, where each band occupies a bandwidth of 4.26 GHz, while the traditional 1-CAP system occupies the entire 25 GHz bandwidth. Furthermore, this 6-CAP system offers the flexibility to adapt the subcarriers with a varying number of bits per symbol. The experimental results of this work depict that the m-CAP system was capable to achieve an aggregated data rate of 102.4 Gbps exhibiting its dominance over 1-CAP system that delivers a data rate of 100 Gbps. The research efforts in [566] report that the authors proposed and experimentally demonstrated a novel WDM-CAP-PON system based on single-side band (OSSB) m-CAP modulation format where a multiuser access network was developed ensuring an aggregate data rate of 9.3 Gbps per user. A number of significant works as delineated in [519] and [473] portray the applicability of m-CAP in fiber infrastructures. The prominence of m-CAP as an advanced modulation format in optical communications can be evidenced from [463]. This work demonstrates the functionality of a three-dimensional (3D) m-CAP system in short reach optical communication systems. Furthermore, the superiority of m-CAP system over the conventional CAP system is apparent from the simulation results where the 3D m-CAP system ensures data rate of 112.5 Gbps in short reach optical communication. Persuaded by these stunning results, m-CAP was adopted as an advanced modulation format for VLC as well.

For the first time, the authors in [199] experimentally demonstrate the performance of m-CAP in VLC systems by considering a fixed bandwidth of 6.5 MHz. It is to be noted that six subcarriers or sub-bands were taken into account and $m = \{1,2,4,6,8,10\}$ with baud rates set as $\{6.50,3.25,1.625,1.083,0.8125,0.65\}$, respectively, and this is illustrated in Fig. 3.45. From this work, it can be inferred that in order to maximize the overall system performance, a bit loading algorithm was exploited. It can be evidenced from the simulation result as illustrated in Fig. 3.46, higher throughputs can be achieved only upon increasing the number of

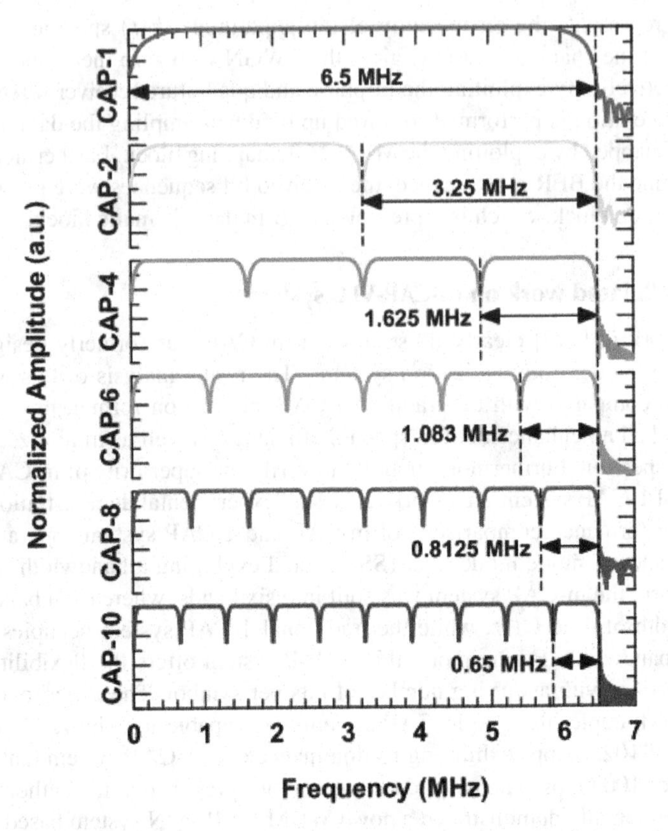

Figure 3.45: Depiction of ideal frequency responses along with the requirement of bandwidth for different orders of m [199].

subcarriers. For instance, in case of 1-CAP system, the achieved throughput and spectral efficiency was 9.04 Mbps and 1.40 b/s/Hz, respectively. Furthermore, when the number of subcarriers were increased from 1 to 10, then the reported throughput and the spectral efficiency were 31.53 Mbps and 4.85 b/s/Hz, respectively. It is also interesting to note that when 10 subcarriers were employed, the achievable BER was 3.8×10^{-3}, thereby ensuring that it is below 7% of FEC limit. Thus, it can be concluded that upon reducing the number of subcarriers, a reduction in throughput as well as the spectral efficiency can be observed due to the wider bandwidths occupied by the subcarriers. Therefore, it can be surmised that it is desirable to exploit more subcarriers to allow for a more accurate flat-band response approximation, thereby preserving the attainment of improved throughput and spectral efficiency. The work in [202] also experimentally demonstrates the performance of m-CAP VLC systems where 8-CAP system illustrates a spectral efficiency of 4.75 b/s/Hz over a realistic distance of 1 m.

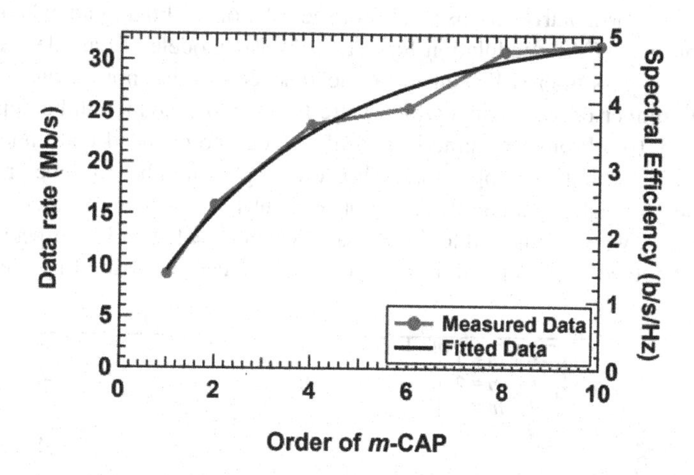

Figure 3.46: Measured data rate and the spectral efficiency for m-CAP system exploiting exponential filtering [199].

The performance analysis of m-CAP system over bandlimited VLC channel can be found in [209]. From this work, it is apparent that realistic system models as well as device models were employed to determine the BER performance of individual subcarriers. The simulated result as emphasized in Fig. 3.47 reveals that the chance for the higher order subcarriers to become vulnerable to the attenuation introduced by the LED is very high. This is clearly illustrated in the simulation result, where the BER performance of 20-CAPVLC system deteriorates for higher order subcarriers. The work in [196] asserts that theoretical investigations were carried out where

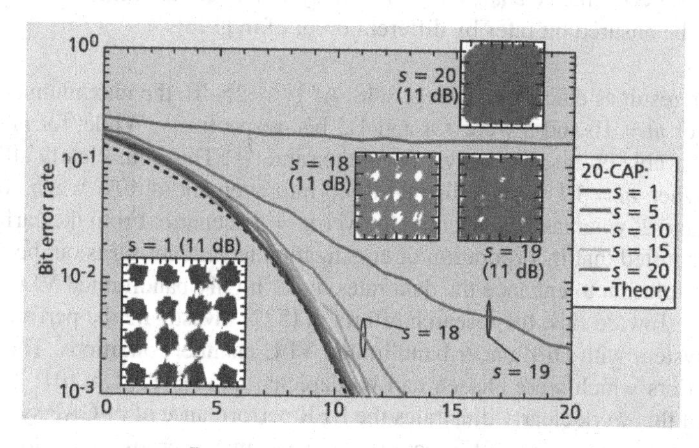

Figure 3.47: BER vs. $\frac{E_b}{N_0}$ performance of 20-CAP modulation format [209].

the authors present numerical results to depict the BER versus SNR performance of m-CAP modulation format pertaining to highly bandwidth limited VLC channel environment. Particularly, in this work, for the sake of facilitating an exhaustive investigation on m-CAP modulation format, LED was modeled as an ideal 1st order analogue low pass filter (LPF) ensuring the absence of other non-linearity aspects. The total number of subcarriers which were taken into account can be depicted as $m = \{1,2,5,10\}$. From the numerical results, it can be deduced that under highly bandwidth constraint, the appropriate choice of higher number of subcarriers, i.e., when $m = 10$ results in a significant improvement in transmission speeds of up to 40% and 17% when compared to the scenario when $m = 1,2$ and 5, respectively for the same targeted BER. The aforementioned aspect can be verified by means of a

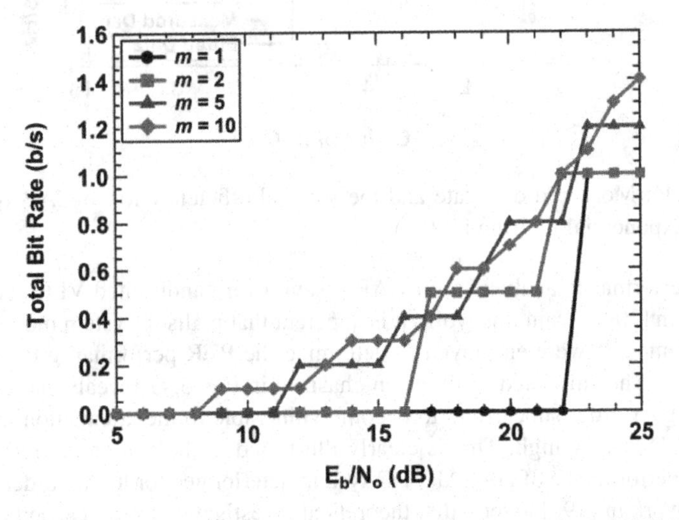

Figure 3.48: Performance analysis of m-CAP system over bandlimited VLC channel depicting the ensured bit rates by different order of m [196].

simulation result as elucidated in Fig. 3.48. At $\frac{E_b}{N_0} = 25$ dB, the maximum attainable bit rates for $m = 10$ and 5 were 1.4 and 1.2 b/s, respectively. While, for $m = 1$ and 2, the achievable bit rate was only 1 b/s at the same $\frac{E_b}{N_0}$. Thus, this clearly affirms the fact that when $m = 10$ offers a distinguished improvement of 40% when compared to $m = 1$ and 2, whereas 17% increment over $m = 5$ scenario. From the earlier studies, it is reported that incorporation of equalization techniques turns out be the most appealing solution to enhance the data rates under highly bandlimited VLC channel conditions. Toward this, the research efforts in [522] investigate the performance of m-CAP system with DFE under bandlimited VLC channel conditions. The number of subcarriers which were chosen can be depicted as $m = \{1,2,5,10\}$. The result analysis of this work clearly illustrates the BER performance of m-CAP system with and without the presence of DFE technique. The primary purpose of exploiting DFE is due to its potential to countervail the effects of ISI which usually emanate on the presently estimated symbols from the previously detected symbols. In general, upon increasing the order of m contributed to a significant improvement in the BER performance due to the assurance of availability of flat-band frequency response. However,

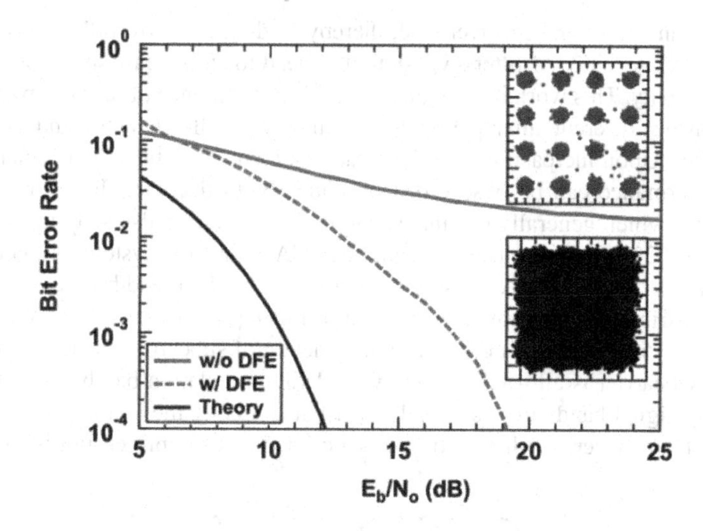

Figure 3.49: Impact of decision feedback equalizer (DFE) on the BER performance when $m = 1$ in m-CAP system under highly bandlimited VLC channel conditions [522].

the trade-off between the system complexity as well as the BER performance exists more exclusively for higher orders of m. In general, the lower orders of m, i.e., $m = 1,2$ suffer from a deterioration in the BER performance due to the prevalence of non-flat frequency response. Therefore, upon incorporation of equalization techniques like DFE, a significant improvement in BER performance can be obtained due to cancellation of ISI. As evident from Fig. 3.49, DFE played a vital role to improve the BER performance of m-CAP system operating over bandlimited VLC channel. When $m = 1$, as portrayed from the simulation result, the BER performance deteriorates without any equalization technique. While, upon enforcing DFE, there is a drastic improvement in BER performance due to the fact that DFE circumvents the ISI that was imposed by the bandlimited conditions. Furthermore, this work analyzes the BER performance for higher orders of m as well. From the simulation results, it can be inferred that for higher orders of m, i.e., when $m = 5,10$, DFE failed to accomplish any striking improvement in the BER performance. Finally, it can be concluded that DFE plays a major role to improve the BER performance for lower orders of m, i.e., when $m = 1,2$. Whereas, for higher orders, the introduction of equalization techniques like DFE results in an additional increase in complexity, as there is no remarkable improvement in BER performance.

Significant research efforts worked in the direction to improve the overall system performance of m-CAPVLC system as well as to attain a reduction in the system complexity. Toward this end, the work in [120] focused on the analysis of pulse shaping filters that play a crucial role in the performance of m-CAPVLC system. As already discussed in case of m-CAP-based VLC systems, it requires $2m$ filters at both the transmitter and receiver side, thereby leading to an overall requirement of $4m$ filters. This choice of filters will definitely lead to an increase in the overall system complexity. This work investigates the BER performance of m-CAP-based VLC system using different filter parameters such as the roll-off factor and the length of the filter. Upon increasing both these factors, a considerable enhancement in the BER performance can be observed. Consequently, to decrease the overall system complexity which generally originates due to the usage of pulse shaping filters, the authors in [523] propose a new concept of m-CAP for VLC systems. In general, in case of conventional m-CAP systems, the total signal bandwidth is divided into m equally distributed subcarriers. In contrast to this equal distribution of subcarriers, the work in [523] follows a different phenomenon of subcarrier distribution, where the first subcarrier is allowed to occupy the LED modulation bandwidth while, the remaining signal bandwidth is equally distributed among the rest of the $m-1$ subcarriers. This concept of distribution of subcarriers is exemplified in Fig. 3.50. The

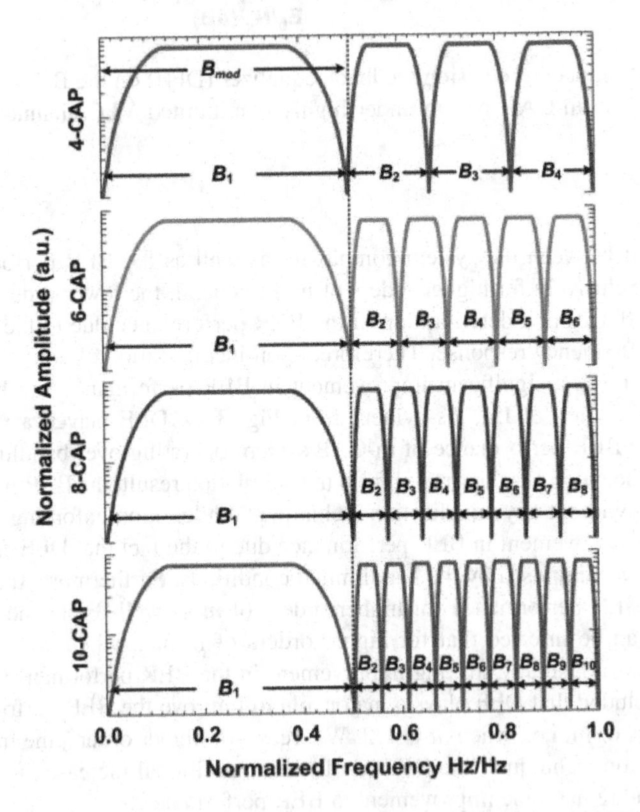

Figure 3.50: Interpretation of unequal subcarrier distribution in m-CAPVLC systems [523].

simulation results of this work confirm the fact that by following such implementation of subcarrier distribution, for $m = \{10, 8, 6, 4\}$ subcarriers, there is a remarkable reduction in overall computational complexity by 80, 75, 67 and 50% when compared with the classical m-CAPVLC system employing $m = \{10, 8, 64\}$ subcarriers. This improvement in computational complexity is possible only at the expense of incurring a power penalty of 1.5 dB. Furthermore, the work in [119] extends the concept of this subcarrier distribution as reported in [523] by performing theoretical investigations on the performance of m-CAP systems by variably splitting the available signal bandwidth. The subcarrier distribution is as follows: Firstly, the total signal bandwidth is split into m subcarriers, where x number of subcarriers were set within the region of LED modulation bandwidth and the remaining subcarriers, i.e., $m - x$ were distributed into the rest portion of the available signal bandwidth as: 2-CAP $(1/1)$, 4-CAP $(1/3, 2/2, 3/1)$ and for 6-CAP $(1/5, 2/4, 3/3, 4/2, 5/1)$, respectively. The simulated result as represented in Fig. 3.51 which is taken from the aforementioned work illustrates the fact that this sort of unequal distribution of subcarriers results in considerable improvement in the transmission speeds of up to 36% in variable 6-CAP system over the traditional 6-CAP system. Additionally, when compared to the conventional m-CAP systems, even under the scenarios of highly bandlimited VLC channel conditions, a significant improvement in the data rate can be witnessed upon employing higher orders of subcarriers in variable m-CAP systems. The work in [120] clearly illustrates that even though the choice of the filter length and the roll-off factor results in improved BER performance, there is an increase in the computational complexity which stems out as the major issue for the implementation of m-CAP on field programmable gate arrays (FPGAs) due to the availability of limited hardware resources. The work in [117] extends the work as signified in [120] by experimentally investigating the performance of m-CAP VLC system over a range of FIR filter conditions with the motive to clearly understand the physical

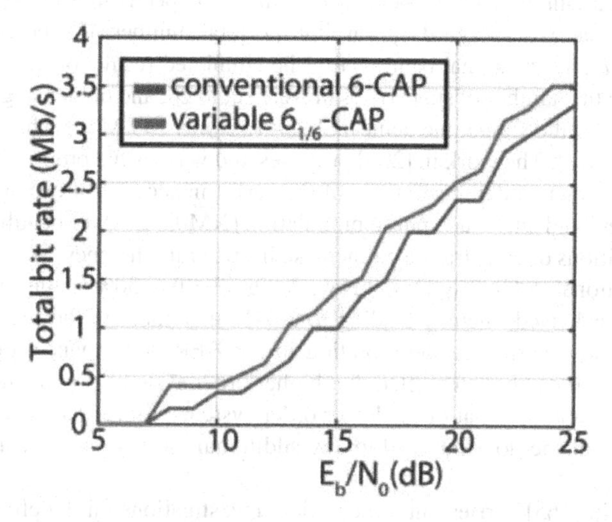

Figure 3.51: Performance comparison of variable CAP with traditional m-CAP system in terms of achievable bit rate [119].

performance of the system. This work gives a clear demonstration of the influence of filter parameters on the measured bit rates or spectral efficiencies and computational complexities. This work also infers that lower order m-CAP system exhibits the same amount of performance as that of the higher orders systems while ensuring lower computational complexity as well. Upon properly optimizing the filter parameters and the order m of a m-CAP VLC link, a significant improvement in the data rate and bandwidth efficiency of 9.69 and 40.43% can be witnessed.

Several other significant research efforts in the recent literature confirm the feasibility of m-CAP modulation format for VLC systems. The work in [521] depicts the experimental demonstration of a 4×4 imaging MIMO VLC system that exploits m-CAP modulation format. Both MIMO and m-CAP techniques independently exhibit their prominence. Consequently, this work employs the combination of these two techniques to attain an improvement in transmission speeds in VLC systems. The link performance was analyzed by varying the total number of subcarriers m, the link distance L and the signal bandwidth. The simulated results of this work ascertain that when the number of subcarriers m were set to 20, and upon using 20 MHz as the bandwidth of the signal, the achievable bit rate was 249 Mbps while attaining a BER of 3.2×10^{-3}. The work in [200] proposes and verifies the proof-of-concept by means of experimental demonstrations of the performance of non-orthogonal multi-band carrierless and amplitude phase modulation (NM-CAP) over bandlimited VLC channel conditions demonstrating an increase in spectral efficiency. When compared to the conventional m-CAP system, this work signifies that 30% savings in the bandwidth can be achieved, thereby leading to a 44% increment in the overall spectral efficiency while ensuring no deterioration in the BER performance. Furthermore, from this work, it can be interpreted that higher order systems provide higher bandwidth compression than that of the lower order systems. Moreover, advantageously, the proposed scheme does not lead to any additional increase in the computational complexity.

The work in [465] carries out a theoretical investigations on the effect of power allocation for subcarriers on the BER performance of m-CAP-based VLC system. With the intent to effectively exploit the passband region of the system's frequency response in order to facilitate for data transmission, this work takes into consideration subcarriers with overlapping as well as subcarriers with different bandwidth allocations. Finally, the simulated results of this work emphasize that the aforementioned methodology results in an improvement in BER performance by 20% when compared with traditional m-CAP system. Much recent literature reports the performance of a non-orthogonal multiband super-Nyquist CAP modulation format in the context of VLC applications [210]. With the intent to ensure an enhancement in spectral efficiency, this work allows for the overlapping of the sub-bands in a manner to disrupt their orthogonality, thereby leading to the origination of interband interference (IBI). From the experimental results, it can be surmised that the proposed system model is capable enough to tolerate the IBI, thereby assuring an enhancement of spectral efficiency without leading to any additional increment in computational complexity at the receiving end. Furthermore, from the simulated results, it can also be deduced that the proposed m-CAP system can withstand up to 30 and 40 compression for 4- and 16-level QAM, thereby exhibiting an improvement in the spectral efficiency of up to 40 and 25% at the expense of reduced BER performance. A much similar work in [201] also investigates the performance of super-Nyquist m-CAP-based VLC systems exploiting polymer LEDs. The result analysis of this work affirms that the proposed system achieves an improvement in spectral efficiency when compared to traditional m-CAP system.

3.2.5.4 Challenging aspects and mitigation techniques associated with CAP-VLC systems

Despite possessing several noteworthy features, CAP modulation is not devoid of its drawbacks. From the earlier studies it can be affirmed that the usage of orthogonal pulse shaping filters and matched filtering operations of CAP transceiver results in an appreciable amount of increase in the sensitivity to timing jitter. Additionally, there is a significant degradation in the overall performance of CAP-VLC system especially in case of non-flat frequency response channels. This is further fueled with the limited bandwidth offered by the LEDs which originates a considerable amount of ISI upon increasing the data throughput of the system. Even though CAP has less PAPR, contrarily, m-CAP much similar to multicarrier modulation formats suffers from high PAPR. Moreover pertaining to m-CAP, even the computational complexity also increases with the increase of the order of m. Thus, this entails for the formulation of sophisticated techniques to enjoy the advantages from both CAP and m-CAP systems. Taking into consideration these aspects, this subsection is directed toward highlighting the possible challenging aspects that hinder the performance of CAP as well as outline certain mitigation techniques to overcome such challenging aspects, and most of these techniques were adopted from [16].

Susceptibility to Timing Jitter

One of the major challenges that hinder the assurance of high data rate communication of CAP-VLC systems is its susceptibility to timing jitter. Principally, timing jitter is referred to as deviation of the receiver clock from the ideal sampling instant which in turn leads to mismatches between the ideal sample and the received sample values. Thus, improper frame misalignments hinder the assurance of high data rate communication. The basic receiver architecture design of CAP systems makes it vulnerable to timing jitter. Practically, the emanation of timing jitter is pictorially depicted in Fig. 3.52. It is imperative from the figure that the ideal sampling instant is portrayed by the thick lines, whereas the actual sampling instant is characterized by the dashed lines, which clearly indicate the emergence of timing jitter τ. As discussed in the previous subsection, the peculiar nature of CAP is that by exploiting the orthogonal set of pulse shaping filters, it avoids the usage of complex carrier modulation and recovery at the receiving end. For example, when compared with QAM, the usage of orthogonal filters in CAP relieves the burden to exploit phase locked loop (PLL) at the receiver to keep track of received phase which is mandatory in QAM.

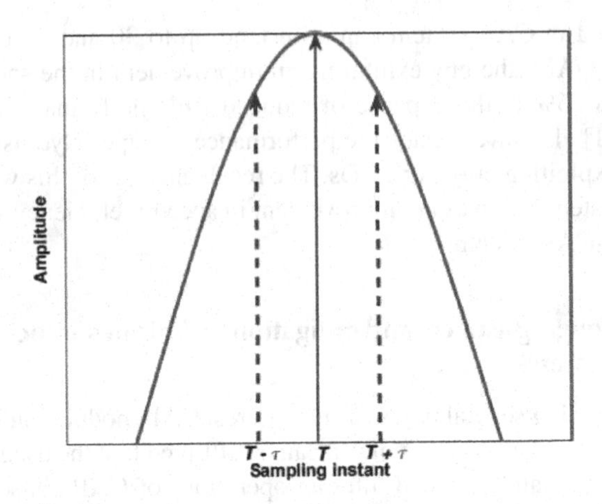

Figure 3.52: Pictorial depiction of the emanation of timing jitter τ [17].

Even though the implementation of CAP is simplified by means of realizing these pulse shaping filters as digital FIR filters, this sort of filter design inherently leads to the occurrence of narrow lobe at the demodulator output as depicted in Fig. 3.52. Consequently, the presence of narrow lobe will definitely enable the CAP system to get susceptible to timing jitter. A comprehensive architectural comparison of CAP and QAM is carried out in [7] highlighting the fact that CAP is severely affected due to the sampling jitter than QAM systems.

The following mathematical analysis verifies the fact that CAP modulation scheme is more sensitive to timing jitter when compared to QAM modulation. This impediment originates due to the receiver architecture of CAP system where the matched filter output constitutes two outputs corresponding to both in-phase and quadrature components. By taking into consideration the receiver schematic of CAP as illustrated in Fig. 3.41, the output of the matched filter pertaining to the in-phase component can be expressed as

$$r_I(t) = x(t) \otimes q(t) \tag{3.108}$$

where $q(t)$ in equation (3.108) corresponds to

$$q(t) = p(T - t) \tag{3.109}$$

By taking into consideration the value of $x(t)$ as depicted in (3.99), equation (3.108) can be rearranged to obtain

$$r_I(t) = \sum_{i=-\infty}^{\infty} u_i p(t - iT) \otimes p(T - t) - \sum_{i=-\infty}^{\infty} v_i \tilde{p}(t - iT) \otimes p(T - t) \tag{3.110}$$

From equation (3.110), the first term manifests the imaginary component, while the second term represents the quadrature component. Further, (3.110) can be simplified as

$$r_I(t) = \sum_{i=-\infty}^{\infty} u_i r_{II}(t - iT) + \sum_{i=-\infty}^{\infty} v_i r_{IQ}(t - iT) \tag{3.111}$$

From (3.111), the desired and interference components can be pronounced as shown by equations (3.112) and (3.113)

$$r_{II}(t) = p(t) \otimes p(T - t) \tag{3.112}$$

$$r_{IQ}(t) = \tilde{p}(t) \otimes p(T - t) \tag{3.113}$$

The desired component can be calculated by taking into consideration that $p(t) = g(t)\cos(2\pi f_c t)$; therefore, the desired component can be solved to attain

$$r_{II}(t) = \int_{-\infty}^{\infty} g(\tau) g(T - t + \tau) \cos(2\pi f_c \tau) \cos(2\pi f_c (T - t + \tau)) d\tau \tag{3.114}$$

By making use of the trigonometric identity $2\cos(A)\cos(B) = \cos(A+B) + \cos(A-B)$ and by assuming $T - t$ as μ, equation (3.114) can be further solved as

$$r_{II}(t) = 0.5\cos(2\pi f_c \mu) \int_{-\infty}^{\infty} g(\tau) g(\mu + \tau) d\tau + \\ 0.5 \int_{-\infty}^{\infty} g(\tau) g(\mu + \tau) \cos(2\pi f_c (\mu + 2\tau)) d\tau \tag{3.115}$$

The second term of (3.115) can be set to zero because it is equivalent to filtering a high frequency signal with a low pass filter, and the high frequency signal corresponds to the signal $g(t)$ modulated over $\cos(2\pi f_c t)$. Thus, by making use of this inference, the expression for the desired signal can be rearranged to yield

$$r_{II}(t) = 0.5\cos(2\pi f_c (T - t)) \underbrace{\int_{-\infty}^{\infty} g(\tau) g(T - t + \tau) d\tau}_{g(t) \otimes g(T-t)} \tag{3.116}$$

Thus, the desired part can be expressed as

$$r_{II}(t) = \cos(2\pi f_c (T - t)) r_{ii}(t) \tag{3.117}$$

Where $r_{ii}(t)$ corresponds to the product of 0.5 with the convolution between $g(t)$ and $g(T - t)$, respectively. Therefore, in a similar fashion, the interference part corresponding to (3.111) can be derived as follows:

$$r_{IQ}(t) = \tilde{p}(t) \otimes p(T - t) \tag{3.118}$$

By substituting the value of $\tilde{p}(t)$ in equation (3.118), the following expression is obtained

$$r_{IQ}(t) = g(t)\sin(2\pi f_c t) \otimes g(T - t)\cos(2\pi f_c (T - t)) \tag{3.119}$$

Furthermore, upon solving equation (3.119), the resultant expression is attained as follows

$$r_{IQ}(t) = \int_{-\infty}^{\infty} g(\tau)\, g(T - t + \tau)\, sin(2\pi f_c \tau)\, cos(2\pi f_c (T - t + \tau))\, d\tau \quad (3.120)$$

Further, (3.120) can be solved to attain

$$r_{II}(t) = 0.5 sin(2\pi f_c \mu) \int_{-\infty}^{\infty} g(\tau)\, g(\mu + \tau)\, d\tau + 0.5 \int_{-\infty}^{\infty} g(\tau)\, g(\mu + \tau)\, sin(2\pi f_c (\mu + 2\tau))\, d\tau \quad (3.121)$$

By using the similar assumptions while deriving the desired component of the signal, the interference part as delineated in (3.121) can be expressed as

$$r_{IQ}(t) = sin(2\pi f_c (T - t))\, r_{ii}(t) \quad (3.122)$$

Figure 3.53, clearly depicts the output of the matched filter pertaining to the in-phase

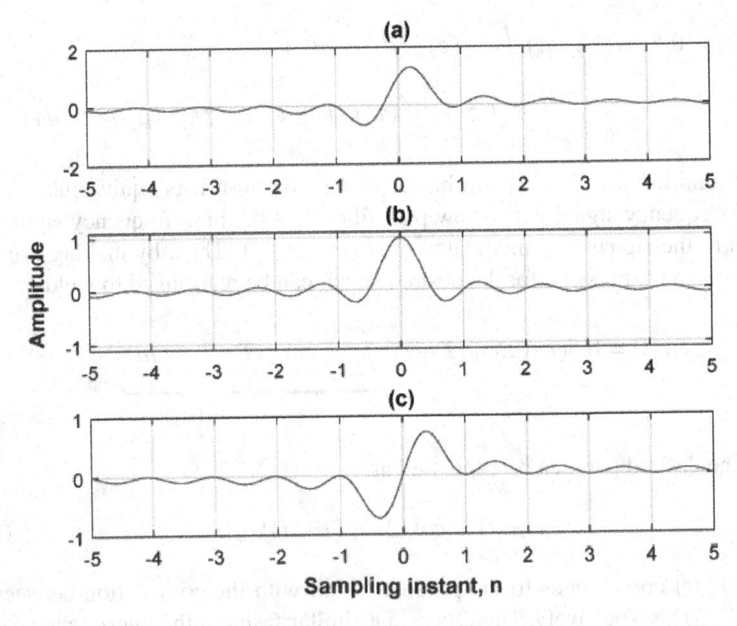

Figure 3.53: (a) Representation of the output of matched filter corresponding to the in-phase component, (b) desired component indicating the presence of peak at sample 0, and (c) the interference part [16].

component which is designated as $r_I(t)$, the desired part of it which is represented as $r_{II}(t)$ and the interference component which is denoted as $r_{IQ}(t)$, respectively. It is clear from the mathematical expressions as delineated in (3.117) that the desired part of the signal has a peak when the sampling time is set to 0, though it has a peak at sampling instant 0, it has a narrow lobe. Even though the interference part of the signal as emphasized by the mathematical expression (3.122) does not lead to any significant amount of distortion at the sampling instant, it has got values between the sampling instants. Consequently, this makes the CAP signal more vulnerable to timing jitter and other channel impediments, since there is a deviation between the ideal sampling instant. In order to demonstrate this fact, this subsection presents a proof of concept taken from the earlier works [17] illustrating the impact of timing jitter on the performance of CAP system. In order to illustrate the deleterious aspects

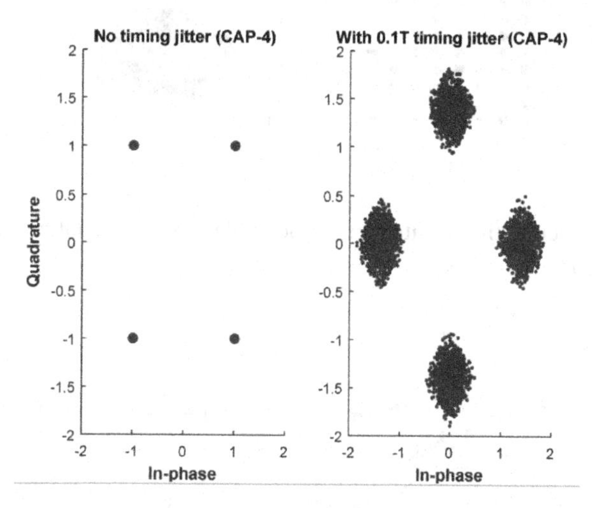

Figure 3.54: Effect of timing jitter on the constellation diagram of CAP-based VLC system [17].

of timing error on the performance of CAP-based VLC system, the authors in [17] carried out simulations to emphasize that timing error has a drastic impact on the performance of CAP-based VLC system. The effect of timing jitter on the constellation of CAP-VLC system with and without the effect of noise is exemplified in Figs. 3.54 and 3.55, respectively. These simulation results underline the fact that a timing jitter of 0.1 T paves the way for the origination of both interference and rotation effect on the received constellation. Furthermore, by taking into consideration a highly noisy environment, the effect of timing jitter on higher order constellation sizes of CAP-VLC system is very detrimental because, under in a noisy environment, it becomes even tougher to distinguish the constellation points pertaining to the received symbols. As evident from the figure, for CAP system comprising of 64 QAM modulation, there are chances for making more inaccurate symbol decisions than that of the CAP-VLC system employing 16 QAM modulation format. Thus, the severity of the impact of timing jitter drastically increases with the increase in the size of the constellation orders. Figure 3.56 clearly illustrates the BER performance of CAP-based VLC system employing 4 QAM modulation scheme under the

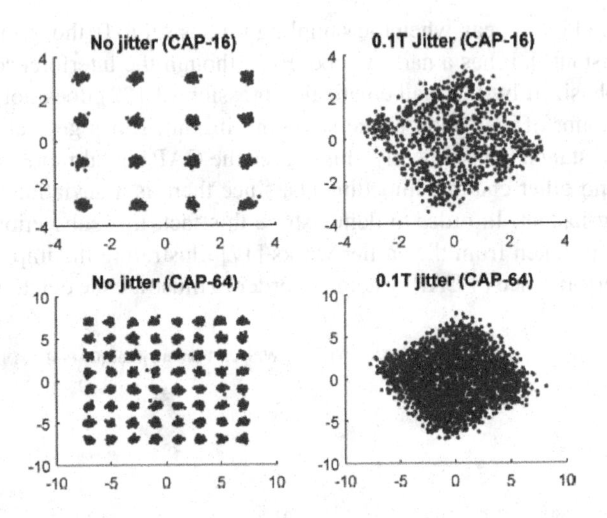

Figure 3.55: Effect of timing jitter on higher orders of constellation in CAP-based VLC system [17].

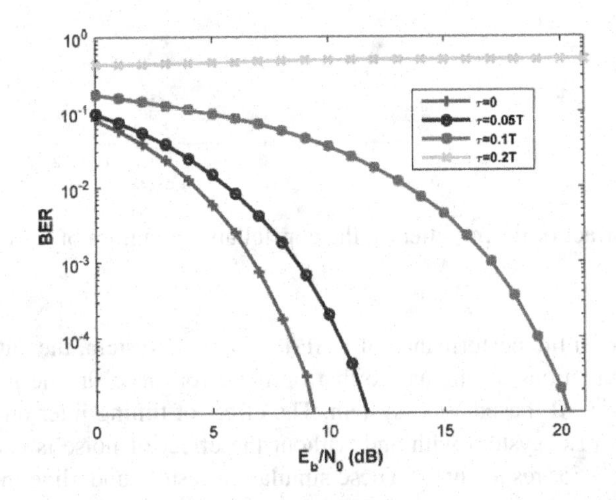

Figure 3.56: Interpretation of the effect of timing jitter on the BER performance in CAP-based VLC system [17].

impact of varying timing jitter. From the simulated result it can be confirmed that upon increasing the timing jitter, there is a degradation in the BER performance. When there is no timing jitter, in order to achieve an error floor of 10^{-4}, it requires SNR of approximately 9 dB, while, when the timing jitter is around 5%, then the amount of SNR required to attain the same probability of error is 12 dB. Thus, this indicates that there is a 3 dB penalty in the amount of SNR required when the timing jitter is just 0.05T. Upon further increasing the timing jitter by 10%, the increment in the amount of SNR is huge. Almost a 10 dB SNR penalty can be observed when the timing jitter is around 0.01T. Furthermore, upon increasing the timing jitter to 20%, it is impossible for CAP-VLC system to reach a reduced error floor. Thus, confirming that the BER performance of CAP-VLC system degrades with the increase in the sensitivity of timing jitter. Additionally, in order to verify the impact of timing jitter on the higher constellation sizes of CAP-VLC systems, the authors in [17] present the BER performance of CAP-VLC system employing different orders of modulation like 4, 16 and 64 QAM, respectively. As depicted in Fig. 3.57, when there is

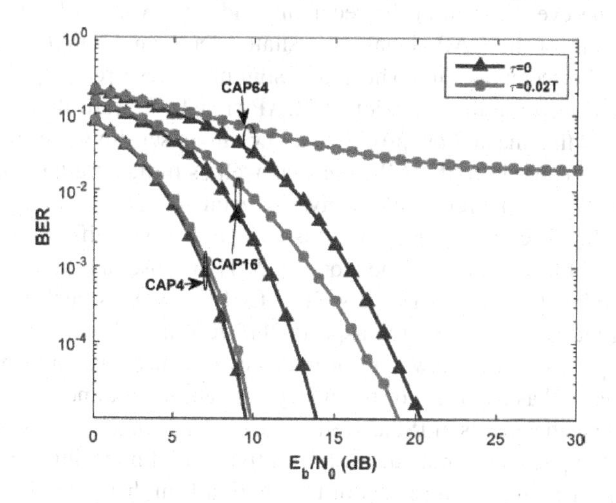

Figure 3.57: Impact of timing jitter on the BER performance of CAP-VLC system exploiting different orders of constellation sizes [17].

an increment in the sensitivity of timing jitter, the BER performance of CAP-VLC system employing different orders of modulation gradually degrades because in order to attain a BER of 10^{-4}, it requires higher amount of SNRs. This figure even infers the certainty that for higher orders of constellation sizes like 64 QAM, it is impossible to attain an irreducible error floor when the timing jitter is around 0.02T. Thus, from the simulation results, it can be surmised that CAP modulation format is severely prone to timing jitter. The constellation points are extremely affected by the presence of timing jitter as the received constellation undergoes interference as well as rotation effect prevails. While, the BER performance degrades drastically when the timing jitter increases leading to huge increase in the SNR penalty. The impact of timing jitter on higher constellation sizes is very detrimental because it hinders the accurate decision capability of the receiver. Therefore, in order to overcome the sensitivity of CAP signal to timing jitter, significant research efforts have been carried out in the literature, among which, enforcing synchronization seems to be the best possible mitigation technique to overcome the timing induced error of CAP system. As depicted from the earlier mathematical analysis, due to the emanation of timing jitter, CAP modulation scheme suffers from severe ISI and cross-channel interference (CCI) as well. Consequently, the research efforts were mainly focused on two principal directions which include either to alter the filter structure of CAP systems or to incorporate a separate synchronization block at the receiving end.

However, from the earlier studies it is revealed that upon modifying the receiver structure, there is a huge increment in the overall system complexity while making the CAP-VLC system sensitive to timing jitter. Much closely related works that are relevant to modifying the receiver structure in order to reduce the impact of timing jitter can be found in a list of references which include [307, 475, 537], etc. When comparing to the classical QAM modulation format, the CAP system has a narrow eye width in the eye diagram at the receiving end. The work in [215] affirms that narrow eye width of the CAP signal necessitates deploying complex synchronization circuit at the receiving end. Therefore, significant research was carried out in this direction to increase the eye width of CAP signals. Toward this end, the work in [475] portrays that the authors proposed a low timing sensitivity receiver structure for enlarging the eye width of CAP signals as well as mitigating the effects of CCI and ISI. Even though this proposed receiver structure is less sensitive to timing jitter, the major drawback associated with it is that there is a significant loss of simple linear structure of CAP receiver. The work in [537] proposes a new set of frequency domain constraints to ensure a ISI as well as CCI free CAP signal. By taking into consideration the aforesaid research aspects, the work in [307] emphasizes that the authors have proposed two new sets of wider eye opened two- dimensional (2D) CAP pulses, as well as a new set of frequency efficient three-dimensional (3D) CAP pulses to mitigate the effects of ISI as well as CCI in CAP-based VLC systems. However, both techniques could not succeed to mitigate the timing jitter in CAP-based VLC systems. The simulation results of this work affirm that 2D pulses can ensure accurate timing synchronization only at the expense of deterioration of BER performance. Meanwhile, the set of 3D CAP pulses demonstrate much higher sensitivity to timing jitter than that of the existing CAP pulses.

It is shown in [18], where the distinct research efforts strive to mitigate the impact of timing jitter on the performance of CAP-based VLC system. The authors in this work propose a novel correlation-based synchronization algorithm to ensure both bit and frame synchronization. It is also to be noted that the synchronization sequence was obtained from the CAP pulse shaping filter. The proposed CAP-filter synchronization technique has been verified through both theoretical as well as simulations. The noteworthy feature of this CAP-filter synchronization technique is that it does not alter the design of receiver structure as well as maintains the average value of the transmitted signal. Furthermore, it also relishes the advantages of Nyquist sampling rate of CAP along with the ISI mitigation capability of RRC filter. Substantially, it is desired to preserve the mean value of the transmitted signal in case of OWC in order to meet the requirements of eye safety regulation and to maintain the quality of the emitted light. The schematic block diagram of CAP receiver which is incorporating a synchronization block is shown in Fig. 3.58. As depicted in the figure, the received signal along with the sequence p are fed as input to the correlator, where the output of the correlator is applied as input to the threshold detector. The main functionality of the threshold detector is to find out whether the exact peak of the correlator is in

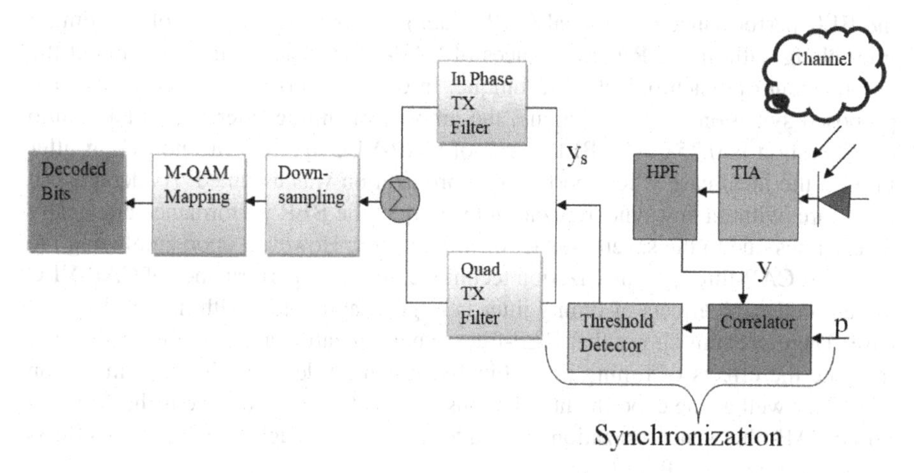

Figure 3.58: Receiver schematic of CAP-VLC system with incorporated bit and frame synchronization block.

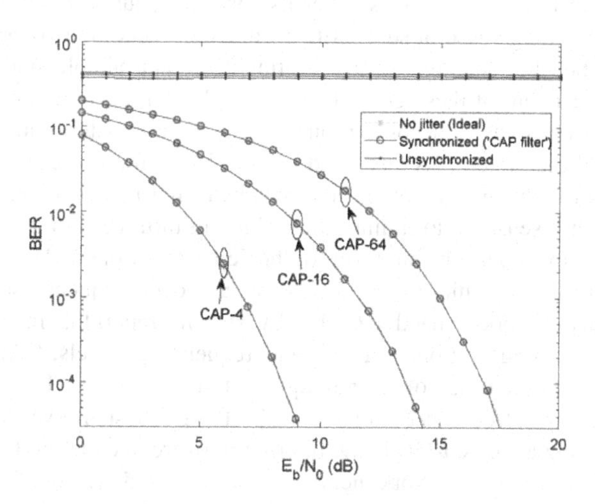

Figure 3.59: BER performance of CAP-VLC system with and without synchronization for different orders of modulation [18].

good alignment with the clock signal. Consequently, in this manner a jitter-free signal as denoted by y_s is subsequently passed through the CAP demodulator to allow for further signal processing. The superiority of CAP filter synchronization scheme has been verified through simulation analysis.

The BER performance of CAP-VLC system with and without synchronization for different orders of modulation is delineated in Fig. 3.59. This simulation result gives the BER performance of an ideal CAP system without the presence of any timing jitter along with the BER performances of CAP-VLC system with and without the enforcement of synchronization technique. In order to verify the superiority of the proposed synchronization technique, the amount of timing jitter that is taken into consideration is 0.25T. The BER curve of CAP-VLC system with no timing jitter implies the ideal case where perfect synchronization was assumed. As depicted in the figure, without any synchronization technique, the BER performance drastically deteriorates due to the severe impact of timing jitter. However, upon employing the proposed CAP-filter synchronization technique, the BER performance of CAP-VLC system under the effects of timing jitter is in good agreement with that of the ideal case. Therefore, this shows that CAP-filter synchronization algorithm is adequate to mitigate the effects of timing jitter. Finally, it can be deduced that the simulation results, as well as the experimental demonstration of this work, unveils that the proposed CAP-filter synchronization technique is capable enough to alleviate the effects of timing jitter in CAP-VLC system.

On the Effect of Limited Modulation Bandwidth of LED

Predominantly, the white phosphorescent LED employed in VLC systems exhibits a non-flat frequency response over its spectrum. Furthermore, it is apparent that the fast switching speed of such LED is hindered by the slow transient response of yellowish phosphor coating, thereby limiting the total available modulation bandwidth. On the account of these grounds, when high data rate communication is desired, VLC systems easily become prone to ISI, thereby deteriorating the overall performance of the system. As discussed in the aforementioned challenges of CAP system, the pulse shaping filters and the matched filtering operations make CAP-VLC system very sensitive to timing jitter which in turn leads to the emanation of ISI. Eventually, this opens the platform for the dependency of CAP-VLC systems on complex equalizers to achieve a good performance over frequency selective channels. A majority of works related to CAP-VLC systems report that in order to ensure high data rate communication over non-flat frequency channels, CAP modulation techniques rely on the usage of several equalization techniques. The work in [512] presents experimental demonstrations of m-CAP-VLC system over a WDM VLC system employing a single RGB LED. In order to overcome the limited modulation bandwidth of the LEDs, this work incorporates a weighted pre-equalization method while at the receiving end, a post-equalization method based upon cascaded multi modulus algorithm (CMMA) was employed. The experimental results of this work reveal that the entire m-CAP-based VLC link was successful to achieve a data rate of 1.35 Gbps over a 30 cm transmission path. The research contribution in [511] reports that the authors employed a Volterra series-based nonlinear equalizer in a WDM-based CAP-VLC system to achieve peak data rates of 4.5 Gbps over a 2 m indoor freespace transmission with a BER of 3×10^{-3}. Similarly, the work in [510] exploits a hybrid post-equalizer in a higher order CAP-based VLC system. The noteworthy feature of this hybrid equalizer is that it employs a linear equalizer, a Volterra-series-based non-linear equalizer and a decision directed least mean squares equalizer to concurrently mitigate the non-linear distortions of VLC systems as well as to accomplish a high data rate communication. In particular, a red-blue-green-yellow LED has been employed and the CAP-based VLC system exploiting such hybrid equalization technique demonstrates aggregate data rate of 8 Gbps.

A list of earlier works in the literature reveal that the equalization techniques were implemented by fixing the tap-spacing of the equalizer weights at the symbol duration, and this type of equalizer is known as symbol spaced equalizer (SSE). In general, this sort of sampling the inputs at symbol rate results in spectrum aliasing since the sampling does not comply with the Nyquist rate. Primarily, the occurrence of spectrum aliasing would lead to the emanation of nulls in the aliased spectrum especially when there is timing jitter, and this in turn degrades the performance of CAP-VLC system employing SSE as the equalization technique due to the enhancement in noise. On the other hand, it is reported from the earlier findings that it is more appropriate to rely on fractionally spaced equalizers (FSE) to counteract the most probable issues like the noise enhancement and performance degradation which were widely encountered when employing SSE in CAP-VLC systems. The most appealing nature of FSE is that it overcomes the drawbacks of SSE by sampling its inputs much faster than the symbol rate. Therefore, considering the high sensitivity of CAP system to timing jitter, SSE cannot be employed as an effective equalization technique owing to its drawbacks. Thus, by considering these aspects, the research contribution in [24] proposed a linear adaptive least mean square (LMS) FSE with the major motive to jointly mitigate the timing jitter as well as the ISI that is usually encountered by CAP-based VLC systems. Furthermore, the authors carried out a comparative analysis of FSE-based CAP-VLC system with its counterpart SSE-based CAP-VLC system in an effort to remove the timing jitter and to address the limited modulation bandwidth of LEDs.

The experimental validations as well as the simulation results of this work unfold that when compared with SSE-based CAP-VLC system, FSE-based CAP-VLC system exhibits its potential to render high data rate as well as imparts a high spectral efficiency along with reduction in computational complexity since there is no requirement of additional synchronization block for removing the impact of timing jitter. In order to understand the differences between FSE and SSE, we present a thorough description about the performance analysis of FSE and SSE-based CAP-VLC systems. The proof of concept along with the experimental validations of FSE and SSE-based CAP-VLC systems is adopted from the remarkable research efforts as articulated in [24]. Figure 3.60 depicts the CAP transceiver employing the FSE equalization technique. The fundamental difference between SSE and FSE is that the matched filtering sampling in case of SSE takes place at the symbol duration, i.e., $\frac{T}{Q} = T$, whereas, in case of FSE, the input samples are taken at much faster rate. It is to be noted that Q signifies how many samples are taken in each symbol duration. It can be inferred that the frequency response of SSE equalizer as stated in [405]

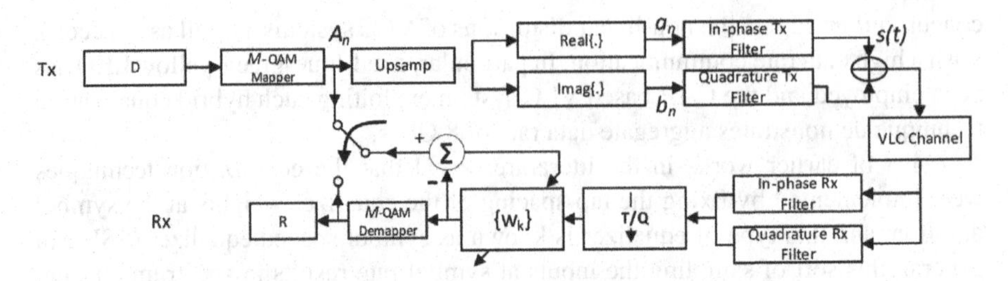

Figure 3.60: Incorporation of fractionally spaced equalizer (FSE) block incorporated into CAP transceiver [24].

can be represented as [24]

$$W_T(f) = \sum_{k=0}^{L-1} w_k e^{-j2\pi fkT} \tag{3.123}$$

From the mathematical expression as depicted by (3.123), w_k and L indicate the weights of the equalizer and the total number of equalizer taps, respectively. Consequently, the equalization spectrum of SSE can be formulated as [24]:

$$H_T(f) = W_T(f) \sum_n X\left(f - \frac{n}{T}\right) e^{j2\pi\left(f-\frac{n}{T}\right)\tau} \tag{3.124}$$

From (3.124), the spectrum of the received distorted signal is signified by $X(f)$, and τ represents the timing delay. Generally, it is obvious for this time delay τ to prevail because ideally it is impossible for the transmitter and receiver clock frequencies to perfectly synchronize with each other. Also from the equalization spectrum of SSE as elucidated by equation (3.124), it can be affirmed that the summation term represents a folded spectrum comprising of the sum of the aliased components. Additionally, the delay component leads to the presence of a phase fact that eventually results in the enhancement of noise. Thus, this problem can be circumvented by means of FSE equalizer, where its inputs are sampled at $T' = \frac{T}{Q}$. Thereupon, the spectrum of FSE can be expressed as

$$H_{T'}(f) = \sum_n W_{T'}\left(f - \frac{n}{T}\right) X\left(f - \frac{n}{T}\right) e^{j2\pi\left(f-\frac{n}{T}\right)\tau} \tag{3.125}$$

Upon comparing the equalized spectrum of the two equalizers, namely, SSE and FSE, the equalized spectrum of SSE as emphasized by (3.124) represents the sum of aliased components, while FSE gives the aliased sum of equalized components, and this is mathematically depicted through equation (3.125). Thus, in this manner, FSE is capable enough to precisely equalize the received signal spectrum, thereby compensating the effects of timing jitter as well as prohibiting the occurrence of noise enhancement. Based upon these grounds, it can be proclaimed that a CAP-VLC system employing FSE equalizer will avoid the degradation of system performance by

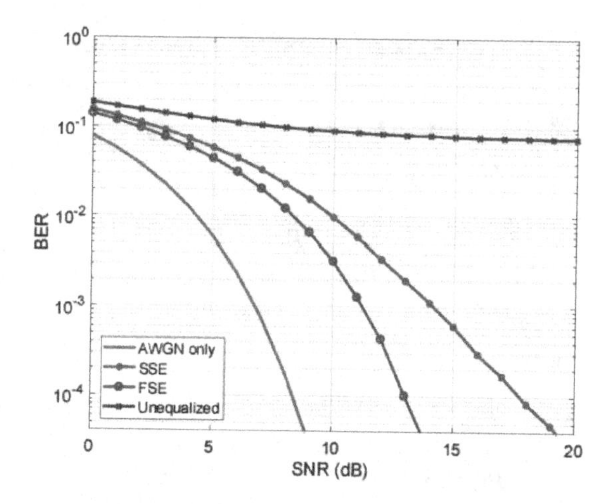

Figure 3.61: Comparison of BER performance of both fractionally-spaced equalizer (FSE) and symbol spaced equalizer (SSE) in a CAP-VLC system over bandlimited VLC channel [24].

counteracting the detrimental aspects of timing jitter. The prominence of the FSE equalization technique over its counterpart, i.e., SSE is clearly illustrated by the simulation result as emphasized by Fig. 3.61. This simulation result pronounces the BER performance of both these equalization techniques in a CAP-based VLC system that is operating over a bandlimited VLC channel. In particular, the BER performance over AWGN channel is exploited as a benchmark to evaluate how well these equalizers work to mitigate the effects of ISI and timing jitter. The dominance of both these equalization techniques is clearly obvious as the CAP-VLC system fails to achieve reduced error floor without enforcing any equalization technique. Meanwhile, the superiority of FSE over SSE can also be evidenced from the simulated result. Furthermore, in order to promulgate the exceptional performance of FSE over SSE in terms of assured data rate as well as the ability to contravene the effects of timing jitter, the authors in [24] carried out Monte Carlo simulations which are depicted in Fig. 3.62 and 3.63. From Fig. 3.62 it can be ascertained that CAP-VLC system exploiting FSE technique exhibits better BER performance as well as offers higher data rates when compared to SSE-based CAP-VLC system. Also upon increasing the SNR from 15 to 20 dB, significant improvement in terms of both BER and assured data rates can be observed. In order to verify the superiority of FSE equalization technique, Fig. 3.63 gives BER performance analysis of both equalization techniques over a range of timing jitter values. It is clear that even with the increase in the range of timing jitter values, FSE is effective to correct them. On the other hand, with the increase in the sensitivity of timing jitter, the performance of SSE-based CAP-VLC system undergoes severe degradation, thereby failing to establish a reliable communication link.

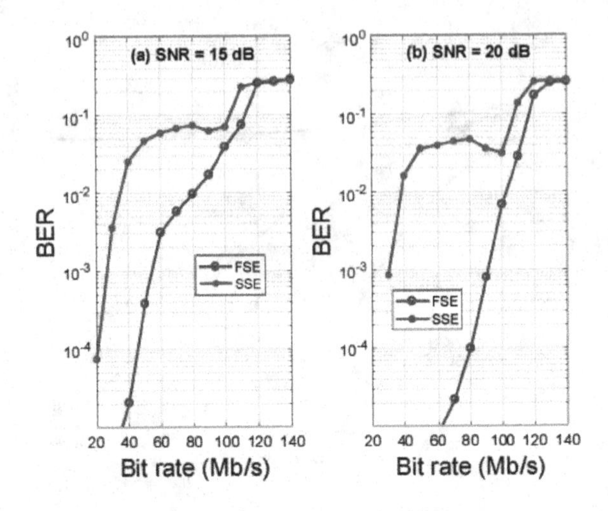

Figure 3.62: Comparison of BER performance of both fractionally spaced equalizer (FSE) and symbol spaced Equalizer (SSE) in a CAP-VLC system for different data rates at SNR of 15 and 20 dB [24].

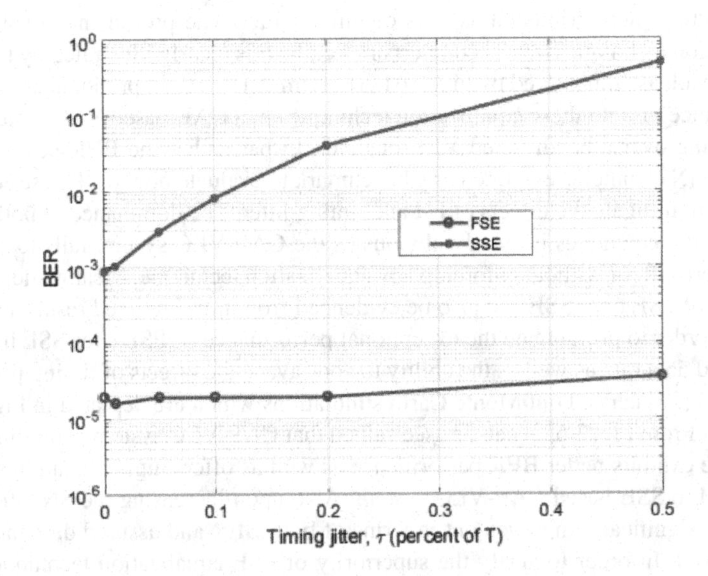

Figure 3.63: Comparison of BER performance of both fractionally spaced equalizer (FSE) and symbol spaced equalizer (SSE) in a CAP-VLC system for different data rates at SNR of 15 and 20 dB [24].

Computational Complexity behind the Implementation of m-CAP Systems

Typically, CAP as well as m-CAP systems result in appreciable amount of computational complexity due to their filter design as well as their exploitation of complex equalization techniques in the process to render high data rate transfer. In particular, in case of CAP and m-CAPVLC systems, the majority of the computational complexity is attributed to the use of matched FIR filters and IFFT/FFT operations. In fact, the appropriate choice of the sub-band count in case of m-CAP and IFFT/FFT sizes plays a significant role in determining the computational complexity. Moreover, in case of m-CAP transceiver, each sub-band at the transmitter terminal requires two FIR filters, and also at the receiving end, matched filters are necessary due to the usage of square root raise cosine (SRRC) operation, thereby incurring a total requirement of $4m$ FIR filters. Look-up tables (LUTs) have been proposed at the transmitting end of CAP and m-CAP systems with the motive to reduce the implementation complexity behind the implementation of filter design. The work in [213] depicts that in order to reduce the computational complexity, the authors employed the Xia pulses in contrast to the traditional RRC filter that is widely employed in classical CAP and m-CAP systems. This work demonstrates a significant reduction in overall computational complexity by 90% even though there is a slight degradation in the BER performance. Particularly, Xia pulses are full Nyquist pulses, thereby avoiding the requirement of matched filtering operation at the receiving end.

In spite of its simple transceiver implementation, the performance of CAP modulation technique is affected due to the limited modulation bandwidth offered by WLEDs. Consequently, the major focus of the literature is directed toward the design of diverse equalization techniques for enhancing the throughput of CAP-VLC system. Nonetheless, these equalization techniques lead to the increase of overall computational complexity of the system. Hence, in order to alleviate such drawbacks, the authors in [21] exploit the concept of spatial domain, i.e., MIMO techniques in order to improve the BER performance of conventional CAP-based VLC system. Generally, with the motive to render high levels of illumination, VLC deploys multiple LEDs, and interestingly this feature paved the way for the realization of several MIMO techniques. Based on these grounds, the research efforts in [21] strived to illustrate the benefits of imposing MIMO techniques (i.e., both spatial multiplexing and repetitive coding) to a CAP-VLC system through theoretical analysis as well as experimental demonstrations. The major contribution of this work is the proposal of spatial multiplexed CAP which has the ability to concurrently transmit different streams of independent CAP signals through multiple LEDs. In this manner, this work witnesses an appreciable amount of improvement in the data rate in the traditional CAP-VLC systems. Additionally, as a second contribution, in order to improve the BER performance of conventional CAP system, this work proposes to use repetitive coded CAP (RC-CAP) which allows for the transmission of multiple identical CAP symbols.

The other technique which can be applied to CAP-VLC systems is spatial modulation. The inherent feature of spatial modulation is that it facilitates only one LED to be active out of a total of N_t LEDs, where the index of this active LED is exploited to encode the data. Particularly, this active LED is responsible for transmitting the CAP signal. Spatial modulation-based CAP system facilitates each active LED to encode additional information bits, thereby contributing to increase the spectral efficiency of traditional CAP system [20]. A much recent work that endeavors to enhance the spectral efficiency of conventional CAP systems while preserving the low complexity of the transceiver design can be found in [20], where the authors propose and investigate the performance of spatial modulation-based CAP-VLC systems. This work reports the derivation of analytical expressions for the BER performance of the aforementioned system and is verified via simulations over a LOS channel environment. Furthermore, the effects of both multipath propagation as well as the mobility on the BER performance of spatial modulation-based CAP-VLC system was also analyzed. In order to circumvent the impairments of channel, power factor imbalance (PFI) and multiple photodetectors were introduced. The simulated results underline the fact that, by exploiting multiple photodetectors, a significant improvement in the overall system performance leading to a 43 dB SNR gain can be witnessed. Thus, spatial modulation-based CAP-VLC system seems to be one of the novel implementations pertaining to MIMO systems.

Susceptibility to PAPR

The main attraction of CAP is its low PAPR. However, in case of m-CAP modulation format, with the increase in the number of sub-bands, the chances for the emergence of high peak values also increase substantially. In order to reduce the deleterious effect of PAPR in m-CAP as well as to improve the spectral and energy efficiency of m-CAP systems, the authors in [22] proposed sub-band index (SI) technique, where only some of the m-CAP sub-bands which are termed as active sub-bands are modulated with data and the remaining unmodulated sub-bands do not result in any PAPR. Consequently, this sort of choice of sub-bands will definitely reduce the amount of PAPR in m-CAP-based VLC systems. In order to make up for the loss in spectral efficiency which generally arises due to modulation of only active subcarriers, the SI-CAP system transmits some additional information bits on the indices of m-CAP sub-bands. The research efforts of this work portray-that the performance analysis of CAP-VLC system exploiting SI has been verified through computer simulations. Additionally, the proposed scheme is validated by means of experimental testing in order to confirm the performance gains exhibited by SI technique. The prominence of SI technique in a short range optical data link using step-index plastic optical fiber (SI-POF) has been verified through experimental demonstrations in [19]. This work confirms the fact that SI technique is very flexible to design as well as considerably improves both spectral and power efficiency of the traditional m-CAP systems. Furthermore, the authors in [23] proposed enhanced sub-band index CAP (eSI-CAP) in order to further improve the spectral efficiency of m-CAP-VLC systems. It is interesting to note that much similar to m-CAP systems, eSI-CAP exploits all of the sub-bands to transport the data and with the intent to improve the spectral efficiency, eSI technique encodes additional bits of information on the indices of the sub-bands. In contrast to m-CAP, where the sub-bands are modulated with data symbols that are drawn from a single constellation, eSI-CAP makes use of dual distinguishable constellations in order to modulate its sub-bands. Thus, in this manner, eSI-CAP system exhibits its potential to enhance the spectral efficiency of traditional m-CAP-VLC systems. It can be illustrated from this work that the performance gains of the proposed eSI technique have been verified by means of theoretical analysis, computer simulations and VLC experimental validations. The simulated and experimental results of this work infer that when compared to traditional m-CAP-VLC systems, the requirement in the amount of SNR for eSI-CAP-VLC systems is relatively less for a fixed data rate. Furthermore, due to its flexible configuration as well as its improvement in both spectral and power efficiency, eSI-CAP system can be conceived as a propitious solution for VLC applications.

3.2.5.5 Research aspects pertaining to CAP-VLC Systems

This section presents in detail several significant research works relevant to the implementation of CAP-VLC systems. More specifically, this section highlights the current state-of-the-art research aspects that have been carried out in the literature to enable CAP to evolve as a potential candidate that is more flexible to be exploited in VLC systems. A list of earlier findings emphasize the superiority of CAP over the multicarrier modulation format like OFDM and the references therein [424,459,472], etc. Additionally, a list of publications show that CAP-based VLC system has the potential to deliver data rates in the range of Gbps. The work in [527] illustrates that a CAP-based VLC system employing a WLED renders an aggregate data rate of 1.1 Gbps with a BER performance of 10^{-3} over a 23 cm air transmission. This work also signifies that in order to circumvent the limited modulation bandwidth of LEDs, the authors have employed optical blue filtering and DFE techniques to accomplish high data transfer of information. The research efforts in [528] delineate that the spectrally efficient CAP and OFDM were exploited to establish a VLC communication link over a bandlimited channel. Initially, the performance of CAP and OFDM-based VLC system employing a blue LED has been analyzed over a bandlimited channel. This work confirms the superiority of CAP-based VLC system over the OFDM-based VLC system, as the former system was capable to deliver a maximum data rate of 1.32 Gbps and the latter could render only 1.08 Gbps. Additionally, with the rationale to further increase the system capacity, WDM scheme has been applied to the aforementioned systems where their performances were analyzed by exploiting a RGB-LED. It can be emphasized that a WDM-based CAP-VLC system making use of RGB-LED exhibits its potential to deliver a maximum aggregate data rate of 3.22 Gbps, while the OFDM-based VLC system delivers only 2.93 Gbps respectively. Thus, from this work it can be presumed that CAP modulation format demonstrates a competing performance and emerges as an alternative spectrally efficient modulation format for the next generation wireless communication systems. Much relevant work in [529] signifies that the authors employed WDM technology along with CAP modulation and evaluated the system performance by making use of RGB-type white LED. Furthermore, the experimental results of this work manifest data rates of 3.22

Gbps over the 25 cm air transmission with the achievable BER of 10^{-3}, respectively.

The research even strived toward providing multiuser access support by exploiting multiband CAP modulation format in a VLC system. In [512], an experimental demonstration of a high-speed WDM-VLC system that employs 3-band CAP64 modulation format can be witnessed. Based upon the combination of different wavelengths and the sub-bands, a dynamic capacity can be accomplished to provide access to 9 users. At the transmitting end, in order to overcome the frequency attenuation of the LEDs, a weighted pre-equalization method has been proposed. Meanwhile, at the receiving end, a post-equalizer based on modified cascaded multi modulus algorithm (CMMA) has been exploited. The experimental results of this work infer that the proposed system was capable to achieve a data rate of 1.35 Gbps over a 30 cm transmission while maintaining a BER of 3.8×10^{-3}, respectively. A high-speed RGB-LED-based WDM-based VLC system that exploits CAP modulation and recursive least square (RLS)-based adaptive equalization imparts higher data rate of 4.5 Gbps over a 1.5 m indoor free space [514]. In addition to rendering high data rate communication, significant research works illustrate that the researchers worked toward mitigating the non-linear aspects of LEDs. Much relevant work in this direction can be found in [514] where the authors experimentally demonstrate the performance of high-speed WDM CAP64 VLC system that employs a Volterra series-based non-linear equalizer to counteract the non-linearity aspects of LED. From this work, it can be inferred that an aggregate data rate of 4.5 Gbps can be achieved successfully over a 2 m indoor free space transmission while the maintaining a BER below 3.8×10^{-3}. The work in [510] verifies through experimental validations the performance of higher order CAP modulation-based VLC system which utilizes a hybrid post-equalizer. This hybrid post-equalizer comprises of a linear equalizer, a Volterra series-based non-linear equalizer and a decision directed least mean square (DD-LMS) equalizer to combat both linear and non-linear distortions of CAP-VLC system that exploits a red-green-blue-yellow (RGBY) LED. The proposed system exhibited its potential to achieve an aggregate data rate of 8 Gbps over a 1 m indoor free space transmission.

Furthermore, the simplicity of CAP enabled it to be exploited as a modulation format in a VLC system operating to render services to people travelling on roads i.e., vehicular communication [331]. From this work, a CAP-VLC system was successful demonstrating 312.5 Mbps over 10 m transmission by exploiting a vehicle LED that has a 3 dB bandwidth of 2.5 MHz. The authors in [521] experimentally demonstrate the performance of a four imaging MIMO VLC system that exploits CAP modulation technique. This work signifies that the aforementioned system was capable of achieving a data rate of 249 Mbps while attaining a BER of 3.2×10^{-3}.

3.2.6 COLOR SHIFT KEYING

So far, the previous subsections gave an elaborate discussion on the baseband modulation techniques. An important aspect to be taken into consideration is that in case of IM/DD-based VLC systems, the modulation frequency does not correspond to the carrier frequency of the optical light sources like LEDs. Pertaining to the practical scenarios, it is very easy to modulate the data by changing the color of LED rather than modulating the carrier frequency of the LED. From the previous literature it can be elucidated that CSK modulation has been proposed by IEEE 802.15.7 standard in order to alleviate the lower date rates and limited dimming support aspects of other multicarrier or single carrier modulation formats that are flexible to be implemented for cost-effective VLC systems. Ever since it originated, CSK has sparked a considerable amount of interest among the research communities which lead to the publication of several research contributions and the references therein [147, 350, 352, 417, 449, 558]. As discussed earlier, in the process of generation of white light by employing blue LEDs with yellowish phosphorus coating, the slow transient response of phosphorus limits the speed at which the LED switches, thereby hindering high data rate communication. Consequently, in order to overcome such drawbacks, the other way to generate white light was to exploit the combination of three different LEDs namely red, green and blue (RGB). Such combination of LEDs is often referred to as TriLED (TLED). Whenever, the data is modulated by varying the intensities of three colors in the RGB or TLED, then the type of modulation can be called color shift keying (CSK). The color space chromaticity diagram as defined by CIE 1931 is a fundamental aspect for the implementation of CSK modulation format. The human eye has got perceives different levels of sensitivity to different colors. Therefore, the basic functionality of the chromaticity diagram is to map all of the colors anticipated by the human eye into two chromaticity parameters, namely x and y.

Human visible wavelength can be divided into seven bands as shown in Table 3.6, where the centers of each band are marked in Fig. 3.64. By taking into consideration this chromaticity diagram, the CSK modulation can be implemented. Firstly, it is vital to find out the RGB constellation triangle, where the center wavelength of the three RGB LEDs play a significant role to determine the constellation triangle. Secondly, it is necessary to map the information bits into chromaticity values. Particularly, the constellation triangle is very prominent to derive the chromaticity values of the symbols. Finally, by varying the intensities of the RGB LEDs, the symbols are transmitted. In general, it is important to solve the optimization problem in order to determine the position of the symbols in the constellation diagram. This is mainly done to maximize the distance between the symbols in order to minimize the ISI. Additionally, it is mandatory for the symbols to be equally distributed in the triangle in a manner to ensure that the amount of light as perceived by the human eye while transmitting different symbols is white light only.

Predominantly, the type of dimming support that is employed in CSK is amplitude dimming. In this scenario, by varying the driving current of the LEDs, a change in the brightness of the resultant white light can be evidenced. In contrast to other modulation formats like OOK and pulse modulations, flickering is not an issue with CSK. Consequently, these benefits of CSK sparked the attention of the researchers to design generalized forms of CSK with arbitrary constellation. The work in [147] exemplifies the fact that the authors employed Billards equivalent disk packing algorithm to present a CSK constellation design technique. Meanwhile, the

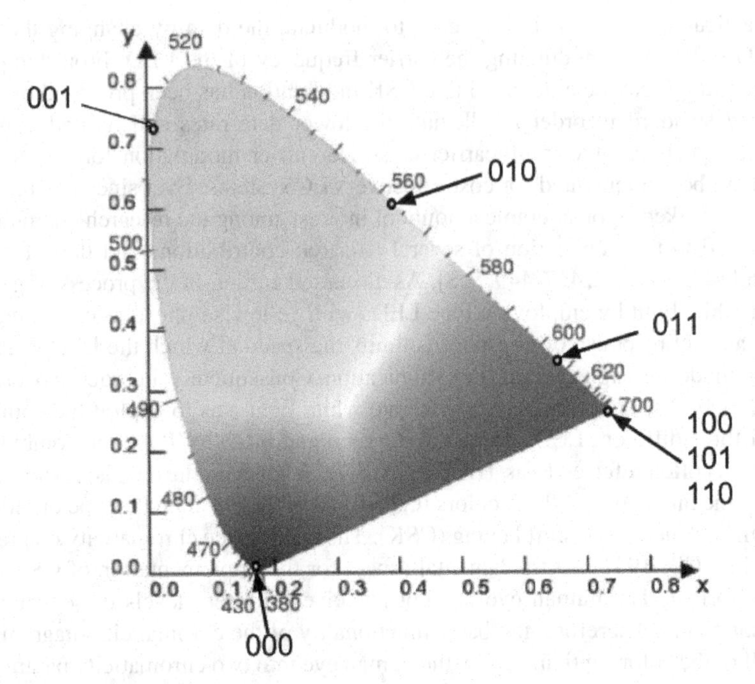

Figure 3.64: Chromaticity diagram as given by CIE 1931. The seven color codes correspond to the seven centers of the seven bands which divide the spectrum of the visible light range [398].

Table 3.6
Illustration of the Seven Bands that are exploited in CSK along with the representation of the center values of codes and chromaticity coordinates [398].

Band (nm)	Code	Center (nm)	(x, y)
380-478	000	429	(0.169, 0.007)
478-540	001	509	(0.011, 0.733)
540-588	010	564	(0.402, 0.597)
588-633	011	611	(0.669, 0.331)
633-679	100	656	(0.729, 0.271)
679-726	101	703	(0.734, 0.265)
726-780	110	753	(0.734, 0.265)

research works in [350] and [351], much similar to [147], developed generalized version of CSK by making use of different optimization algorithms like the interior point methods. All of the proposed constellation design techniques were designed in such a manner to comply with the color balance requirement for the TLED source to generate any desired color for the purpose of illumination. Much significant work in [449] suggests the usage of four LEDs, namely, blue, cyan, yellow and red where the combination of these four colors provides a flexibility to achieve a quadrilateral constellation shape in a manner that resembles QAM-like constellation design. When compared with traditional CSK modulation format, i.e., with three LEDs, the former modulation format that is achieved by combining four different colors was shown to be more energy efficient as well as more reliable in terms of reduction of ISI. As stated from the earlier findings in the literature, the RGB TLED can be exploited to implement wavelength division multiplexing (WDM), a widely deployed multiplexing technique for fiber optics communication. The work in [502] signifies that separate data streams were modulated on three colors, where the aggregate data rate of 803 Mbps was witnessed by the usage of RGB LED. There are still several other related works that exploit different modulation formats like CAP, OFDM, etc., along with RGB LED to render high data rate communication.

3.3 CONCLUSION

This chapter presents a comprehensive in-depth discussion about several varieties of modulation techniques that adhere to the requirements of IM/DD-based VLC systems. Substantially, all of these modulation formats must comply with both illumination and communication prerequisites. The Base-band modulation techniques offer low to moderate data rates and are more suitable to be employed for frequency flat channels. From the earlier research findings it can be surmised that even though base band modulation formats are very straightforward to implement for VLC systems; however, it is mandatory for these schemes to rely on complex equalization techniques in case of frequency-selective channels. Contrarily, the multicarrier modulation formats offer the most viable solution for IM/DD-based VLC systems since they exhibit the potential to render high data rate communication.

In particular, the widely deployed multicarrier modulation technique like OFDM is proven to be as the most suitable candidate for the cost-effective realization of IM/DD systems, especially when adaptive bit and power loading algorithms were incorporated. The major emphasis of this chapter is to underline different variants of optical OFDM with much emphasis on power efficiency, spectral efficiency and computational complexity. Additionally, this chapter elaborates several significant research efforts that strived to propose modulation formats that are meeting the important illumination aspects like dimming and flickering. Accordingly, this chapter categorizes the diverse variants of optical OFDM as the earliest unipolar variants, superposition variants and hybrid variants and details about the pros and cons associated with each category of optical OFDM. Furthermore, several aspects which include the mathematical analysis involved to attain a real and non-negative valued signal, spectral efficiency, power efficiency and computational complexity aspects were also highlighted.

By taking into account the detrimental aspects like the high power consumption and computational inefficiency that emerged due to the enforcement of Hermitian Symmetry constraint in case of classical IFFT-based optical OFDM, this chapter focuses on the performance analysis of several other real transform techniques like DHT, DCT and DST-based optical OFDM systems in terms of BER, power efficiency, spectral efficiency and computational complexity. Moreover, a brief comparative analysis is carried out among the real transformation techniques and their counterpart like complex valued IFFT transform. The simulated results and experimental validations carried out in the literature suggest that the real trigonometric transform-based optical OFDM variants manifest more distinguished advantages than those of the Hermitian Symmetry imposed IFFT-based optical OFDM systems. A brief review on the special modulation format like HCM, which is based on FWHT as well as the WPDM and CSK, is also the scope of this chapter.

Besides base band modulation formats and multicarrier modulation format like OFDM, the major focus of this chapter is directed towards providing insights on the spectrally efficient modulation format like CAP and m-CAP that has been proposed to handle the limited modulation bandwidth challenge of WLED-based VLC systems. This chapter underlines certain distinct features of CAP modulation format that are more relevant to VLC applications. In addition, several challenging aspects that limit the performance of CAP-VLC systems were stressed along with certain notable research efforts carried out to mitigate them. Finally, the main emphasis of this chapter is to point the reader to a comprehensive analysis of diverse modulation formats that can be exploited in VLC systems to ensure reliable high data rate transfer of information.

4 NON-LINEARITIES OF OPTICAL SOURCES

Despite being superior in terms of high data rate transmission and possessing several attractions, the certainty of the individual subcarrier signals to add up coherently results in high peaks in the time-domain signal of optical OFDM, which in turn leads to high peak to average power ratio (PAPR) and stems out as the most vital aspect which needs to be addressed especially for VLC. The emergence of high peaks indicates that the optical source will have to operate outside its linear region for the sake of accommodation of the entire amplitude swings of the corresponding time-domain signal. If this scenario prevails, this is very unappealing, as the levels of distortion present in the time-domain transmitted signal increase significantly and, ultimately, the overall performance of the system deteriorates.

4.1 NON-LINEARITY IN OPTICAL SOURCES

An optical communication system is built with an optical source; in the scenario of IM/DD systems, the key optical source is LED. This optical source is used for generating an optical power which is a function of modulated input electrical signal where the information is converted into optical beam. Generally, the optical sources which are used in OWC systems have a peak power limit which enforces that the optical time-domain signals should have limited peak amplitudes. Therefore, this peak power constraint results in distortion of the optical OFDM signal. LED is the foremost source of non-linearity in VLC. As delineated in Fig. 4.1, LED has a turn on voltage (TOV) as minimum threshold value which is the onset of current flow and emission of light. For voltages below TOV, the LED is considered to be in a cut-off region, i.e., it does not conduct any current. However, above TOV, the LED conducts the amount of current flows and the light output increases exponentially with voltage. LED outputs light power that is linear with the drive current [159]. Thermal aspects cause a drop in the electrical-to-optical (E/O) conversion efficiency, precisely, light output of the LED decreases and then approaches a steady-state value; hence, it is vital to consider such aspects. Therefore, in order to prevent the degradation in output light or, in the worst case, the total failure of the LED chip, it necessitates to adjust the dc and ac/pulsed currents accordingly.

The transfer function of the LED distorts the signal amplitude which forces the signal peaks below the TOV of the LED to be clipped, as well as the upper peaks are clipped purposefully before modulating the LED for the sake of avoiding chip overheating. Hence, the transmitted time-domain signal of optical OFDM should be constrained to a limited range due to the non-linear characteristics of LEDs. As a result, the high PAPR of optical OFDM signals gives rise to a large dynamic power

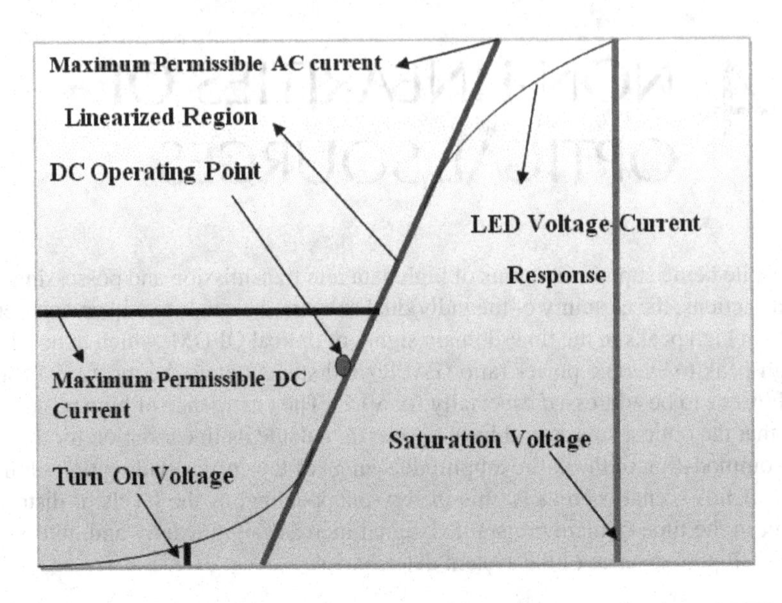

Figure 4.1: Depiction of non-linear and linearized transfer function of LED.

excursion which harms the energy efficiency at both the transmitting and receiving ends. Thus, the optical OFDM signal with high PAPR is subsequently clipped by the LED transmitter originating serious clipping distortion.

To circumvent this major drawback, there are certain solutions proposed in the literature. Because of the structure of the LEDs, the output optical power and the forward current are related by a non-linear function. Therefore, the non-linearities of LEDs can be overcome by linearizing the non-linear region of LED, and the second approach is to adopt PAPR reduction techniques. One such technique is predistortion which is proposed in [161] to linearize the relationship between the output optical power and the forward current. However, this technique requires an accurate model for the design of predistortion and linearization over the dynamic range of optical source. It is reported in [136] that a polynomial model is employed for the purpose of describing the non-linear transfer function of the optical sources from which a predistortion function can be designed to linearize the non-linear region. The problem encountered while employing the predistortion technique is that the non-linear transfer function of the optical sources can change due to many reasons, one of which is the temperature of the transmitter. The optical sources dissipate a portion of the input energy as heat, and this leads to the increase of the temperature of the device which consequently changes the non-linear relation between the output power and forward current. Hence, the predistortion function is not sufficient enough to maintain the linearity of the device and, subsequently, this scenario compels the necessity of dynamic feedback for modifying the instantaneous non-linear transfer function of the optical sources [420].

4.2 PAPR REDUCTION TECHNIQUES FOR IM/DD SYSTEMS

This section highlights the research efforts done by several researchers to reduce the detrimental aspect like PAPR in optical OFDM. Among the variants of optical OFDM, DCO-OFDM system suffers from high PAPR due to the superimposition of huge number of subcarriers as well as for assuring a positive-valued signal, a certain amount of DC bias is added. Therefore, there is a necessity to reduce the amount of PAPR in such system. Hence, it is vital to rely on a few such PAPR reduction techniques which include clipping with filtering [531], selected mapping (SLM) [293, 294, 509] partial transmit sequence (PTS) [358], discrete Fourier transform (DFT)-Spread [563] and signal companding [49, 192]. However, the traditional PAPR reduction schemes cannot be directly implemented in case of optical domain because of the real and positive signal transmission.

The aforementioned methods result in certain correlation between the subcarrier symbols of the OFDM blocks, and such methods can be classified into deterministic and probabilistic approaches. Firstly, clipping with filtering falls into the category of deterministic approach, i.e., signal distortion technique and in general it is referred to as a non-linear process which causes serious in-band and out-of band distortion because it limits the overshooting signal amplitude to a pre-defined level. The second category of PAPR reduction techniques like SLM, PTS, signal companding and DFT-Spread comes under multiple signalling and probabilistic approaches which tend to reduce the probability of high PAPR. SLM and PTS are most widely employed PAPR reduction techniques, but the major drawback associated with them is that they require the side information which significantly reduces the bandwidth-efficiency of the system. Companding is a non-linear transformation technique which tends to destroy the orthogonality among the OFDM subcarriers. Meanwhile, DFT-Spread technique is a type of precoding method which exploits the DFT operation before the IFFT transformation module to negate the levels of high PAPR.

While dealing with the optical domain, several techniques have been proposed to reduce the peaks of the time-domain optical OFDM signal, among which are the following:

In [565], the authors addressed the performance-limiting factor like high PAPR in DCO-OFDM system by applying semi-definite relaxation approach to tone injection scheme. From the simulated results there is a significant reduction in PAPR by approximately 5 dB when compared to traditional DCO-OFDM system. The work in [219] employs branch-and-bound method (BBM)-based tone injection scheme for the purpose of reducing the amount of PAPR in DCO-OFDM system. Tone injection is an effective way for reducing the amount of PAPR when compared to PTS and SLM techniques because it does not transmit any side-information. The simulated results emphasize that the proposed BBM-based tone injection method is more superior than the existing methods in terms of PAPR reduction. However, the aforementioned techniques as used in [565] and [219] result in the increase of power in order to achieve a distinct improvement in BER performance in the presence of LED non-linearities. Additionally, the amount of computational complexity incurred is also high.

In [146], in order to resolve the problem of high PAPR, the authors proposed to apply a modified active constellation extension (ACE) method or tone reservation as a power derating reduction technique in DCO-OFDM VLC systems. In this work, the authors take into account cubic metric (CM) because CM focuses on the cubic non-linearities which generally emanate due to several types of distortions including both in-band and out-of-band distortion. The research in [45] signifies that for the purpose of reducing the PAPR effectively with faster convergence and lower complexity, a tone reservation scheme which is based on the combination of signal-to-clipping noise ratio (SCR) procedure and the least squares approximation (LSA) approach. In [218], a tone reservation technique which is based on the time-domain kernel matrix (TKM-TR) scheme is utilized for reducing the PAPR in DCO-OFDM system. The computational complexity of these aforesaid methods is extremely high since they need to search over all possible combinations of the expanded constellation, i.e., peak cancelling signals.

An iterative clipping method, as proposed in [562], reduces the upper and lower PAPR of the oversampled VLC-OFDM signals without introducing in-band distortions. The work in [48] proposes an exponential non-linear companding method to reduce the PAPR in VLC-OFDM systems, and this method exploits the advantages of exponential companding function to compress large signals and to expand small signals at the same time. One major drawback associated with iterative clipping and exponential companding is that the property of the transmitted signal is destroyed. Pilot-aided PAPR reduction schemes were proposed in [383, 384, 404] where the reduction in PAPR depends upon the density of pilot sequences. But, these methods lead to data rate loss, and there is a degradation in bandwidth efficiency due to the introduction of pilot sequences.

One of the popular techniques for reducing the PAPR in RF-based OFDM system is SLM. For exploiting the SLM concept in DCO-OFDM VLC systems, several studies have been conducted and few can be reported in [165, 227, 536]. The work in [536] combines chaos with SLM technique so that the generation of phase factors can be controlled with the help of chaotic sequences. The result analysis emphasizes that chaotic SLM (CSLM) technique attains better BER performance than traditional SLM technique in OFDM IM/DD system. Specifically, the performance of SLM technique was investigated in [165] which employs five different families of phase sequences, namely, chaotic, Shapiro-Rudin, pseudo-random interferometry code (PRIC), Walsh-Hadamard and random sequences. These different phase sequences result in different levels of PAPR reduction. By exploiting SLM technique, it is obligatory to utilize the side information for indicating the transmitted candidate signal and this subsequently reduces the bandwidth efficiency. The work in [227] employs a special set of symmetric vectors as phase sequences in SLM technique for the purpose of reducing PAPR in DCO-OFDM-based VLC system. In this approach, the magnitude difference between the received signal and the pre-defined phase sequences in the frequency domain was used to detect the side information blindly. Nevertheless, these methods have extremely high computational complexity as they need multiple IFFT operations for candidate generation. The deleterious aspect like PAPR is addressed even in recent times where the research in [226] proposes a novel PAPR reduction scheme using optimized even and odd sequence combination (OEOSC) technique for DCO-OFDM VLC systems to avoid side information transmission and high computational complexity while preserving the PAPR reduction capability.

Having looked at the drawbacks associated with Hermitian Symmetry imposed IFFT-based optical OFDM system, this chapter focuses on analyzing the PAPR performance of real trigonometric transform techniques-based optical OFDM systems. Based upon the previous research efforts, it is clear that among the PAPR reduction techniques, spreading technique exhibits its prominence. Hence, the major focus of this chapter is to present an elaborate discussion on spreading technique along with the comparison with multiple signalling and probabilistic technique like PTS, and signal distortion technique like clipping and filtering. Firstly, it is vital to statistically characterize the DCO-OFDM system exploiting DCT, i.e., FOOFDM system, the same phenomenon holds well for DST-DCO-OFDM system. For any choice of subcarrier number, an FOOFDM symbol (in general, a DCO-OFDM symbol) can be treated as the sum of a group of independent and identically distributed (i.i.d.) samples. By utilizing central limit theorem, the FOOFDM symbol approximately becomes Gaussian. The mean of the FOOFDM symbol is equal to zero. The FOOFDM signal $s(n)$ can be modeled as an i.i.d. Gaussian process with probability density function (PDF) $P_{DCO}(s)$ which is given as

$$P_{DCO}(s) = N\left(s;\beta_{DC},\sigma^2\right) \tag{4.1}$$

Where $N\left(s;\beta_{DC},\sigma^2\right)$ is Gaussian and is given as

$$N\left(s;\beta_{DC},\sigma^2\right) = \frac{1}{\sqrt{2\pi\sigma^2}}e^{-\frac{(s-\beta_{DC})^2}{2\sigma^2}} \tag{4.2}$$

The optical power is given as

$$\underset{optical}{P_{DCO}(s)} = \int_{-\infty}^{\infty} sP_{DCO}(s)\,ds \tag{4.3}$$

Upon substitution of (4.2) into (4.3)

$$\underset{optical}{P_{DCO}(s)} = \int_{0}^{\infty} s\frac{1}{\sqrt{2\pi\sigma^2}}e^{-\frac{(s-\beta_{DC})^2}{2\sigma^2}}ds \tag{4.4}$$

Solving for the terms in integral by assuming $\frac{s-\beta_{DC}}{\sigma} = t$, then equation (4.4) reduces to

$$\underset{optical}{P_{DCO}(s)} = \frac{\sigma}{\sqrt{2\pi\sigma^2}}\int_{-\frac{\beta_{DC}}{\sigma}}^{\infty}(\sigma t+\beta_{DC})\,e^{-\frac{t^2}{2}}dt \tag{4.5}$$

(4.5) can be further solved to attain

$$\underset{optical}{P_{DCO}(s)} = \frac{\sigma}{\sqrt{2\pi}}\int_{-\frac{\beta_{DC}}{\sigma}}^{\infty} te^{-\frac{t^2}{2}}dt + \frac{\beta_{DC}\sigma}{\sqrt{2\pi\sigma^2}}\int_{-\frac{\beta_{DC}}{\sigma}}^{\infty} e^{-\frac{t^2}{2}}dt \tag{4.6}$$

The integral in the first term of (4.6) can be solved by assuming $t^2 = u$

$$\underset{optical}{P_{DCO}}(s) = \frac{\sigma}{\sqrt{2\pi}} \int_{\left(\frac{\beta_{DC}}{\sigma}\right)^2}^{\infty} \frac{1}{2} e^{-\frac{u}{2}} du + \beta_{DC} \frac{1}{\sqrt{2\pi}} \int_{-\frac{\beta_{DC}}{\sigma}}^{\infty} e^{-\frac{t^2}{2}} dt \quad (4.7)$$

Upon solving the integral in the first term and by writing the integral in the second term by means of Q-function as

$$Q(z) = \frac{1}{\sqrt{2\pi}} \int_{z}^{\infty} e^{-\frac{x^2}{2}} dx \quad (4.8)$$

Thus, the optical power in (4.7) can be obtained as

$$\underset{optical}{P_{DCO}}(s) = \frac{\sigma}{\sqrt{2\pi}} e^{-\frac{\beta_{DC}^2}{2\sigma^2}} + \beta_{DC} Q\left(-\frac{\beta_{DC}}{\sigma}\right) \quad (4.9)$$

When less amount of bias β_{DC} is added, i.e., when β_{DC} is small, then it results in clipping of lots of signal power, thereby leading to the emanation of clipping noise. Therefore, in order to overcome clipping noise, β_{DC} should be large enough. So, when β_{DC} is sufficiently large enough, then the optical power as represented in (4.9) is approximately equal to β_{DC}. Then,

$$\underset{optical}{P_{DCO}}(s) \sim \beta_{DC} \quad (4.10)$$

From (4.10), it is obvious that the optical power increases with the increase in added bias value. Thereby, this corresponds to reduction in power efficiency. Thus, it is vital to reduce the amount of PAPR in such systems.

4.3 PAPR ANALYSIS IN DCT/DST-BASED MULTICARRIER SYSTEM

This section deals with the performance analysis of the real trigonometric transform-based, i.e., DCT or DST-based spreading technique for PAPR reduction along with the presentation of PTS technique and clipping and filtering techniques. Followed up by mathematical derivations, the proof of concept is evaluated by means of simulation results which will be illustrated in the subsequent sections.

4.3.1 PERFORMANCE ANALYSIS OF DCT/DST-BASED SPREADING TECHNIQUES FOR PAPR REDUCTION

As evident from the schematic representations in Figs. 4.2 and 4.3, DCT-Spread (DCT-S) FOOFDM system along with the DST-Spread (DST-S) optical OFDM system employ much simpler and real signal processing techniques like the DCT and DST abandoning the deployment of Hermitian Symmetry imposed complex signal processing technique like IFFT. Moreover, these trigonometric transform techniques stem out as the most appealing solution to yield a pure real-valued signal. Especially,

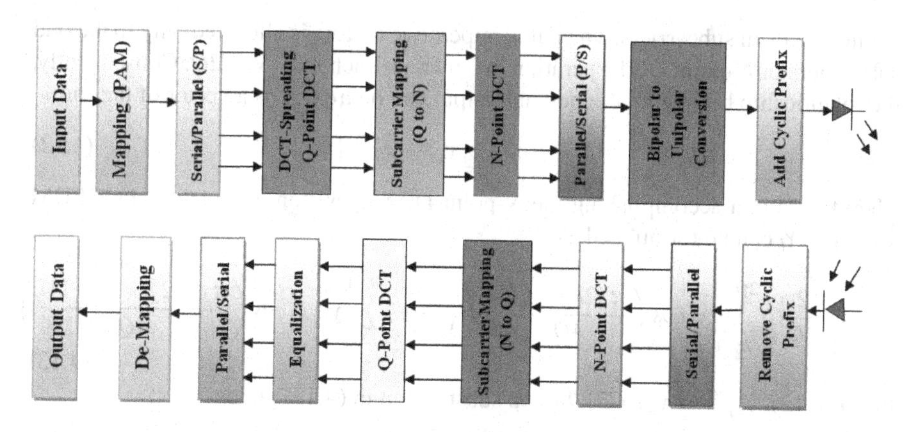

Figure 4.2: DCT-Spread FOOFDM for VLC [499].

DCT transforms possess excellent energy concentration property and exhibit a remarkable trait that works well with the principalities of IM/DD systems. From a list of earlier renowned works [60, 541], etc., it can be proclaimed that the major motive to spread an input sequence is to dramatically subside the high amount of PAPR much similar to the capacity of single carrier transmission systems.

As stated earlier and throughout, the fundamental aspect which needs to be taken into account while designing or formulating any modulation scheme, signal processing technique, etc., is that it is imperative for the transmission signal statistics to obey the signal processing characteristics of IM systems. This implies that the transmitted signal statistics in the time domain should be both real-valued and non-negative in nature. Thus, as delineated from the transmitter and receiver schematics of the DCT-S-FOOFDM system, the input data signal is converted into a set of symbols drawn from the real-constellation mapping like M-ary PAM. Without relying on complex mapping techniques like M-ary QAM, a much simpler and one-dimensional real-mapping format like M-PAM is sufficient to fetch a real-valued signal. Nonetheless, it is not necessary for the yielded real time-domain signal to be non-negative-valued as well. Therefore, this necessitates to adopt the ACO-OFDM principalities in order to realize a unipolar signal transmission as well. Consequently, the set of real-mapped symbols are transmitted in parallel with the aid of serial to parallel converter. Thereupon, in order to realize a DCT-Spread FOOFDM system, the set of parallel data symbols are spread by means of a Q-point DCT transform technique.
Predominantly, the mathematical expression for the output signal upon enforcing Q-point DCT operation can be defined as

$$x_m = \sqrt{\frac{2}{Q}} \sum_{q=0}^{Q-1} C_q X_q cos\left(\frac{\pi(2m+1)q}{2Q}\right), \; 0 \leq m \leq Q-1 \tag{4.11}$$

As discussed in Chapter 3, the fundamental aspect that is associated with ACO-OFDM system is that only the odd subcarriers are exploited for data mapping leaving behind the even subcarriers. Thus, it is imperative that x_m is allocated only to the odd subcarriers of N-point DCT operation in a manner such that, $N = 2Q$. Consequently, the relationship between the input and output can be treated as a vector of the form

$$Y = [0, x_0, 0, x_1, \cdots x_{m-1}] \tag{4.12}$$

Therefore, upon accomplishing the N-point DCT operation, the yielded FOOFDM sequence Y_i can be formulated as:

$$Y_i = \sqrt{\frac{2}{2Q}} C_i \sum_{h=0}^{2Q-1} y_h cos\left(\frac{\pi(2h+1)i}{4Q}\right) = \sqrt{\frac{1}{Q}} C_i \sum_{j=0}^{Q-1} y_{2j+1} cos\left(\frac{\pi(2(2j+1)+1)i}{4Q}\right) \tag{4.13}$$

From (4.12), $y_{2j+1} = x_j$ [577]. So, on substitution in (4.13) yields

$$Y_i = \sqrt{\frac{1}{Q}} C_i \sum_{j=0}^{Q-1} x_j cos\left(\frac{\pi(4j+3)i}{4Q}\right) \tag{4.14}$$

$$x_j = \sqrt{\frac{2}{Q}} \sum_{q=0}^{Q-1} C_q X_q cos\left(\frac{\pi(2j+1)q}{2Q}\right) \tag{4.15}$$

Upon substituting the expression for the input signal as portrayed in (4.15) into the mathematical expression as elucidated by equation (4.14), the output sequence attained is as follows [499]

$$Y_i = \frac{1}{\sqrt{2}Q} C_i \sum_{j=0}^{Q-1} \sum_{q=0}^{Q-1} C_q X_q \left[cos\left(\frac{4\pi(i-q)j + \pi(3i-2q)}{4Q}\right) + cos\left(\frac{4\pi(i+q)j + \pi(3i+2q)}{4Q}\right)\right] \tag{4.16}$$

The negative components in Y_i are clipped to zero to yield a unipolar real-valued signal $\underset{unipolar}{y}$ given as

$$\underset{unipolar}{y} = \begin{cases} Y_i, & Y_i > 0 \\ 0, & Y_i \leq 0 \end{cases} \tag{4.17}$$

In particular, with the major emphasis to overcome the ISI, which generally prevails due to the dispersive nature of the channel, it is vital to add a sufficient amount of cyclic prefix. In general, the amount of cyclic prefix is appended at a rate of $\frac{1}{4}$th of the subcarrier's size. Thereupon, time attained real-valued and unipolar time-domain signal is propagated through the optical channel. Further, at the receiving end, inverse operations such as removal of cyclic prefix, N-Point DCT, P-Point DCT, de mapping, etc., are implemented in order to ensure a reliable recovery of the transmitted data stream. Generally, in OFDM systems, the cyclic prefix/guard interval is widely employed with the primary reason to convert the linear convolution of the OFDM/FOOFDM signal with the channel into circular convolution; hence, a fundamental frequency domain equalizer like zero-forcing can be employed.

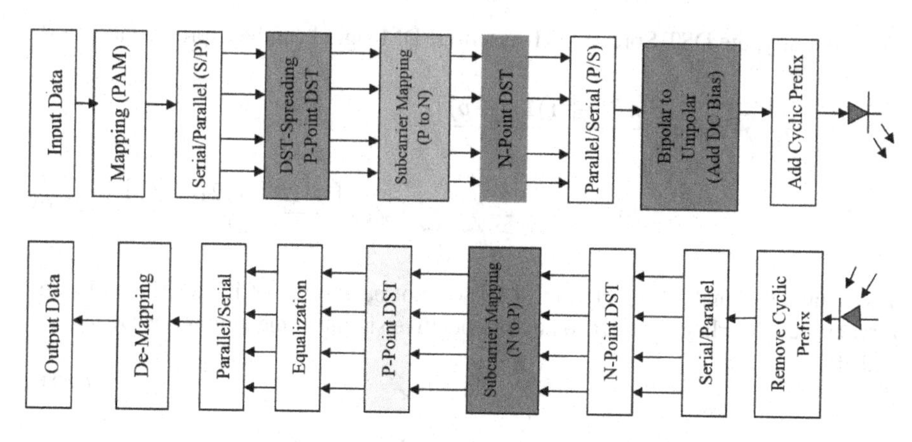

Figure 4.3: DST-Spread-DCO-OFDM system for VLC [494, 495].

Complementary Cumulative Distribution Function (CCDF) for PAPR: The PAPR is generally the ratio between the maximum peak power and the average power of the discretized OFDM signal and is represented as

$$PAPR = 10log_{10}\left(\frac{Max\left\{\left|\underset{unipolar}{y}\right|^{2}\right\}}{E\left\{\left|\underset{unipolar}{y}\right|^{2}\right\}}\right) \tag{4.18}$$

From (4.18), $E\{\}$ denotes the expectation operator. One is familiar with, CCDF being used for evaluating the probability that PAPR of a particular OFDM time-domain symbol exceeds a certain value of threshold PAPR_0 which is given as

$$CCDF = Pr(PAPR > PAPR_0) \tag{4.19}$$

Since the addition of DC bias in DCO-OFDM system has resulted in a power-inefficient scheme, there is a necessity to reduce the amount of PAPR in DCO-OFDM system. Consequently, this work enforces the PAPR reduction techniques to DST-based DCO-OFDM system. The block diagram of DST-based spreading in DST-based DCO-OFDM system is represented in Fig. 4.3. In contrast to Hermitian Symmetry imposed IFFT and FFT, this system incorporates the DST transformation module for both multiplexing and de-multiplexing. The M-PAM mapped data is loaded into the P-point DST block to accomplish the DST-Spreading operation.

Consequently, the DST-Spread DST-based-OFDM signal can be expressed as [495]

$$Y_i = \frac{1}{P}\sum_{h=0}^{P-1}\sum_{p=0}^{P-1} X_p cos\left[\frac{\pi(2h+1)2(i-p)}{4P}\right] - \frac{1}{P}\sum_{h=0}^{P-1}\sum_{p=0}^{P-1} X_p cos\left[\frac{\pi(2h+1)2(i+p)}{4P}\right] \quad (4.20)$$

The obtained signal as depicted in (4.20) cannot be assured of its positivity. Therefore, as stated earlier, in order to attain a positive signal, some amount of DC bias is added.

$$\underset{Tx}{Y_i} = Y_i + \beta_{DC} \quad (4.21)$$

Finally, PAPR is computed to this transmitted time-domain signal which is denoted as $\underset{Tx}{Y_i}$. The PAPR of the time-domain signal is given as the ratio of the maximum peak power to average power.

$$PAPR = 10log_{10}\left(\frac{Max\left\{\left|\underset{Tx}{Y_i}\right|^2\right\}}{E\left\{\left|\underset{Tx}{Y_i}\right|^2\right\}}\right) \quad (4.22)$$

4.3.2 EXPLOITATION OF PTS TECHNIQUE IN DCT/DST-BASED OPTICAL OFDM

In general, PTS is one of the multiple signalling and probabilistic techniques which is widely employed in RF-based traditional OFDM systems to reduce the amount of PAPR. The basic principal methodology involved behind the implementation of PTS technique is that the incoming huge stream of data is partitioned into sub-blocks before getting applied to the transformation module, i.e., the transformation operation is computed individually to each sub-block and then weighted by a phase factor. However, this technique is incorporated to reduce the levels of PAPR in DCT and DST-based optical OFDM system in accordance with IM/DD systems and is elucidated in Figs. 4.4 and 4.5. In order to show the difference between the mathematical expressions, we have enforced PTS technique to DCT-based ACO-OFDM and DST-based DCO-OFDM system. The schematic representation of FOOFDM VLC transmitter employing conventional PTS technique for PAPR reduction is illustrated in Fig. 4.4. The serialized BPSK/M-PAM mapped data stream is parallelized and then partitioned into L disjoint subblocks of equal size in a manner as depicted in (4.23).

$$X = \left[X^0, X^1, X^2, X^3, \cdots X^{L-1}\right]^T \quad (4.23)$$

Similar to ACO-OFDM methodology, the arrangement in each sub-block X^l for $0 \leq l \leq L-1$ is made in such a way that only odd subcarriers are modulated setting the

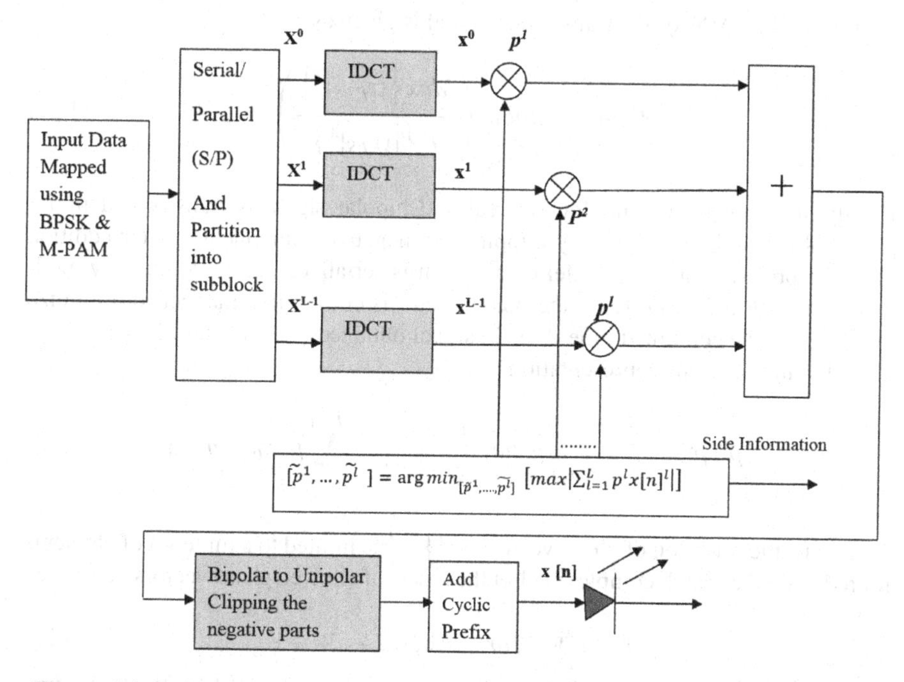

Figure 4.4: PAPR reduction in FOOFDM using partial transmit sequence (PTS).

even subcarriers to zero. So, if there are N subcarriers only $\frac{N}{2}$ are utilized due to inclusion of odd subcarriers. Hence, the representation looks like

$$X^{l}_{PTS} = \left[\left\{X^{l}_{PTS}[1]\right\}, 0, \left\{X^{l}_{PTS}[3]\right\}, 0, \cdots \left\{X^{l}_{PTS}\left[\frac{N}{2}-1\right]\right\}\right]^{T} \tag{4.24}$$

Therefore, this signal is fed to the input of the IDCT to yield the time-domain signal which can be expressed as

$$\underset{DCT-ACO}{x^{l}_{PTS}[n]} = \sum_{l=0}^{L-1}\left\{\sqrt{\frac{2}{N}}\sum_{k=0}^{\frac{N}{2}-1} C^{l}[k] X^{l}_{PTS}[k] \cos\left(\frac{\pi(2n+1)k}{2N}\right)\right\} \tag{4.25}$$

The obtained time-domain signal $x^{l}_{PTS}[n]$ is phase rotated independently by getting multiplied with the corresponding complex phase factor $p^{l} = e^{j\phi_{l}}$ where $l = 0, 1, 2, 3, \cdots L-1$ and $\phi_{l} = [0, 2\pi)$ to yield the combined time-domain signal with the lowest PAPR as follows:

$$\widetilde{x_{PTS}} = \sum_{l=0}^{L-1} p^{l} \underset{DCT-ACO}{x^{l}_{PTS}[n]} \tag{4.26}$$

Therefore, the PAPR of the transmitted signal is given as

$$PAPR = 10log_{10}\left(\frac{Max\left\{|\widetilde{x_{PTS}}|^2\right\}}{E\left\{|\widetilde{x_{PTS}}|^2\right\}}\right) \tag{4.27}$$

Finally, the composite time-domain real and bipolar signal is transformed into a pure real and unipolar signal by clipping the negative components. After addition of cyclic prefix, then this parallel data stream is serialized and then intensity modulated through the LED. Here, the phase vector is chosen in a manner to minimize the PAPR of the combined time-domain signal obtained from all sub-blocks. This is given by the following representation

$$\left[\tilde{p}^1, \tilde{p}^2, \cdots \tilde{p}^l\right] = \underset{\left[\tilde{p}^1, \tilde{p}^2, \cdots \tilde{p}^l\right]}{arg\ min}\left(\underset{0,1,2,\cdots N-1}{max}\left|\sum_{l=0}^{L-1} p^l\, x^l_{\substack{PTS\\DCT-ACO}}[n]\right|\right) \tag{4.28}$$

Generally, the selection of phase vectors $\{p^l\}_{l=0}^{L-1}$ is limited to a finite set of elements for reducing the search complexity. Let these sets of allowed phase vectors be defined as

$$p^l = e^{\frac{j2\pi q}{W}}, for\ q = 0, 1, 2, \cdots W-1 \tag{4.29}$$

In (4.29), the total number of allowed phase vectors is denoted by W. If the first phase factor p^1 is set to 1, without any loss of information, then there are a total of $L-1$ phase factors to be found by exhaustive search. Therefore, a total of W^{L-1} sets of phase factors are to be searched to find the optimum one. It is to be taken into consideration that the amount of PAPR reduction has a direct relationship between the phase factors W and the number of sub-blocks L. So here, with the increase in the number of sub-blocks, the PAPR of the overall system decreases significantly but at the expense of exponential increase in search complexity. Moreover, L number of IDCTs are involved for each data sub-block, so it requires log_2W^{L-1} bits of side information to be transmitted.

In order to reduce the amount of PAPR as well as to substantially reduce the search complexity of the phase vectors, a sub optimal way of choosing the phase factors is proposed in the literature. Accordingly, all the phase factors are initially set to 1 followed by the computation of PAPR. p^1 remains 1 and the values of other phase factors are chosen among W possible sets. Then in the next iteration, the second phase factor is changed to p^2 and then PAPR is computed. Therefore, if the PAPR computed upon choosing p^2 is less than the previous one, then this phase factor is chosen as the part of the final set of phase factors. This procedure repeats until all the phase factors are explored. Moreover, for PAPR reduction using PTS, every sub-block should imbibe the Hermitian Symmetry criteria to meet the requirements of IM/DD systems. Eventually, this incurs a lot of additional complexity overhead.

Hence, this necessitates to stress on real transformation techniques like DCT and DST. As evident from Fig. 4.5, the incoming huge stream of mapped data is partitioned into L disjoint sub-blocks as delineated in (4.23). In particular, these sub-blocks X^l where l lies in the interval $0 \leq l \leq L-1$ are of equal size and are sequentially located. Now, IDST operation is computed individually for each sub-block to

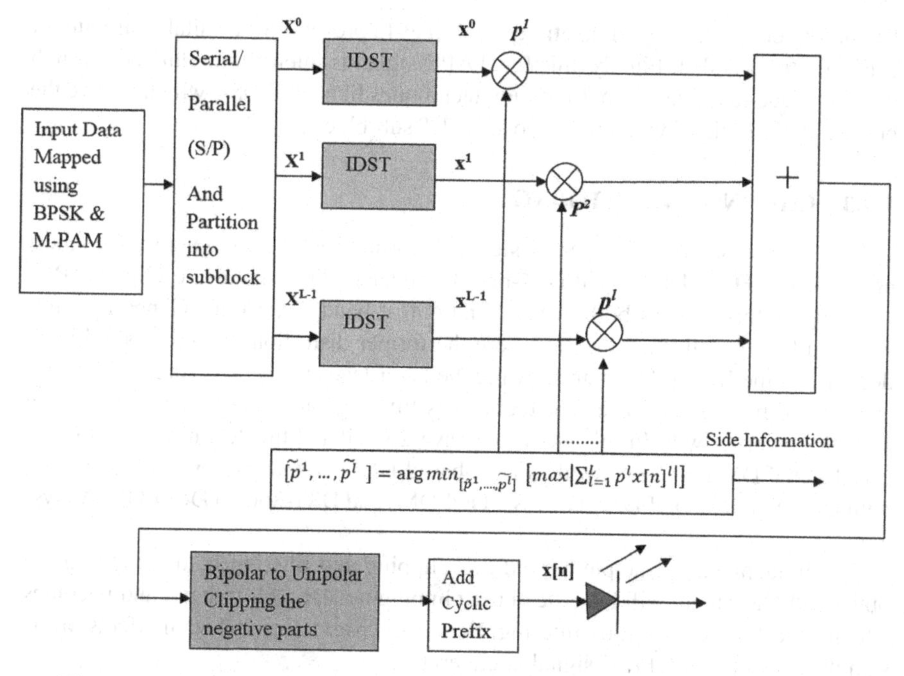

Figure 4.5: PAPR reduction in DST-DCO-OFDM system using PTS [495].

attain the time-domain signal and the mathematical representation looks as follows:

$$\underset{DST-DCO}{x^{l}_{PTS}[n]} = \sum_{l=0}^{L-1} \left\{ \sqrt{\frac{2}{N}} \sum_{k=0}^{N-1} X^{l}_{PTS}[k] \sin\left[\frac{\pi(2n+1)(2k+1)}{4N}\right] \right\} n = 0,1,2,3,\cdots N-1 \tag{4.30}$$

Now, signal scrambling is applied individually to each sub-block, i.e., each block of time-domain signal as depicted in (4.30) is individually weighted by a phase factor. Therefore, the representation looks like

$$\widetilde{x_{PTS}} = \sum_{l=0}^{L-1} p^{l} \underset{DST-DCO}{x^{l}_{PTS}[n]} \tag{4.31}$$

The primary motive of PTS technique is to reduce the PAPR of the combined time-domain signal as represented in (4.31), by carefully selecting the phase factors which are given mathematically as

$$\left[\tilde{p}^{1}, \tilde{p}^{2}, \cdots \tilde{p}^{l}\right] = \underset{[\tilde{p}^{1}, \tilde{p}^{2}, \cdots \tilde{p}^{l}]}{arg\ min} \left(\underset{0,1,2,\cdots N-1}{max} \left| \sum_{l=0}^{L-1} p^{l} \underset{DST-DCO}{x^{l}_{PTS}[n]} \right| \right) \tag{4.32}$$

The selection of the phase vectors is detailed in the previous subsection, i.e., PTS technique for DCT-ACO-OFDM system. In the rest of the operations like addition of

DC bias to ensure the signal positivity as well as to prevent ISI, a suitable amount of cyclic prefix is added. Finally, this low PAPR signal is intensity modulated through the LED. Due to the real transformation techniques like DCT/DST which relieve the burden of Hermitian Symmetry imposed IFFT sub-blocks.

4.3.3 CLIPPING AND FILTERING

This is one of the fundamental signal distortion techniques where the highest peaks of the FOOFDM signal/DST-based optical OFDM signal are clipped. But, in doing so, this induces both in-band and out-of-band distortions. Generally, clipping can be a non-linear process, where the former distortion leads to a significant decrease in the BER performance, while the latter leads to spectral spreading. However, out-of-band distortion can be reduced by filtering, but at the expense of increase in peak power growth. In this work, we have assimilated this technique in conventional FOOFDM system as well as DST-based DCO-OFDM system for comparing with DCT-Spread FOOFDM (DCT-S-FOOFDM) and DST-Spread DCO-OFDM system.

The fundamental principle behind this clipping and filtering is that clipping the optical OFDM signal will lead the entire clipping noise to fall in-band and becomes a tedious task to remove it by filtering. Hence, in order to combat such effects, oversampling the optical OFDM signal is imperative.

$$x_m^{'} = \underset{FOOFDM}{x_m^{'}} = \sqrt{\frac{2}{PN}} \sum_{k=0}^{PN-1} C_k X_k^{'} cos\left(\frac{\pi\,(2m+1)\,k}{2PN}\right) \tag{4.33}$$

Meanwhile, the time-domain signal corresponding to DST-based DCO-OFDM system is given as

$$x_m^{'} = \underset{DST-DCO}{x_m^{'}} = \sqrt{\frac{2}{PN}} \sum_{k=0}^{PN-1} X_k^{'} sin\left[\frac{\pi\,(2m+1)\,(2k+1)}{4PN}\right], \quad m = 0,1,2,3,\cdots PN-1 \tag{4.34}$$

where $x_m^{'}$ in (4.33) and (4.34) represents the discretized time domain of FOOFDM and DST-DCO-OFDM signal and P denotes the oversampling factor. Primarily, oversampling is implemented by affixing $N.(P-1)$ zeros in the frequency domain.

Since the physical characteristics of opto-electronic devices oblige the VLC system to exploit IM to drive the LED, this necessitates the transmitted signal to be both real and unipolar. Hence, in this scenario, a desired amount of upper clipping and lower clipping followed by the addition of DC bias to fetch a unipolar signal is illustrated mathematically as given below.

$$\underset{clip}{x_m^{'}} = \begin{cases} A_u, & if\ x_m^{'} \geq A_u \\ x_m^{'}, & -A_l < x_m^{'} < A_u \\ -A_u, & if\ x_m^{'} \leq -A_l \end{cases} \tag{4.35}$$

where A_u and A_l in (4.35) denote the upper and lower clipping levels, respectively. Therefore, the upper and lower clipping ratios can be defined as

$$Upper\ Clipping\ Ratio = \frac{A_u^2}{\sigma^2} \tag{4.36}$$

$$Lower\ Clipping\ Ratio = \frac{A_l^2}{\sigma^2} \tag{4.37}$$

where σ^2 is the variance of the time-domain signal. The simulated results evidence a significant decrease in PAPR by incorporation of this technique, but the decrease is not predominant when compared with DCT-Spreading in FOOFDM system as well as DST-Spreading in DST-DCO-OFDM system.

4.3.4 PERFORMANCE ANALYSIS OF PAPR REDUCTION TECHNIQUES IN A DCT/DST-BASED MULTICARRIER SYSTEM

For the mathematical analysis performed in the earlier sections, this section provides the proof-of-concept in order to determine the PAPR performance of different PAPR reduction techniques in DCT/DST-based multicarrier system. The significance

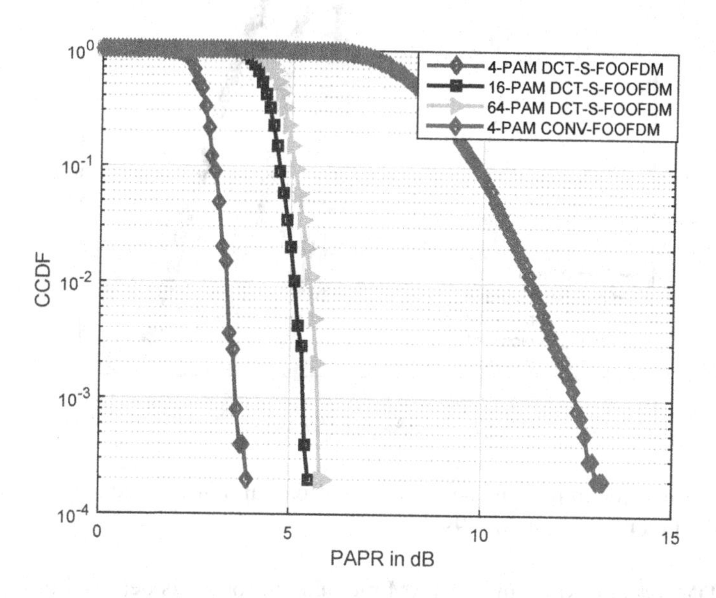

Figure 4.6: Performance of DCT-based spreading in DCT-based optical OFDM, i.e., (FOOFDM) system exploiting M-PAM modulation format [499].

of imposing spreading techniques for PAPR reduction in FOOFDM system is clearly evident in Fig. 4.6. In order to determine the prominence of spreading technique, comparisons are carried out with the traditional FOOFDM system using 4-PAM modulation format. Upon imposing DCT-based spreading technique, the PAPR reduces drastically. For instance, at a probability of 2×10^{-4}, the attained PAPR for DCT-Spread-FOOFDM system is around 3.9 dB, while for the traditional FOOFDM system without any spreading technique, the attained PAPR is around 13 dB. This analysis reveals that spreading technique plays a vital role in reducing the amount of PAPR greatly, where a gain of around 10.9 dB is achieved. Furthermore, in order to depict the superiority of spreading technique in FOOFDM system, comparisons were carried out for higher orders of modulation which include the 16-PAM and 64-PAM, respectively. With the increase in orders of modulation, there is an increment in PAPR. However, due to the enforcement of spreading technique, the amount of PAPR reduction in FOOFDM system using 16-PAM and 64-PAM is much better when compared with the traditional FOOFDM system without any PAPR reduction technique. This is because at CCDF of around 2×10^{-4}, DCT-S-FOOFDM system using 16-PAM and 64-PAM modulation formats yields a PAPR reduction of around 7.5 and 7.2 dB, respectively. The effect of PAPR reduction by using PTS technique

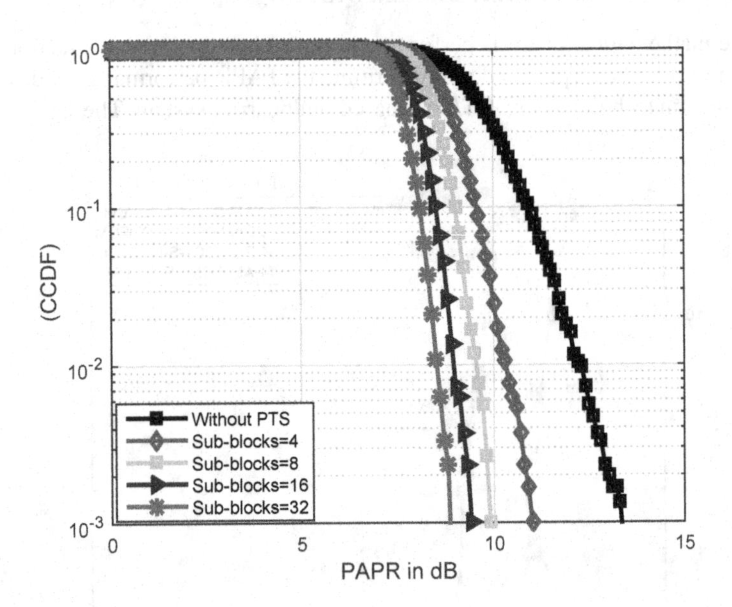

Figure 4.7: Interpretation of the performance of partial transmit sequence for PAPR reduction in FOOFDM system [499].

in FOOFDM system exploiting 16-PAM modulation format is depicted in Fig. 4.7. For the purpose of simulation, the total number of subcarriers employed were 512. As delineated, with the increase in the number of sub-blocks, the PAPR reduces drastically. The total number of sub-blocks employed are 4, 8, 16 and 32, respectively. In order to highlight the amount of PAPR reduction by using PTS technique, comparisons were carried out with the traditional FOOFDM system without enforcing any PTS technique. When the number of sub-blocks employed were 32, at a CCDF of approximately 10^{-3}, there is a significant reduction in PAPR and when compared with conventional FOOFDM system without the application of PTS technique, a gain of around 4.5 dB is achieved. The same inference can be drawn with other sub-blocks as well, because, when 16, 8 and 4 sub-blocks were employed, then when compared with traditional FOOOFDM system, PTS technique achieves a gain of around 3.9, 3.4 and 2.1 dB, respectively. Predominantly, this analysis clearly affirms that PAPR reduces by a significant amount upon exploiting PTS technique in FOOFDM system. However, DCT-Spreading technique outperforms PTS technique for PAPR reduction in FOOFDM system. The PAPR reduction by using the signal

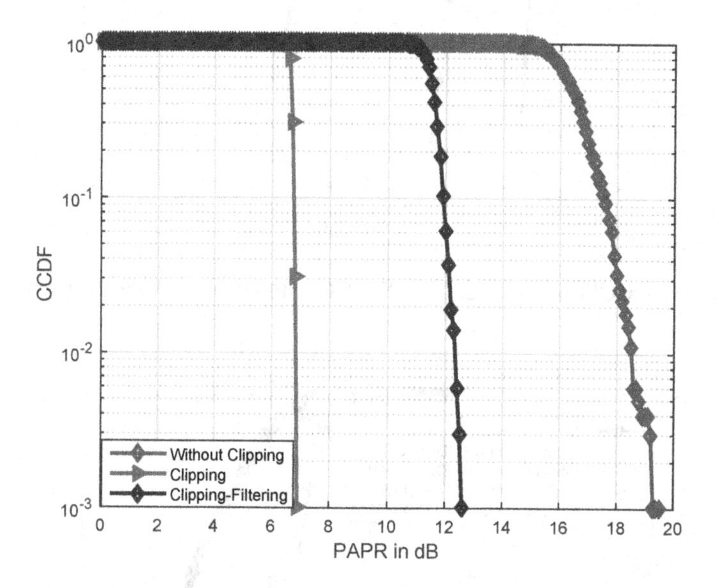

Figure 4.8: PAPR performance of FOOFDM system using clipping and filtering technique [499].

distortion and probabilistic technique in FOOFDM system is delineated in Fig. 4.8. From the simulated result, it is evident that when compared with filtering technique, the reduction in PAPR in FOOFDM system is much better upon incorporating clipping technique. For FOOFDM system employing 4-PAM modulation format, the achieved PAPR for clipping and clipping with filtering operations is around 6.9, 12.6 dB, respectively. Whereas, for a traditional FOOFDM system, the attained PAPR is around 19.3 dB. The amount of PAPR incurred in traditional FOOFDM system is high due to the inherent fact of addition of DC bias. Individually, clipping and filtering techniques exhibit their prominence in reducing the PAPR in optical OFDM system, but the amount of PAPR reduction when compared with a spreading technique is not that impressive. In a similar manner, DST-based spreading is incorporated in DST-based DCO-OFDM system in order to reduce the amount of PAPR. The simulated result as portrayed in Fig. 4.9 signifies that for DST-based DCO-OFDM system

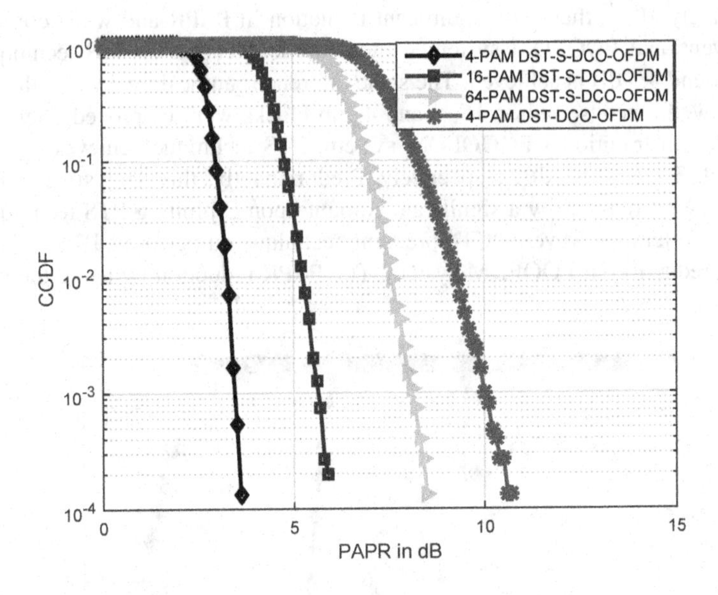

Figure 4.9: CCDF curves for DST-S-DCO-OFDM using M-PAM [495].

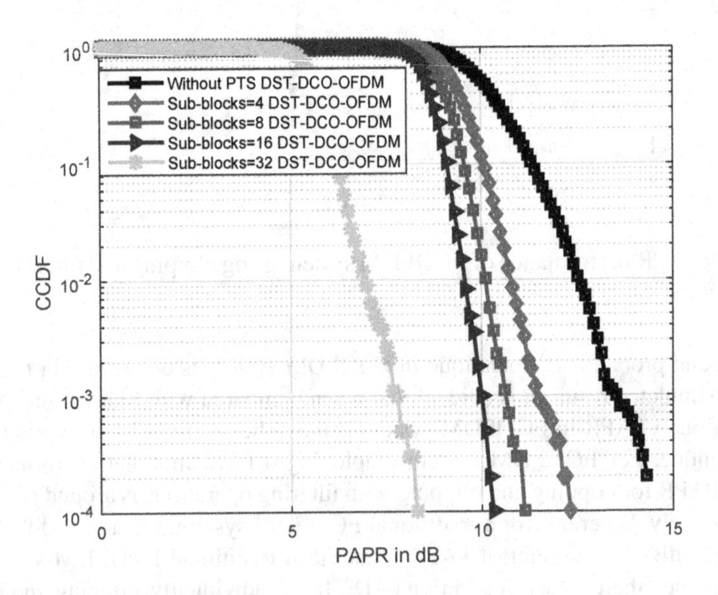

Figure 4.10: CCDF vs. PAPR for PAPR reduction using partial transmit sequence (PTS) in DST-DCO-OFDM system [495].

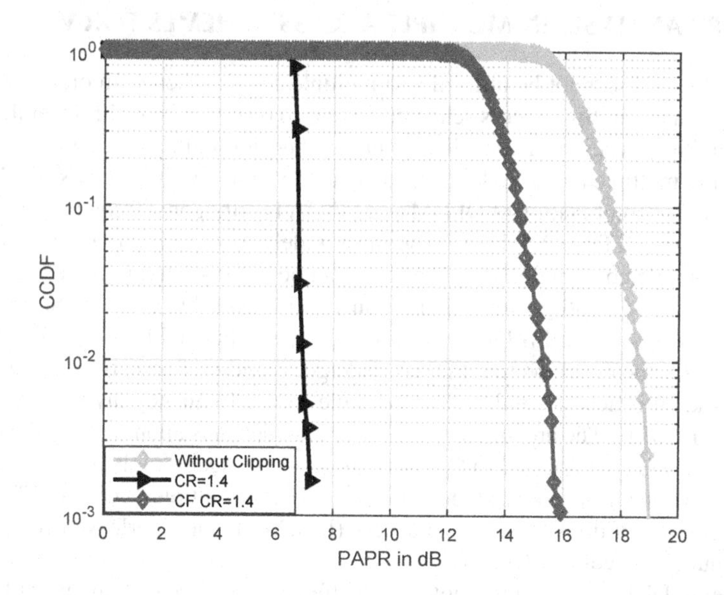

Figure 4.11: CCDF vs. PAPR for PAPR reduction using clipping and filtering in DST-DCO-OFDM system [495].

exploiting 4-PAM modulation format and upon imposing DST-based spreading, the reduction in PAPR is significant when compared with traditional DST-based DCO-OFDM system without any spreading operation. For the same order of modulation, i.e., 4-PAM, at a CCDF of approximately 10^{-4}, the attained PAPR in DST-S-DCO-OFDM system is 3.6 dB, while the traditional DST-based DCO-OFDM system achieves a PAPR of around 10.7 dB. This clearly confirms that a gain of around 7.1 dB can be obtained from the spreading technique. Furthermore, upon increasing the orders of modulation to 16-PAM and 64-PAM, the achieved PAPR was around 5.9 and 8.5 dB, respectively. Thus, when compared with traditional DST-based DCO-OFDM system using 4-PAM modulation format, DST-S-DCO-OFDM system employing 16-PAM and 64-PAM modulation formats, achieve a gain of around 4.8 and 2.2 dB, respectively.

In order to depict the superiority of spreading technique, the CCDF vs. PAPR performance of PTS and clipping and filtering techniques in DST-based DCO-OFDM system were analyzed, and this is illustrated in Figs. 4.10 and 4.11. Pertaining to PTS technique, upon increasing the number of sub-blocks to 32, the amount of reduction in PAPR is significant. At a CCDF of 10^{-4}, DST-based DCO-OFDM system with PTS technique achieves a gain of around 6 dB when compared with the traditional DST-based DCO-OFDM system without enforcement of any PAPR reduction techniques. Dealing with the scenario of clipping and filtering technique, at a probability of 10^{-3}, for a clipping ratio of 1.4, clipping technique achieves a PAPR of 7 dB, while clipping and filtering together attains a PAPR of 16 dB.

4.4 PAPR ANALYSIS IN MULTIPLE ACCESS SCHEMES FOR VLC

OFDM by itself cannot be utilized as a multiple access scheme. In order for it to be employed as a multiple access scheme, it needs to be interfaced with several other existing multiple access schemes like frequency division multiple access (FDMA), time division multiple access (TDMA), and code division multiple access (CDMA). The well-known combination of OFDM and FDMA is orthogonal frequency division multiple access (OFDMA), which is the most widely used multiple access scheme in cellular mobile communication. The same can be incorporated in optical domain complying with real and positive signal transmission and this type of multiple access scheme employed in IM/DD systems is called as optical OFDMA (OOFDMA). However, since the envelope in OFDMA undergoes random fluctuations giving rise to high peaks, it is necessitated to rely on single carrier frequency division multiple access (SC-FDMA). The amount of PAPR reduction depends upon the way the number of subcarriers are allocated to the users/subscribers. Accordingly, there are three different ways, namely, distributed frequency division multiple access (DFDMA), localized frequency division multiple access (LFDMA), and interleaved frequency division multiple access (IFDMA).

In general, DFT spreading is employed to effectively reduce the amount of PAPR of a multicarrier system to the level of a single carrier system. Therefore, M point DFT is employed for spreading, and N point IDFT is employed for fetching a time-domain signal. In case of DFDMA, the M DFT outputs are distributed over the entire band of N subcarriers with zeros filled in the unused $(N-M)$ subcarriers. While, if DFDMA distributes the M DFT outputs with equidistant as $\frac{N}{M} = Q$, where Q is the spreading factor, then it is referred to as IFDMA. Whereas, LFDMA distributes the DFT outputs to M consecutive subcarriers out of the N subcarriers. Relevant works pertaining to the exploitation of the aforesaid multiple access schemes to IM/DD systems can be reported in [128, 299, 300]. The research in [128] compares the PAPR performance of optical OFDM interleaved division multiple access (O-OFDM-IDMA) with OOFDMA. The simulation results in this work suggest that the O-OFDM-IDMA is more power-efficient than O-OFDMA, especially for higher throughput at the cost of higher computational complexity. However, the CCDF curves for O-OFDM-IDMA and OOFDMA do not show much difference. The superiority of SC-FDMA over OOFDMA in WLED-based communication system is outlined in [299]. The work in [300] proposes an improved frequency-domain decision feedback equalizer (FD-DFE) for SC-FDMA-based VLC system. This proposed structure remarkably reduces the decision error of FD-DFE and at the same time improves the BER performance as well as the transmission performance. Figure 4.12 interprets an uplink transmission scenario using visible light in an indoor environment emphasizing that the channel consists of both LOS as well as NLOS components. This growing technology like VLC provides the flexibility for creation of a small-scale communication network in an indoor room environment where each installed LED can act as a base station rendering services to multiple mobile terminals which are within the vicinity of LED lighting fixtures. In this regard, it is vital to exploit multiple access schemes for imparting high data rate communication. OFDM

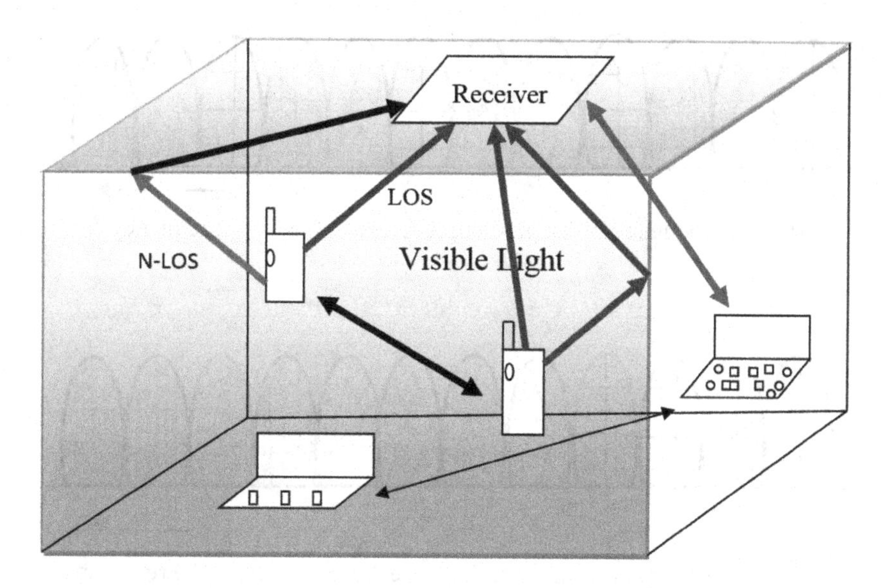

Figure 4.12: Illustration of uplink scenario using visible light.

is one straightforward technique which is easily reconcilable with multiple access schemes giving rise to OFDMA. Meanwhile, exploiting OFDMA to optical domain, it can be called optical OFDMA (OOFDMA).

It is imperative that, in case of OFDMA system, the subcarriers are partitioned and are assigned to multiple mobile users. Precisely, OFDMA facilitates the subcarriers belonging to each OFDM symbol to be divided orthogonally among multiple users. However, OFDMA waveform exhibits rapid envelope fluctuations resulting in high amount of PAPR. Therefore, in order to reduce the amount of PAPR in a multicarrier system, SC-FDMA can be expedited for the reduction of PAPR. This is implemented by enforcing M-point DFT spreading before the N-point IDFT operation at the transmitting end. In case of SC-FDMA, each of the terminals in the uplink employs a certain set of subcarriers to transmit its own data. Meanwhile, the subcarriers which are not involved in data transmission will be filled with zeros. Further, upon incorporating DFT-Spreading, there exist different ways of allocating subcarriers to a specific user. Based upon the way of this assignment of subcarriers to each terminal, there is a profound effect in PAPR reduction. Accordingly, the most remarkable ones are

- IFDMA
- LFDMA

Figure 4.13 portrays the allocation of subcarriers to different users. In order to clearly illustrate the subcarrier mapping strategies, this figure employs three users, the sizes of N and M are 12 and 4, respectively. While, the spreading factor S is 3. The detailed derivations of the mathematical expressions relevant to these multiple access

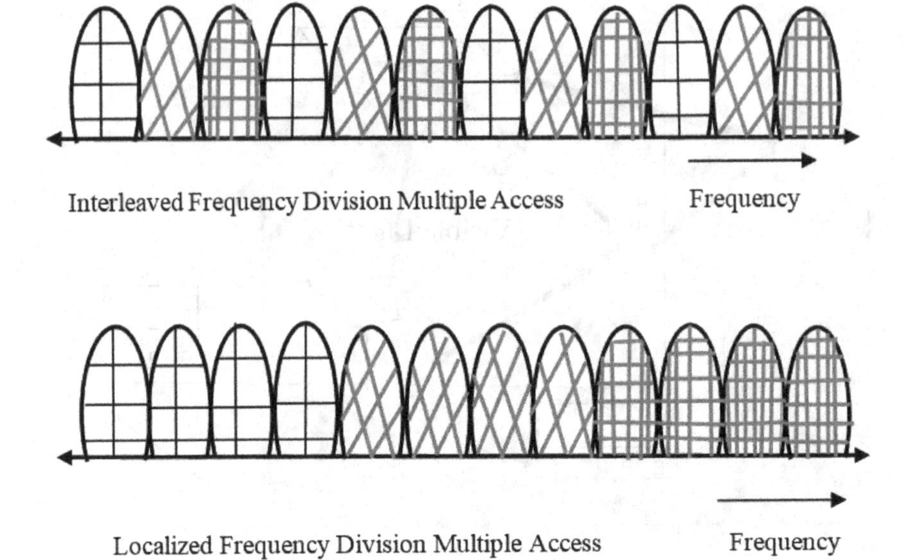

Figure 4.13: Different mapping strategies.

strategies will be given in the subsequent sections. In order to emphasize the benefits of enforcing real trigonometric transforms like DCT and DST to the aforesaid multiple access schemes, we carried out derivations of mathematical expressions pertaining to the Hermitian Symmetry imposed IFFT-based multiple access schemes like DFT-OIFDMA, DFT-OLFDMA.

4.4.1 DST-BASED MULTIPLE ACCESS SCHEMES

This subsection is mainly directed toward the analysis of uplink scenario in the optical domain. The performance analysis of different multiple access schemes like the optical IFDMA, LFDMA and OFDMA were compared. In particular, the uplink transmission systems were designed with the aid of trigonometric signal processing transform techniques like DCT and DST. As accentuated in Fig. 4.14, a set of real mapped data symbols are transmitted in parallel and then undergo the phenomena of spreading by exploiting DST transform technique. In general, it can be articulated from the figure that M-point DST is incorporated for enabling the spreading operation in order to yield the frequency domain representation of the transmitted data points. At the later stage, different subcarrier mapping techniques like the IFDMA and LFDMA are applied and then followed by passing the signal to a IDST transformation block in order to yield a time-domain representation of the signal. Typically, the different methodologies behind the mapping strategies are underlined in Figs. 4.15 and 4.16, respectively. More precisely, the output of the DST is mapped to the N-point IDST in a manner that $(N > M)$. Upon attaining a time-domain signal, in

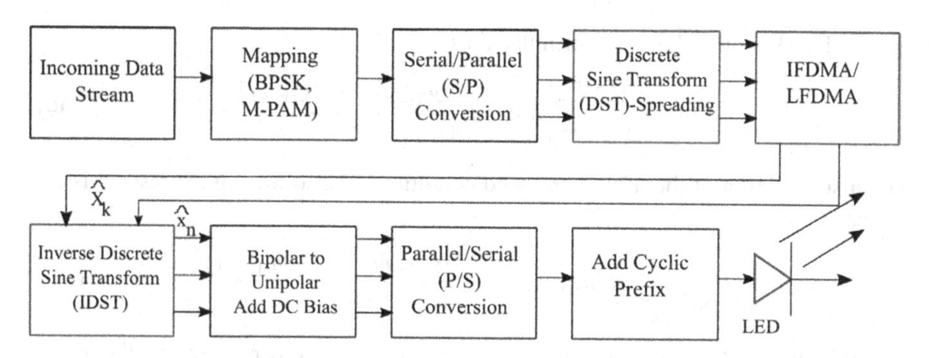

Figure 4.14: Illustration of uplink transmission employing DST-Spreading in optical IFDMA/LFDMA (OIFDMA/OLFDMA) [494, 495].

order to facilitate a unipolar signal transmission, the earliest variant of optical OFDM, i.e., the DCO-OFDM system methodology is assimilated, and then followed up by inclusion of an appropriate amount of cyclic prefix with the major emphasis to circumvent the detrimental aspects of ISI. Eventually, the attained time-domain signal is then intensity modulated by means of opto-electronic device like LED.

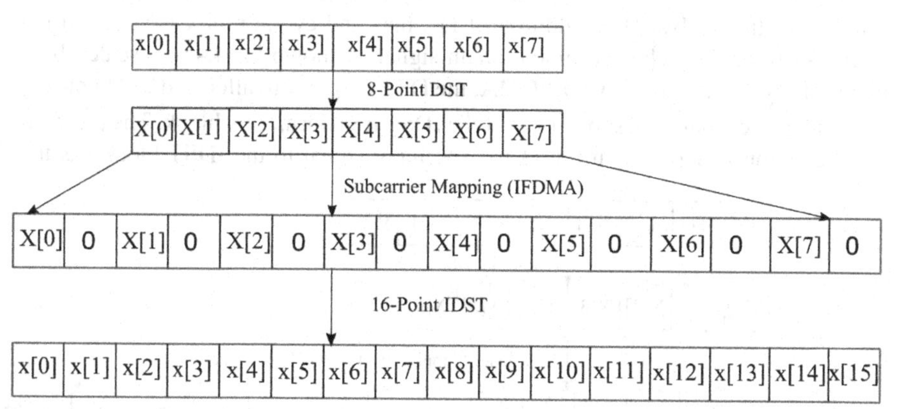

Figure 4.15: Signal format depicting optical IFDMA (OIFDMA) mapping strategy [494, 495].

Figure 4.15 represents the subcarrier mapping utilizing 8-point DST and 16-point IDST. The outputs of DST are allocated in an equidistant manner as $\frac{N}{M} = S$, where S represents the bandwidth spreading factor. Therefore, mathematically this can be represented as [111]

$$\tilde{X}[k] = \begin{cases} X\left[\frac{k}{S}\right], & k = Sm_1, \quad m_1 = 0, 1, 2, 3, \cdots M-1 \\ 0, & Otherwise \end{cases} \tag{4.38}$$

where $\tilde{X}[k]$ specifies the input of the IDST module and the output sequence of IDST which is denoted as $\tilde{x}[n]$ with $n = Ms + m$ for $s = 0, 1, 2, 3, \cdots S-1$ and

$m = 0, 1, 2, 3, \cdots M-1$ can be formulated as

$$\tilde{x}[n] = \sqrt{\frac{2}{N}} \sum_{k=0}^{N-1} \tilde{X}[k] \sin\left[\frac{\pi(2n+1)(2k+1)}{4N}\right] \tag{4.39}$$

Upon substitution of the aforementioned conditions, the following expression is attained

$$\tilde{x}[n] = \sqrt{\frac{2}{SM}} \sum_{m_1=0}^{M-1} X[m_1] \sin\left[\frac{\pi(2Ms+2m+1)(2Sm_1+1)}{4MS}\right] \tag{4.40}$$

Accordingly, the time-domain output signal can be derived by rearranging the mathematical expression as illustrated by (4.40) with the aid of trigonometric identities in order to attain the resultant expression as follows:

$$\tilde{x}[n] = \begin{cases} \frac{1}{\sqrt{S}} \frac{2}{M} \sum_{m_1=0}^{M-1} X[m_1] \sin\left[\frac{\pi(2m+1)m_1}{2M} + \frac{\pi(2(m+Ms)+1)}{4MS}\right](-1)^k, \\ \quad if\ k\ is\ odd \\ \frac{1}{\sqrt{S}} \frac{2}{M} \sum_{m_1=0}^{M-1} X[m_1] \sin\left[\frac{\pi(2m+1)m_1}{2M} + \frac{\pi(2(m+Ms)+1)}{4MS}\right], \\ \quad if\ k\ is\ even \end{cases} \tag{4.41}$$

The rest of the analysis like addition of DC bias and cyclic prefix is stated in the earlier section. The achieved time-domain signal as shown in (4.41) is scaled by a factor of $\frac{1}{\sqrt{S}}$. In case of DST-OLFDMA, the DST outputs are allocated to M consecutive subcarriers out of N subcarriers of the IDST transformation block. The pictorial representation is depicted in Fig. 4.16. The input signal to the IDST block has the

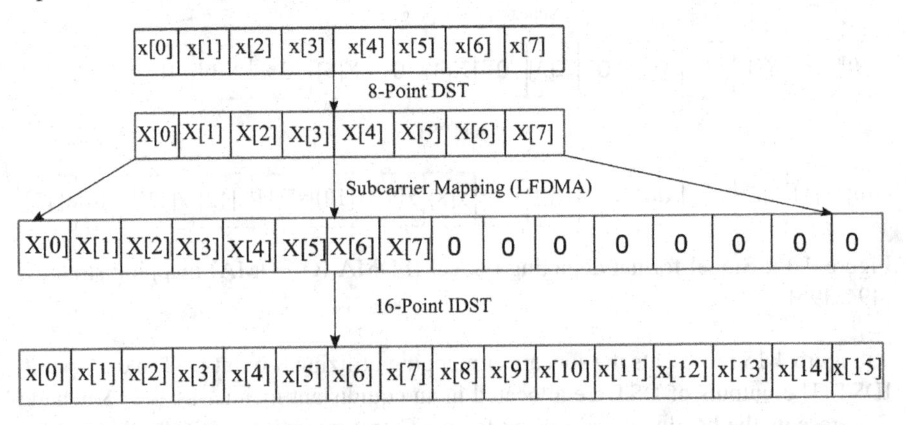

Figure 4.16: Signal format depicting optical LFDMA (OLFDMA) mapping strategy [494, 495].

following representation as [111]

$$\tilde{X}[k] = \begin{cases} X[k], & k = 0, 1, 2, 3, \cdots M-1 \\ 0, & k = M, M+1, \cdots N-1 \end{cases} \tag{4.42}$$

The time-domain output sequence of the IDST transformation block which is denoted as $\tilde{x}[n]$ with $n = Sm + s$ for $s = 0, 1, 2, 3, \cdots S-1$ can be represented as:

$$\tilde{x}[n] = \tilde{x}[Sm+s] = \sqrt{\frac{1}{S}}\sqrt{\frac{2}{M}}\sum_{k=0}^{M-1} X[k]\, sin\left[\frac{\pi S\left(2m+\frac{2s}{S}+\frac{1}{S}\right)(2k+1)}{4MS}\right] \tag{4.43}$$

Typically, the output expression is analyzed under two scenarios, i.e., when $s = 0$ and $s \neq 0$ as follows:

Case a: when $s = 0$, the output sequence can be manifested as follows:

$$\tilde{x}[n] = \tilde{x}[Sm] = \frac{1}{\sqrt{S}}\sqrt{\frac{2}{M}}\sum_{k=0}^{M-1} X[k]\, sin\left[\frac{\pi(2m+1)(2k+1)}{4M}\right] cos\left[\frac{\pi\left(1-\frac{1}{S}\right)(2k+1)}{4M}\right] - \frac{1}{\sqrt{S}}\sqrt{\frac{2}{M}}\sum_{k=0}^{M-1} X_k cos\left[\frac{\pi(2m+1)(2k+1)}{4M}\right] sin\left[\frac{\pi\left(1-\frac{1}{S}\right)(2k+1)}{4M}\right] \tag{4.44}$$

Case b: when $s \neq 0$,

Let $X[k] = \sqrt{\frac{2}{M}}\sum_{p=0}^{M-1} x[p]\, sin\left[\frac{\pi(2p+1)(2k+1)}{4M}\right]$, then (4.43) can be solved to attain the following time-domain signal as follows:

$$\tilde{x}[n] = \tilde{x}[Sm+s] = \frac{1}{\sqrt{S}}\frac{2}{M}\sum_{k=0}^{M-1}\sum_{p=0}^{M-1} x[p]\, sin\left[\frac{\pi(2p+1)(2k+1)}{4M}\right] \times \left[sin\left(\frac{\pi\left(2m+\frac{1}{S}\right)(2k+1)}{4M}\right) cos\left(\frac{\pi s(2k+1)}{2MS}\right) - \left[cos\left(\frac{\pi\left(2m+\frac{1}{S}\right)(2k+1)}{4M}\right) sin\left(\frac{\pi s(2k+1)}{2MS}\right)\right]\right. \tag{4.45}$$

Therefore, the time-domain DST-OLFDMA signals as represented by (4.44) and (4.45) depict that the time-domain signal is multiplied by additional weighing factors, thereby contributing to the increase in PAPR when compared with DST-OIFDMA. This mathematical analysis is confirmed through simulation results.

4.4.2 FAST OPTICAL IFDMA AND FAST OPTICAL LFDMA

Much similar to the earlier section which illustrates a DST-based multiple access system, this section briefs the mathematical representation of the uplink fast optical IFDMA (F-O-IFDMA) and fast optical LFDMA (F-O-LFDMA) systems. The schematic representation of the transmitter block diagram is shown in Fig. 4.17. This figure elucidates the uplink transmitter employing DCT-Spreading for both the subcarrier mapping schemes, i.e., interleaved and localized FDMA. The M-PAM mapped data is spread using DCT and then allocated (subcarrier mapping) using

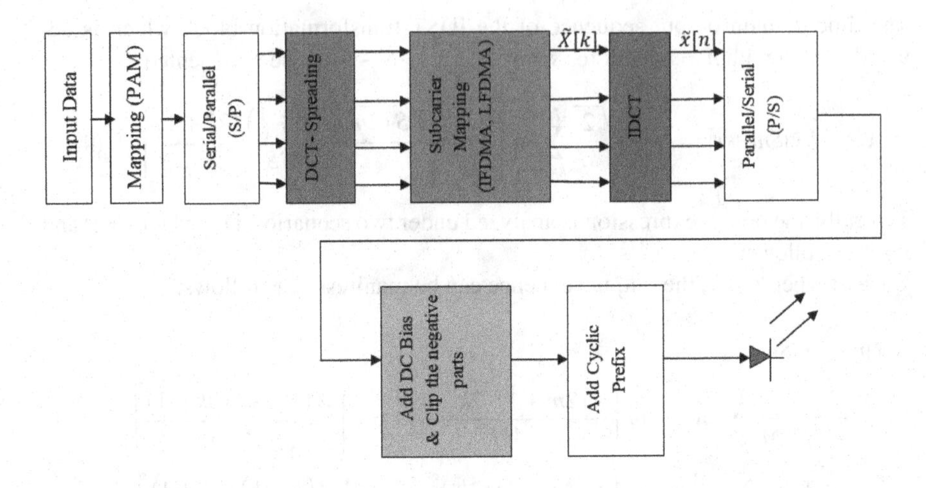

Figure 4.17: Uplink transmitter utilizing DCT-Spreading for F-O-IFDMA and F-O-LFDMA [499].

(4.38). After subcarrier mapping, this sequence $\tilde{X}[k]$ is fed to the IDCT to yield the time domain signal $\tilde{x}[n]$. The IDCT output sequence $\tilde{x}[n]$ with $n = Ms + m$ for $s = 0, 1, 2, 3 \cdots S-1$ and $m = 0, 1, 2, 3, \cdots M-1$ can be solved to attain the following: Upon substitution of the subcarrier mapped sequence, $\tilde{X}[k]$ as represented in (4.38) into (3.77) obtains

$$\tilde{x}[n] = \sqrt{\frac{2}{SM}}\left[\sum_{m_1=0}^{M-1} C\left[\frac{k}{S}\right] X\left[\frac{k}{S}\right] cos\left(\frac{\pi(2Ms+2m+1)}{2N}k\right)\right] \tag{4.46}$$

From (4.38), since, $k = Sm_1$ so, on substitution in (4.46), the following time-domain F-O-IFDMA signal is attained as follows:

$$\tilde{x}[n] = \begin{cases} \frac{1}{\sqrt{S}}\sqrt{\frac{2}{M}}\sum_{m_1=0}^{M-1} C[m_1] X[m_1] cos\left(\frac{\pi(2m+1)m_1}{2M}\right)(-1)^k, \\ \quad if\ k\ is\ odd \\ \frac{1}{\sqrt{S}}\sqrt{\frac{2}{M}}\sum_{m_1=0}^{M-1} C[m_1] X[m_1] cos\left(\frac{\pi(2m+1)m_1}{2M}\right), \\ \quad if\ k\ is\ even \end{cases} \tag{4.47}$$

By taking into consideration the signal representation for LFDMA mapping scheme as dealt with in the previous section which is given by (4.42), the corresponding IDCT output sequence $\tilde{x}[n]$ with $n = Sm + s$ for $s = 0, 1, 2, 3, \cdots S-1$ can be expressed as follows:

$$\tilde{x}[n] = \tilde{x}[Sm+s] = \sqrt{\frac{2}{MS}}\sum_{k=0}^{M-1} C[k] X[k] cos\left(\frac{\pi(2(Sm+s)+1)k}{2SM}\right) \tag{4.48}$$

Accordingly, the time-domain signals for the two scenarios, i.e., when $s = 0$ and $s \neq 0$ can be derived as [499]

$$\tilde{x}[n] = \tilde{x}[Sm] =$$
$$\frac{1}{\sqrt{S}}\left\{\sqrt{\frac{2}{M}}\sum_{k=0}^{M-1} C[k]X[k]\cos\left(\frac{\pi(2m+1)k}{2M}\right)\cos\left(\frac{\pi\left(1-\frac{1}{S}\right)k}{2M}\right)\right\} \quad (4.49)$$

If $s \neq 0$,
then let us consider $X[k] = \sqrt{\frac{2}{M}}C[k]\sum_{q=0}^{M-1} x[q]\cos\left(\frac{\pi(2q+1)k}{2M}\right)$ and upon substituting in (4.49) yields [499]

$$\tilde{x}[n] = \tilde{x}[Sm+s] = \frac{1}{\sqrt{S}}\frac{2}{M}\sum_{k=0}^{M-1} C[k]^2\left\{\sum_{q=0}^{M-1} x[q]\cos\left(\frac{\pi(2q+1)k}{2M}\right)\right\}\times$$
$$\cos\left(\pi\left(\frac{Sm+s}{SM}\right)k+\frac{\pi k}{2SM}\right) \quad (4.50)$$

(4.49) and (4.50) reinforce that the time-domain F-O-LFDMA signal represents that the input signal is scaled by a factor of $\frac{1}{S}$ and, moreover, they are getting multiplied with the real weighting factors. Hence, this results in significant increase in PAPR for F-O-LFDMA when compared with F-O-IFDMA. This is verified through simulation results.

Furthermore, an inference can be drawn by making use of (4.47), (4.49) and (4.50), where the real time-domain signals are fetched with good spectral efficiency as all of the subcarriers are utilized for data transmission. Based on the aforementioned arrangement of subcarriers, PAPR reduces accordingly and the simulation results are detailed next.

4.4.3 OPTICAL INTERLEAVED FREQUENCY DIVISION MULTIPLE ACCESS

As discussed in the aforesaid section, this section highlights the mathematical analysis of the time-domain signal for DFT-based multiple access system. Moreover, due to the constraint of Hermitian Symmetry criteria, even the subcarrier mapping strategies are varied when compared with conventional RF domain. As depicted in Fig. 4.18, the subcarrier mapped data is constrained to satisfy HS criteria before getting applied to the IDFT transformation block. The detailed description of the subcarrier mapping is outlined in Fig. 4.19. For the input sequence $\tilde{X}[k]$, the N-point IDFT output sequence can be represented as $\tilde{x}[n]$ with $n = Ms+m$ for $s = 0,1,2,3,\cdots S-1$ and $m = 0,1,2,3,\cdots M-1$ can be obtained as

$$\tilde{x}[n] = \frac{1}{N}\sum_{k=0}^{N-1}\tilde{X}[k]e^{\frac{j2\pi nk}{N}} \quad (4.51)$$

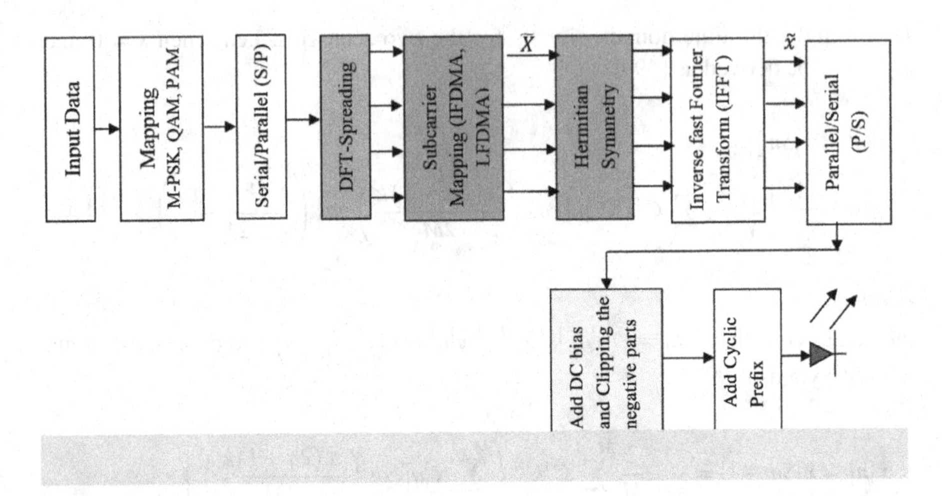

Figure 4.18: Uplink transmission employing DFT-Spreading in optical IFDMA/LFDMA (OIFDMA/OLFDMA) [495].

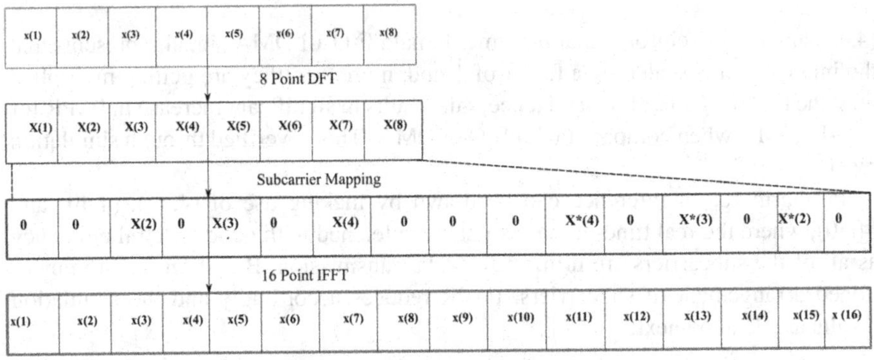

x(1)	x(2)	x(3)	x(4)	x(5)	x(6)	x(7)	x(8)

8 Point DFT

X(1)	X(2)	X(3)	X(4)	X(5)	X(6)	X(7)	X(8)

Subcarrier Mapping

0	0	X(2)	0	X(3)	0	X(4)	0	0	0	X*(4)	0	X*(3)	0	X*(2)	0

16 Point IFFT

x(1)	x(2)	x(3)	x(4)	x(5)	x(6)	x(7)	x(8)	x(9)	x(10)	x(11)	x(12)	x(13)	x(14)	x(15)	x (16)

Figure 4.19: DFT-OIFDMA subcarrier mapping format [499].

Upon substituting the aforementioned constraints as well as rearranging, (4.51) can be reduced as

$$\tilde{x}[n] = \frac{1}{SM} \sum_{m_1=0}^{M-1} X[m_1] e^{\frac{j2\pi n m_1}{M}} \tag{4.52}$$

However, $X[m_1]$ is constrained to satisfy Hermitian Symmetry criteria. As already seen in the prior chapter, according to Hermitian Symmetry criteria, the first and the middle subcarriers are set to zero as

$$X[0] = X\left[\frac{M}{2}\right] = 0 \tag{4.53}$$

In order to avoid the presence of any complex part of the signal at the output of IDFT block, the arrangement is made in such a manner that all of the inputs are not utilized for data transmission, i.e., out of N subcarriers only $\frac{N}{2}$ are exploited for data transmission and the rest $\frac{N}{2}$ are flipped complex conjugate versions of the previous ones. Therefore, the arrangement looks like

$$X\left[M-k\right]=X^{*}\left[k\right],\ \ k=1,2,3,\cdots\frac{M}{2} \tag{4.54}$$

Accordingly, (4.52) can be rearranged as

$$\tilde{x}\left[n\right]=\frac{1}{S}\frac{1}{M}\Bigg[X\left[0\right]e^{\frac{j2\pi(0)m}{M}}+\sum_{m_{1}=1}^{\frac{M}{2}-1}X\left[m_{1}\right]e^{\frac{j2\pi m_{1}m}{M}}+ X\left[\frac{M}{2}\right]e^{\frac{j2\pi\left(\frac{M}{2}\right)m}{M}}+\sum_{m_{1}=\frac{M}{2}+1}^{M-1}X\left[m_{1}\right]e^{\frac{j2\pi m_{1}m}{M}}\Bigg] \tag{4.55}$$

Since, the first and the middle subcarriers are set to zero, (4.55) can be reduced to

$$\tilde{x}\left[n\right]=\frac{1}{S}\frac{1}{M}\left[\sum_{m_{1}=1}^{\frac{M}{2}-1}X\left[m_{1}\right]e^{\frac{j2\pi m_{1}m}{M}}+\sum_{m_{1}=\frac{M}{2}+1}^{M-1}X\left[m_{1}\right]e^{\frac{j2\pi m_{1}m}{M}}\right] \tag{4.56}$$

Upon applying change in variable transformation as $M-m_{1}^{'}=m_{1}$

$$\tilde{x}\left[n\right]=\tilde{x}\left[Ms+m\right]=\frac{1}{S}\frac{1}{M}\left[\sum_{m_{1}=1}^{\frac{M}{2}-1}X_{m_{1}}e^{\frac{j2\pi m_{1}m}{M}}+\sum_{m_{1}^{'}=1}^{\frac{M}{2}-1}X\left[M-m_{1}^{'}\right]e^{\frac{-j2\pi\left(M-m_{1}^{'}\right)m}{M}}\right] \tag{4.57}$$

Therefore, by employing (4.54), as well as rearranging (4.57), the following expression can be obtained

$$\tilde{x}\left[n\right]=\tilde{x}\left[Ms+m\right]=\frac{1}{S}\frac{1}{M}\left[\sum_{m_{1}=1}^{\frac{M}{2}-1}X\left[m_{1}\right]e^{\frac{j2\pi m_{1}m}{M}}+\sum_{m_{1}=1}^{\frac{M}{2}-1}X^{*}\left[m_{1}\right]e^{\frac{-j2\pi m_{1}m}{M}}\right] \tag{4.58}$$

By using Euler inequalities, (4.58) can be rearranged as

$$\tilde{x}\left[n\right]=\tilde{x}\left[Ms+m\right]=\frac{1}{S}\left[\frac{2}{M}\sum_{m_{1}=1}^{\frac{M}{2}-1}\underset{RC}{X\left[m_{1}\right]}cos\left(\frac{2\pi m_{1}m}{M}\right)-\underset{IC}{X\left[m_{1}\right]}sin\left(\frac{2\pi m_{1}m}{M}\right)\right] \tag{4.59}$$

where $\underset{RC}{X\left[m_{1}\right]}$ and $\underset{IC}{X\left[m_{1}\right]}$ in (4.59) represents the real and imaginary components. Equation (4.59) clearly illustrates that there is a loss in spectral efficiency as all of the subcarriers are not utilized for data transmission because, out of N subcarriers, only $\frac{N}{2}$ are involved in transporting data.

4.4.4 OPTICAL LOCALIZED FREQUENCY DIVISION MULTIPLE ACCESS

This subsection illustrates the uplink transmission scenario of ACO-LFDMA system and the pictorial representation of it is elucidated in Fig. 4.20. In case of DFT-Spreading scheme for OLFDMA, the input to the IFFT has the following representation as [111]

$$X[k] = \begin{cases} X[k], & k = 0,1,2,\cdots M-1 \\ 0, & k = M, M+1, M+2, \cdots N-1 \end{cases} \tag{4.60}$$

Elaborate signal format representation is shown in Fig. 4.21. The IFFT output sequence $x[n]$ with $n = Sm + s$ for $s = 0,1,2,3,\cdots S-1$ can be expressed as follows [499]:

Case 1: $s = 0$

$$x[n] = x[Sm] = \frac{1}{S}\left\{\frac{2}{M}\left[\sum_{k=1}^{\frac{M}{4}-1} X[k]_{Real} cos\left(2\pi\frac{m}{M}k\right) - \sum_{k=1}^{\frac{M}{4}-1} X[k]_{Imag} sin\left(2\pi\frac{m}{M}k\right)\right]\right\} \tag{4.61}$$

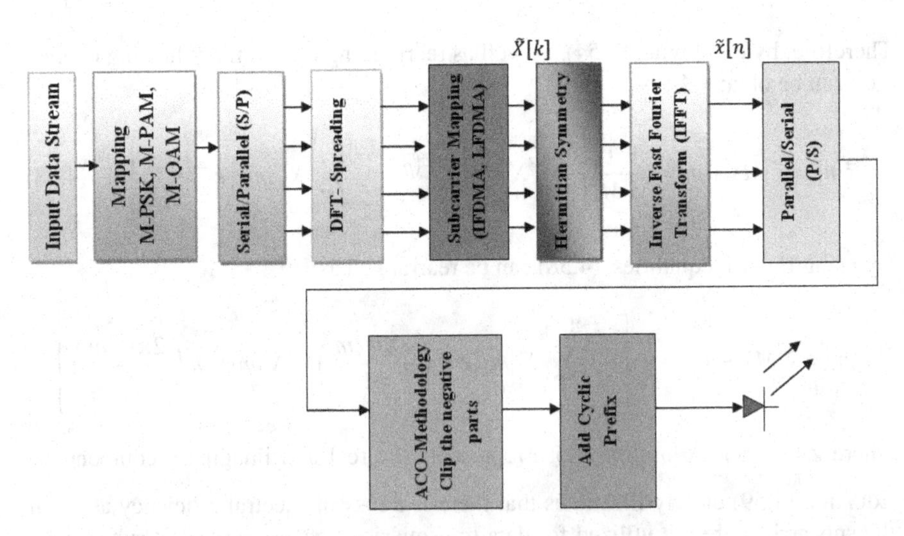

Figure 4.20: Uplink transmission employing DFT-Spreading in ACO-LFDMA.

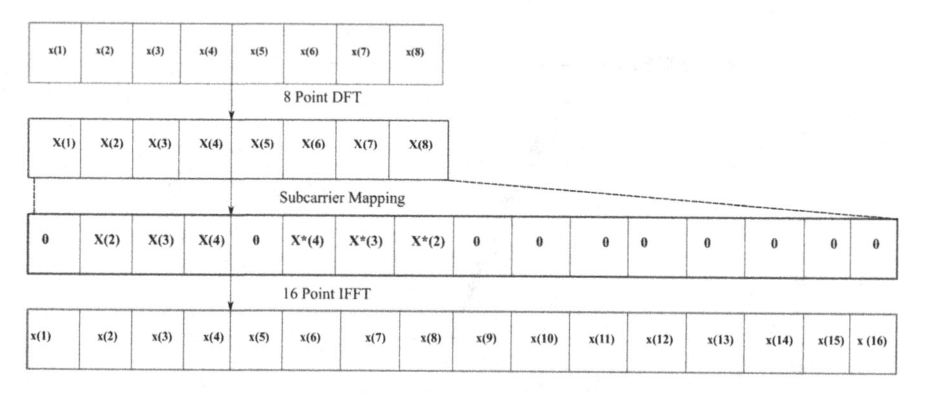

Figure 4.21: Signal format representation for DFT-OLFDMA [499].

Case 2: $s \neq 0$

$$x[n] = \frac{1}{S} e^{\frac{j2\pi(M-1)(Sm+s)}{MS}} \left[\sum_{p=0}^{M-1} \left(\frac{2}{M} \sum_{k=1}^{\frac{M}{4}-1} \underset{Real}{X[k]} cos\left(\frac{2\pi pk}{M}\right) - \underset{Imag}{X[k]} sin\left(\frac{2\pi pk}{M}\right) \right) \right] \left[\frac{sin\pi\left(m+\frac{s}{S}-p\right)}{M sin\pi\left(\left(\frac{Sm+s}{M}-\frac{pS}{M}\right)\right)} \right] e^{\frac{j\pi p}{M}} \tag{4.62}$$

Therefore, comparison of (4.59), (4.61) and (4.62) with (4.41), (4.44) and (4.45) confirms the fact that DFT-based multiple access strategies result in a decrease in throughput due to the inherent fact that all of the subcarriers cannot be exploited for data transmission.

4.4.5 PERFORMANCE ANALYSIS OF DCT/DST-BASED MULTIPLE ACCESS SCHEMES COMPATIBLE WITH IM/DD SYSTEMS FOR VLC

This section presents the performance analysis of DCT/DST-based multiple access schemes that are compatible with IM/DD systems for VLC. The system models along with the detailed mathematical analysis of the attained time-domain signals for each multiple access scheme are detailed in the previous subsection.

In order to illustrate the PAPR performance of different DCT-based multiple access schemes, i.e., F-O-IFDMA, F-O-LFDMA and F-OOFDMA, Figs. 4.22 and 4.23 present the CCDF vs. PAPR curves for the aforesaid multiple access schemes by varying the number of subcarriers. From the CCDF curves, it is ascertained that the reduction in PAPR is dependent on the way the subcarriers are allocated to the mobile terminals. Based upon the number of subcarriers M that are appropriated to each subscriber, the PAPR performance varies. The PAPR performance deteriorates with the increase in the number of subcarriers from $M = 4, 8, 16, \cdots 128$. In order to illustrate this fact, we have presented the CCDF vs. PAPR curves for $M = 16$ and 64 subcarriers.

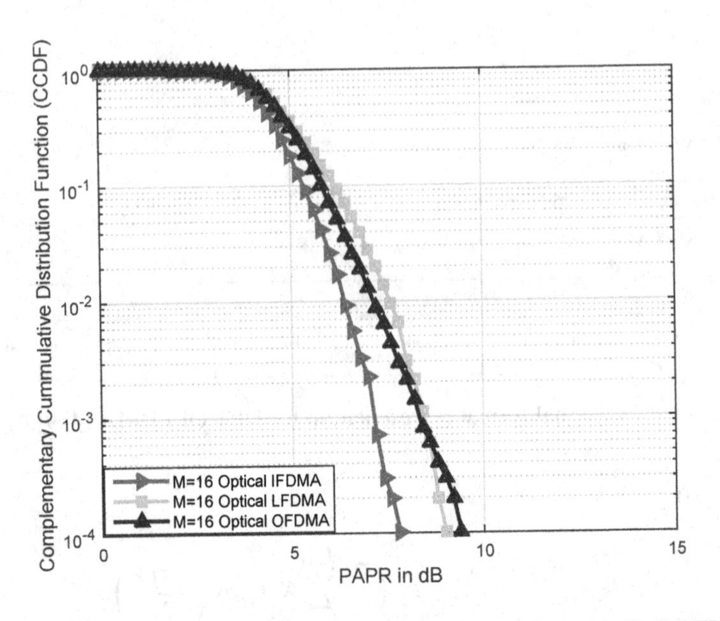

Figure 4.22: CCDF vs. PAPR employing M=16 for F-O-IFDMA, F-O-LFDMA and F-OOFDMA [499].

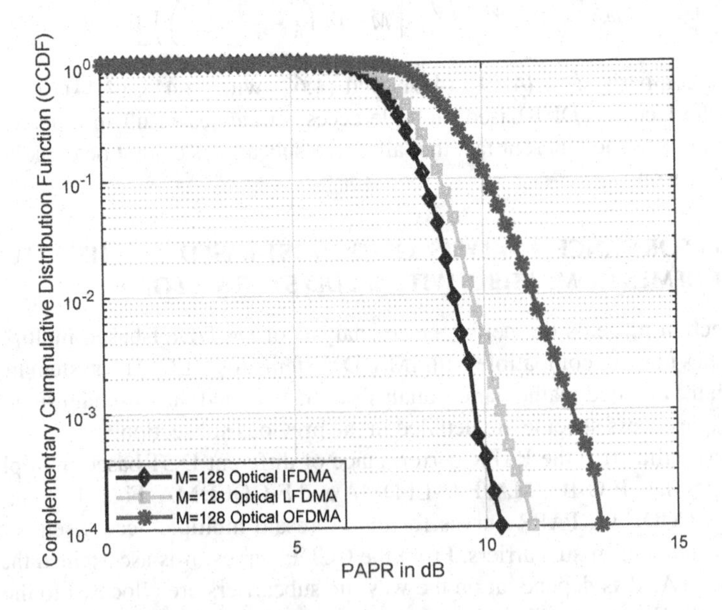

Figure 4.23: Comparison of PAPR performance for OIFDMA, OLFDMA and OOFDMA for higher order of M employing 16-PAM [499].

Furthermore, upon taking into consideration the PAPR performance analysis of DST-based multiple access schemes as portrayed from the work [495] (please see figures 21, 22 and 23 from reference [495]), it can be deduced that the amount of reduction in PAPR for both the multiple access schemes like DCT-based IFDMA, DCT-based LFDMA, DST-based OIFDMA and DST-based OLFDMA is superior when compared to DCT/DST-based OFDMA. However, IFDMA-based multiple access strategy yields a better performance when compared with LFDMA.

Listing 4.1: MATLAB code to illustrate the PAPR performance of PTS technique in conventional optical OFDM system. (Note: This MATLAB code works with MATLAB R2014b version).

```
1  clc;
2  clear all;
3  close all;
4  N=256; %Size of the Transform employed
5  Os_f=4; %Oversampling Factor
6  NOs_f=N*Os_f;  %Product of the transform and the
      oversampling factor
7  b=4; %Total Number of bits employed for the
      constellation
8  M=2^b; %Emphasizes M-ary Modulation i.e., M=16
9  Es=1; A=sqrt(3/2/(M-1)*Es); % Normalization factor
      for M-QAM
10 No_sb=[4,8,16,32]; %Total Number of subblocks
      employed in PTS PAPR reduction scheme
11 dBs = [0:0.1:15]; dBcs = dBs+(dBs(2)-dBs(1))/2;
12 Nblk = 5000; % Total Number of OFDM blocks for
      iteration
13 mod_object=modem.qammod('M',M,'SymbolOrder','gray');
      %Mapping using QAM modulator
14 for nblk=1:Nblk
15    mod_sym = A*modulate(mod_object,randint(1,N,M));%
         Modulation of the incoming data using M-QAM
16    [Nr,Nc]=size(mod_sym);
17    zero_pad_sym=zeros(Nr,Nc*Os_f);%Zero Padding
18    for k=1:Nr          % zero padding for
         oversampling
19 %%%Enforcing Hermitian Symmetry Criteria
20       zero_pad_sym(k,1:Os_f:Nc*Os_f)=mod_sym(k,:);
21       datamat1=zero_pad_sym;
22       datamat1(1)=0;
23       datamat1(513)=0;
24       datamat1(:,514:1024)=flip(conj(datamat1
            (:,2:512)));
25    end
```

```
26     ifft_sym=ifft(datamat1, NOs_f);%Performing IFFT
          operation
27     sym_pow=abs(ifft_sym).^2;%calculating the symbol
          power
28     mean_pow(nblk)=mean(sym_pow); max_pow(nblk)=max(
          sym_pow);
29  end
30  PAPR=max_pow./mean_pow;  %determining PAPR as the
       ratio of maximum peak power to average power
31  PAPRdB=10*log10(PAPR); % measure PAPR in dB
32  dBcs = dBs + (dBs(2)-dBs(1))/2;   % dB midpoint
       vector
33  count = 0;  N_bins = hist(PAPRdB,dBcs);
34  for i=length(dBs):-1:1
35     count = count+N_bins(i); CCDF(i) = count/Nblk;
36  end
37  %%Plotting CCDF curve without enforcing any PAPR
       reduction technique
38  semilogy(dBs,CCDF,'ks-','LineWidth',2), hold on
39  %%Plotting CCDF curves by enforcing PTS-based PAPR
       reduction technique
40  %%%Plotting CCDF curve for subblocks=4
41     k1 = 1;
42     Nsb=No_sb(k1);
43     CCDF1=PAPR_PTS(N,Os_f,Nsb,b,dBs,Nblk);
44  %%%Plotting CCDF curve for subblocks=8
45     k2 = 2;
46     Nsb1=No_sb(k2);
47        CCDF2=PAPR_PTS(N,Os_f,Nsb1,b,dBs,Nblk);
48     %%%Plotting CCDF curve for subblocks=16
49      k3 = 3;
50     Nsb2=No_sb(k3);
51        CCDF3=PAPR_PTS(N,Os_f,Nsb2,b,dBs,Nblk);
52  %%%Plotting CCDF curve for subblocks=32
53     k4 = 4;
54     Nsb3=No_sb(k4);
55     CCDF4=PAPR_PTS(N,Os_f,Nsb3,b,dBs,Nblk);
56  semilogy(dBs,CCDF1,'rd-','LineWidth',2);
57  semilogy(dBs,CCDF2,'gs-','LineWidth',2);
58  semilogy(dBs,CCDF3,'b>-','LineWidth',2);
59  semilogy(dBs,CCDF4,'m*-','LineWidth',2);
60  axis([dBs([1 end]) 1e-3 1]);
61  grid on;
62  xlabel('PAPR in dB');
```

```
ylabel('Complementary Cummulative Distribution
    Function (CCDF)');
legend(3,'Without PTS','Sub-blocks=4','Sub-blocks=8',
    'Sub-blocks=16','Sub-blocks=32');
```

Listing 4.2: MATLAB code for the function *PAPR_PTS* as depicted in the previous program.

```
function CCDF=PAPR_PTS(N,Os_f,Nsb,b,dBs,Nblk)
% This function indicates the CCDF function for the
    Partial Transmit Sequence
% N     : Number of Subcarriers
% Os_f  : Indicates the Oversampling factor
% Nsb   : Number of subblocks
% b     : Number of bits per QAM symbol
% dBs   : dB vector
% Nblk : Number of OFDM blocks for iteration
NOs_f=N*Os_f;  % FFT size
M=2^b; Es=1; A=sqrt(3/2/(M-1)*Es); % Normalization
    factor for M-QAM
mod_object=modem.qammod('M',M,'SymbolOrder','gray');
for iter=1:Nblk
   w = ones(1,Nsb);         % Phase (weight) factor
   mod_sym = A*modulate(mod_object,randint(1,N,M)); %
       Modulation M-QAM
   [Nr,Nc] = size(mod_sym);
   zero_pad_sym = zeros(Nr, Nc*Os_f);
   for k=1:Nr                          % zero padding
      for oversampling
      zero_pad_sym(k,1:Os_f:Nc*Os_f) =mod_sym(k,:);
   end
   sub_block=zeros(Nsb,NOs_f);
   for k=1:Nsb  % Disjoint Subblock Mapping
      kk = (k-1)*NOs_f/Nsb+1:k*NOs_f/Nsb;
      sub_block(k,kk) = zero_pad_sym(1,kk);
      datamat=sub_block;
      datamat(1)=0;
      datamat(513)=0;
      datamat(:,514:1024)=flip(conj(datamat(:,2:512))
         );
   end
   ifft_sym=ifft(datamat.',NOs_f).';  % IFFT
   % --  Phase Factor Optimization  -- %
   for m=1:Nsb
      x = w(1:Nsb)*ifft_sym;
      sym_pow = abs(x).^2;
```

```
34        PAPR = max(sym_pow)/mean(sym_pow);
35        if m==1, PAPR_min = PAPR;
36         else if PAPR_min<PAPR, w(m)=1; else PAPR_min =
               PAPR; end
37        end
38        w(m+1)=-1;
39     end
40     x_tilde = w(1:Nsb)*ifft_sym; % The lowest PAPR
         symbol
41     sym_pow = abs(x_tilde).^2; % Symbol power
42     PAPRs(iter) = max(sym_pow)/mean(sym_pow);
43  end
44  PAPRdBs=10*log10(PAPRs); % measure PAPR
45  dBcs = dBs + (dBs(2)-dBs(1))/2;    % dB midpoint
       vector
46  count=0;  N_bins=hist(PAPRdBs,dBcs);
47  for i=length(dBs):-1:1
48     count=count+N_bins(i); CCDF(i)=count/Nblk;
49  end
50  plot_or_not = 0;
51  if plot_or_not>0
52    figure(1), clf
53    semilogy(dBs,CCDF,'-s'); axis([dBs([1 end]) 1e-4
         1]); grid on; hold on
54    title(['16 QAM CCDF of OFDMA PAPR, ',' ',num2str(N)
         ,'-point ' ,num2str(Nblk) '-blocks']);
55    xlabel('PAPR_0 [dB]'); ylabel('Pr(PAPR>PAPR_0)');
56  end
```

Listing 4.3: MATLAB code to depict the PAPR performance of different subcarrier mapping strategies using different orders of M-PAM. The simulation result attained upon executing this code is elucidated in Fig. 4.24. Note: This MATLAB code works with MATLAB version R2014b.

```
1   clc;
2   clear all;
3   close all;
4    N_Trans=256; %Size of the Transform employed (i.e.,
       DCT/DFT/DST)
5   SDLB=128; % Size of the Data block (i.e., total
       number of subcarriers appropriated to user)
6   bb=2;
7   N_b=length(bb);
8   dBs = [0:0.2:15];
9   dBcs = dBs+(dBs(2)-dBs(1))/2;
10  Nblk = 1000; % Total Number of OFDM blocks employed
```

```
    for iteration
11 for i=1:N_b
12     b=bb(i);
13     M=2^b; %Size of the constellation
14    CCDF_OFDMA = SubcarrierMapping('OF',N_Trans,b,
        N_Trans,dBcs,Nblk); % Computation of CCDF of
        OFDMA
15    CCDF_LFDMA = SubcarrierMapping('LF',SDLB,b,N_Trans
        ,dBcs,Nblk); % Computation of CCDF of LFDMA
16     CCDF_IFDMA = SubcarrierMapping('IF',SDLB,b,
         N_Trans,dBcs,Nblk); % Computation of CCDF of
         IFDMA
17 end
18 %%%%%%
19 %Computation of PAPR for 8PAM
20 N_Trans=256; %Size of the Transform employed (i.e.,
     DCT/DFT/DST)
21 SDLB=128; % Size of the Data block (i.e., total
     number of subcarriers appropriated to user)
22 bs1=3;
23 N_b1=length(bs1);
24 dBs = [0:0.2:15];
25 dBcs = dBs+(dBs(2)-dBs(1))/2;
26 Nblk = 1000; % Total Number of OFDM blocks employed
     for iteration
27 for i=1:N_b1
28    b1=bs1(i);
29    M=2^b1;
30    CCDF_OFDMA1 = SubcarrierMapping('OF', N_Trans,b1,
        N_Trans,dBcs,Nblk); % Computation of CCDF of
        OFDMA
31    CCDF_LFDMA1 = SubcarrierMapping('LF',SDLB,b1,
        N_Trans,dBcs,Nblk); % %Computation of CCDF of
        LFDMA
32    CCDF_IFDMA1 = SubcarrierMapping('IF',SDLB,b1,
        N_Trans,dBcs,Nblk);
33  %Computation of CCDF of IFDMA
34 end
35 %%%%
36 %Computation of PAPR for 16PAM
37
38 N_Trans=256; %Size of the Transform employed (i.e.,
    DCT/DFT/DST)
39 SDLB=128; % Size of the Data block (i.e., total
```

```
      number of subcarriers appropriated to user)
40  bs2=4;
41  N_b2=length(bs2);
42  dBs = [0:0.2:15];
43  dBcs = dBs+(dBs(2)-dBs(1))/2;
44  Nblk = 1000; % Total Number of OFDM blocks employed
      for iteration
45  for i=1:N_b2
46     b2=bs2(i);
47     M=2^b2;
48     CCDF_OFDMA2 = SubcarrierMapping('OF', N_Trans,b2,
         N_Trans,dBcs,Nblk); % Computation of CCDF of
         CCDF of OFDMA
49     CCDF_LFDMA2 = SubcarrierMapping('LF',SDLB,b2,
         N_Trans,dBcs,Nblk); % Computation of CCDF of
         LFDMA
50      CCDF_IFDMA2 = SubcarrierMapping('IF',SDLB,b2,
          N_Trans,dBcs,Nblk); % Computation of CCDF of
          IFDMA
51  end
52  %%%%%Plotting the CCDF curves for different
      subcarrier mapping strategies using different
      orders of modulation
53  semilogy(dBs,CCDF_IFDMA,'yd-','LineWidth',2);hold on;
54  semilogy(dBs,CCDF_IFDMA1,'bs-','LineWidth',2);
55  semilogy(dBs,CCDF_IFDMA2,'k>-','LineWidth',2);
56  semilogy(dBs,CCDF_LFDMA,'c>-','LineWidth',2);
57  semilogy(dBs,CCDF_LFDMA1,'ks-','LineWidth',2);
58  semilogy(dBs,CCDF_LFDMA2,'r>-','LineWidth',2);
59  semilogy(dBs,CCDF_OFDMA,'g<-','LineWidth',2);
60  semilogy(dBs,CCDF_OFDMA1,'yd-','LineWidth',2);
61  semilogy(dBs,CCDF_OFDMA2,'m*-','LineWidth',2);
62  axis([dBs([1 end]) 10^-3 1]);
63  grid on;
64  xlabel('PAPR in dB');
65  ylabel('CCDF');
66   legend(3,'OIFDMA-4PAM','OIFDMA-8PAM','OIFDMA-16PAM',
         'OLFDMA-4PAM','OLFDMA-8PAM','OLFDMA-16PAM','
       OOFDMA-4PAM','OOFDMA-8PAM','OOFDMA-16PAM');
```

Listing 4.4: MATLAB code for the function SubcarrierMapping which is exploiting DCT transform for the main code depicted above.

```
1  function [CCDF,PAPRs]=SubcarrierMapping(type_of_fdma,
      size_datablock,b,N,dBcs,N_block)
```

```
2
3  %type_of_fdma suggests whether it is OFDMA, IFDMA or
      LFDMA
4  %size_datablock infers the size of the data block
5  %b is the total number of bits employed
6  %N represents the size of the FFT transform
7  %dBcs is the dB vector
8  %N_block is the total number of OFDM blocks employed
      for iteration
9  M=2^b; %Size of the constellation
10 Es=1; A=sqrt(3/2/(M-1)*Es); % Normalization factor
      for QAM
11 mod_object=modem.pammod('M',M,'SymbolOrder','gray');
      %Real constellation mapping
12 S=N/size_datablock; % Spreading factor portrays the
      division of the FFT transform to that of the size
      of the data blocks.
13 for iter=1:N_block
14    mod_sym = A*modulate(mod_object,randint(1,
         size_datablock,M)); %Generation of random bits
         and then modulating them using M-PAM
15    switch upper(type_of_fdma(1:2))
16      case 'IF', dct_sym = zero_insertion(dct(mod_sym,
           size_datablock),N/ size_datablock); %
           Indicates interleaved frequency division
           multiple access subcarrier mapping strategy
17      case 'LF', dct_sym = [dct(mod_sym,
           size_datablock) zeros(1,N- size_datablock)];
           % Indicates the localized frequency division
           multiple access subcarrier mapping strategy
18      case 'OF', dct_sym = zero_insertion(mod_sym,S);
           % Oversampling, No DCT spreading
19      otherwise  dct_sym = mod_sym; % No oversampling,
           No DCT spreading
20    end
21     idct_sym = idct(dct_sym,N);  %Computation of
         Inverse Discrete Cosine  transform operation
22    idct_sym1=idct_sym+3.25; %Addition of Bias value i
        .e., by finding out the maximum negative value
        in the OFDM symbol and then adding that value
        to ensure a real and positive value is
        transmitted.
23    if nargin>7, idct_sym = zero_insertion(idct_sym1,
        Nos); end
```

```
24 %    if nargin>6, idct_sym = conv(idct_sym1,psf); end
25     sym_pow = idct_sym.*conj(idct_sym); %Calculation
         of Symbol power
26     PAPRs(iter) = max(sym_pow)/mean(sym_pow); %
         Calculation of PAPR which is defined as the
         ratio of maximum signal power to that of the
         average signal power
27 end
28 PAPRdBs = 10*log10(PAPRs); % PAPR in dB
29 N_bins = hist(PAPRdBs,dBcs);  count = 0;
30 for i=length(dBcs):-1:1
31     count = count+N_bins(i); CCDF(i) = count/N_block;
32 end
```

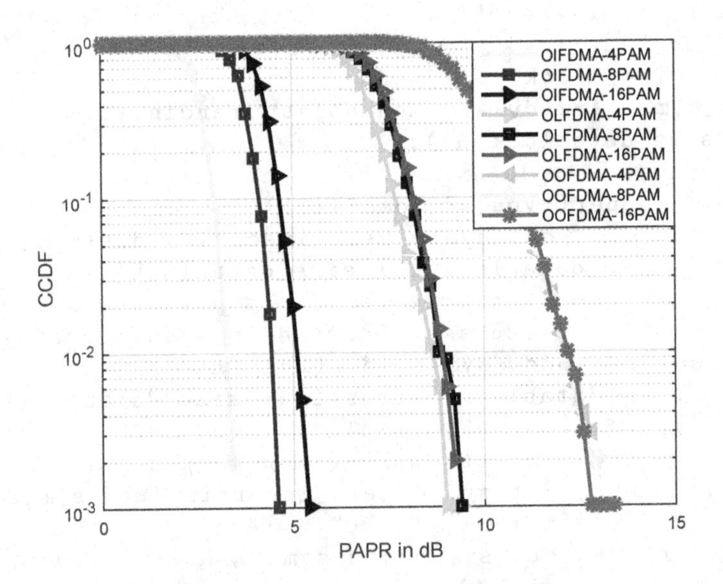

Figure 4.24: PAPR performance comparison for different subcarrier mapping techniques like OIFDMA, OLFDMA and OOFDMA.

The simulation result as emphasized in Fig. 4.24 clearly depicts that OIFDMA achieves a superior performance in terms of PAPR reduction when compared with OLFDMA and OOFDMA. The dominant performance of OIFDMA when compared to OLFDMA is due to the fact that time-domain signal corresponding to OLFDMA results in additional weights, thereby leading to the deterioration in the PAPR performance. Meanwhile, OOFDMA results in higher amount of PAPR than the former subcarrier mapping techniques due to the rapid fluctuations in the envelope of the time-domain signal.

4.5 CONCLUSION

This chapter presents an in-depth analysis of the current state-of-the art research aspects that were carried out in the literature to address the detrimental aspect of PAPR in both multicarrier and multiple access schemes that are reconciling with the requirements of IM/DD systems for VLC. Taking into consideration the significant advantage offered by DCT/DST-based multicarrier systems when compared with DFT-based optical OFDM system, for the transmission of the same bit rate, both DCT/DST-based multicarrier systems support double the constellation sizes. Moreover, without the requirement of Hermitian Symmetry criteria, a real-valued signal is attained, thereby relieving the burdern to rely on additional signal processing techniques and hardware. Furthermore, when compared with DCT and DST, the computational complexity in terms of the requirement of number of additions and multiplications is more for DFT. Thus, this chapter gives a proof of the concept demonstrations of the PAPR analysis in DCT/DST-based multicarrier and multiple access systems that are flexibile to be implemented for VLC. It is apparent that the DCT/DST-DCO-OFDM system, in spite of being superior in terms of enhanced throughput when compared with DCT/DST-ACO-OFDM system, the addition of DC bias enables it to evolve as a power-inefficient scheme. Consequently, this aspect emerges to be more pronounced in IM/DD systems; this is due to the fact that the limited dynamic range of LEDs leads to clipping of the discretized time-domain signal which does not fit within its linear range, thereby leading to the occurrence of deleterious non-linear distortion like clipping distortion.

Therefore, to significantly reduce the amount of PAPR in DCT/DST-DCO-OFDM systems, this chapter presents a DCT/DST-based spreading technique as a PAPR reduction scheme and furthermore compares with other PAPR reduction techniques like PTS, clipping and filtering for the same developed system models (i.e., DCT/DST-based optical OFDM systems). The simulation result emphasizes that the reduction in PAPR upon imposing DCT/DST-based spreading technique in DCT/DST-based optical OFDM system is significant when compared with conventional DCT/DST-based optical OFDM systems. Upon imposing the other PAPR reduction schemes like PTS, clipping and filtering to DCT/DST-DCO-OFDM systems, PAPR reduces by an appreciable amount. From the mathematical analysis carried out in the literature, it is evident that even though other PAPR reduction techniques like PTS, clipping and filtering exhibit their remarkable nature to reduce the PAPR, but spreading dominates the former PAPR reduction techniques.

Particularly, the major attraction of this chapter is its presentation of elaborate mathematical analysis for the time-domain signal formats pertaining to different multicarrier and multiple access schemes that are complying with the requirements of IM/DD systems for VLC. Due to the constraints imposed in case of optical domain, where the transmitted signal should be both a real and non-negative valued signal, adaption of the RF-based PAPR reduction techniques to VLC is not straightforward. Consequently, this chapter gives a thorough literature on several PAPR reduction techniques that have been exploited in multicarrier systems. Moreover, taking into consideration the sophisticated nature of this technology where, with the already installed lighting fixtures, a small scale cellular network can be formulated, thereby making the deployment of multiple access schemes mandatory. In this regard, this chapter deals with the PAPR analysis in multiple access systems. Accordingly, this chapter presents the performance of different subcarrier mapping strategies like DCT/DST-OIFDMA and DCT/DST-OLFDMA as well as derives the analytical expressions for transmitted signal formats for DFT-OIFDMA and DFT-OLFDMA. The simulated results evidence that among the subcarrier mapping strategies, DCT/DST-OIFDMA is superior in terms of PAPR reduction than those of DCT/DST-OLFDMA. However, both schemes achieve a better reduction in PAPR when compared with DCT/DST-OOFDMA. Its ease of simplicity as well as enhancement in the throughput will definitely enable DCT/DST-based multicarrier and multiple access systems to be considered as attractive techniques for the cost-effective realization of IM/DD systems for VLC.

5 MULTIPLE ACCESS SCHEMES AND VLC FOR SMART CITIES

5.1 MOTIVATION

Dated back to the early 1980s and 1990s, where the second generation (2G) cellular communication systems such as the well-known global system for mobile (GSM) and the interim standard (IS)-95, i.e., IS-95, were introduced with the intent to support low date rate services such as voice calls and short message services (SMS). With the passing of time, the third and the fourth generation cellular systems, i.e., 3G and 4G, adopted suitable physical layer solutions such as the adaptive modulation, turbo coding, single-user point to point communication, multiuser multiple input multiple output (MIMO) transmission and reception techniques, OFDM, coordinated multipoint transmission schemes, etc., for the purpose of guaranteeing reliable and high data rate communication in the orders of Mbps. In recent times, due to the emanation of the IoT era, the entire world has witnessed an explosive intensification of connected devices. Moreover, the rapid proliferation of mobile broadband subscribers which reached around 8.6 billion in 2020 will assuredly impose an unprecedented growth in mobile traffic demand. In this context, in order to meet this enormous growth, it is anticipated for the future wireless communication networks to offer projected data rates in the order of several Gbps. Eventually, the advent of many new applications addressing the requirements of low-latency and high spectrum efficiency evolves as a significant challenge for the fifth generation (5G) and beyond-5G communication networks.

Taking into consideration this huge surge in mobile data, the ongoing research is approaching toward two main directions. Firstly, the enrichment of spectral efficiency of the available RF spectrum can be attained by adopting advanced modulation formats, multiple access techniques, etc. Secondly, the potentials of the unlicensed portion of the electromagnetic spectrum can be exploited to render high data rate communication. Facilitated by the tremendous demand for high data rates, improved security and increased energy efficiency geared the move from the congested RF bands to the optical spectrum which assures abundant unregulated bandwidth, low interference, secure, reliable and cost-effective communication as compared to the RF-based wireless communication. This has spurred an interest in the VLC which can be viewed as potential candidate to complement as well as to offload the RF-based wireless communication systems more particularly for the indoor user dense scenarios like the workplaces, homes, multinational companies, exhibition and conference halls, train cabins and aircrafts. As already stated in the previous chapters, the data in case of VLC is transmitted by modulating and demodulating the intensities of LEDs in a process called IM and DD. Furthermore, a typical VLC system can be realized by exploiting the off-the-shelf standard LEDs with the sophisticated trait of accomplishing simultaneous 'high data rate transfer' and 'indoor or outdoor illumination'. Thus, this remarkable nature of VLC creates an opportunity for the exploitation of the already available illumination infrastructural units to render not only illumination, but also wireless communication support. In addition to this, the most stunning feature of VLC is its inherent communication security since light cannot penetrate through several objects like walls, thus paving the way to be suitable for small cell design, thereby imparting high quality services without any intercell interference.

In this respect, VLC has expanded as a small cell technology which can be integrated as a part of future ubiquitous 5G communication systems. Despite offering great advantages, VLC systems have certain shortcoming that need to be fully addressed for the sake of the realization of the full potential of this growing technology. Among many such issues, the achievable data rates of VLC systems are restrained by the limited modulation bandwidth of white phosphorescent LEDs ranging up to a few megahertz. Consequently, in order to realize the full potential of such beneficial VLC systems in terms of its ubiquitous connectivity and assurance of high data transfer, it is mandated to rely on advanced optical modulation formats which are compatible with IM/DD systems. The detailed overview of different modulation formats are summarized in Chapter 3. Furthermore, with the motive to evolve as a key technology enabler for 5G and beyond-5G communication networks, multiple access schemes have been proposed in the literature addressing several arresting aspects like enabling massive ubiquitous connectivity, low latency and effective utilization of the spectrum resources. To this end, in an effort to focus VLC on a multiuser, scalable and fully networked wireless communication technology, there has been a growing literature on multiple access schemes in VLC systems.

The main objective of this chapter is to elaborate an exhaustive overview of the current state-of-the-art of different, multiple access schemes that can be exploited in the context of VLC systems. Explicitly, we provide a thorough review of multiple access schemes for RF-based wireless communications followed by the in-depth technical discussions on optical orthogonal frequency division multiple access (OOFDMA), optical code division multiple access (OCDMA), optical spatial division multiple access (SDMA) and wavelength division multiple access (WDMA). Finally, the major attraction of this chapter is a thorough review and analysis of a promising multiple access scheme like non-orthogonal multiple access (NOMA) which is envisaged to renders high data rate communication support for 5G and beyond-5G communication networks. From the earlier studies it is evident that NOMA outperforms other conventional multiple access schemes, and offers remarkable benefits when exploited in VLC systems. Additionally, the distinguished nature of NOMA is that it can coexist with other multiple access schemes, thereby rendering high spectral efficiency. Besides providing deep technical insights about different multiple access schemes in VLC systems, this chapter gives a brief description of exploiting VLC technology for smart city applications with major stress on design of VLC-based smart lighting systems.

5.2 REVIEW ON CONVENTIONAL AND EMERGING RF-BASED MULTIPLE ACCESS SCHEMES

Predominantly, RF-based multiple access schemes manifest an effective utilizations, i.e., sharing of the available network resources among a large number of users. In general, this has been realized by employing frequency and time division which constitutes the formulation of frequency division multiple access (FDMA) and time division multiple access (TDMA), the well-known multiple access techniques which were exploited in the first generation (1G) and second generation (2G) cellular technology. On the other hand, the third generation (3G) cellular communication technology adopted code division multiple access (CDMA) as the multiple access strategy where a set of subscribers/users exploit either spreading or signature sequences to allow for orthogonal multiple access, while facilitating the sharing of the same frequency and time. In particular, the detection methodology in CDMA is either through single user or multiuser detection mechanisms. In case of single user detection, the other users, interference is treated as an additional noise, while pertaining to multiuser detection mechanism, all the multiple users, signals are detected either through the maximum likelihood (ML) detection or minimum mean square error (MMSE) detection.

The fourth generation (4G) cellular technology employed OFDMA where the subcarriers are appropriated to the users either in interleaved or random manner. Whereas, fifth generation (5G) and beyond-5G networks require ensuring seamless services to the high data rate subscribers that are sharing the same network resources. From this perspective, NOMA can be viewed as a potential candidate multiple access scheme that provides the flexibility for multiple users to share the time and frequency resources concurrently through power domain multiplexing. Therefore, in contrast to orthogonal multiple access (OMA) schemes like TDMA, FDMA and CDMA, NOMA allows multiple users to utilize the entire time and frequency resources simultaneously, thereby emerging as a key multiple access technique for both RF and OWC to render a considerable amount of improvement in achievable system throughput as well as to reinforce ubiquitous connectivity. The underlying principle behind the power domain multiplexing is that there is a provision for the multiple users to simultaneously transmit and receive over the same frequency band interval by exploiting varying power levels.

Principally, there are three principal versions of NOMA: the power domain NOMA (PD-NOMA), code domain NOMA (CD-NOMA) and multiplexed version of NOMA, which is available by multiplexing either the power domain, code domain or the spatial domain. Particularly, among the available domains, PD-NOMA is the most widely adopted. In case of PD-NOMA, the multiplexed signals which belong to multiple users are separated by employing channel gain differences at the receiving end. In line with the interest in both research and industrial units, there has been a growing literature on exploitation of NOMA for both RF and OWC systems.

NOMA offers several remarkable benefits of which, its assurance of high cell-edge throughput is the most significant; additionally, it even guarantees a high spectral efficiency, user fairness as well as attainment of low-latency transmission. Nonetheless, in spite of its distinct advantages, there are certain drawbacks associated with NOMA like the increase in the complexity of the receiver implementation and the mutual interference among the users who are sharing the same resources also increase. Consequently, to alleviate such drawbacks, extensive research efforts have been carried out. The receiver implementation complexity can be reduced by exploiting user-grouping or user-clustering techniques and the mutual interference among the users can be reduced by employing superposition coding (SC) at the transmitting end and successive interference cancellation (SIC) algorithm at the receiver.

Owing to its remarkable performance offered by NOMA, it is considered a potential candidate for various standardization activities. NOMA has been exploited with the name of multiuser superposition transmission (MUST) in LTE-A systems i.e., 3rd Generation Partnership Project (3GPP) release 13. Meanwhile, the 3GPP releases 13 and 14 make use of a new category of user terminals that are capable enough to cancel out the interference giving rise to a method called network assisted interference cancellation and suppression (NAICS). It is also interesting to note that NOMA schemes were adopted even in the 3GPP releases 15 and 16. Even though a wide variety of multiple access schemes have been exploited in RF-based wireless communications, all of them cannot be utilized in the optical domain in a direct manner without any essential modifications. This is imperative because the primary requirement of IM and DD systems is a pure real and positive valued signal. Therefore, as a result, this enforces the application of necessary intermediate signal processing at transmitting end for the purpose to transform complex symbols into real and unipolar symbols. In this context it is vital to have a clear understanding on the key design aspects which need to be taken into consideration when exploiting the RF-based multiple access schemes to VLC.

5.3 MULTIPLE ACCESS SCHEMES FOR VLC

This section outlines the comprehensive overview of these multiple access schemes that are proficient to render high data rate communication and bandwidth efficiency in the context of VLC systems. More specifically, this section focuses on current state-of-the-art technical advancements, as well as gives essential insights about the challenges and constraints that are associated with each multiple access technique when employed for IM/DD systems. It is vital to take into consideration such aspects because these are of paramount importance in the design and deployment of cost-effective and high data rate communication systems. The major emphasis of this section is that it gives deep insights about the illustration of the realization of VLC system that supports multiple users by making use of different multiple access techniques. Before proceeding with the different multiple access schemes, a thorough emphasis is laid on the distinctive features of VLC that need to be taken into account in order to realize a fully networked VLC system.

5.3.1 DESIGN ASPECTS OF VLC SYSTEMS

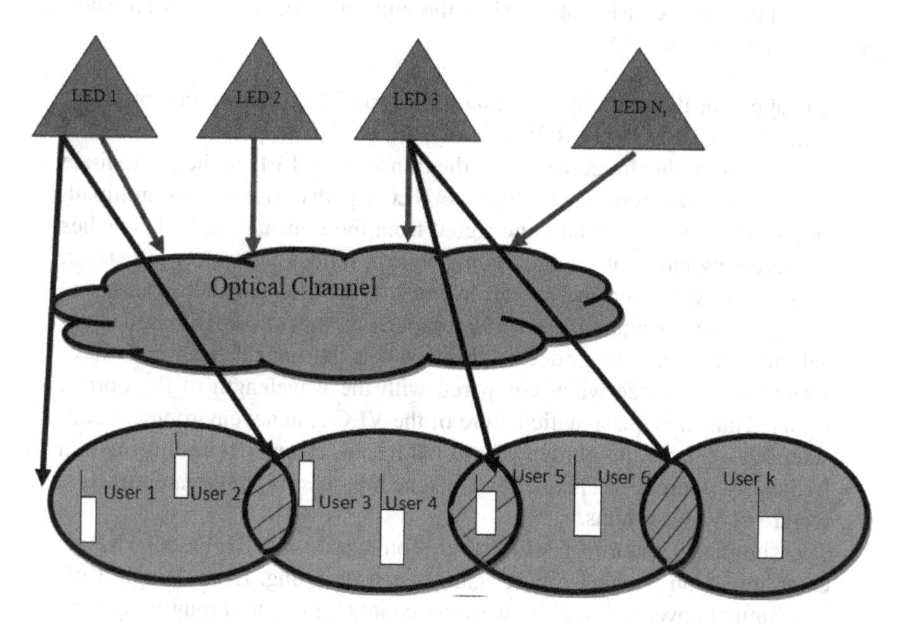

Figure 5.1: Typical multiuser VLC system model.

A general schematic model of a typical downlink VLC scenario is delineated in Fig. 5.1, where LEDs are installed on top of the ceiling and are transmitting information to the receiver terminals which are being equipped with photodiodes. Particularly, the FOVs of both the transmitting and receiving opto-electronic elements are designed in a manner such that they satisfy the requirements of simultaneous 'illumination' and 'communication'. As elucidated from the figure, there is a seamless transmission of information to the users which are within the vicinity of the installed lighting fixtures. Precisely, within the VLC channel environment, the signal does not undergo fading; meanwhile, once the signal is outside the range of the installed LED lighting fixtures, there is a degradation in the signal to interference noise ratio (SINR). In general, there is a significant deterioration of SNR at the edge of the interference-free region as well when the user moves away from the LOS path. The shaded area, as represented in the figure, specifies the ICI arising from the neighbouring cell. Generally, In case of a multiuser scenario, it is vital to take into consideration the distribution of resource units, namely, time, wavelength, frequency or space. Based upon the allocation of these resources, the various performance metrics like the user fairness, spectral efficiency, latency, energy efficiency, etc. are satisfied. Consequently, such types of practical scenarios, especially in case of the downlink and uplink, harnessed several research efforts on multiple access schemes which are appropriate for VLC systems. Even though several of the multiple access schemes have been adopted from the RF literature, it is vital to contemplate the peculiar characteristic features of VLC systems.

Some of the remarkable features of VLC along with the system design which needs to be taken into consideration when the multiple access schemes are adopted to VLC are listed below:

1. It is apparent that the impulse response of the VLC channel comprises of both LOS and NLOS paths. Principally, the LOS path specifies the shortest distant path or the direct path from the transmitting LED to the photodiode installed receiver terminal. While the NLOS paths arise due to multipath propagation scenario, where the signal from the transmitting LED reaches the receiving terminal by undergoing several reflections from walls, floor, ceiling, etc. However, in general, in case of the VLC channels, the signal fading can be considered less when compared to that of the RF-based channel environment. This is due to the fact that the optical front ends have a large aperture size when compared with the wavelength of the optical signal. Thus, this quasi-static nature of the VLC channel environment considerably relieves the burden of estimating the channel state information frequently. Moreover, channel-aware scheduling is no longer an attractive feature of VLC systems.
2. The significant nature of VLC when compared with RF-based wireless communication is being robust against eavesdropping. Even though it offers limited coverage area, light waves cannot penetrate through any non-transparent objects like walls and many such objects. Therefore, this intrinsic property of light waves not only guarantees high amount of security as there is no co-channel interference between VLC systems in adjacent rooms, but also offers the opportunistic feature of efficient spatial reuse of optical resources.
3. The non-coherent emission characteristics of LEDs allow the signal transmission to be both real and positive valued. Thus, this makes the design of optimal and suboptimal transmission schemes pertaining to the multiuser VLC systems more complex when compared to RF-based communications. Furthermore, it is to be remembered that 'illumination' is the primary task of LED and 'communication' is used as an add-on service. Therefore, it is of paramount importance to consider both the requirements as well as constraints of illumination which include adequate dimming control, suitable mechanisms to circumvent the flickering problems, etc. in the design of VLC multiuser systems. This can be elaborated with an example where the design of multiuser VLC systems necessitates the dependency of highly directional LEDs to reduce the multiuser interference; nonetheless, this choice might not fulfill the illumination requirements. Therefore, it is indispensable to facilitate a joint design where both lighting and communication requirements are met when VLC is intended to be used for multiuser scenarios.
4. Despite offering abundant unregulated bandwidth, the offered data rates of VLC systems are restricted by the limited modulation bandwidth of the off-the-shelf standard LEDs. Eventually, this accelerated the emergence of

new optical technologies or manipulation of RF-based advanced wireless communication concepts like MIMO, spectrally efficient multiple access schemes, etc. in order to realize the full potential of VLC systems.

5.3.2 OPTICAL FREQUENCY DIVISION MULTIPLE ACCESS

As already discussed, multiple access schemes were adopted in every generation of the cellular telecommunication systems as FDMA, TDMA and CDMA which were a core component in 1G, 2G and 3G cellular mobile communication standards. In the same manner, upon moving to the higher generations, i.e., 4G, communications exploited OFDMA as a multiple access technique to furnish the high data rate transfer. It is widely accepted that the High-dimensional multicarrier schemes like OFDM by itself cannot be exploited as a multiple access scheme. In order for it to be a suitable multiple access technique, it is required for OFDM to be integrated with the existing multiple access schemes like TDMA, FDMA and CDMA. OFDM creates the flexibility, adaptability and simplicity for the establishment of a multiuser access scenario. The straightforward integration of OFDM with FDMA resulted in the emergence of a well-known and robust multiple access scheme like OFDMA. Taking into consideration the distinctive features offered by OFDMA in RF-based wireless communications, it has been adopted as an effective multiple access technique in optical communication systems. In the context of VLC systems, it is of immense prerequisite to take into account the design aspects when adopting OFDMA technique in the optical domain.

This growing technology like VLC facilitates the ease of formulation of a small scale cellular communication network that can be energetically embedded into an indoor room environment which comprises of multiple lighting fixtures. Primarily, each installed and spatially separated illuminating device behaves as a base station (BS) or as an access point (AP) rendering seamless services to the users underneath it. Such kind of cellular network ensures a wide coverage area to multiple mobile terminals/user equipments which are within the proximity of the installed lighting fixtures. From the earlier research findings, it is evident that a system of such kind offers a considerable improvement in wireless system capacity when compared to RF-based wireless communication systems. Particularly, the scenario where multiple APs communicate with the multiple user equipments signifying the downlink scenario and generally, in the uplink scenario, one or more users can communicate with the AP and this sort of network is referred to as optical attocell. Considering the benefits of OOFDM, it has been deemed as one of the prime candidates for enabling signal modulation in VLC systems. Additionally, OOFDM endeavors the straightforward realization of optical OFDMA (OOFDMA). Most of the previous research work in the literature adopt the spectrally efficient DCO-OFDM methodology in the downlink scenario. In addition to the assurance of spectral efficiency, the added DC bias in DCO-OFDM is harnessed for the purpose to support adequate levels of illumination.

Significant research efforts have been directed toward the implementation of OFDMA-VLC system to provide high data rate communication to multiple users which are within the vicinity of the installed illumination fixtures. The majority of the earlier works portray the system implementation of downlink OOFDMA system, and very few works witness the realization of uplink OOFDMA system. Firstly, it is vital to emphasize on the system implementation of OOFDMA-based VLC system. A DCO-OFDMA-based VLC system which depicts the downlink scenario is elucidated in Fig. 5.2. For the purpose of understanding, we limit our discussion to

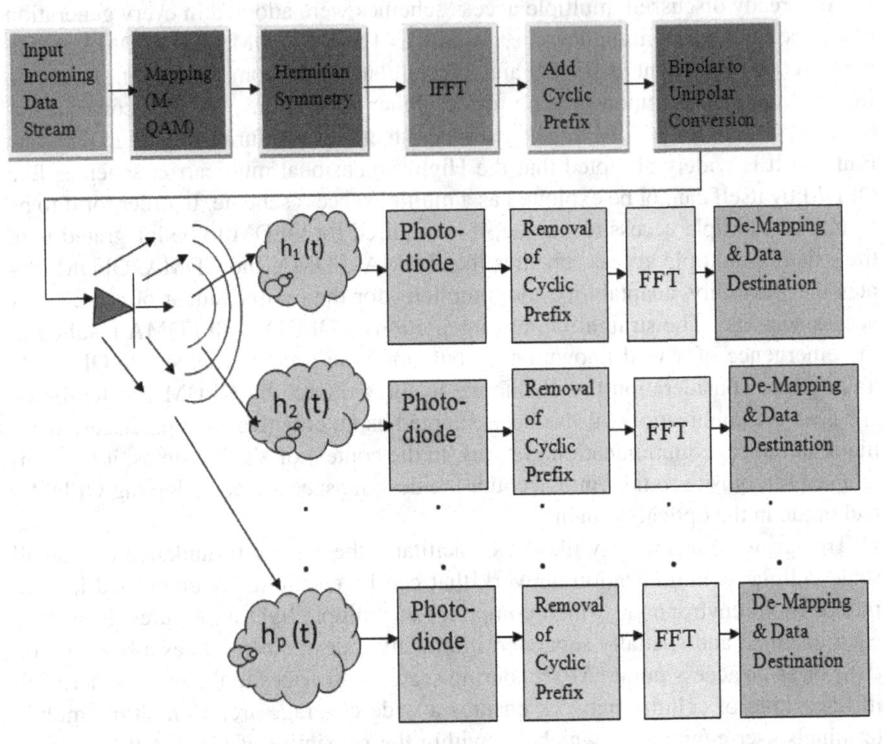

Figure 5.2: Schematic representation of downlink scenario in DCO-OFDMA system for VLC.

a single attocell. This is also valid because light signals cannot penetrate through other non-transparent walls, doors and partitions; therefore, we can focus on single attocell scenario where the interference emanating from the neighboring attocells in the adjacent rooms can be ignored. It is to be taken into account, the transmitter and receiver schematic as represented in the figures are designed in accordance with the requirements of IM/DD systems. As illustrated, a set of P user terminals in the attocell receive their services from the installed lighting fixtures. At the transmitting terminal, the incoming high stream of data is transmitted into several low data rate streams by employing a serial to parallel converter and then mapped by using higher order complex constellation like M-QAM. Since Fourier signal processing implies complex signal processing where the phase carries fundamental information; therefore, it is vital to constrain the input to the IFFT transformation block to satisfy the

Hermitian Symmetry criteria to yield a real-valued signal, or it is also vital to design a downlink multiple access system by using real trigonometric transformation techniques like DHT, DCT, DST, etc.

Mathematically, the aforesaid discussion can be expressed as follows:
Let the total number of subcarriers in each OFDM symbol be denoted as N. Correspondingly, S_k conveys the complex mapped data on the kth subcarrier. As stated, in order to obtain the real-valued time-domain OFDM signal, the OFDM frame structure needs to be expanded in such a manner that it accomplishes the criteria of Hermitian Symmetry. According to Hermitian Symmetry criteria, the first and the middle subcarriers are set to zero and the remaining subcarriers are flipped complex conjugate versions of the previous ones. This sort of arrangement is generally done to ensure the absence of any complex component at the output of IFFT block. Thus, this can be represented as

$$S_0 = S_{\frac{N}{2}} = 0 \tag{5.1}$$

$$S_{N-k} = S_K^*; \quad k = 1,2,3,\cdots \frac{N}{2} \tag{5.2}$$

Therefore, by taking into consideration the aforementioned equations (5.1) and (5.2), the input to the IFFT can be represented in the form of a vector which can be expressed as

$$S_k = \left[0,\ S_1,\ S_2,\ \cdots\cdots\ S_{\frac{N}{2}-1},\ 0,\ S^*_{\frac{N}{2}-1},\ \cdots\cdots\ S_1^*\right] \tag{5.3}$$

Further, the time-domain format of the IFFT signal can be proclaimed as

$$s_n = \frac{1}{N}\sum_{k=0}^{N-1} S_k e^{\frac{j2\pi nk}{N}} \tag{5.4}$$

Thus, upon substitution of (5.3) into the time-domain signal format for IFFT as given by (5.4), the following expression is attained as shown below

$$s_n = \frac{1}{N}\left[S_0 e^{\frac{j2\pi n(0)}{N}} + \sum_{k=1}^{\frac{N}{2}-1} S_k e^{\frac{j2\pi nk}{N}} + S_{\frac{N}{2}} e^{\frac{j2\pi n\left(\frac{N}{2}\right)}{N}} + \sum_{k=\frac{N}{2}-1}^{1} S_k e^{\frac{j2\pi nk}{N}}\right] \tag{5.5}$$

Upon enforcing the aforementioned conditions and solving, the expression for the time-domain signal format in DCO-OFDMA system can be put as follows

$$s_n = \frac{2}{N}\sum_{k=1}^{\frac{N}{2}-1}\left[\underset{RC}{S_k cos}\left(\frac{2\pi nk}{N}\right) - \underset{IC}{S_k sin}\left(\frac{2\pi nk}{N}\right)\right] \tag{5.6}$$

As a result, from (5.6), it can be inferred that when the total number of subcarriers is set to $2N$, the effective subcarriers will be only $N-1$ because only half are exploited for data transmission due to the Hermitian Symmetry criterion. It is to be noted that the set of the effective subcarriers is denoted by $N = \{1,2,3,\cdots N-1\}$ and the set of the subcarriers which are appropriated to the user equipment p are denoted by N_p. The time-domain signal as represented by equation (5.6) needs to be intensity modulated through the LEDs. Hence, this necessitates the transmitted signal to be both real and positive valued. So far, the attained time-domain signal as represented by (5.6) ensures only for the real nature. Consequently, in order to accomplish a positive valued signal transmission, it is necessary to enforce bipolar to unipolar conversion strategies. In this regard, since DCO-OFDM offers a straightforward realization of OFDMA, it is easy to add a suitable amount of DC bias to the non-negative valued signal in order to fetch a positive valued signal. Accordingly, the DC bias added positive valued time-domain signal can be expressed as

$$s_{n,Dc} = s_n + \beta_{DC} \tag{5.7}$$

$s_{n,DC}$ in (5.7) specifies the DC bias added time-domain signal. Thereupon, this time-domain signal is intensity modulated with the help of LED. From now on, at the receiving end, we confine our discussion to the subscriber/user equipment p, where transmitted signal is sensed by means of a photodiode. Thereupon, inverse operations are incorporated at the receiving end, such as removal of cyclic prefix, followed by translation of the time-domain signal into frequency domain signal with the help of FFT transform technique. Accordingly, the received frequency domain signal can be formulated as

$$\hat{S}_{k,p} = H_{k,p} S_{k,DC} + N_{k,p} \tag{5.8}$$

From (5.8), $H_{k,p}$ denotes the frequency response of the channel corresponding to the subscriber p at subcarrier k, whereas the parameter $N_{k,p}$ signifies the amount of AWGN and $S_{k,DC}$ represents the DC bias added transmitted signal in the frequency domain. Finally, each subscriber or the user equipment will demodulate the received signal by exploiting suitable demapping techniques in order to restore the useful information.

5.3.2.1 Related work on OOFDMA-VLC systems

An optical attocell network is formed by carefully installing the LEDs termed as APs on the ceiling with an appropriate alignment. Each installed LED or the AP offers services to the users or mobile stations which are within the coverage area or illumination region of the AP. In general, whenever a mobile station or the user equipment comes outside the vicinity of the AP, then handover techniques can be employed to offer the most reliable means of wireless services by taking into account that the mobile station should be served by the most appropriate AP. Exploiting the same frequency resources in the adjacent cells results in the emergence of co-channel interference (CCI) in the optical attocell networks. The existence of CCI will substantially degrade the performance of the optical attocell network, as there is a severe degradation in the SINR of the users located at the cell edges. Consequently, the offered data rates will be limited. Thus, this entails to circumvent the CCI in order to maximize the system throughput as well as to maximize the quality of the signal within the whole coverage region. With this motive, the work in [97] introduces the concept of multipoint joint transmission (JT) to a VLC cellular network. JT refers to contemporaneous transmission of data from multiple cooperating BSs to MS. Thus, by employing this means of coordinated transmission of data mitigates the strong CCI in VLC optical attocell network. From this work, it can be emphasized that the SINR of the cell-edge user improves significantly. The simulation results affirm that when compared to full frequency reuse system, the JT scheme results in considerable improvement in the median SINR by 16.4 dB. Besides, when compared to a static resource partitioning system, the JT scheme achieves a 67.6% improvement in the system throughput.

This section presents in detail the research work carried out in the literature that are relevant to OOFDMA-based VLC systems. The work in [95] proposes a framework of an indoor VLC cellular network which is often referred to as optical attocell network and employs the multiple access scheme like OFDMA which is based on DCO-OFDM methodology. Particularly, this work develops an analytical approach to calculate the statistics of SINR of a user who is randomly located in the optical attocell. The analytical analysis of the derived SINR was verified by means of Monte Carlo simulations. Furthermore, with the motive to estimate the wireless capacity of the downlink scenario, this work studies the average spectral efficiency of the optical attocell network. The simulated results affirm the strong dependency of the spectral efficiency on the radius of the optical attocell as well as on the half-power semi-angle of the light transmission profile. By appropriate selection of attocell parameters, the numerical results demonstrate an average spectral efficiency of 5.9 bits/s/Hz. In order to ensure an enhancement in system throughput by providing seamless services to multiple users which are within the coverage area of the APs/LEDs, it is vital to stress on several important aspects like the user mobility, assurance of duplex communication and handover mechanisms. With this motive, the research work in [50] provides a hybrid network model which exploits the combination of VLC and OFDMA where the VLC channel is exploited only for enabling the downlink transmission, while OFDMA is chosen for uplink transmission. In order to resolve the mobility of the users among different hotspots, this work proposes a novel protocol that takes into consideration the joint combination of access, horizontal and vertical handover mechanisms for the mobile terminal. Furthermore, it is also interesting to witness a new VLC network scheme and its corresponding frame format to address the multiuser access problems that usually occur in every hotspot. The authors even define a new metric ρ for analyzing the capacity of this hybrid network. The analytical and simulation results of this work reveal the superiority of the hybrid network in terms of capacity when compared with traditional OFDMA system.

The work in [513] studies the issues that are associated with resource allocation in VLC-OFDMA systems. The authors in this work propose two mechanisms for the resource allocation, namely, the optimization-based resource allocation scheme and the low complexity resource allocation scheme. Furthermore, the performance of these two schemes in terms of data rate, computational complexity and user fairness was evaluated and compared with that of the resource allocation strategy in case of TDMA system. The simulated result as shown in Fig. 5.3 depicts the average user data rate variation over the baseband modulation bandwidth where the proposed low

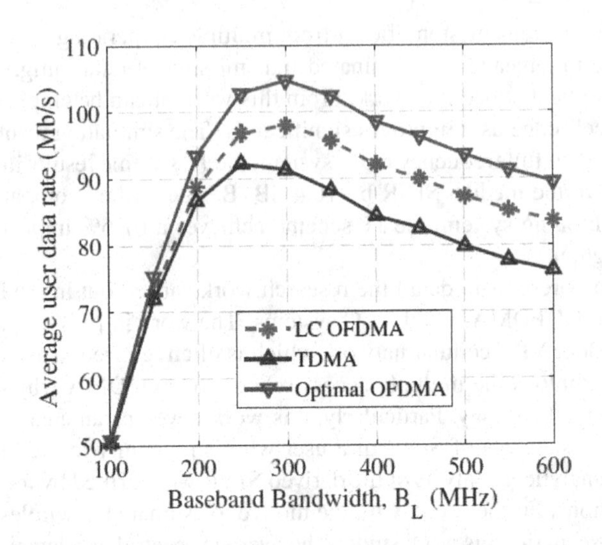

Figure 5.3: Performance analysis of user data rate with respect to the baseband modulation bandwidth [513]. LC-OFDM denotes the low complexity resource allocation-based OFDMA, optimal OFDMA implies the optimization-based resource allocation-OFDMA.

complexity resource allocation-based OFDMA and optimization-based resource allocation schemes were compared with the resource allocation in TDMA. From the simulated result, it is clear that both resource allocation schemes in OFDMA system outperform the TDMA system. Upon increasing the modulation bandwidth, the user data rate significantly decreases due to the fact that the first subcarrier remains unused due to the Hermitian Symmetry criteria; as a result, there is a substantial increase in the unused bandwidth on the first subcarrier, thereby leading to the inefficient usage of the total bandwidth. Furthermore, upon observation of the simulation results of this work, it can be concluded that OFDMA system exploiting the aforementioned resource allocation schemes exhibits a superior performance than TDMA system in terms of both data rate as well as user fairness. Also, the low complexity resource allocation scheme attains a near-optimal performance by resulting in 90% reduction in the overall computational complexity.

The work in [96] presents the interference coordination in optical wireless cellular networks by exploiting different frequency reuse techniques. Particularly, this work introduces the concept of fractional frequency reuse (FFR) technique and compares with the other two benchmark techniques, like full frequency reuse and cluster-based static resource partitioning techniques. From the earlier findings, it is evident that full frequency reuse technique results in higher throughput only at the expense of poor performance of the users located at the cell edges. Meanwhile, the second benchmark technique facilitates the cell edge users, to experience a better performance only at the cost of reduced system throughput. Thus, contrarily, FFR technique has been exploited to balance between the cell edge users, performance and system throughput along with offering minimum system complexity. By making use of suitable power control factors, it can be delineated from the simulated results that FFR technique achieves a guaranteed user throughput of 5.6 Mbps with an average spectral efficiency of 0.3389 bps/Hz/m^2. When compared with the other benchmark techniques, significant improvement in the achievable spectral efficiency and user throughput can be witnessed by using FFR technique. Therefore, with this motive to ensure higher throughputs, there are several significant research contributions in the literature. The work in [138] proposes two soft handover mechanisms to increase the data rates of OFDMA-VLC systems. The simulation results of this work infer that by exploiting the proposed soft handover techniques, the overall OFDMA-VLC system yields higher data rates as well as enhances the performance of the users located at the cell boundaries.

The work in [261] proposes a downlink cooperation for indoor VLC cellular networks to enhance the spectral efficiency. This work adopts OFDMA as the multiple access strategy, and FFR technique is also considered with the intent to enhance the experience of the users located at the cell boundaries. A relay assisted transmission is accomplished between the neighboring BSs where a non-orthogonal amplify-and-forward (NAF) protocol is implemented to achieve the downlink cooperation. This work gives the analysis for the SINR as well as derives the spectral efficiency expressions for users situated at random locations in an optical attocell. Furthermore, the derived expressions were validated by means of Monte Carlo simulations. A much similar research contribution in [94] analyzes the downlink performance of optical attocell networks. Primarily this work derives the mathematical expressions for SINR, outage probability, and the achievable data rates of OFDMA-VLC system. In general, the combination of optical attocell with the existing lighting infrastructures, as well as the imposition of other physical constraints, fails to achieve optimized regular HEX cell deployments. In this regard; this work takes into account several other network topologies like the square and random cell deployments, etc. Since the Poisson point process (PPP) represents a close association with the real-world scenario; this work takes into consideration such deployment process and the simulation result infers that the hexagonal and PPP random cell deployments offer the best and the worst-case performance for the realization of practical optical attocell deployments. Typically, optical attocells offers very high data rate density due to the inherent nature that they can be deployed densly in a room. Consequently, in order to proclaim this advantage offered by the optical attocell networks, this work further carries out comparison between the performance of optical attocell network, RF-small cell networks which can be termed as femtocells and indoor mmWave systems in terms of achievable data rate. The simulation results show that the well-designed optical attocell dominates the femtocell network or mmWave systems in terms of offered data rate.

In [304] and [305], an indoor VLC positioning system exploiting OFDMA scheme was presented, where the proposed system could provide both indoor positioning as well as data communications. The feasibility of the proposed system model was analyzed in a room environment comprising of dimensions $10 \times 10 \times 9$ cm, where the simulated results determine that the proposed scheme offers a mean positioning error (PE) of 1.32 cm and the error vector magnitude (EVM) was 10 dB over a 12 cm free space transmission span. The work in [305] extends the work as portrayed in [304]; in order to enable indoor positioning, three subcarriers with maximum received signal intensity pertaining to three LEDs were employed. The simulation results validate the sophisticated nature of the proposed system, where it is capable to determine the position of the receiver without any (ICI) (INTER CARRIER INTERFERENCE) as well as exhibit its potential to render data communication as well. An indoor room environment comprising of dimensions $20 \times 20 \times 15$ cm was employed to determine the viability of the proposed scheme. For a 20 cm free space transmission span, the proposed system employing QPSK modulation scheme guarantees a mean PE of 1.68 cm and the EVM is more than 15 dB. The research efforts in [308] report that the authors address the multiuser access scenario in a small cell by exploiting OOFDMA. In order to facilitate high throughputs in downlink transmission scenario, this work focuses on resource allocation problem for OOFDMA-VLC system. Additionally, this work gives an approach for the joint design of the bias level, power and subcarrier allocation in the aforementioned system. Furthermore, this work depicts the proposal of efficient algorithms for optimizing the bias value along with presentation of its convergence analysis. Low-complexity dual decomposition method and a fast constant power allocation method for both power and subcarrier allocation were proposed. The simulated results of this work clearly illustrate the characteristics of different algorithms as well as highlight the dominance of the joint design approach over other allocation methods.

OFDMA has been exploited even in the field of vehicular communication as well. The work in [546] portrays that a vehicular VLC system has been developed where OFDMA was exploited as a multiple access scheme, and the link was developed by means of commercial blue LEDs and avalanche photodiodes (APDs). The experimental results demonstrat a 43.1 Mbps uplink transmission over a 30 m long VLC link. Furthermore, both experimental and simulation results signify that both the non-linear effects of LED as well as the shot noise of the APD induce serious effect on the EVM due to the emanation of out-of-band interference. This eventually brings forth a conclusion that digital compensation of non-linear induced distortion as well as reduction of background light are vital when deploying higher multicarrier modulation formats. Furthermore, the viability of implementation of OFDMA as a multiple access technique in underwater VLC applications can be found in the recent literature [155]. This work illustrates the downlink scenario of an underwater sensor network (USN), where the communication aid between the sensor nodes and the central command units is facilitated by means of VLC. This proposed system relies on DCO-OFDMA-based multiple access strategy where the subsets of subcarriers were appropriated to different sensor nodes. With the motive to furnish an identical data rate at each sensor node in USN as well as to fulfill a targeted BER, this work proposes a joint subcarrier allocation and bit loading algorithms. Finally, the significance of these methods was validated by means of Monte Carlo simulations.

The alarming increase in the rate of mobile devices imposes a challenge on the current RF spectrum. Toward this end, the wireless fidelity (Wi-Fi) networks fail to provide reliable services to the end users. On these grounds, it is vital to exploit the unlicensed portion of the spectrum such as the visible light and the infrared. Much recent work in the literature provides a hybrid light fidelity (LiFi)/Wi-Fi network (HLWN) as an entrancing solution for future wireless communications, where the Wi-Fi network is reinforced with the significant nature of LiFi like its capability to ensure ultra-high speeds as well as low-latency wireless connectivity. This work studies a dynamic load balancing (LB) with handover mechanism in HLWNs. Furthermore, in order to evaluate the performance of the proposed HLWNs, the authors take into account an orientation-based random waypoint (ORWP) mobility model as well as propose OFDMA-based resource allocation method in LiFi systems [564]. In addition, this work proposes an enhanced evolutionary game theory (EGT)-based LB scheme with handover mechanism in HLWNs. In the proposed EGT approach, each user is given a provision to adapt their strategy with the purpose to enhance the payoff till load balancing is accomplished across the LiFi and Wi-Fi networks. Thereupon, the Wi-Fi system exploits the carrier sense multiple access with collision detection (CSMA/CA) and LiFi system makes use of the OFDMA-based resource allocation method. From the simulation results, it can be surmised that, upon comparing with TDMA scheme, OFDMA-based resource allocation method in LiFi system shows an outstanding performance in terms of user data rate as well as user fairness. Meanwhile, While, pertaining to LB in HLWNs, the proposed game theory approach like the EGT scheme illustrates significant improvement in the throughput over the other standard schemes like hard threshold (HT) scheme and random access point assignment (RAA) scheme.

5.3.3 OPTICAL CODE DIVISION MULTIPLE ACCESS

The same underlying principle of CDMA as used in RF-based wireless communications implies the optical domain. In case of OCDMA technique, each user is assigned a dedicated code to allow for simultaneous transmission and reception over the same frequency band. While, at the receiving end, the user correlates the received signals with its respective designated code and accordingly makes a decision. Pertaining to CDMA communications, the design of signature codes is tremendously foremost aspect and the same is applicable to VLC-CDMA systems. In case of traditional RF wireless communications, CDMA systems generally rely on bipolar codes due to their distinguished salient features like their excellent correlation functions that ease the ability to distinguish different users in an effective manner by combating the detrimental issues like multiuser interference and multipath interference. However, it is not that easy to exploit these bipolar codes in VLC systems that use IM and DD. This is due to the fact that the particularities involved in the design of VLC transmitter compel to rely on relatively complex modulation format while adopting these codes for cost-effective VLC systems. Eventually, this in turn increases the overall system complexity. Thus, this drawback can be overcome in a propitious manner either by modifying bipolar codes that well suit the VLC systems, or by entirely redesigning the unipolar codes for optical CDMA systems that are flexible to be used for VLC.

Modification of bipolar codes that well fit the VLC systems is not devoid of disadvantages. The transformation of "– 1" to "0" in code elements practically turns out to be an infeasible solution, as there is a deterioration in the original orthogonality of the signature codes that are based on "– 1" and "+ 1" chips [412]. As a result, multiple access interference is inevitable. Nonetheless, significant research efforts have been carried out to exploit bipolar codes for OCDMA-based VLC systems. Whereas, pertaining to the unipolar codes, a variety of designs of unipolar codes can be witnessed, such as random optical codes (ROC), prime codes (PM), optical orthogonal codes (OOC), zero cross-correlation codes (ZCC), and bipolar to unipolar (BTU) codes. The rest of this subsection is structured in such a manner that the different types of VLC-CDMA systems comprising of unipolar and bipolar codes were introduced along with a major emphasis of the performance evaluation of CDMA-VLC systems employing unipolar and bipolar codes.

5.3.3.1 CDMA-based VLC systems comprising unipolar codes

Before moving forward with the elaborate discussion on unipolar codes, it is essential to have an understanding on the system model of VLC-CDMA that is exploiting unipolar codes. The transmitter and receiver schematic representation of VLC-CDMA system with unipolar codes is illustrated in Figs. 5.4 and 5.5. In such type

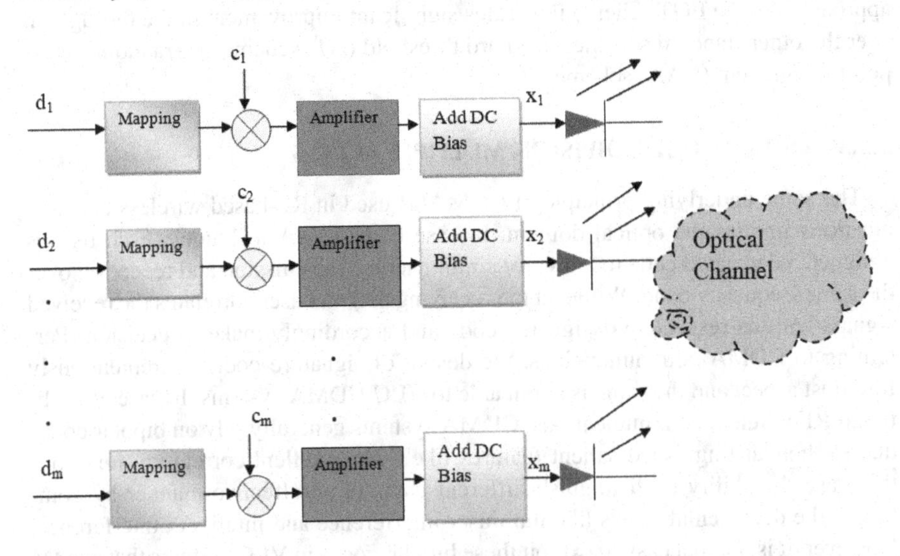

Figure 5.4: Transmitter schematic of VLC-CDMA system exploiting unipolar codes.

of systems, the modifications in the design of transmitter and receiver need to take into account several factors like the types of LEDs employed, the channel models, types of modulation schemes, etc. From the transmitter schematic, the data stream

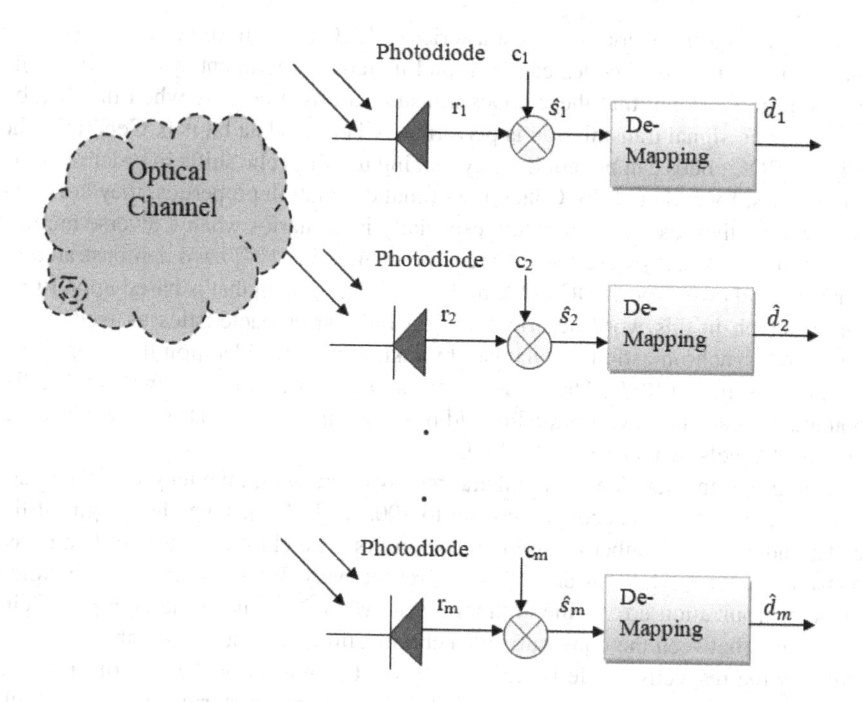

Figure 5.5: Receiver schematic of VLC-CDMA system exploiting unipolar codes.

$\{d_1, d_2, d_3, \cdots d_m\}$ is modulated into a set of symbols $\{s_1, s_2, s_3, \cdots s_m\}$ by using an appropriate mapping technique and then followed by multiplication of signature codes given by the pattern $\{c_1, c_2, c_3, \cdots c_m\}$, respectively. Furthermore, this signal is amplified and then an appropriate amount of DC bias is added to attain unipolar signals. Thereupon, these sets of unipolar signals are intensity modulated through the LED. Particularly, the input signals to the LEDs drive the LEDs only in their linear range to avoid clipping induced distortions. Further, these intensity modulated signals are propagated through the indoor VLC channel environment, where at the receiving end, the signals are detected by means of a photodiode to yield the set of received signals $\{r_1, r_2, r_3, \cdots r_m\}$ followed by multiplication with the same set of signature codes to yield the corresponding data symbols $\{\hat{s}_1, \hat{s}_2, \hat{s}_3, \cdots \hat{s}_m\}$, respectively. Finally, suitable demapping schemes are enforced in order to yield the data $\{\hat{d}_1, \hat{d}_2, \hat{d}_3, \cdots \hat{d}_m\}$. The details of performance analysis of CDMA-VLC systems exploiting different unipolar codes will be discussed below:

A. Performance Analysis of CDMA-VLC Systems with Random Optical Codes

The research efforts show that ROC works well in scenarios where a large number of users share a common channel. This sort of scenario is very similar to wireless sensor networks, where a set of sensor nodes commonly communicate to the cluster head. The remarkable trait of ROC is that in spite of the correlation functions of ROC not being optimal, it offers the flexibility for users to share a common VLC channel. The earlier works state that a series of ROC codes have been proposed and the application of ROC codes can be found in audio experiment systems. It should be taken into account that these codes transmit information only when the data bit is 1 while no signal transmission is performed when the data bit is 0. Generally, the original ROC codes can be modified by making use of cyclic shifts in order to attain more codes. Even though ROC lack in optimal correlation properties, they are characterized by their ease of generation especially in scenarios when a diverse number of simultaneous users were considered. The authors in [194] have demonstrated an experimental prototype of OCDMA-multiuser-VLC system that is based upon ROC. The research in this work describes the operational characteristics as well as the functional synchronization mechanism by taking into consideration the poor correlation properties of ROC. This work portrays that the developed framework has the potential to transmit medium-quality audio data from several LEDs through different optical channels by making use of ROCs.

Going into the detailed experimental results in this work, the length of the codes which were taken into account were set to 100, while 10 denotes the weight of the code, and 8 is the number of simultaneous users. The chip rate and the data transmission rates were 1 Mcps and 10 Kbps, respectively. For the purpose of avoiding the synchronization errors, the authors in this work introduced the concept of circular shifts between the consecutive synchronization bits, and the number of cyclic shifts of the respective code is signified by m. The impact of the size of m on the BER and synchronization error rate (SER) on the overall performance of the developed system were also analyzed. From the result analysis for lower orders of m, i.e., when $m = 1$ and 2, there is a degradation in the BER performance of the system, while upon increasing m to 3 and 4, the obtained BERs for both experimental and theoretical results were in close association with each other, i.e., the attained BERs for theoretical and experimental results were 3×10^{-5} and $\sim 2 \times 10^{-5}$, respectively. Pertaining to the SER analysis when $m = 3$, the achieved SER for both theoretical and experimental results was 4.9×10^{-6} and 9×10^{-6}, while, upon increasing the size of cyclic shifts, i.e., when $m = 4$, an error-free communications can be illustrated where the SER for the theoretical and experimental results were 2.9×10^{-8} and zero. Thus, it can be deduced that the higher the size of the cyclic shifts, the greater the achievement in the error-free communications. Meanwhile, a similar analysis can be reported in [193] where experimental validations of an OCDMA-based VLC system that was interfaced to an Ethernet network was presented. Much similar to the previous work as stated in [194], this work proposes as well as evaluates the synchronism mechanism to mitigate the relative poor correlation properties of ROCs. For the same code parameters and the same number of simultaneous users as stated in [194], the work in [193] considers a chip rate and data transmission rate of 2 Mcps and 20 Kbps, respectively. The attained SER for $m = 3$ was 5×10^{-6} while for $m = 4$, zero SER can be reported.

Much relevant to the research works as stated in [194] and [193], the work in [186] takes into consideration a new parameter like the number of applied code extensions E. The main reason for using this additional parameter is to extend the size of the family of base codes by making use of differently cyclic-shifted code replicas, thus facilitating the accommodation of more users. Thus, in this work, a new mechanism called cyclic code-shift extension (CCSE) has been proposed to facilitate the asynchronous multiuser transmissions, especially for the scenarios which requires to accommodate large number of users accomplishing seamless asynchronous transmission. The simulated results of the work as portrayed in [186] emphasized that with the increase in parameter E, the total number of users supported by the system also increase.

B. Performance Analysis of CDMA-VLC Systems with Prime Codes

From the earlier works it can be inferred that prime codes (PC) exhibit their simplicity in terms of code generation process and offer a comparatively less complexity in the design aspects of both encoder and decoder and hence widely deployed in non-coherent OCDMA and wavelength hopping OCDMA systems. Despite its simplicity, its correlation functions not being optimal and at the same time the large autocorrelation side lobes hinder the capability of achievement of accurate synchronization between the transmitter and the receiver, thus, making it unfit for asynchronous VLC-CDMA systems. Consequently, in order to address this drawback, it is mandatory to enforce necessary modifications and accordingly lead to the design of generalized modified prime code (GMPC) and inverted GMPC (IGMPC) by making use of the modified prime sequence code (MPSC). By exploiting an improved design of PM, the performance of OCDMA-based VLC systems was analyzed in [341]. This work signifies the performance comparison of the improved version of PC-based OCDMA-VLC system and the conventional OCDMA-VLC system in terms of average light intensity and average number of users. Furthermore, comparisons of the aforesaid two systems were carried out in terms of the relationship between the normalized light intensity fluctuations and the average number of the users. The code length of GMPSC and the maximum number of users taken into account are 64 and 32, and when the total number of simultaneous active users are less than 32, then there is no chance for the error to occur in both systems because they are capable enough to tolerate the multiuser interference. On the other hand, when compared to the traditional OCDMA-VLC systems, the system with improved PC exhibits higher average light intensity and lower normalized light intensity fluctuations [412].

C. Performance Analysis of CDMA-VLC Systems with Optical Orthogonal Codes

Optical orthogonal codes (OOCs) are one of the mature signature codes for OCDMA and generally are exploited in optical fiber communication systems. OOC comprises of (0, 1) sequences. The design aspect of OOC is complex as the positions of "1" are not regular when compared with PC. Even though the correlation function of OOC are better than that of the PC, the ability to generate as many codes as possible is not quite large like ROC. Thus, these factors limit the capacity of OOC-based CDMA-VLC systems. Consequently, to improve the capacity of OCDMA-VLC systems, research work concentrated mainly either to improve the design aspects of OCC or to jointly combine the usage of OCC with other codes. Pertaining to the design aspects, there are two related works that report the improvement in design aspects of OOC. A relationship between the three parameters like the code length, code weight and the correlation level were studied in [93]. This relationship can be expressed as $(n, w, 1) = (q^2+q+1, q+1, 1)$ where q^2+q+1, $q+1$ and 1 denote the code length, code weight and the correlation level limitation respectively, while q specifies a constant. The improved OOC has a ZCC, thus minimizing the multiple access interference. However, the number of users that can be supported is less and this family of OOC is power inefficient.

The simulated results in [93] give a performance comparison in terms of BER and SNR between the OOC and the improved version of OOC, which are obtained by using the relationship $(n, w, 1) = (q^2+q+1, q+1, 1)$. Upon assuming the value of q to be 3, the corresponding code set is given by $(13, 4, 1)$ which was the improved version of OOC and upon comparing with the traditional OOC code set $(13, 3, 1)$, the improved version can support more users, i.e., a maximum of 6 users can be supported when 2-ary modulation was employed. However, from the simulated results, it can also be observed that even though the improved version of the OOC can accommodate more users, it has low power efficiency. Furthermore, the research efforts in [278] also strive to improve the OOC, where two-dimensional OOC have been proposed based upon the three colored wavelengths, namely, red, green and blue for the purpose to construct the code word for each user present inside the aircraft. Notably, this work is mainly directed toward the proposal of aeronautical architecture which exploits VLC as the prime technology for dispensing the in-flight entertainment services. In order to accomplish this task, the authors propose two different wavelength assignment approaches in the corresponding VLC cells. The first method is directed toward the combined usage of wavelength division multiplexing (WDM) and CDMA techniques to mitigate the effects of intracell and intercell interferences. The second approach, in order to ensure adequate sharing of resources among users, two-dimensional OOCs have been proposed. Furthermore, from the simulated results it can be observed that the BER performance of two-dimensional OOC was better than that of the one-dimensional OOC.

In addition to the improvements made for OOC, the other way to increase the capacity of OOC was to combine with other codes. Therefore, it can be evidenced from [375] and [376] where these OOC codes were jointly used with balanced incomplete block design (BIBD) codes. Notably, in this work, the M-ary data signal corresponding to each user was encoded using the OOC codeword that has been assigned to that specific user and then followed by getting applied to a BIBD encoder to fetch a multilevel signal. Thus, based upon the ratio of code length and code weight of the BIBD, the amount of PAPR of the overall transmitted signal can be controlled. Generally, this sort of approach is facilitated to support multiuser access and the reason behind combining BIBD with OOC was to distinguish different users, and this sort of system was named coded-multilevel expurgated PPM (C-MEPPM) which was compared with code cycle modulation (CCM)-based OCDMA system with respect to the supported number of users and the BER performance. The OOC designated by $(101, 11, 2)$ combined with BIBD with the code set $(101, 256, 6)$ of C-MEPPM system was compared with CCM-OCDMA system comprising of OOC with code set $(101, 25, 7)$, where the BER performance of the former system was better than that of the latter for the same number of users.

5.3.3.2 CDMA-based VLC systems comprising bipolar codes

This subsection gives a critical discussion about different types of bipolar codes that are feasible to be applicable for CDMA-based VLC systems. Furthermore, stressing on the performance of CDMA-VLC systems exploiting different bipolar codes is also the focus of this subsection. Predominantly, the bipolar codes that are well suitable and that are widely employed in case of RF-based wireless CDMA systems can be applied to VLC systems only with the precaution that the transmitted signal must be assured of its real and positive nature. Firstly, in this subsection we briefly introduce the well-known and well-applicable bipolar codes like the Walsh Hadamard sequences, Gold sequences, m-sequences, modified Gold sequences and modified Walsh-Hadamard sequences.

In synchronous channels, the autocorrelation and cross-correlation functions of Walsh-Hadamard sequences are ideal and hence are more suitable and can be applied to VLC systems. Even though several variants of unipolar codes are existing, the cross-correlation functions of such codes do not perform better and, hence, it is appropriate to rely on bipolar codes. Much significant works which are relevant to the application of Walsh-Hadamard sequences for CDMA-VLC systems can be found below.

The research in [445] reveals that the performance of a multicarrier CDMA-VLC system has been analyzed by taking into consideration OFDM platform. In order to overcome the light-dimming issue, the authors in this work exploited RPO-OFDM methodology, and the CDMA part makes use of the Walsh-Hadamard sequences as signature codes to enable efficient resource sharing in synchronous channels. Furthermore, analysis emphasizing the derivations of closed-form expressions for the SINR and BER is also the major contribution of this work. The simulation results presented in this work employ the Hadamard codes of length 32 that are capable to support up to 32 synchronous users in the room. Figure. 5.6 illustrates the comparison of analytical and simulated results of multicarrier CDMA system employing Walsh-Hadamard sequences. In this figure, the performance of BER versus the number of users by employing three different modulation formats, like BPSK, 4 QAM and 16 QAM at two bit rates 5 and 20 Mbps, was analyzed. The simulated results closely match with the derived analytical expressions, thereby demonstrating the accuracy of the derived mathematical expressions. Furthermore, the higher the orders of modulation, the lower the supported number of users. At a forward error correction (FEC) limit, for the bit rate of 20 Mbps scenario, the number of users that can be supported by exploiting 16 QAM, 4 QAM and BPSK were 0, 4 and 9, respectively. On the other hand, when the data rate is 5 Mbps, at a BER of approximately 10^{-5}, the supported number of users by using these modulation formats like 16 QAM, 4 QAM and BPSK were 6, 16 and 20, respectively. Therefore, from this analysis it can be deduced that at the same data rate upon increasing the orders of modulation, fewer users are supported.

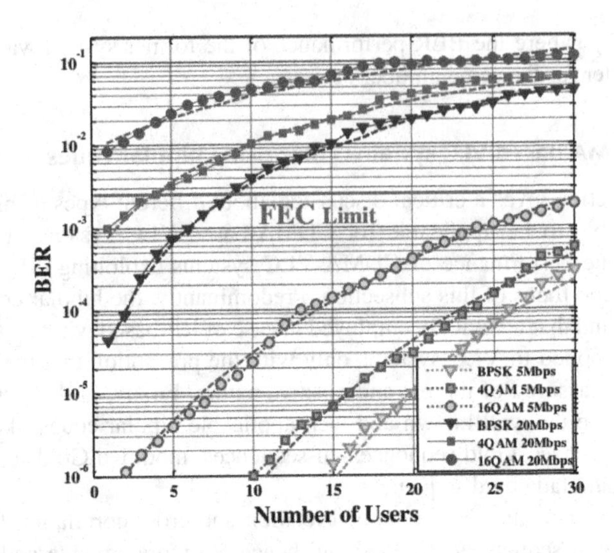

Figure 5.6: BER performance highlighting the analytical and simulation results for multicarrier CDMA system using Walsh Hadamard codes. [445].

The other significant work making use of Walsh-Hadamard sequences can be exemplified in [574] where a femtocell VLC system that is based upon CDMA technology was experimentally demonstrated. In this work, for the purpose of preserving the orthogonality among different users within a same cell, unique orthogonal codes like Walsh-Hadamard codes were employed and with the intent to distinguish different cells, phase-shifted maximum-length pseudo-noise (PN) sequences, which are also called as m-sequences, were adopted. The experimental demonstrations of this work illustrate that upon exploiting orthogonal Walsh-Hadamard codes, the intercell interference can be eliminated. In this experimental work, 4 femtocells were employed and 4 commercially available phosphorescent white LED modules that were fixed to the ceiling were taken into account, and the distance between each adjacent base station was 30 cm. In general, these LEDs were regarded as base stations rendering services to multiple users that are within the vicinity of the installed lighting fixtures. Within the femtocell, each user was assigned a Walsh-Hadamard code of length 16. Furthermore, all the users within each femtocell can simultaneously enjoy a data rate of 3 Mbps, and since the power of the system was limited, each femtocell was capable enough to support fewer than 10 users. Moreover, when receivers were placed away from the transmitting LED, the number of supported users was less.

The work in [433] uses CDMA as a technology for a VLC system that makes use of PWM technique. Primarily, illumination which is the foremost task of LEDs should not be hindered because of the data communication point of view. Consequently, this entails that the type of modulation technique to be employed for VLC should not affect the average light output of the illuminating sources. Additionally, the type of modulation formats to be employed for VLC should take into consideration dimming and flickering aspects of the light sources. As a result, in order to address these issues, the authors considered PWM technique for embedding data in the light output. In the developed system model, Walsh-Hadamard sequences were used as spreading codes. The simulated results therein emphasize the BER and mean square error (MSE) performance for CDMA-based VLC system using PWM technique in which three color lighting LEDs were employed. Also from the simulated results it can be evidenced that the communication link is almost error-free within the ranges of 10.7, 9.3 and 8.1 m for the three color LEDs, namely, red, green and blue LEDs, respectively. However, upon increasing the distance between the transmitting LEDs and the receiver, degradation in the overall performance of the system can be witnessed. Different color LEDs exhibit different variations in the performance of error-free regions due to the differences in the LED light output power and different properties of photodiode responsivity.

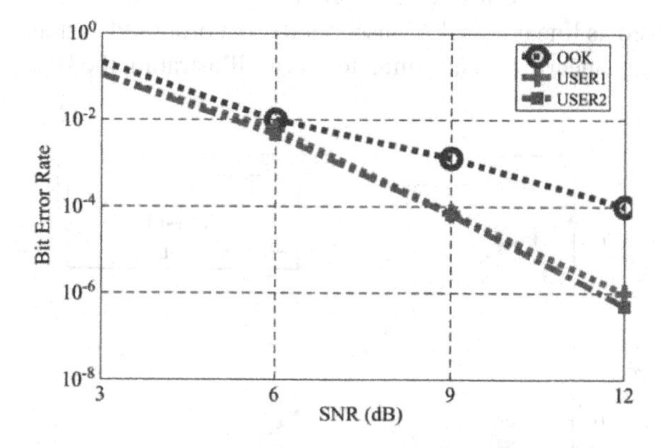

Figure 5.7: Comparison of experimental results for BER performance of CSK-CDMA VLC systems versus OOK modulation [107].

The authors in [107] for the first time proposed and demonstrated a VLC-CDMA system where CSK was used as a modulation scheme that exploits the red, green and blue light waves simultaneously with the orthogonal spreading codes like Walsh-Hadamard codes. Meanwhile, mobile phone cameras were utilized as the receivers to capture the information from different types of light waves. In this work, the capacity of the VLC system can be enhanced by making use of CSK, as well as the single color light interference can be mitigated and CDMA technology provides the flexibility for multiple users to access the network. The experimental set-up of CSK-CDMA-VLC system uses a 4×4 array of RGB LEDs, and the transmission distance is set to 50 cm. Figure 5.7 delineates an experimental result of the CSK-CDMA-VLC system and compares the BER performance of this system with conventional OOK modulation method. The proposed system was capable to serve two users; however, upon increasing the orthogonal spreading codes, there is a scope to support more users. From the result analysis, it is clear that both the users achieve a low error transmission of 10^{-6}. Upon comparison with OOK modulation, in order to achieve a

BER of 10^{-4}, the values of SNR required for CSK modulation and OOK modulation were 9 and 12 dB, respectively. This confirms the fact that CSK-based CDMA-VLC system achieves a 3 dB gain in BER performance when compared to OOK modulation.

The work in [411] proposes a CDMA-VLC system that is capable to solve the problem of non-linearity of LEDs. For the purpose of simulation, micro-LED array of size 8×8 was employed for enabling digital transmission, for spreading the users' data, the length of the Walsh-Hadamard code used was 64 and the total number of supported users were 37. Unlike the traditional mechanism of VLC-CDMA system, where the multiplexed user's data is transmitted using a single LED, in this work, the data belonging to different users is transmitted by employing separate LED elements. Thus, in each LED, the input stream of data that is modulated by employing different mapping schemes manifests a constant envelope. Thus, even if the LED is driven into saturation region, also, it is highly probable for the binary output from each LED to be viewed as linear signal. Consequently, the proposed system mitigates the effect of LED non-linearities. The simulated result illustrating the BER performance

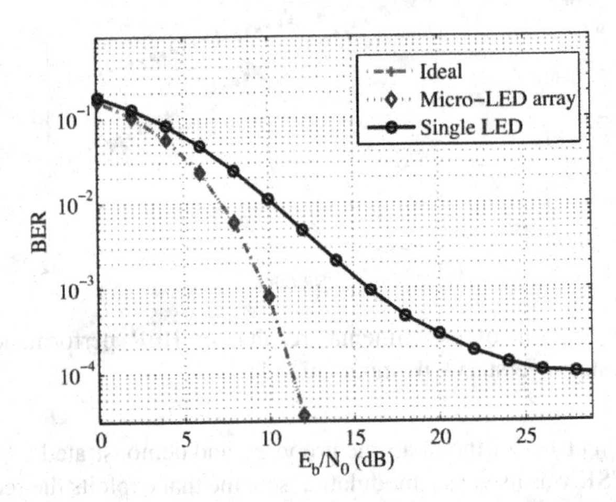

Figure 5.8: Comparison of simulated result of micro-LEDs-based VLC-CDMA and single LED-based VLC-CDMA system employing Walsh-Hadamard codes [411].

of Micro-LEDs-based VLC-CDMA system and a single LED-based VLC-CDMA system is depicted in Fig. 5.8. From the figure it is evident that the BER performance of VLC-CDMA system comprising of micro-LEDs imputes a good consistency with that of the ideal scenario, and this performance analysis substantiates the fact that the non-linearities of the LEDs were overcome by the aforesaid proposed system. Contrarily, the BER performance of single LED-VLC-CDMA architecture still indicates the presence of non-linearities of LEDs. This is attributed to the reason that BER performance fails to achieve a desired error floor of 10^{-4}, which clearly affirms that the non-linear distortion dominates the performance loss at high ranges of

SNRs. Eventually, if the PAPR is very large for CDMA signal, then there is a huge degradation in the BER performance of the overall system.

The aforementioned research work portrays the exploitation of Walsh-Hadamard codes for CDMA-VLC systems; however, there are other types of bipolar codes, namely, Gold sequences, m-sequences, modified Gold sequences and modified Walsh-Hadamard sequences. Unlike the Walsh-Hadamard sequences, which exhibits excellent correlation properties, the Gold sequences and m-sequences do not have ideal cross-correlation and autocorrelation properties. In spite of this drawback, they can be exploited in VLC-CDMA systems due to their large code family. The applications of these sets of bipolar codes is reported in the following literature: From the research efforts in [132] and [133], it is evident that in order to enforce Gold sequences and m-sequences for VLC-CDMA systems, it is mandatory to add a DC bias unit at the transmitting end, and at the receiving end it is vital to make use of a high pass filter (HPF) for removing out the induced DC levels. The prominent performance of these codes for positioning and localization applications can be found in references [132] and [133]. Simulated and experimental results therein underline the fact that these bipolar codes outperform the unipolar codes. This is mainly due to the fact that pertaining to bipolar codes, the HPF at the receiving end of VLC-CDMA systems greatly reduces the effect of noise emanating from the surrounding light. Besides the applications of m-sequences in positioning applications, they can be employed for vehicle to vehicle (V2V) communications as well, and a much relevant reference is [538] which signifies that for the purpose to broadcast high priority event-driven messages, VLC-CDMA is used as a key technology. Owing to the exploitation of HPF at the receiving end, the system can get rid of the background noise, and also the results in this work demonstrate the suitability of m-sequences for different sorts of applications. However, the poor correlation properties of m-sequences might not allow their applicability for multiple access communications. Moreover, from the simulated result analysis of this work it can be interpreted that the orthogonality nature of Walsh- Hadamard sequences, as well as the poor cross-correlation properties of m-sequences enables Walsh-Hadamard sequences-based VLC-CDMA system to offer better performance than that of the m-sequences-based CDMA-VLC system in the multi user scenarios.

From the performance analysis of different bipolar codes which were discussed above, it can be encapsulated that these codes, namely, the Walsh-Hadamard sequences, Gold sequences, m-sequences, etc. can be found applicable for CDMA-VLC systems provided they comply to the requirements of IM and DD, where it is mandatory to rely on bias units to ensure the bipolar to unipolar conversion strategy. Among the bipolar codes, Walsh-Hadamard sequences possess good auto-correlation and cross-correlation properties as well the orthogonal nature of Walsh-Hadamard sequences enables the CDMA-VLC systems to offer an improved BER performance in the case of multiuser communications. Even though Gold sequences and m-sequences lack good auto correlation and cross-correlation properties, their ease of implementation and their large code family enable their exploitation in CDMA-VLC systems.

5.3.4 OPTICAL SPACE DIVISION MULTIPLE ACCESS

By harnessing the frequency-reuse gains, a cellular communication network can yield high data density. Particularly, when dealing with optical cells, these are comparatively small than those of the RF cells, this is due to the LEDs offering a limited coverage area. Upon comparing with small cell RF wireless networks, optical cellular network expedites much higher frequency gains and consequently exhibits higher data density. It is evident from the literature that optical cellular networks offer substantially improved performance than that of the RF femtocell networks. Generally, in a multiuser cellular system, TDMA is the widely used multiple access scheme. The underlying principle behind TDMA is that the information signals which are associated for different users are transmitted in different time slots in such a manner that each user can access the entire system bandwidth. In spite of its remarkable benefits, there are some limitations which are associated with TDMA. Firstly, since only one user can be served in a time slot, the throughput offered by TDMA system is limited, and therefore this sort of scenario fails to effectively exploit the common bandwidth resource. Secondly, in order to ensure the full cell coverage, it is vital for the transmitter to radiate signals uniformly in all directions which in turn leads to the occurrence of intercell interference (ICI), notably affecting the users located at the cell edges. Thus, in order to overcome these limitations, it is convenient to add spatial dimension to conventional TDMA system, thereby leading to the emergence of space division multiple access (SDMA), a renowned multiple access scheme that has been integrated with 4G wireless communication standards like LTE and IEEE 802.11ac.

In SDMA, an antenna array is employed as the transmitter for the purpose of generation of multiple narrow beams that are pointing toward different active users. Thus, in this manner, SDMA creates a flexibility to serve multiple users in the same time slot concurrently. Even though SDMA in RF-based wireless communications exhibited distinguished performance, its advantages cannot be directly utilized in case of VLC due to the particularities involved in the design of VLC transmitter. It is because in case of RF-based wireless communications exploiting SDMA, directional narrow beams are generated by varying the amplitude and phase of the signals that are transmitted by the antenna array [109]. Nonetheless, this similar approach cannot be adopted in a straightforward way in case of VLC because VLC relies on IM and DD. Thus, inevitably, this implies that the transmitted signal has to ensure of non-negative and real nature as well as at the same time should comply with the constraints of the illumination devices. However, in case of VLC, in order to generate directional narrow beams, the inherent characteristic feature of LEDs like its confined emission semi-angle, i.e., the limited FOV of the LEDs can be employed. Consequently, in case of optical SDMA (OSDMA), an angle diversity transmitter comprising of multiple directional narrow beams of LED elements turns out to be the most appropriate choice as an optical transmitter. Thus, upon activating different transmitter elements, multiple users situated at different locations can be served simultaneously by the angle diversity transmitter.

Precisely, the angle diversity transmitter is able to generate narrow beams of light intended for multiple active users located at different positions. The difference between TDMA and SDMA is highlighted in Fig. 5.9 where it can be inferred that by using TDDMA, only one user can be served within a time slot. Whereas, by making use of SDMA, four different users can be served simultaneously within the same timeslot. This nature of SDMA enables it to exhibit a remarkable performance than that of the conventional TDMA system. Furthermore, the implementation strategy of

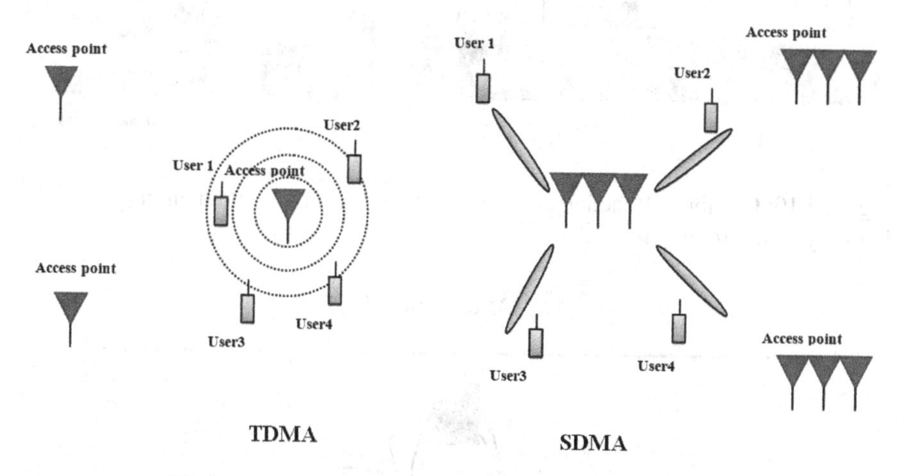

Figure 5.9: Differences between TDMA and SDMA.

SDMA and TDMA can be elaborated in Fig. 5.10. As evident from the figure, multiple active users are grouped within one timeslot in case of SDMA, while pertaining to the scenario of TDMA, each active user occupies one time slot. Thus, this enhances the throughput of the system at the expense of certain challenges. Grouping of the multiple users within the same timeslot obliges the system to rely on multielement transmitters to support parallel and directional transmissions. In addition to this, if the grouping of the users is performed incorrectly, then there is a chance for the emanation of multiuser access interference which in turn leads to the deterioration of the data rate of each user belonging to a certain group. Consequently, taking into consideration the aforesaid aspects, an angle diversity transmitter along with optimal spatial grouping strategy is essential while dealing with OSDMA systems.

5.3.4.1 System model of optical SDMA (OSDMA)-based VLC system exploiting angle diversity transmitter

A general downlink transmission scenario is considered where the installed LED lighting fixtures act simultaneously as APs and illumination devices. In particular, it is assumed that the APs are positioned on the ceiling and the receiver terminals are placed at a desktop height. The information from the LED/AP is transmitted to the active user by means of LOS path as delineated in Fig. 5.11. Therefore, the DC gain

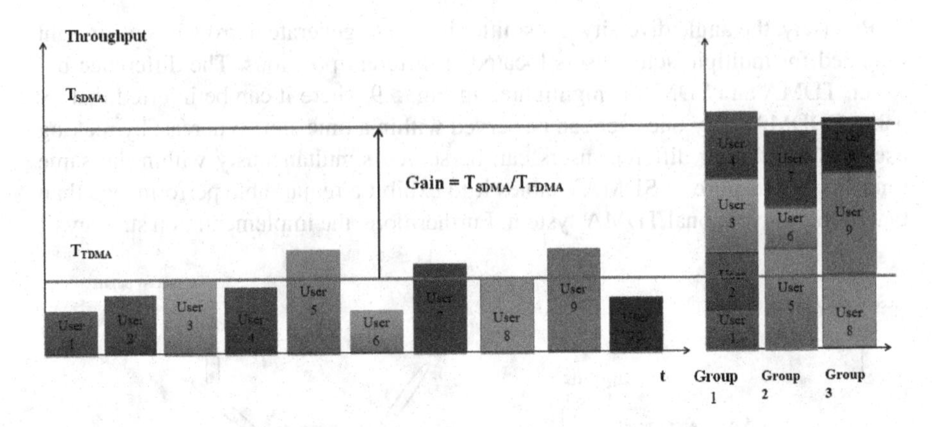

Figure 5.10: Grouping 10 active users into 3 spatial groups to attain an improvement in the system throughput.

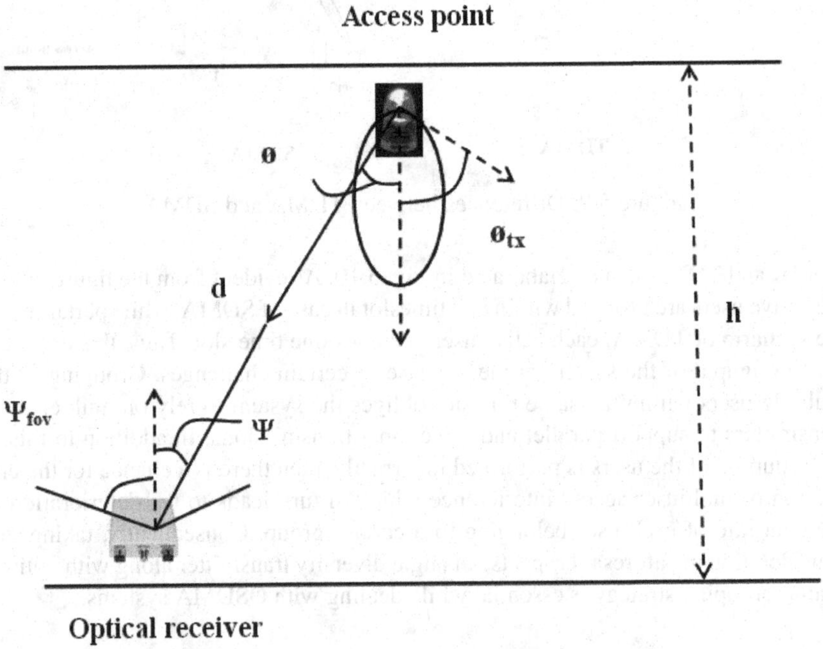

Access point

ø

øtx

d

h

Ψfov

Ψ

Optical receiver

Figure 5.11: VLC LOS propagation channel model.

of the corresponding LOS link can be formulated as follows:

$$H_0 = \frac{(m+1)\,A_{pd}}{2\pi d^2} cos^m(\phi)\, cos(\psi)\, rect\left(\frac{\psi}{\psi_{FOV}}\right) \tag{5.9}$$

From (5.9), the distance between the transmitting LED and the receiver is denoted by d, while the angles ϕ, ψ and ψ_{FOV}, represent the irradiance angle, incident angle at the receiver and the FOV of the receiver, respectively. Since the type of the LED employed is of Lambertian type, the Lambert's mode number is denoted by m. In order to highlight the performance of SDMA-based VLC system, it is vital to have a clear-cut understanding about the optical TDMA system as in the latter part of this subsection, the performance of OSDMA is compared with TDMA-VLC system.

Optical TDMA

As already stated, the information intended for different users is transmitted in different time slots, where each time slot is of identical length. Only one user can be served within a time slot. Thus, the received signal to interference noise ratio (SINR) of the active user k can be put up as follows:

$$SINR_k = \frac{\left(\gamma P_{TX} H_{(\hat{b},\, k)}\right)^2}{N_0 B + \sum_{b' \neq \hat{b}} \left(\gamma P_{TX} H_{(\hat{b},\, k)}\right)^2} \tag{5.10}$$

The parameter γ in (5.3.4.1) specifies the responsivity of the photodiode in $\frac{A}{W}$, P_{TX} represents the amount of optical power that is transmitted by the AP and is assumed to be similar for all the APs. The channel response between the corresponding AP which is denoted by $\hat{b}$ and the user k is denoted by $H_{(\hat{b},\, k)}$ and the square term in the denominator, i.e., $\sum_{b' \neq \hat{b}} \left(\gamma P_{TX} H_{(\hat{b},\, k)}\right)^2$ specifies the interference signal that emanates from the interfering APs b'. While, the terms N_0 and B symbolize the AWGN power spectral density and the optical communication, bandwidth, respectively.

Recalling that in optical TDMA, each active user can access the entire system bandwidth within their allocated time interval, i.e., upon dividing the transmission frames into different time slots of identical length, each user has a provision to access the entire bandwidth. Accordingly, by supposing a simple round robin (RR) schedular, the overall throughput/spectral efficiency is given by

$$\Omega_{TDMA}(k) = \frac{1}{K} \sum_{k=1}^{K} B log_2 (1 + SNR_k) \tag{5.11}$$

K in (5.11) represents the active users.

Drawbacks associated with TDMA-VLC system

In case of TDMA, the users are allocated with fixed time slots, which implies that only one user can access the bandwidth within a time slot. Consequently, since the available bandwidth cannot be effectively shared, this limits the overall system performance as well as has a direct impact on the data rate pertaining to each active user. This issue is more conspicuous in case of the optical domain (i.e., optical attocell network), where the bandwidth is limited by the slow transient response of the yellowish phosphor coating of white LEDs. The other major aspect is that a single illuminating device (LED) employed in case of optical attocell network should radiate uniformly in all directions to accomplish uniform illumination. However, when multiple light fixtures are employed in a room, the chance of occurrence of overlapping light cones will be high and this scenario is more pronounced at the optical cell edges. Eventually, strong co-channel interference (CCI) is inevitable in such regions.

Optical SDMA-based VLC system

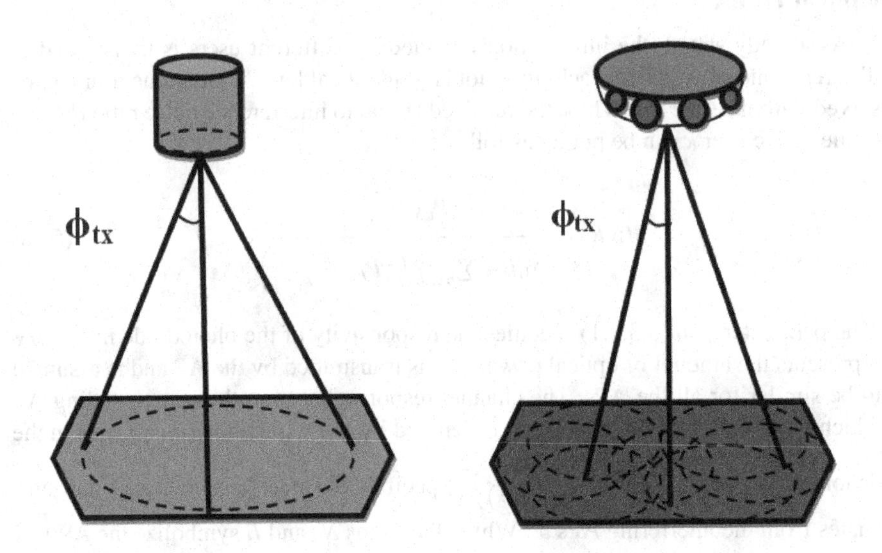

Figure 5.12: Optical TDMA with single element optical transmitter (left) and OS-DMA comprising of angle diversity transmitter (right).

The aforementioned drawbacks of TDMA-based VLC system can be overcome by optical SDMA. In recent times, SDMA has been proposed to provide multiple access support for RF-based wireless communication. Meanwhile, dealing with the RF domain, each AP comprises of an antenna array that is capable enough to generate simultaneous multiple directional narrow beam signals. Thus, in contrast to TDMA system where only one user can be served within a time slot, in case of OS-DMA system, multiple users can be served simultaneously within a single time slot. The interference emanating between different narrow beams can be mitigated by exploiting the location information of active users; thus, when compared with TDMA, SDMA exhibits superior potential to combat the CCI. Deploying SDMA in case of the optical domain enforces a constraint on the design of optical transmitter. In order for the optical transmitter to fulfill the methodology of SDMA, it should have the potential to originate multiple narrow beams of light. Thus, this entails to rely on a multi element angle diversity transmitter. The schematic of angle diversity transmitter is delineated in Fig. 5.12. From the figure, it is clear that this angle diversity transmitter comprises of multiple LED elements where each LED element is directed toward a different direction. In general, these multiple LED elements are driven by different electronics. As evident from the figure, the first LED is installed at the center of the semi-sphere base, while the remaining LED elements are installed around this central LED element.

Angle diversity transmitter can facilitate parallel data transmission by activating the LED elements that are within the coverage area of the active users. The rest of the LEDs that are not meant for communication purposes can be appropriated to render constant light for illumination purposes. Thus, by deploying angle diversity transmitter, both indoor illumination as well as communication support to multiple users can be ensured. In order for SDMA to be exploited in RF-based wireless communication, it is mandatory for the transmitters, to rely on multiple RF chains as well as complex beam-steering algorithms for the purpose of ensuring pointed beam signals. However, this burden can be overcome in case of optical domain, thanks to its opto-illuminating devices which are having the potential to generate narrow FOV optical signals. Thus, when SDMA is exploited for VLC, there is no necessity to stress on complex beamforming algorithms.

Furthermore, the methodology of OSDMA in case of VLC can be effectively accomplished by making use of the angle diversity transmitter which is potential enough to effectively mitigate the effects of ICI in case of optical attocell network. The difference between conventional TDMA-based VLC system and OSDMA-based VLC system is outlined in Fig. 5.13. Additionally, the difference between single-element transmitter and angle-diversity transmitter is also highlighted in this figure. From the captured snapshot of optical TDMA system, there are two neighboring cells, i.e., cell 1 and cell 2, in which the active users 3 and 5 were served by them. Since the semi-angle of the single-element transmitter is quite large, it is probable for user 3 which is located at the cell edge to experience a high amount of ICI. Whereas, upon referring to the snapshot of OSDMA-based optical attocell network, since angle diversity transmitter is employed, all the active users are simultaneously served by means of directional beams. As depicted from the figure, since the coverage area of LEDs is confined only within the area of active users, a significant amount of reduction in ICI can be obtained.

In case of OSDMA, it is vital to enforce the phenomenon of spatial grouping because it is not appropriate to group users arbitrarily into different time slots, as the chances of some users being spatially close will be high. This kind of scenario definitely leads to overlapping beams, and hence as a result leads to the emergence of high multiuser access interference. Specific to arithmetical terms, spatial grouping is referred to as partition $P = \{G_1, G_2, G_3 \cdots\}$ comprising of a set of active users U, where each spatial group is referred to as G_j. In order to ensure user fairness, it is necessary to take into consideration that spatial groups are mutually exclusive as well as collectively exhaustive. Briefly, this implies that each spatial group should comprise of only one active user, and none of the active users should remain unserved.

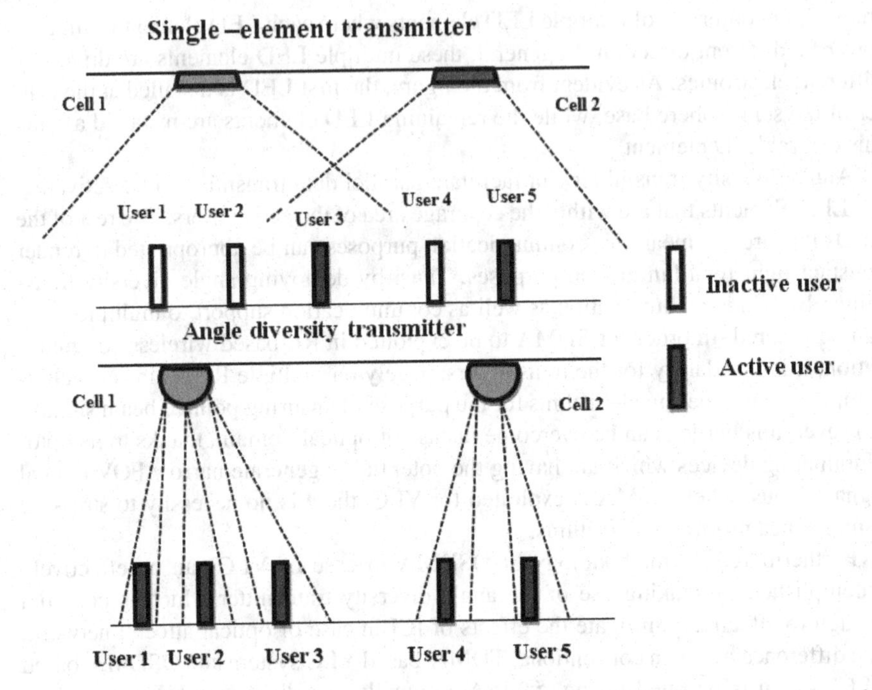

Figure 5.13: Single-element optical transmitter in optical TDMA system and angle diversity transmitter in OSDMA system.

Performance evaluation of OSDMA-VLC system

Interference plays a major role evaluating the performance of OSDMA-based VLC system. We need to take into consideration two important interferences like the intracell interference, which generally emanates from the active LED elements belonging to a desired optical cell. The other interference is the intercell interference which occurs from other LED elements belonging to neighboring optical cells. Consequently, the SINR of the active user u_i belonging to the spatial group G_j can be calculated as follows:

$$SINR_{(u_i,G_j)} = \frac{\left(\gamma P_{TX} H_{(b,\, u_i)}\right)^2}{N_0 B + \sum_{b' \in B_{inter}^{(b,G_j)}} \left(\gamma P_{TX} H_{(b',\, u_i)}\right)^2} \tag{5.12}$$

From (5.3.4.1) the term $\sum_{b' \in B_{inter}^{(b,G_j)}} \left(\gamma P_{TX} H_{(b',\, u_i)}\right)^2$ refers to the interference power, and the set of active interfering LEDs that serve other users in the group G_j is specified through $B_{inter}^{(b,G_j)}$. As already stated in case of OSDMA, each spatial group generally occupies one time slot and this is evident from Fig. 5.10. Furthermore, within each time slot, multiple active users are served concurrently, thereby the overall spectral efficiency in case of OSDMA system can be expressed as

$$\Omega_{SDMA}(k) = \frac{1}{J}\sum_{j=1}^{J}\sum_{u_i \in G_j}^{max} Blog_2\left(1 + SINR_{(u_i,G_j)}\right) \tag{5.13}$$

J represents the total number of spatial groups where each spatial group can take one time slot for data transmission. The average spectral efficiency of OSDMA scheme

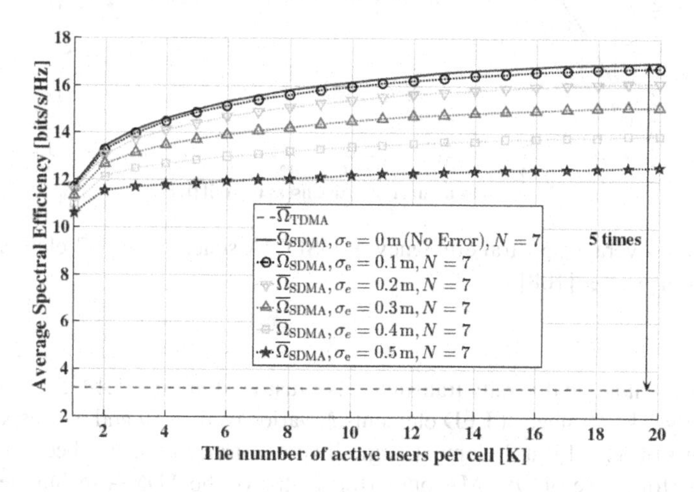

Figure 5.14: Average spectral efficiency of OSDMA scheme with 7-element angle diversity transmitter [108].

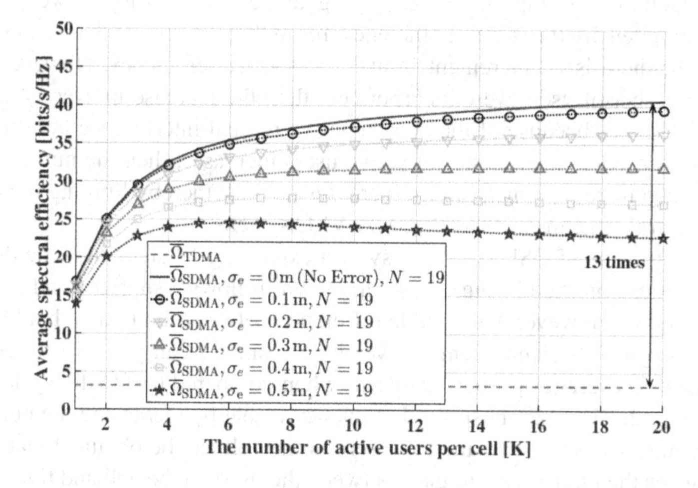

Figure 5.15: Average spectral efficiency of OSDMA scheme with 19-element angle diversity transmitter [108].

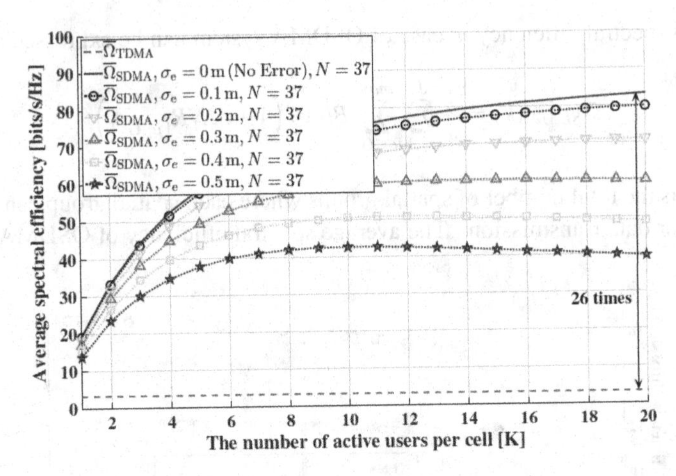

Figure 5.16: Average spectral efficiency of OSDMA scheme with 37-element angle diversity transmitter [108].

with N-element angle diversity transmitter is illustrated in Figs. 5.14, 5.15 and 5.16, respectively. The number of LED elements N varies from 7, 19 and 37, respectively. As shown in Fig. 5.14, a 7-element angle diversity transmitter has been employed and the performance of OSDMA outperforms that of the TDMA technique, where the achievable spectral efficiency of OSDMA is 6 times higher when compared with TDMA. The underlying reason behind this increase in spectral efficiency is that an angle diversity transmitter exhibits the potential to generate narrow directional beams of light, which play an important role to mitigate the effects of ICI as well as at the same time to intensify the spectral efficiency. From the figure, note that for less number of users, there is an increment in average spectral efficiency. However, when increased number of users were incorporated, then the increase in average spectral efficiency saturates because of the occurrence of mutual interference between each link. Moreover, when the number of active users increases, then the number of active LEDs also increase which in turn increases the ICI, thereby limiting the offered average spectral efficiency of the OSDMA-VLC system.

The performance of OSDMA-VLC system exploiting 19 element angle diversity transmitter is exemplified in Fig. 5.15. This figure follows a similar trend as that of the previous one. However, the notable difference between the two is that OSDMA system employing 19-element angle diversity transmitter can yield 13 times better improvement in average spectral efficiency when compared to TDMA system. The reason behind this increase in spectral efficiency is that by increasing the number of LED elements, a greater separation of users in the cell can be obtained that significantly reduces the mutual interference between the users in the cell and thus ensures a better performance. Upon further increasing the number of LED elements to 37, the increase in spectral efficiency is more significant when compared to the former systems (i.e., 7, 19 element angle diversity transmitter) and this is portrayed in Fig.

5.16. A 26-fold increase in spectral efficiency can be witnessed by using 37-element angle diversity transmitter-based OSDMA VLC system when compared with conventional TDMA system. Thus, the aforementioned analysis reveals the fact that OSDMA-based VLC system exploiting angle diversity transmitter significantly enhances the average spectral efficiency when compared with that of a conventional optical TDMA-based VLC system.

5.3.4.2 Research efforts pertaining to OSDMA-based VLC system

The work in [110] investigates the performance of VLC system tailored by optical SDMA. Furthermore, this work adopts an angle diversity transmitter as an optical transmitter to serve multiple active users contemporaneously within the same time-slot. From the simulated result analysis it can be affirmed that OSDMA-based VLC system significantly manifests higher system throughput than that of the TDMA-based VLC system. Notably, OSDMA-based VLC system can obtain an improvement in system throughput of over 10 times that of the TDMA-based VLC system. Upon exploiting angle diversity transmitter comprising of a larger number of transmitting elements, the improvement in system throughput is even higher.

The work in [108] proposes OSDMA technique for optical attocell networks. The framework developed in this work emphasizes that an angle diversity transmitter replaces a conventional single-element transmitter in each optical cell with the motive to enable simultaneous transmissions to multiple users located at different positions. Such type of system configuration mitigates the ICI between optical cells. In this work, optical TDMA-VLC system is used as a benchmark for the purpose of comparison of the obtained performance of OSDMA-VLC system. The simulated result analysis shows that OSDMA unveils more superior performance than that of the conventional TDMA scheme. For a 37-element LED angle diversity transmitter, OSDMA technique significantly improves the average spectral efficiency of the system by a factor of 26. Moreover, this study also takes into account the user position errors. From the Monte Carlo simulations, it is clear that the proposed frame work is robust to user position errors. The same authors extend this work by deriving upper and lower bound for average spectral efficiency for the aforesaid system [109]. The simulated results of this work demonstrate that OSDMA system is robust to user position errors and upon considering the practical state-of-the-art indoor position techniques, the performance of the system is only compromised at most by about 14%.

In the same context, in order to provide seamless services to multiple users, Yin et al. [555] studied the performance of coordinated multipoint VLC (CoMP-VLC) system which is enabled by SDMA. Research in [322] states that an optimal linear precoding transmitter, which is based upon the MMSE criterion, has been proposed for the VLC system, where power line communication (PLC) was exploited to coordinate multiple LEDs for the purpose of facilitating signal broadcasting. The system of such kind is referred to as a CoMP-VLC system. In order to achieve a better trade-off between the system performance and fairness throughput, a low complexity SDMA grouping algorithm has been proposed. This SDMA algorithm, which is also called random pairing (RP) algorithm, can be exploited for improving the area spectral efficiency (ASE) for different user loadings. The fairness throughput in this work was evaluated in terms of Jain's index of fairness (JIF). The simulated result illustrating the achieved ASE against different user loadings is delineated in Fig. 5.17. From the figure, it can be inferred that the simulation is carried out against three

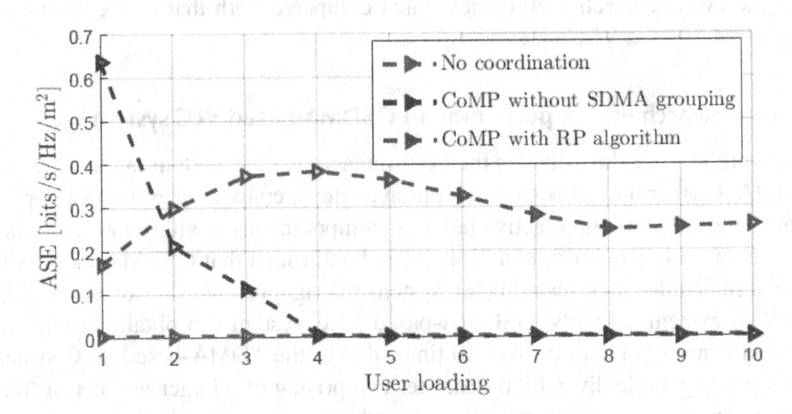

Figure 5.17: Simulation result elucidating area spectral efficiency (ASE) for different user loadings [555].

distinct baselines, namely, a VLC system employing FDMA, COMP-VLC system without SDMA grouping and COMP-VLC system exploiting RF algorithm. Among the three scenarios, the COMP-VLC system employing RF algorithm demonstrates high gains in the achievable ASE across different user loadings. On the contrary, COMP-VLC system without SDMA grouping achieves higher ASE for low user loadings, and upon increasing the user loadings, the performance of ASE deteriorates due to grouping of user equipments that are not spatially compatible. Meanwhile, the VLC-FDMA system has lower ASE than that of the other two systems, which this is due to the fact that there is no frequency reused. The Figure 5.18 indicates the variation of JIF against different user loadings. As depicted from the figure, a high fairness throughput can be achieved for a VLC system employing FDMA technique without any frequency reuse. Pertaining to CoMP-VLC system without SDMA grouping, the achievable throughput is low when compared with CoMP-VLC system which is making use of RP algorithm. Moreover, the achievable throughput in case of CoMP-VLC system without SDMA grouping decreases with the increase in user loadings. However, this poor performance of decreasing fairness throughput can be improved by making use of RP algorithm. A very similar research work can be witnessed in [556] which can be an extended work of [555] where the authors have investigated a CoMP-VLC downlink system in which the time and frequency resources were shared among a group of users relying on SDMA. With the intent to combat the interuser interference and at the same time to make use of the linear operating region of LED, a linear zero-forcing (ZF) transmit precoding (TPC) was adopted to the CoMP-VLC system. Similar to [555], the work in [556] employs the low-complexity suboptimal SDMA user-grouping algorithm like RP and

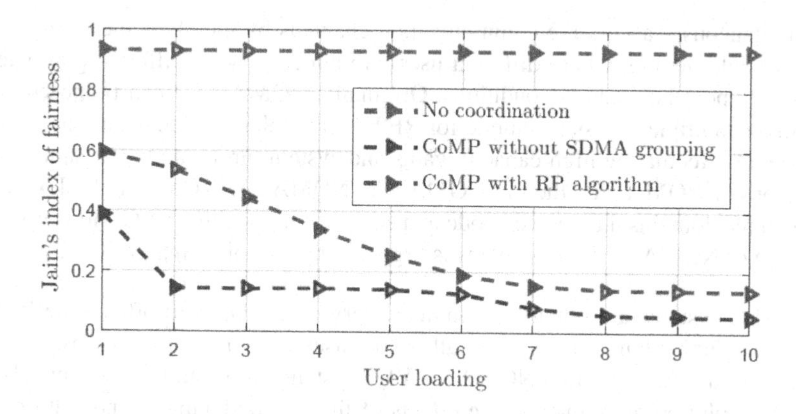

Figure 5.18: Simulation result indicating the variation of Jain's index of fairness (JIF) for different user loadings [555].

evaluates the system in terms of both ASE and JIF. Additionally, this work compares the performance of RP algorithm with the other existing algorithms like tree-based (TB) grouping algorithm [170], random grouping (RG) algorithm and best fit (BF) grouping algorithm [437]. Simulation results emphasize that considerable amount of improvement in achievable ASE gains can be noticed upon enforcing intelligent SDMA user-grouping algorithms, rather than blindly grouping all the users in the cell. Among the other user-grouping algorithms, RP algorithm yields distinguished performance and is more robust to different user loadings.

5.3.5 OPTICAL NON-ORTHOGONAL MULTIPLE ACCESS

The previous sections provide a thorough emphasis on the exploitation of different orthogonal multiple access schemes like OOFDMA, OCDMA and OSDMA for multi user VLC systems. The present section focuses on the non-orthogonal multiple access schemes like NOMA and its use for VLC system. Starting with the introduction of the underlying principle of NOMA-based VLC systems, this subsection is directed toward the presentation of performance analysis of NOMA-VLC systems by exploiting different power allocation algorithms. Additionally, this subsection gives a comprehensive review about the interface of OMA and NOMA in the context of VLC systems along with thorough research aspects pertaining to NOMA-based VLC systems. Finally, this subsection is concluded by briefly listing out the applications of VLC-NOMA systems and the associated challenges of such systems.

NOMA has been proposed as a potential multiple access technique for VLC to increase the system throughput as well as to ensure ubiquitous connectivity in VLC systems. The most stupendous nature of NOMA is that it allows multiple users to contemporaneously exploit the entire frequency and time resources, thus offering superior enhancements in spectral efficiency when compared with that of the OMA schemes. The underlying principle behind NOMA is that, it facilitates multiple users to simultaneously access the communication channels by the phenomena of power domain multiplexing, where different users are allocated with different power levels based upon the channel conditions. Originally, NOMA has been proposed as a propitious multiple access technique for RF-based wireless communications for the purpose of effectuating high capacity gains and system throughput. The same classifications of NOMA, i.e., the PD-NOMA, CD-NOMA and NOMA multiplexing in all the three domains, i.e., power, code and spatial, hold well for VLC. However, the adoption of NOMA for VLC systems is triggered by the following grounds:

1. VLC systems are potential complementary counterparts to offload the RF communication systems especially under user dense indoor scenarios.
2. Particularly, exploiting NOMA for VLC systems turns out to be an appealing solution to counteract the effects of limited modulation bandwidth of the current off-the-shelf LEDs as it ensures high data rate communication.
3. The fundamentally quasi-static nature of the VC channel enables the NOMA receiver to accurately estimate the channel gain for the subsequent power allocation.

These aforementioned advantages motivated the exploitation of NOMA for VLC systems and a significant literature has been evidenced addressing numerous aspects, which finally enhances the high data rate communication when compared with the existing OMA counterparts.

5.3.5.1 Underlying principle of NOMA technology

To have a clear cut understanding of PD-NOMA, let's consider a typical indoor room environment which depicts a downlink scenario where the LEDs are installed on top of the ceiling and are rendering services to the mobile terminals or user equipments which are within the vicinity of the LEDs. Particularly, each receiver terminal has a photodiode installed in it to allow for the conversion of light signal into suitable electrical signal. Figure 5.19 illustrates a two-user downlink NOMA scheme where User 1 is located at a close distance to the LED, and User 2 is located at a farther distance from the LED. Here each installed LED acts as a BS or as an AP. Therefore, User 1 can be categorized as a weak user since it is situated far away from the BS/AP. As a result, User 1 suffers from high path loss and, consequently, it has weak channel gain. Whereas User 2, which is located close to the AP, can be termed as a strong user and considerably has more amount of channel gain. At the transmitting end, the signals corresponding to both the weak and strong users are superimposed by allocating different values of power for each user and are transmitted simultaneously. Apparently, more amount of power is assigned to the weak User 2 that has high path loss and low channel gain. Meanwhile, less power is allocated by the transmitter to the strong User 1. Predominantly, this mechanism of power allocation at the transmitting end is referred to as SC. At the receiving end, a simple multi user detection is performed by relying on SIC. The detailed overview of SC and SIC is presented below:

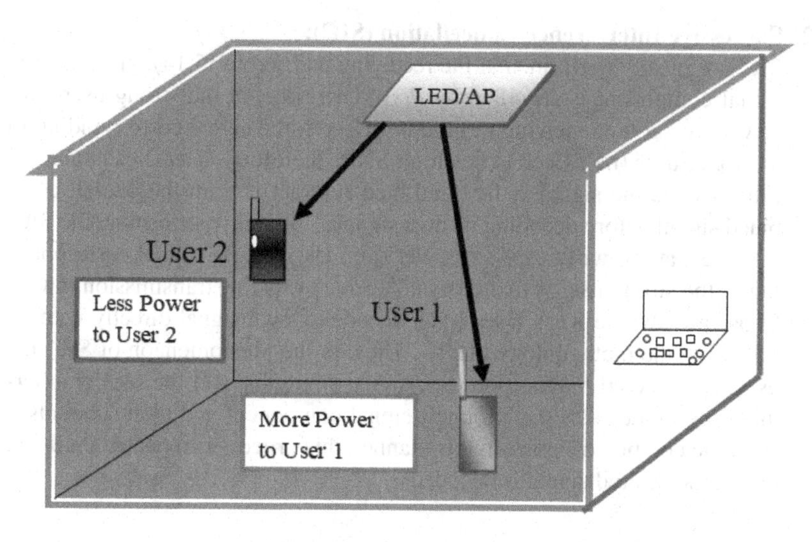

Figure 5.19: Schematic representation of downlink NOMA-VLC system with 2 users.

1. **Superposition Coding (SC):**
For the first time, the basic idea of the usage of SC was started in [124], where different information signals can be transmitted to several receivers in a downlink broadcast channel. As stated above, the basic phenomenaon of SC is to allocate high values of power to users with adverse channel conditions, and correspondingly a low amount of power is assigned to users with favorable channel conditions. As interpreted in the figure, two real and positive-valued signals s_1 and s_2 are transmitted by the LED/AP to the two user terminals/equipments User 1 and User 2, respectively. Since, User 2 is positioned at a close proximity to the transmitting LED/AP, it possesses high channel gain; accordingly, the AP appropriates lower values of power P_2 to s_2. Subsequently, the signals s_1 and s_2 corresponding to Users 1 and 2 are superimposed and are transmitted concurrently with the following representation

$$X = P_1 s_1 + P_2 s_2 \tag{5.14}$$

From the above discussion, it is obvious that in (5.14), the power value $P_1 > P_2$. The sum total of the allocated power values to the users equals the total transmitting power of the LED/AP. The same underlying phenomenon holds well even for the multiple user scenario where the power allocation is done upon sensing the channel conditions of the corresponding users.

2. **Successive Interference Cancellation (SIC):**
   SIC is typically performed at the receiving end. From (5.14), the strongest signal component is specified by $P_1 s_1$. Therefore, at the strong user's receiver, i.e., at the receiving terminal of User 2, the signal corresponding to the weak user (i.e., User 1) has high SNR; therefore, User 2 can successfully decode the signal s_1 first and then subtract it from the overall combined signal before decoding its own signal s_2, thereby performing the SIC. Whereas, at the weak user's receiver (i.e., User 1), the signal corresponding to the strong user is treated as noise due to its low transmission power. Consequently, the weak User 1 can decode its own signal directly without following the methodology of SIC. Thus, in the phenomenaon of SIC, the users are ordered based upon their signal strengths, and the receiving terminal firstly decodes the strongest signal component and then subtracts it from the combined signal. In this manner, this process is repeated until and unless the desired signal is decoded.

5.3.5.2 Power allocation mechanisms in ONOMA

As already seen in the previous section, the principle methodology of NOMA is to allocate more amount of power to the users with the worst channel conditions and less values of power to the users with exceptional channel conditions. Therefore, in this regard, the vital aspect which needs to be considered is to allocate appropriate levels of power to different users for the purpose of expediting SIC as well as to obtain better trade-off between throughput and fairness. This motivated the proposal of power allocation mechanisms to attain a better system performance with improved coverage probability. Thus, the different power allocation techniques are:

1. Fixed Power Allocation [270]
2. Gain Ratio Power Allocation [335]
3. Exhaustive Search Power Allocation [554]
4. Normalized Gain Difference Power Allocation [98]
5. Sum Rate Maximization Power Allocation [548]
6. Sum Rate Maximization under Quality of Service (QoS) Constraints Power Allocation [570]
7. Max-Min Fairness with Sum Rate Maximization Power Allocation [440]

The schematic representation of downlink DCO-OFDM-based NOMA VLC system is elucidated in Fig. 5.20. As evident from the NOMA transmitter schematic, the information signals corresponding to the two users are processed in parallel to yield the two time-domain real and positive-valued DCO-OFDM signals $s_1(t)$ and $s_2(t)$. These two signals are meant for the two user equipments/terminals i.e., $UE1$ and $UE2$ are transformed to M-QAM complex-valued symbols and further these complex-valued symbols are constrained to satisfy the Hermitian Symmetry criteria to yield a real-valued signal. Thereupon, an appropriate amount of DC bias is added to enassure both real and positive-valued signals as denoted by $s_1(t)$ and $s_2(t)$. Eventually, at the transmitting end, SC is employed to intensity modulate the overall

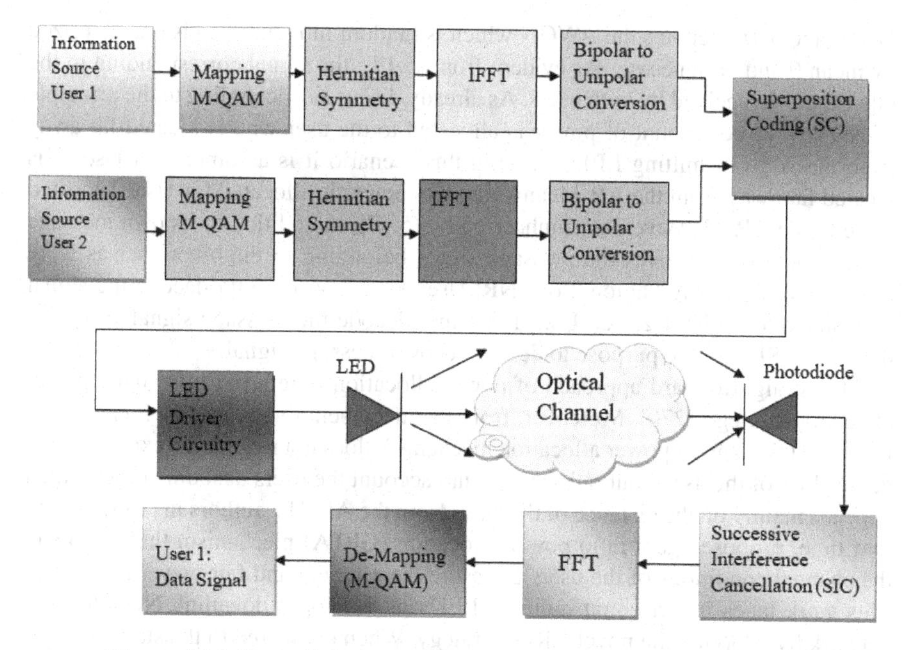

Figure 5.20: Downlink DCO-OFDM-based NOMA-VLC system.

combined signal which is given as

$$s(t) = \sqrt{P\alpha_n} s_1(t) + \sqrt{P(1-\alpha_n)} s_2(t) \tag{5.15}$$

From the transmitted signal as shown in (5.15), the parameter P specifies the total power that is available for the transmitted signal. The level of power to be appropriated to each user's information signal is specified by the power splitting factor α_n. Also, $E\left\{|s_1(t)|^2\right\}$ and $E\left\{|s_2(t)|^2\right\}$ are normalized to 1. Finally, the real and positive-valued signal $s(t)$ is intensity modulated through the LED luminaire which generally comprises of Q number of LED chips. At the receiver terminal of user equipments UE_l, $l \in 1, 2$, the electrical signal can be expressed as

$$y_l(t) = \gamma \sum_{i=1}^{Q} h_l(t) \otimes s(t) + n_l(t) \tag{5.16}$$

Upon substitution of (5.15) into (5.16), the final expression of the received signal is obtained as

$$y_l(t) = \gamma \sum_{i=1}^{Q} h_l(t) \otimes \left[\sqrt{P\alpha_n} s_1(t) + \sqrt{P(1-\alpha_n)} s_2(t)\right] + n_l(t) \tag{5.17}$$

From (5.17), the γ specifies the responsivity of the photodiode in $\left(\frac{A}{W}\right)$, the electrical channel impulse response between the AP, i.e., the source and the user l is given by

$h_l,(t)$, and $n_l(t)$ denotes the AWGN which is random in nature and is characterized by mean 0 and variance N_0. As evident from (5.17), the signal corresponding to the other users is received by both users. As already discussed, according to the principle of NOMA, more amount of power is allocated to the user who is located far away from the AP/transmitting LED. Here, in this scenario it is assumed that User 2 is located far away from the AP. Meanwhile, less power is allocated for User 1 located close to the AP. Moreover, as emphasized by (5.17), in addition to its own message signal, each user receives the message signal belonging to the other user as well. Since, User 2 has high amount of SNR, User 1 can successfully decode the signal corresponding to User 2, i.e., User 1 can first decode the message signal $s_2(t)$ and then apply SIC for the purpose to decode its own message signal.

This straightforward approach of power allocation is referred to as fixed power allocation strategy [270]. Moreover, from the aforementioned analysis it can be inferred that this sort of power allocation mechanism does not require the exact channel gain values of the users, but rather takes into account the users decoding order which depends mainly on the distance of the users from the AP. The authors in [335], for the first time, proposed gain ratio power allocation (GRPA) mechanism that considers the channel conditions of the users to ensure the efficient and fair power allocation. This work takes into account multiple LEDs to develop a downlink NOMA-VLC network by adopting the novel GRPA strategy. When compared to the static or fixed power allocation technique, this approach of channel dependent power allocation strategy, i.e., GRPA significantly enhances the performance of the system by maximizing the sum rate of the users. The authors in this work considered a realistic indoor room environment comprising of multiple LEDs as delineated in Fig. 5.21. As evident from the figure, the beams emitted by the two LEDs are overlapped in

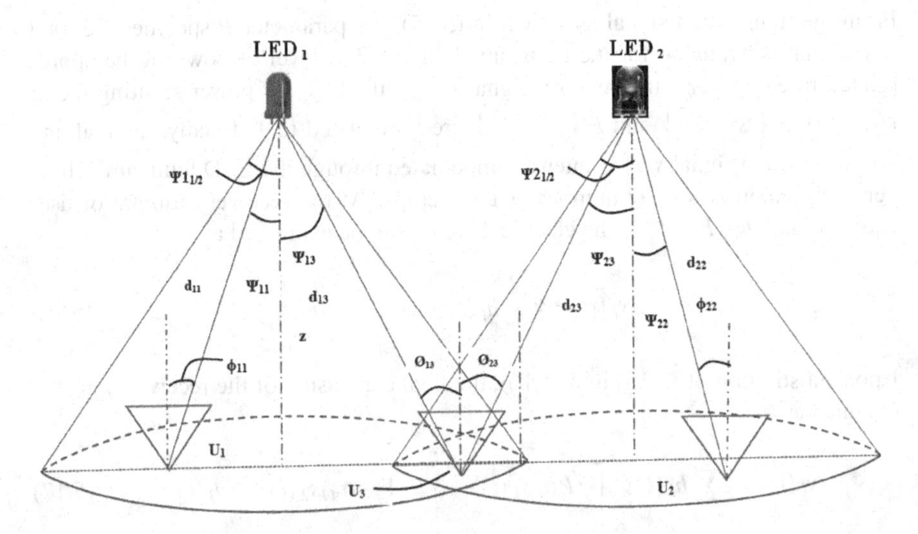

Figure 5.21: Schematic representation of downlink-indoor NOMA-VLC system with 2 LEDs and 3 users.

order to ensure the user located at the cell boundary to receive the data streams belonging to the two adjacent LEDs. As depicted from the figure, there are 2 LEDs and 3 UEs denoted by U_1, U_2 and U_3, respectively. The UEs U_1 and U_2 are located within the vicinity of the two LEDs which are signified as LED_1 and LED_2, while U_3 is located at the intersection area of the two LEDs. Employing the principle of NOMA, LED_1 transmits two real and unipolar signals with power values P_{11} and P_{13} to the two UEs, U_1 and U_3. Particularly, the two signals $s_1(t)$ and $s_2(t)$ are meant for the two users U_1 and U_3. Similarly, the transmitting LED_2 transmits two real and positive-valued signals $s_2(t)$ and $s_3(t)$ with power values P_{22} and P_{23} for users 2 and 3, respectively. Further, it is interesting to note that the $UE\ U_3$ exhibits the potential to achieve the diversity gain because it receives double the copies of the same symbol $s_3(t)$ which is transmitted from both the LEDs.

Therefore, by employing the principle of SC, the transmitted signal from each LED is the superposition of the signals contemplated for its users. For ease of understanding, let us specify the users belonging to each LED by C_q. Accordingly, the signal transmitted from LED_q, $q = 1, 2$ can be expressed as

$$s_q(t) = \sum_{l \in C_q} P_{ql} s_l(t) \tag{5.18}$$

where the total transmitted power is given by

$$P_q = \sum_{l \in C_q} P_{ql} \tag{5.19}$$

In general, the signal received at each UE, i.e., UE_l, $l = 1,2,3$ is the combination of all the signals transmitted by the LEDs within the networks and can be put up as

$$y_l(t) = \sum_{q=1}^{Q} \sum_{l \in C_q} h_{ql} P_{ql} s_l(t) + n_l(t) \tag{5.20}$$

From (5.20), the total number of LEDs is represented by Q, where in this scenario Q is taken as 2 and $n_l(t)$ specifies the AWGN. The path gain corresponding to the LOS component from the transmitting LED_q to the UE U_l is given by

$$h_{ql} = \frac{A_l(m+1)}{2\pi d_{ql}^2} T_s(\phi_{ql})\, g(\phi_{ql})\, cos^m \psi_{ql} cos\phi_{ql},\ 0 \leq \phi_{ql} \leq \Phi_l \tag{5.21}$$

From the channel impulse response as shown by equation (5.21), A_l exemplifies the photodiode area pertaining to the lth user, d_{ql} denotes the distance between the qth LED, i.e., LED_q to the UE U_l, ψ_{ql} specifies the irradiance angle with respect to the transmitter perpendicular axis, the incidence angle is denoted by ϕ_{ql}, the FOV of the UE U_l is denoted by Φ_l. The gains of the optical filter and the optical concentrator are given by $T_s(ql)$ and $g(\phi_{ql})$, respectively. The impulse response of the channel is zero for $\phi_{ql} > \Phi_l$, which implies that if the incident light falls outside the range of the FOV of the receiver, then the optical LOS channel gain is zero. By exploiting SC strategy, the multi user interference is eradicated. Furthermore, the decoding phenomenon is accomplished based upon the order of increasing channel gain. Accordingly, the UE U_l can successfully decode the signals belonging to the other users who are comprising of lower decoding order. Meanwhile, the interference emanating from the users with higher decoding order, i.e., from U_r with $r > l$ can be treated as noise. Consequently, the instantaneous SNR of UE U_l is given as

$$\gamma_l = \sum_{q=1}^{Q} \frac{h_{ql} P_{ql}}{\sum_{r>l} h_{ql} P_{ql} + \sigma_l^2} \tag{5.22}$$

This aforementioned analysis exemplifies that the allocation of the transmitted power among the users plays a vital role to enhance the performance of the VLC system. In contrast to the fixed power allocation strategy, the GRPA mechanism takes into account the actual channel path gains of all the users and the power which is assigned to the lth sorted user is given by

$$P_l = \left(\frac{h_{q1}}{h_{ql}}\right)^l P_{l-1} \tag{5.23}$$

Therefore, according to GRPA, the power allocation mechanism depends not only on the user gain compared to the gain of the first sorted user, but also on the decoding order l. From (5.23), with the increase of the channel gain h_{ql}, the amount of allocated power decreases, and this is also valid because less amount of power is sufficient for the users with favorable channel conditions to decode their signals after subtracting the signals of the other users with lower decoding order. The simulated results in this work infer that GRPA strategy of power allocation outperforms the fixed power allocation mechanism in downlink NOMA-VLC system. Furthermore, the authors studied the effect of tuning the transmission angles of the LEDs and the FOVs of the receivers for maximizing the throughput of the system.

With the intent to further enhance the performance of NOMA-based VLC system, the authors in [554], optimized the power values based upon the channel conditions of the users. Primarily, exhaustive search (ES) method is employed for determining the optimum set of power allocation coefficients, and this sort of approach maximized the coverage probability of the NOMA-VLC system. Further, the research efforts in [98] emphasize that NOMA technique is exploited in order to improve the sum rate of MIMO-based VLC system. The authors in this work for the first time proposed a normalized gain difference power allocation (NGDPA) algorithm to warrant efficient and low-complexity power allocation based upon the channel conditions of the users in MIMO-NOMA-based VLC system. The performance analysis of an indoor 2×2 MIMO-VLC system which is exploiting NOMA technique is verified through numerical simulations. The simulation results reveal the fact that upon imposing NGDPA technique, the achievable sum rate of the aforesaid system improves significantly. Furthermore, when compared with GRPA strategy, NGDPA accomplishes an improvement in sum rate by 29.1% in the same system, i.e., 2×2 MIMO-VLC exploiting NOMA for 3 users scenario. In the same manner, in order to further improve the performance of NOMA-VLC system in terms of maximization of sum throughput of the multiuser VLC system, the authors in [548] further optimized the allocated power coefficients. The original non-convex optimization problem was equivalently transformed into a convex one by the introduction of auxiliary variables and by the application of variable transformation. By employing the Karush-Kuhn-Tucker (KKT) conditions for optimality, an optimal power control algorithm has been proposed which exhibits an improvement in sum throughput when compared with the GRPA and static/fixed power allocation algorithms.

In [570], the research efforts signify that for the purpose of reducing the interference for VLC multi cell networks, grouping based on user locations is exploited where the residual interference from the SIC algorithm in NOMA is considered, and the power allocation in each cell is optimized for the sake of achieving an improvement in the user rate under user quality of service constraint. The authors in [440] consider a 2 users scenario, where they optimize the power allocation strategies under both sum-rate maximization and max-min fairness criteria, where practical optical power and quality of service constraints were included. The main contribution of this work is the attainment of semi-closed forms of the optimal power allocation solutions via the mathematical analysis. Moreover, in the context of VLC downlinks, the simulated results demonstrate the remarkable superiority of NOMA over OMA schemes.

5.3.5.3 MIMO-NOMA-based VLC system

The pictorial representation of transmitter block diagram of indoor 2×2 MIMO-VLC-based multi user system exploiting the power domain NOMA technique is illustrated in Fig. 5.22. In general, DCO-OFDM methodology is incorporated to convert the bipolar signal to unipolar signal. Going into the details of the transmitter schematic, the M-QAM mapped complex symbols are constrained to satisfy the Hermitian Symmetry criteria to ensure for a real-valued signal at the output of the IFFT, followed by power domain superposition and addition of DC bias. Consequently, the signal which is fed as input to the qth LED where $q = 1,2$ can be formulated as

$$s_q(t) = \sum_{l=1}^{K} \sqrt{P_{q,l}} s_{q,l}(t) + \beta_{DC} \tag{5.24}$$

From (5.24), $s_q(t)$ represents the time-domain signal belonging to the qth LED, $s_{q,l}(t)$ signifies the signal intended for the lth user, $(l = 1,2,\cdots K)$ which is confined to the qth LED, β_{DC} denotes the added DC-bias which is sufficient enough to drive the LED and $P_{q,l}$ specifies the assigned power level for the lth user in the qth LED. Upon propagation through the freespace, the electrical signal vector at the receiving terminal of the lth user can be put up as

$$y_l = \gamma P_{opt} \xi H_l s + n_l \tag{5.25}$$

As already discussed several times throughout this book, γ is the responsivity of the photodiode, and the output optical power of the LED is denoted by P_{opt}, and ξ

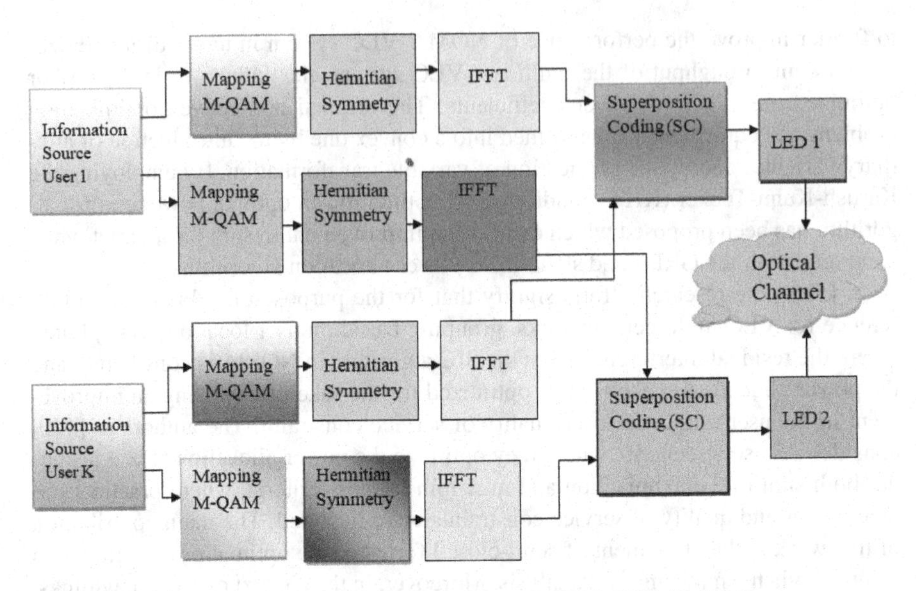

Figure 5.22: Transmitter schematic of 2 × 2 MIMO-NOMA-VLC system.

specifies the modulation index. The 2 × 2 matrix pertaining to the lth user is given by H_l, while the transmitted electrical signal is determined by $s = [s_1, s_2]^T$ and n_l represents the AWGN vector of the lth user. Note that, [.] enumerates the transpose operation. If the type of the LED employed is of Lambertian emission (i.e., LED radiates uniformly in all the directions), then the LOS optical channel gain of the lth user in accordance to the qth ($q = 1,2$) LED and the jth ($j = 1,2$) photodiode can be interpreted as

$$h_{jq,l} = \frac{A_l(m+1)}{2\pi d_{jq,l}^2} T_s(\phi_{jq,l})\, g(\phi_{jq,l})\, cos^m \psi_{jq,l} cos\phi_{jq,l},\ 0 \leq \phi_{jq,l} \leq \Phi_l \tag{5.26}$$

The parameters stipulated in (5.26) are mentioned in the previous subsections. From the receiver block diagram as portrayed in Fig. 5.23, for the purpose of attaining perfect reception of the transmitted data, MIMO demultiplexing and low complexity ZF channel equalization can be performed. Thereupon, the estimated electrical signal vector in relevance to the lth user can be manifested as

$$\tilde{s}_l = s + \frac{1}{\gamma P_{out}\xi} H_l^{-1} n_l \tag{5.27}$$

In (5.27), H_l^{-1} indicates the inverse of H_l. Furthermore, in order to facilitate multi-user detection, SIC is enforced at each user with respect to each LED. Prior to accomplishing the SIC operation, it is essential to determine the decoding order of the users with respect to each LED. In contrast to single-LED NOMA, where a single LED is employed as an AP to render services to multiple users, MIMO-NOMA involves multiple LEDs. Eventually, this means to rely on a new way to develop the

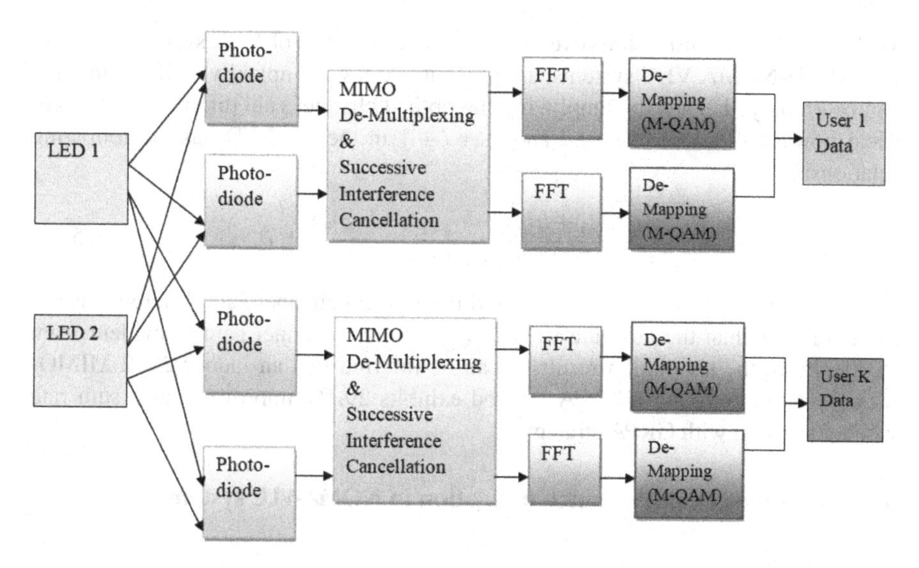

Figure 5.23: Receiver schematic of 2×2 MIMO-NOMA-VLC system.

decoding order of the users. Unlike the single-LED NOMA VLC systems, where individual optical channel gain of each user is considered to determine the decoding order, in case of MIMO-NOMA VLC systems, the sum of the optical channel gains of each user belonging to each LED is taken into account to sort the users. Thus, without any loss of generality, the l users with respect to the qth LED are sorted according to the sum of their optical channel gains in the decreasing order to attain the following expression

$$h_{1q,1} + h_{2q,1} > h_{1q,2} + h_{2q,2} > \cdots\cdots h_{1q,l} + h_{2q,l} \tag{5.28}$$

Therefore, the decoding order with respect to the qth LED is given by

$$D_{i,1} < D_{i,2} < D_{i,3} \cdots\cdots D_{i,l} \tag{5.29}$$

Thereupon, at the receiving end, followed up by DCO-OFDM demodulation, the data corresponding to the two users is obtained. In order to ensure the effective power allocation, the authors in [98] imposed GRPA and NGDPA strategies for NOMA-based MIMO VLC system. As already stated above, the power allocation in case of GRPA strategy exploits the optical channel gain of each user. Meanwhile, in case of the MIMO-based VLC system, the sum of the optical channel gain is employed. Thus, in case of a 2×2 MIMO-NOMA-VLC system, by considering the decoding order as given by (5.29), the correspondence between the powers allocated to the lth user and $l+1$th user in the qth LED is determined as

$$P_{q,l} = \left(\frac{h_{1q,l+1} + h_{2q,l+1}}{h_{1q,1} + h_{2q,1}} \right)^{l+1} P_{q,l+1} \tag{5.30}$$

With the intent to further improve the achievable sum rate of VLC systems employing MIMO-NOMA VLC systems, an efficient and low-complexity NGDPA method can be employed. Thus, by employing the optical channel gain difference, the electrical powers allocated to user l and user $l+1$ in the qth LED has the following relationship as

$$P_{q,l} = \left(\frac{h_{1q,1} + h_{2q,1} - h_{1q,l+1} - h_{2q,l+1}}{h_{1q,1} + h_{2q,1}} \right)^{l} P_{q,l+1} \tag{5.31}$$

Therefore, from (5.31), it can be inferred that optical channel gain difference is exploited rather than the absolute values of the optical channel gain as evident from (5.30). From the simulation results it can be inferred that an indoor 2×2 MIMO-VLC system exploiting NGDPA method exhibits 29.1% improvement in sum rate when compared with GRPA strategy.

5.3.5.4 Inter-cell interference mitigation in NOMA-VLC systems

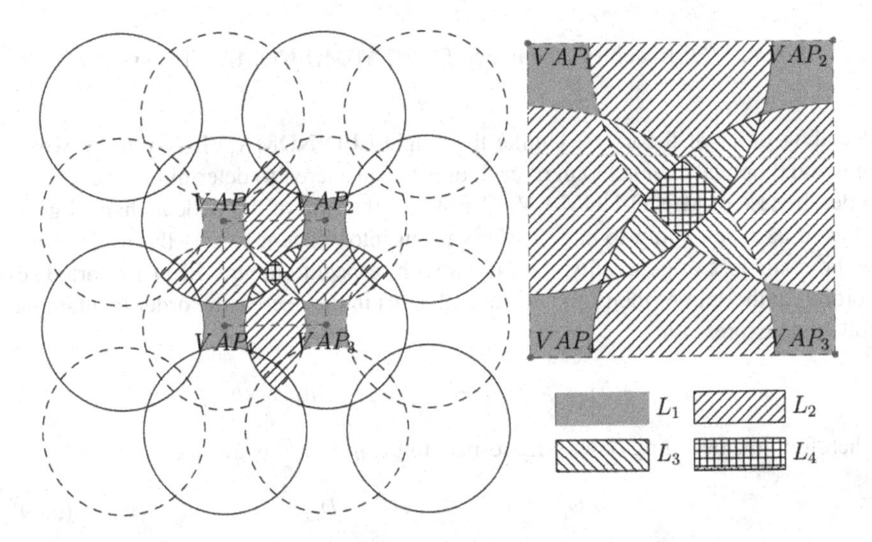

Figure 5.24: Illustration of multicell NOMA-VLC network [570].

With the motive to mitigate the inter cell interference in multi cell NOMA-based VLC networks, user grouping based upon the location of users was adopted in [570]. Figure. 5.24 delineates a typical NOMA-based VLC network. From the figure, it is clear that transmitting LEDs act as visible light access points (VAPs) to impart simultaneous high-speed data transfer and illumination. The VAPs are identified by the red dots in the figure and based upon the degree of cell overlapping, the coverage areas were classified into four types. Particularly, the FOV of the user, vertical distance between the transmitting VAP and the receiver, attenuation coefficient, etc., determine the cell size. Primarily, the interference in the multicell VLC network as represented in the figure is specified by several factors as follows:

1. FOV of the Users: Generally, if the receiver FOV of a certain user is adequately small, then it implies that only one VAP (i.e., transmitting LED) exists in the FOV of the corresponding user. This sort of situation entails the assurance to eradicate the inter cell interference. However, the absence of LOS optical signal in the coverage holes ultimately hinders the assurance of communication coverage in that specific zone.
2. Distribution of User: The other way to eliminate the intercell interference is to ensure that the co-frequency VAPs do not work. However, this type of condition is dependent upon the time-varying user distribution.
3. Impact of Frequency Reuse (FR) factor: Even though upon setting $FR = 1$ achieves the highest spectrum efficiency, users located in the regions L_2, L_3, L_4 experience severe interference which in turn obliges to rely on advanced interference cancellation or user scheduling techniques. Nonetheless, upon using $FR = 2$, the situation can be significantly improved as the users located in the region L_4 can undergo co-frequency interference.

In order to balance the trade-off between the interference cancellation and frequency efficiency, a $FR = 2$ was implemented in the aforementioned research work. Taking into consideration the location of the users, the user grouping for NOMA-VLC networks was carried out as stated below: The users located in the region L_1 received from a single LED source the so-called VAP without any sort of interference. The users located in the region L_2 can select either of the VAP; nonetheless;, since each LED employed a different portion of the spectrum, so there is no interference. There is a probability for the users located in the region L_3 to receive from two different LEDs operating in the same bandwidth; hence they need to be scheduled for load balancing. The chances for the users to be located in the region L_4 is quite small; therefore, the users are assigned with a dedicated part of the bandwidth to combat the interference.

Similarly, in [440], the problem encountered in the scenario of overlapping VLC cells was analyzed. This scenario emphasizes that for a user located in the coverage region of two adjacent LEDs, the probability of receiving two signals superimposed in the power domain is very high. Generally, the amount of power allocation depends upon the observed channel conditions. At the receiving end, the user performs SIC to decode the individual signals from the composite constellation. Further, upon performing the phase pre-distortion on the transmitted signals, the performance of the system can be enhanced because the optimal phase difference between the two channel gains was attained. Further, with the intent to overcome the cell overlapping, the research work in [335] emphasizes that the authors have exploited cell zooming technique, where the transmitting angles of the LEDs were adjusted to control the cell sizes. From this work, it can be inferred that the proposed framework exploits LED with two different transmission angle settings. Particularly, in order to collect the relevant information about the location of the users and their corresponding channel gains, a central control unit (CCU) was employed for the purpose of configuring the transmission angles. From this work, it can be deduced that it is always not appropriate to shrink the coverage areas of the LEDs, as it might lead to loss of coverage.

Additionally, the adjustment of the transmission angles of the LEDs has an impact on the width and intensity of the light beams, as there is a chance to yield undesired illumination inconsistency across the indoor room environment.

5.3.5.5 Interface of OMA schemes with NOMA-VLC systems

The remarkable peculiarity of NOMA is that it allows for the multiplexing of different users in the power domain by facilitating different users to share the same time and frequency resources. However, since interference is inevitable, it becomes difficult to multiplex a large number of users. This emerges due to the fact that the transmitting LED/VAP can be considered as a small cell offering a limited coverage area to the users within its vicinity. Generally, in case of small cell size comprising of a large number of users, the channel conditions might not differ among the users. Therefore, under such scenarios, OMA techniques stem out to be a more preferred option. Subsequently, in order to accomplish a better trade-off between capacity and reliability, it is vital to allow for the co-existence of OMA techniques with NOMA in a VLC network. Thereupon, under this hybrid multiple access strategy, it is possible for users to be divided into different groups which can be multiplexed by exploiting the OMA techniques like OOFDMA, TDMA, etc., while the users in each group can be multiplexed in the power domain by means of ONOMA. In view of the peculiar advantages offered by both the OMA techniques and NOMA, it is indispensable to furnish appropriate insights on their hybrid applicability for VLC systems. Much pertinent literature in close association toward the joint interface of OMA techniques and NOMA in VLC systems is briefed below.

The authors in [270] analyzed the performance of NOMA-based VLC downlink system for a two-user scenario as well as compared the performance of NOMA-VLC system with OOFDMA-based VLC system. Primarily, for the conversion of the bipolar signal into unipolar signal, DCO-OFDM methodology is incorporated in the aforementioned systems. Figure. 5.25 illustrates the boundaries of the achievable rate region comparisons for both OOFDMA and NOMA in downlink VLC system, where perfect interference cancellation is assumed for NOMA-based VLC system. As evident from the figure, NOMA-based VLC system attains higher data rates for both users than those of the OOFDMA-VLC system except at the corner points. The other way around, the performance of NOMA-VLC system outperforms OOFDM-VLC system when the cancellation error is 1%. However, for higher cancellation errors of up to 2%, the performance of NOMA system gradually deteriorates. The impact of tuning the transmission angles of the LEDs and the receiver FOV on the performance of NOMA-based VLC system is investigated in [335] and [554]. In [335], the effect of tuning of both the transmission angles of the LEDs and the receiver FOV on the overall NOMA-VLC system performance was verified through the simulation results for the following scenarios:

1. No tuning of the transmission angles and the FOV of the receiver
2. Tuning of the transmission angles
3. Tuning of the FOV of the receiver

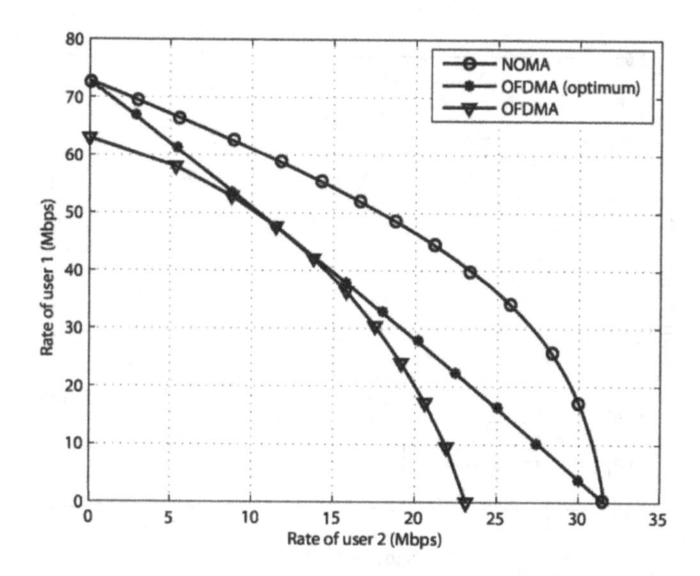

Figure 5.25: Comparison of boundary rate pairs (Mbps) for both OOFDMA and ONOMA in downlink VLC system where perfect interference cancellation is assumed for ONOMA [270].

For the simulation set-up, the authors have employed a $6 \times 6 \times 3\ m^3$ indoor room environment comprising of two transmitting LEDs. The tunable transmission angles of the LEDs which were taken into account are $(45, 60)$, and the FOVs of the receiver considered are $(15, 30, 50)$, respectively. For a small number of users, there is an increase in the sum data rate upon tuning both the transmission angles and the FOVs of the receiver. This is due to the fact that the interference between the two light beams can be eliminated. Nonetheless, it is obvious from the simulation results that with the increase of number of users, since less amount of power is allocated to each user, it enables the users located at the celledges to receive their signals from both LEDs. Thereupon, the tuning of the transmission angles will eventually deteriorate the system throughput, since it hinders the cell-edge users to receive the signal from only one LED. Besides, the tuning of the FOV of the receiver attains the best performance since each user is facilitated to optimize the reception according to their respective position.

From the simulated results as delineated in Figs. 5.26 and 5.27, it can be deduced that when compared with fixed/static power allocation, GRPA strategy gives an improvement in achievable sum rate when tunable FOVs were taken into consideration. Additionally, the number of handovers can be reduced by employing the tunable FOVs of the receivers. NOMA-based VLC system is compared with TDMA-based VLC system in [554]. As depicted in Fig. 5.28, for a large number of users, NOMA achieves a higher sum rate when compared with the other OMA scheme like TDMA. Furthermore, the effect of varying different LED semi-angles on the

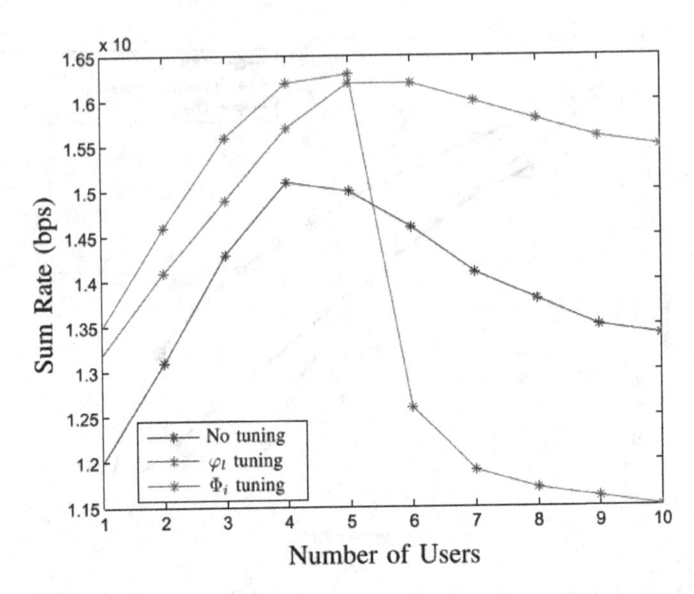

Figure 5.26: Performance of NOMA-VLC system employing fixed power allocation strategy for the three scenarios: no tuning of transmission angles and FOV, tunable transmission angles and tunable FOVs of the receiver [335].

performance of the system was analyzed for both TDMA and NOMA based VLC systems. When employing LED with semi-angles of 59.9 and 44.9 degrees, a TDMA-based VLC system yields the same amount of performance in terms of achievable sum rate in *bpcu*. Meanwhile, reducing the LED semi-angle to 30 degrees will decrease the spectral efficiency in case of TDMA-VLC system. On the other hand, reducing the LED semi-angle of from 59.9 to 44.9 degrees drastically improves the achievable sum rate of NOMA-VLC system. Upon further reducing the LED semi-angle to 30 degrees, even while increasing the number of users to 20, the achievable sum rate of NOMA-VLC system increases by 125% when compared with TDMA-VLC system.

5.3.5.6 State-of-the-art research aspects associated with NOMA-VLC systems

NOMA is adopted as a potential multiple access technique for VLC with the motive to enhance the throughput, ameliorate the experience of the cell-edge users, augmenting the user fairness and connectivity and ensuring seamless services with decreased latency. The matter-of-fact associated with NOMA is that a single resource component namely subcarrier, time slot or a spreading code can be accessed or shared by multiple users. In this context, there exist different types of NOMA: the PD-NOMA, pattern division multiple access (PDMA), and sparse code multiple access (SCMA), and these are represented as excellent candidates for the 5G

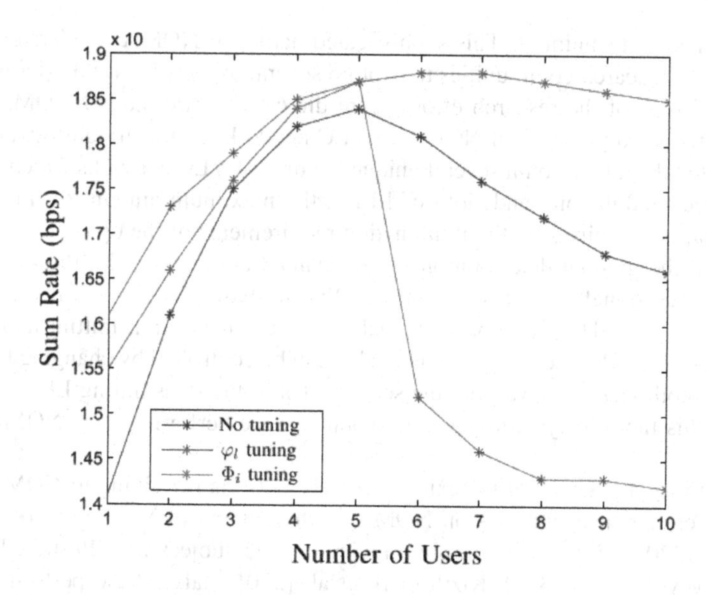

Figure 5.27: Performance of NOMA-VLC system employing GRPA strategy for the three scenarios: no tuning of transmission angles and FOV, tunable transmission angles and tunable FOVs of the receiver [335].

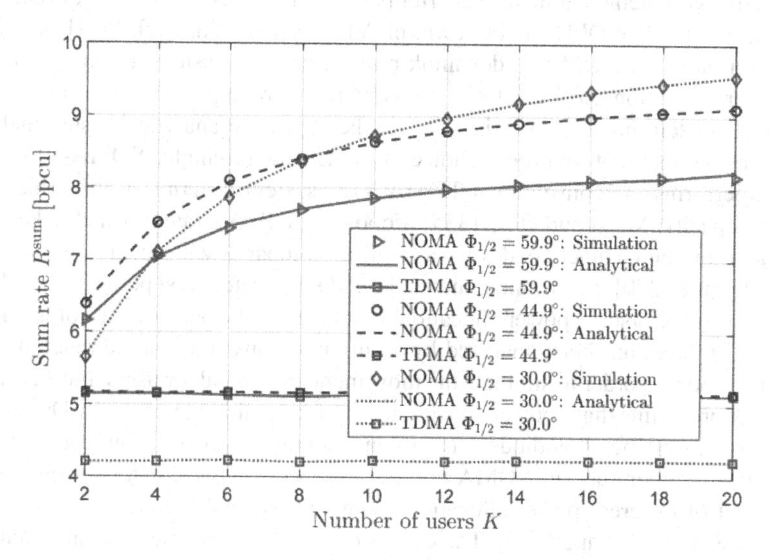

Figure 5.28: Comparison of ergodic sum rate for a different number of users in both TDMA and ONOMA-based VLC systems [554].

multiple access techniques. This sophisticated nature of NOMA accelerated an interest in the research communities to explore several aspects in NOMA-based VLC systems. Much of the research efforts were directed toward the PD-NOMA-based VLC systems. Application of NOMA to VLC networks is not straightforward, as it requires to take into account several unique features of VLC networks which include the limited modulation bandwidth of LEDs, the maximum amount of transmitted power that is restrained by the illumination requirements of the opto-electronic devices, and the gradual deterioration in the channel conditions with the increase in distance. Additionally, the performance of NOMA-based VLC systems can be significantly improved by incorporating tunable opto-electronic transmitting and receiving components. Precisely, the channel value can be controlled by changing the FOV of the photodiodes or by varying the semi-angles of the transmitting LEDs. Consequently, this factor plays a major role to enhance the performance of NOMA-VLC systems.

Going into the details of the existing research efforts pertaining to NOMA-VLC systems, earlier investigations on NOMA in the context of VLC networks can be found in [270, 335, 554]. A thorough review on the subject was illustrated in few such surveys like [26, 336]. Kizilirmark et al. [270] analyzed the performance of DCO-OFDM-based NOMA-VLC system as well as compared it with DCO-OFDM-based OOFDMA system. For the first time, the authors have proposed fixed/static power allocation strategy in power-domain NOMA-VLC systems. The simulated results of the work demonstrate the superior performance of NOMA over conventional OMA technique like OFDMA in downlink VLC system. Yin et al. [554] evaluated the performance of NOMA in downlink multi user VLC system. Upon optimizing the power allocation coefficient set, the system coverage probability of the downlink VLC system has been maximized. Furthermore, the analytical result analysis ascertains the fact that by proper choice of the LED semi-angle, NOMA-VLC system outperforms the conventional TDMA-VLC system in terms of the achievable system capacity. Marshoud et al. [335] proposed a GRPA strategy which takes into consideration the channel conditions as well as compared with fixed power allocation technique [270], where the simulated results indicate the superiority of GRPA strategy over fixed/static power allocation. In this work, the performance of the multi user VLC system has been analyzed by taking into consideration the random walk mobility model in order to simulate the movements of the indoor users. Furthermore, this work proclaims that upon adaptively tuning the semi-angle of the LEDs and the FOV of the receiving photodiodes, the system sum rate improves significantly. Much of the existing literature on NOMA in multi user VLC networks stresses on the development of different power allocation strategies. Related literature can be found in [98, 548, 554, 570] and [440]. The details of the aforementioned PA mechanisms are given in Subsection 5.3.5.2.

The authors in [553] extend the work as signified in [554] by developing a theoretical framework to interpret the performance of a multi user VLC system for two different scenarios. This proposed framework formulates user pairing strategies to enhance the system throughput. In order to provide guaranteed quality of service

(QoS), analytical expressions were derived pertaining to the system coverage probability. Meanwhile, in the scenario to ensure opportunistic best effort service, closed-form expressions for ergodic sum rate were formulated. Under high SNR regime, this work derives the upper bound of the sum rate gain of NOMA over OMA technique like OOFDMA. Besides, the theoretical and the simulated result analysis of this work reveals the fact that the performance gain of NOMA system can be enhanced upon pairing users with distinctive channel conditions. Also, the effect of the proper choice of LED semi-angles on the performance of the system has been analyzed. The choice of LEDs with larger semi-angles is an appropriate option to ensure guaranteed QoS. Meanwhile, the LEDs with 35-degree semi-angle are the most suitable choice to provide opportunistic best effort service.

From the majority of the previous literature, it can be reported that most of the contributions on optical NOMA take into account perfect knowledge of the channel fading coefficients. Nonetheless, pertaining to the real time applications, it is of utmost importance to first estimate the co efficients of the channel and then exploit them while performing the detection process. It is always not appropriate to assume perfect channel state information in the real time scenarios, as it might lead to subsequent degradation in the overall system performance. Consequently, motivated by this fact, the researchers in [557] and [334] exploited the channel state information to analyze the performance of MIMO VLC systems. In [557], the authors exploited the VLC channel estimation errors for the purpose of joint optimization of precoder and equalizer in optical MIMO systems. Whereas, the contribution in [334] investigates the impact of noisy channel state information on the performance of different MIMO precoding schemes. Driven by this motivation, the research contribution in [323] considered the effect of channel state errors on the performance of a multi user VLC downlink network. Interestingly, this work takes into account two channel uncertainity models such as the noisy channel state information model and the outdated channel state information model for evaluating the performance of the system.

Taking into consideration the aforementioned aspects, the contribution in [337] exploits NOMA as a promising multiple access technique for 5G wireless networks in the context of VLC systems, where the corresponding system performance was evaluated under different channel uncertainity models. This work reports that closed-form expressions for the BER performance of NOMA-VLC system were derived for an arbitrary number of users for the scenario of perfect channel state information. Furthermore, the impact of two channel uncertainity models like the noisy and the outdated channel state information models on the performance of NOMA-VLC system were analyzed, where closed-form expression for the BER under noisy channel state information was derived along with the derivation of tight upper bound on the BER under outdated channel state information. The analytical framework of this work was substantiated with the aid of comprehensive Monte Carlo simulations which reinforced to thrive pragmatic intuition on the impact of imperfect channel state information on the overall system performance. Thereupon, the simulated result analysis affirms that noisy channel state information results in slight degradation in the BER performance of NOMA-VLC system; whereas, while dealing with outdated channel state information, an appreciable amount of degradation in the BER performance can be witnessed especially when there is a change in the channel gains of the users which generally prevails due to the involved mobility.

The work in [306] signifies an experimental demonstration of bidirectional NOMA-OFDMA VLC system, where the proposed system imparted a high throughput and a higher system capacity for a large number of users. Furthermore, this work investigated the effects of power allocation and channel estimation on the BER performance of the system. From the experimental results it can be deduced that inter-user interference can be eliminated by incorporating efficient channel estimation algorithms. Additionally, from the experimental results, it can be surmised that the proposed bidirectional NOMA-OFDMA-VLC system can be envisioned as a potential solution for both downlink and uplink VLC networks. From [191], it can be emphasized that the authors have proposed a phase pre-distorted joint detection (JD) method in contrast to SIC algorithm in uplink NOMA-VLC system. Based upon the channel information of different users, the optimal phase-distortion term was computed and then the users' transmitted signals were pre-distorted for attaining a better improvement in the overall system performance. The experimental results underline the fact that the aforesaid (i.e., phase pre-distorted JD) method exhibits a more superior performance than that of the SIC-based NOMA system.

Despite the noteworthy advantages offered by NOMA-VLC system, its performance is limited because of the implicit non-linear behavior of the voltage-current characteristics of the LEDs. Thus, the existing literature focuses on the exploitation of pre-distortion or post-distortion techniques to linearize the limited non-linear region of the LEDs or to employ the PAPR reduction techniques to mitigate the non-linear effects of the LEDs. However, there remains a necessity to understand the impact of non-linearities on the performance of NOMA-based VLC system. The authors in [348] have taken into account the non-linear characteristics of LED in the design process of pre-distorter for cognitive radio inspired NOMA-based VLC system. Moreover, this work proposes singular value decomposition-based Chebyshev precoding for the purpose of improving the performance of non-linear MIMO-NOMA-VLC system. This work infers that a generalized power allocation strategy is derived which can work well even in the scenarios where the users encounter similar channel conditions. Thorough analytical analysis is also performed where derivations on upper bounds for the BER of the proposed detector using M-QAM can be found. A similar work addressing the LED non-linearity issue can be reported in [303] where the maximum driven current of optical PD-NOMA was more significantly reduced than that of the conventional PD-NOMA. The research in this work shows that by means of experimental investigations, the proposed system achieves an improvement in terms of its BER performance over that of the conventional system.

The contribution in [296] signifies that a hierarchical pre-distorted layered ACO-OFDM (HPD-LACO-OFDM) scheme has been proposed for NOMA that offers remarkable benefits of assurance of superior spectral efficiency and high optical power efficiency when compared with that of a DCO-OFDM-based NOMA-VLC system. The enthralling advantage of HPD-LACO-OFDM-based NOMA-VLC system is that the inter layer interference of LACO-OFDM which generally occurs because of the clipping distortion at the transmitting end can be eliminated by exploiting pre-distortion technology. By employing the phenomenaon of pre-distortion technique, there is no scope for the occurrence of interference between the layers and the signals corresponding to all the layers can be recovered significantly without relying on SIC demodulation. Eventually, the receiver complexity is gradually reduced and the error propagation (EP) problem no longer exists. Thus, a better utilization of the spectrum can be achieved by means of HPD-LACO-OFDM-based NOMA-VLC system than that of the DCO-OFDM-based NOMA VLC system. Moreover, the former system exhibits a more superior BER performance than that of the latter system.

The impact of users' mobility on the performance of NOMA-VLC system has been investigated in [551] and [327]. Particularly, VLC networks employing NOMA based transmission technique is not devoid of its drawbacks. The major drawback associated with VLC-NOMA is that it is required for the NOMA-VLC transmitter to sort the users based upon their channel gains. Primarily, it is mandatory to estimate the channel gain at the user side, and then it is vital to incorporate the feedback mechanism to the transmitter. Therefore, it is obvious that while estimating the unknown channel gains, the computational complexity of the system increases. Moreover, sending the channel quality information to the transmitter results in link overhead. Additionally, a majority of the transmission scenarios in VLC is highly dependent on LOS links and pertaining to the real-time applications, this might not be feasible at all times, especially when the receiving direction toward LED is outside the FOV of the receiver. The list of works in [135, 163, 164, 407, 451, 452, 504, 505] portrays that the mobile users, especially due to the random receiver orientations in VLC networks, encounter the problem of unavailability of LOS link.

The contribution in [551] proposed a realistic multiuser VLC network, where the mobile users with random vertical orientations were served based upon the NOMA strategy by employing individual and group-based user ordering techniques with various feedback schemes. By employing the mean value of the vertical angle and the distance information, feedback mechanism on the channel quality was computed with the intent to reduce the computational complexity and link overhead. Additionally, for the group-based user scheduling, two-bit feedback scheme was proposed, which takes into account both the distance and the vertical angle in contrast to the conventional one-bit feedback which considers only distance. Analytical expressions for the outage probability and the sum-rate expressions were derived where the analytical analysis exhibits a good agreement with the simulated result analysis. Also, from the numerical results, it can be inferred that the practical feedback scheme with mean vertical angle attains a near-optimal sum-rate performance, while the two-bit feedback mechanism depicts a more superior performance than that of the one-bit feedback.

Similar to contribution [551], the research work in [327] also focuses on user mobility where the authors have investigated fundamental aspects on NOMA-VLC networks with major emphasis on the achievable sum rates and the optimal power allocation for both static and mobile users. Firstly, in this work, for the static user scenario, closed-form expressions of both the lower and upper bounds of the achievable rates were derived. In particular, this derived lower bound plays a vital role in determining the optimal power allocation scheme for static users in NOMA-VLC networks. Additionally, this derived lower bound is helpful in minimizing the transmit power under the minimum rate requirements and individual LED power constraints which turns out to be NP-hard. Consequently, by making use of semi-definite relaxation (SDR) technique, the optimal power allocation scheme can be obtained by solving the convex semi-definite program (SDP). Secondly, this work is directed toward the optimal power allocation for mobile users. In order to cope with the imperfect channel state information, which generally emerges due to the users' mobility, these uncertainties in channel state information were characterized as ellipsoidal regions. Then a robust power allocation scheme was developed by taking into consideration the ellipsoidal channel state information model for minimizing the total LEDs power, at the same time fulfilling both the minimum rate requirements of all the users and per-LED power constraints; generally, this turns out to be as a non-convex problem. The superiority of the proposed power allocation schemes can be affirmed through the simulation results.

To address the explosive augmentation in the demand for seamless wireless connectivity, it is an immense prerequisite to cater for efficient and reliable management of the available wireless resources. Toward this end, NOMA can be regarded as an excellent candidate due to its noteworthy feature of assurance of enhancement in the achievable data rates. Despite its advantages, the performance of NOMA-VLC system can be deteriorated under specific conditions especially in scenarios where a large number of users exist in highly symmetrical locations. Therefore, OMA techniques are proven to be as viable candidates, as they guarantee better link reliability; but this comes out at the expense of decreased spectral efficiency. Thus, in this regard, it is an immense prerequisite to facilitate a certain degree of intelligence in the transmitting LED/AP to assist in the real-time configuration of the multiple access protocol. The research efforts in [12] strive for the design of a framework that is capable enough to offer for dynamic multiple access selections (DMAS) in intelligent APs. The proposed DMAS is the intended to expedite for the adaptive configuration of the available multiple access modes like OMA, NOMA or hybrid OMA/NOMA based upon the computational evaluation of the sum data rate, average outage probability and fairness criteria. The simulated results draw a conclusion that DMAS is the most feasible solution that accomplishes better satisfaction of the system requirements when compared to the static configuration of a single multiple access technique.

Experimental validations of NOMA-based VLC systems have been investigated in [442], where the authors have proposed and demonstrated real-time software reconfigurable dynamic power-and-subcarrier allocation scheme for OFDM-based NOMA-VLC system for the multi user scenario. The flexibility and the reliability of the aforementioned system are validated by employing Xilinx-7 field programmable gate array (FPGA) to accomplish real-time transmissions. In order to optimize the overall BER performance of the system, the power allocation for both user and sub-band level has been incorporated. With the aid of software reconfigurable technique, based upon the channel conditions, the power ratios of different sub-bands can be dynamically adjusted, thereby resulting in superior user fairness and optimal transmission performance. Additionally, from this work it can be witnessed that in order to enhance the system flexibility as well as to satisfy various capacity demands of users, subcarrier allocation for the OFDM-NOMA-VLC system has been proposed. Experimental results reveal that OFDM-NOMA-VLC system achieves 1.84 Gbps real-time transmission, wherein the obtained BER is below 3.8×10^{-3} forward error correction (FEC) limit.

In spite of the rich research investigations of exploitation of NOMA with VLC, there is a limited literature on the joint interface of VLC/RF networks with NOMA. Intuitively, these two networks, i.e., VLC and RF, exhibit several distinctive features, as they operate at different regimes of the electromagnetic spectrum. This fact evolves two vital aspects, like the discrepancies in the capacities of the two subsystems along with their respective offered coverage area. Apparently, VLC offers a limited coverage area with better capacity. Therefore, due to this rate asymmetry, there occurs a trade-off between the achievable rate and the fairness, where this issue is more pronounced while dealing with VLC/RF networks. The investigations in [397] confirm the fact that the authors for the first time proposed a hybrid VLC/RF network in a practical indoor scenario, where the two sub networks, i.e., namely RF and VLC, performed NOMA. Furthermore, the peculiarities of NOMA make optimal user grouping an open research problem. Consequently, the asymmetry in the achievable rates of the users, as well as the different nature of the two subsystems, leads to the origination of several challenges in a hybrid VLC/RF network. More particularly, fairness stems out as a major problem because there is a scope for users being served by VLC network to increase their capacity far beyond the respective users that are being served by the RF AP. Thus, in such a hybrid network it is indispensable to allow for appropriate user selection or user grouping to utilize the advantages offered by such hybrid network as well as to overcome the congestion aspects. Since the users in such hybrid networks experience different channel conditions at their respective APs, it becomes very cumbersome to classify users' as strong or weak. Eventually, in such hybrid networks, user grouping stems out as an important challenge.

Subsequently, in order to attain a balance between the fairness as well as to maximize the individual rates, the research in [397] proposes a novel utility function that takes into account the additional complexity of the NOMA scheme. With the intent to model the interactions between the users and the respective APs of the hybrid VLC/RF network, the authors exploited a coalitional game theory where each coalition is appropriated to a specific AP, and users are given the flexibility to join a coalition that best suits them. The research efforts in [535]. much similar to [397], address certain aspects in hybrid RF/VLC-NOMA system, with the motive to extend the coverage area as well as to augment the QoS. By taking into account both advantages as well as the vulnerabilities of the technologies like VLC, RF, NOMA, the work in [535] is directed toward the design of a downlink hybrid VLC/RF relaying network with NOMA in a two-user scenario. More particularly, both users can gain access through the VLC technology while as a complementary technology, RF, is employed for the purpose of reducing the outage probability for the far user; specifically, RF, is used for forwarding another copy of message to the far user. Hence, the far user has a flexibility to receive two messages, i.e., one from the VLC AP and the other is through the RF. In this work, the concept of cross-band selection combining (CBSC) has been introduced, which states that the far user was continuously served either through the hybrid VLC/RF link or solely via the VLC link. Moreover, considering VLC particularities, accordingly where the transmitted signal has to be non-negative as well as constrained by the illumination requirements, it can be inferred that this work derives the closed-form expressions for the achievable rates as well as the outage probability of each user. Furthermore, for the sake of confirming the effectiveness as well as accuracy of the proposed VLC/RF cooperative NOMA network, Monte-Carlo simulations have been presented. From the result analysis it can be deduced that the proposed CBSC imparts significant improvement in the outage probability of the weak user, and an increment in the sum throughput of the system can be witnessed.

5.3.5.7 Challenges associated with NOMA-VLC systems

Until now, the previous sub sections presented elaborate research efforts about the state-of-the-art aspects that are in close association toward the exploitation of NOMA in VLC systems. This subsection portrays some of the enthralling research challenges pertaining to the practical amalgamation of NOMA in VLC systems. Nevertheless, the adoption of this non-orthogonal multiple access technique like NOMA has sparked a significant amount of momentum in the research community, yet there are several aspects that need to be paid special attention to ever since the interface of this multiple access scheme with VLC systems.

A. Optical MIMO-NOMA VLC System

With the motive to render a sufficient amount of illumination, indoor spaces like conference halls, libraries, museums, multi national companies, train cabins, aircrafts, hospitals, homes, offices, etc. are furnished with multiple LEDs. This scenario gave rise to the realization of MIMO configurations in VLC systems. The exploitation of MIMO for NOMA-based VLC system is the most appealing solution to guarantee both capacity and reliability improvements. However, MIMO-NOMA VLC system is not free from challenges. It is apparent that one of the spatial diversity techniques of MIMO is spatial multiplexing (SMP); therefore, in MIMO system exploiting SMP, users are broadly divided into subgroups; it is intended that a single transmitting LED or VAP serves the users belonging to each subgroup. Meanwhile, the signals belonging to the users confined to a particular subgroup will be superimposed in the power domain. This phenomenon confirms the fact that there are two different levels of interference cancellation that need to be associated for the multi user detection. Firstly, transmit precoding can be used to separate the MIMO subchannels [334]. Secondly, the interference cancellation methodology like SIC can be performed to ensure the multi user detection for users belonging to the same LED. Therefore, this naturally implies that it is vital to ensure an exhaustive design of SMP-MIMO-NOMA-based VLC systems to replenish a remarkable enhancement in capacity by augmenting the number of users that are to be served concurrently. The other diversity technique like repetition coding (RC) can be employed to improve the link reliability by ensuring diversity gains. Despite its noteworthy advantage offered by RC, its interface with NOMA-VLC systems is not straightforward, as the design aspects are very complex. This is due to the fact that, for a MIMO-NOMA-VLC system, the assignment of power should be done across the overall combined signal and, on top of it, there is necessity to exploit the channel gains of the users belonging to different LEDs. Thus, this process entails intensifying the overall complexity associated in the design of such systems especially, when the users' mobility is taken into account.

B. Emphasis on VLC/RF-NOMA Networks

A limited coverage area is offered by a VLC network, and it is also imperative for the VLC receiver to face the VAP which stems out as an important challenging aspect for VLC systems. To overcome such drawbacks, it is a major prerequisite to exploit RF and VLC technologies together. Precisely, VLC technology can be used as a complementary technology to off-load the data especially in user-dense indoor environments. Even though there is a growing literature on the joint exploitation of RF/VLC networks, the joint interface of RF/VLC-NOMA systems is still in its infancy. Particularly, the work on the interface of VLC/RF based NOMA networks has recently started and can be witnessed in [397] and [535]. Intensive research efforts on downlink NOMA were studied, and limited attention is paid to uplink NOMA-VLC system. As of now, one study which emphasizes the experimental demonstrations on the BER performance of uplink NOMA-VLC system can be reported in [306]. Furthermore, relay-assisted VLC systems play a prominent role to enhance the overall system performance, where secondary lights like the desk lamps, task lights, etc. were employed to direct the ceiling LED-based illumination infrastructural units. It is worthwhile to stress on relay assisted NOMA VLC system to augment the coverage area of the cell edge users.

C. Exploiting CD-NOMA for Multi user VLC Systems

As already mentioned earlier, NOMA occurs in three different forms like PD, CD and NOMA in the multiplexed format. However, most of the research focused on PD NOMA and limited literature reports on CD NOMA. Sparse code multiple access (SCMA) is one such class of CD-NOMA which can be exploited to increase the bandwidth efficiency. Much significant work pertaining to its usage in VLC can be found in the research work as stated in [30], where SCMA has been applied to VLC systems for the purpose of maximizing the throughput of the system. This work contributes to the monitoring of long-range health information. Generally, to ensure simultaneous transmission as well as effective management of various significant medical data with real-time monitoring, it is required to rely on multiple data transmission techniques. Apparently, CDMA exhibits its prominence in enabling multiple connections where the data symbols are spread over a sequence of orthogonal codes for the purpose of mitigating the effects of interference. Nonetheless, its assurance of maximum connection is confined to this codebook option. Contrarily, the availability of a novel codebook for NOMA, SCMA exhibits the potential to ensure around 50 and 200% more connections when compared with traditional CDMA or other OMA techniques.

The research in this work elucidates the fact that a bandwidth effective multiple transmission technique like SCMA has been exploited for a VLC system that is compatible for RF-free indoor healthcare monitoring. In this work, SCMA technique has been employed for the purpose of improving the bandwidth efficiency and as an add-on, long-message passing algorithm (MPA) was used to reduce the decoding complexity that is associated with SCMA. This work focuses on the design of a practical multiple medical data transmission and monitoring system where SCMA has been used for transferring multiple medical data by using a single LED. Thereupon, the collected data was monitored by means of a smart phone or a PC in the real time. By exploiting this information, the medical professional provides the necessary medical care. The experimental results of the proposed system reveal the superiority of SCMA-based VLC system where six different medical data were transferred concurrently over a distance of 5 m with a reliable packet error rate (PER) of 10^{-5}.

D. Impact of LED Non-Linearities on the Performance of NOMA-VLC Systems

The performance of VLC systems is limited due to the non-linearities of the optical sources especially the LEDs. Accordingly, this demands a thorough investigation of the impact of non-linearities of LEDs on the performance of ONOMA-based VLC system. Apparently, it is mandatory to have such an investigation in order to enable the actual realization of ONOMA-VLC system and to determine the achievable spectral efficiency and error rate analysis. Mitigating the effects of non-linearities of LEDs and analyzing the performance of NOMA-VLC systems by taking into consideration such non-linear effects is still in its infancy, and a limited literature can be witnessed. Therefore, this constitutes an important challenging aspect for the realization of future high-speed NOMA-based VLC systems.

E. Effect of Symmetrical VLC Channel Conditions on NOMA-VLC Systems

Particularly, when dealing with NOMA, the existence of dissimilar channel conditions is mandatory to ensure successful detection of users in case of a multiuser scenario. Nonetheless, VLC channel exhibits high correlations, and in fact this sort of scenario implies that the channel gains of the users will be in close association with one another. Furthermore, the positions of the receiving terminals also have a significant impact on the performance of NOMA-VLC systems. In case of indoor room environments, when the LEDs are positioned at the center of the room this enables the VLC configurations to undergo high symmetrical channel gains. Until now, the best feasible solution to address the strong symmetrical VLC channel gains ,is to tune the transmission angles and FOV of the LEDs and the photodiodes. The future research can be directed toward the enhancement of the performance of NOMA-VLC system under highly symmetrical VLC channel gains, as well as evaluation of the performance of the system by dynamically adjusting the transmitting and receiving angles by taking into consideration the location of users.

F. Future Directions of NOMA-VLC Systems

Toward the realization of high speed, spectrally efficient and power-efficient NOMA-VLC networks, future research should focus on exploitation of massive MIMO VLC systems and amalgamating with ONOMA technique, enforcing proper user pairing as well as stressing on optimization of power allocation algorithms, laying emphasis on the formulation of ONOMA-uplink VLC systems, analyzing the performance of O-NOMA in asynchronous communications, and stressing on the detrimental aspect like PAPR analysis and reducing it in NOMA-VLC systems.

5.4 SMART CITIES EXPLOITING VISIBLE LIGHT COMMUNICATION TECHNOLOGY

The foremost rationale of a smart city is to facilitate seamless amalgamation of information and communication technology (ICI) with the IoT era in order to efficiently supervise the assets of the city. This in turn plays a vital role in enhancing the well-being of the people as well. The assets of a city or a nation mainly include transportation units, hospitals, educational institutes such as schools and colleges, libraries, local departments, information systems, power plants, water supply networks, waste management, law enforcement and other such community services. The extensive emergence of both information and communication technologies will definitely play a significant role in helping a city to evolve as a smart city. Predominantly, the major task of the smart city is to employ different types of IoT sensors to gather information and to gain insights in order to manage the city's assets, resources and services effectively. In this manner, the gathered data will definitely enable to accomplish uninterrupted services within the city. Figures 5.29 and 5.30 delineate the features and overview of the smart city.

Despite RF-based technology being widely deployed in many applications in the smart city, from a glimpse of earlier findings, it is apparent that RF-based technology will no longer ensure the desired quality of service for the evolving smart city applications. It is valid because the fascinating demands of the subscribers throughout the world and the explosive increase in the mobile appliances result in an unprecedented increase in the mobile data traffic. As a consequence, the present RF spectrum is not capable to fully comply with this enormous increase in the demand for data. Thus, to relieve the burden of RF-based wireless communications, research efforts have been shifted toward the unlicensed portion of the electromagnetic spectrum. Toward this end, VLC has evolved as a potential candidate affirming the most

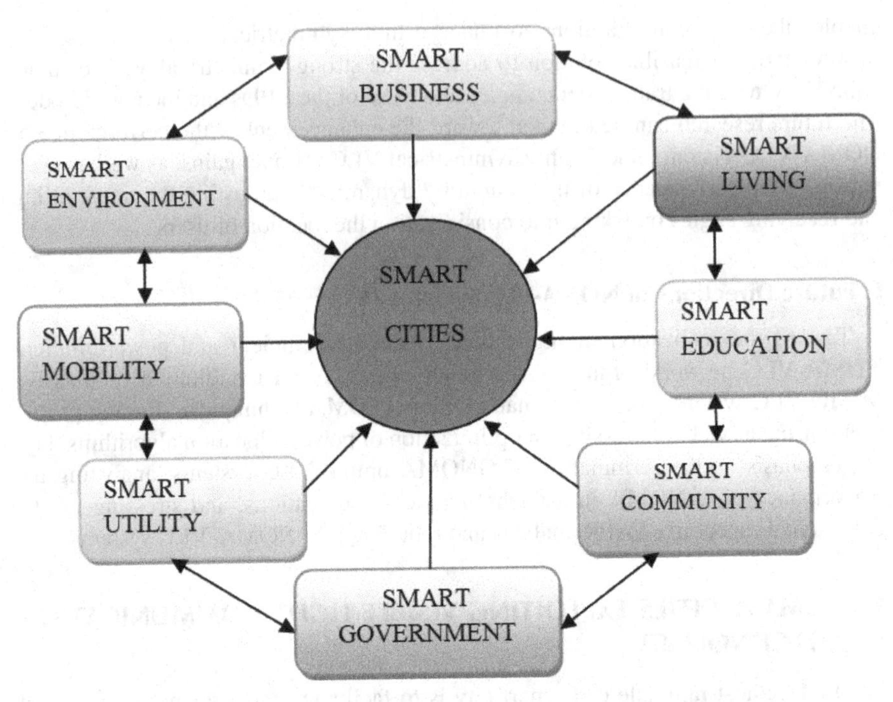

Figure 5.29: Depiction of various features of a smart city [497].

Figure 5.30: Interpretation of the outline of a smart city [497].

reliable services for bandwidth intensive applications. Thus, this sort of communication can be exploited to render seamless connectivity among the people, infrastructural units, transportation units, government sectors, health care services, etc. in a smart city. VLC offers the most feasible solution in enabling seamless coverage through the city due to its capability to support simultaneous high data rate communication and illumination which is aesthetically pleasing to the human eye.

VLC exhibits several noteworthy advantages, among which the most arresting benefit is its ability to not to interfere with other existing technologies like RF units, etc. Moreover, the LEDs that were exploited in VLC technology provide an ample amount of support to realize a small-scale communication cellular network in a manner that each installed lighting fixture ensures uninterrupted services to the end users which are within the coverage area of the LEDs. Consequently, this technology offers the flexibility to be easily integrated with several units present in the smart city which include the road side units, traffic lights, street lighting systems, transportation infrastructural units, dash boards, sign boards, etc. As already discussed, multiple access provision is also possible to render services to the subscribers. In this regard, a variety of multiple access schemes like OOFDMA, OCDMA, OTDMA, OSDMA and NOMA play a dominant role to enhance the high data rate communication to users. Furthermore, VLC technology can go hand-in-hand with RF communications without causing any forms of interference. Thus, in this regard, a cost-effective smart lighting, smart intelligence to automotive applications, etc. can be realized in a smart city. On these grounds, much relevant work that is closely associated with this approach to design a light-based communication architecture for enabling intelligent transportation applications in smart cities can be portrayed from [68]. Figure 5.31 outlines the general architecture illustrating the functionality of several communication layers. Typically, as per [68], the allocation of these communication layers can be listed below as:

1. *LED-to-LED Communication Layer*
2. *LED-to-VLC subgateway*
3. *VLC-Subgateway-to-Free Space Optical Wireless Communication (FSO) Gateway*
4. *Inter-FSO Gateway*

The detailed summary interpreting the functionality of each layer in an effort to furnish connectivity in a smart city is presented below:

1. The omnipresence of LEDs, aided by their exceptional prevalence like their cost-effective and energy-efficient nature, sustainability, etc., enabled all the illumination appliances to become LED oriented. The work in [68] claims LED to be used as both light source as well as a sensor. This choice is to provide the flexibility for both data modulation as well as data gathering. More particularly, when LED is used as a light source, it accomplishes the criteria of illumination, and the data signal is modulated by its fast switching speeds. As a sensor, LED-to-LED communication creates a provision for the formulation of a wireless optical sensor network that plays a significant role in gathering the real time data relevant to high traffic prone areas, distance between two moving vehicles, high accident prone areas, emergency aspects like fire accidents, ambulance aids, etc. Thereupon, this sensed information is then disseminated to the *VLC subgateway*.
2. The primary task of *subgateway* is to supervise and to control the LED-based communication devices which are within the vicinity of it.

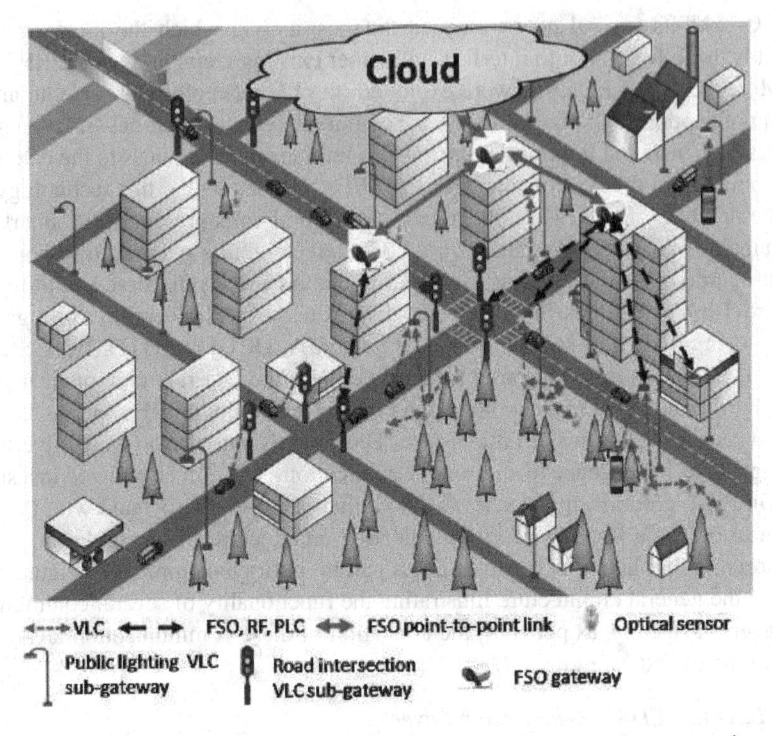

Figure 5.31: Smart lighting-based communication architecture for smart city applications [68].

Furthermore, it even creates the flexibility for providing access to smart devices, mobile phones, vehicles, etc. The other prominent role of the *subgateway* is to collect all relevant information from all the optical sensors and based upon the gathered information, it takes critical decision regarding the emergency situations. It is clearly evident from the figure that there are two subgateways, namely, the public lighting VLC sub gateway and the road-intersection VLC subgateway. The former subgateway is deployed in the lighting poles that are located alongside the road. The latter subgateway is necessitated to provide seamless coverage of the road intersection units.

3. In order to ensure seamless connectivity in the smart city, the *VLC subgateway* should be capable enough to possess highest storage capacity as well as it should be computationally efficient. Typically, the foremost role of the subgateway is to facilitate communication between smart city and other network service providers. This work exploits the FSO as a medium of communication between all the gateways present in the city. In order to establish this sort of communication, even the RF-based wireless communication technologies can work hand-in-hand with VLC to enhance high data rate transfer.

4. By employing FSO point-to-point links, data can be communicated across several *VLC subgateways* that span throughout the city. Thus, in this manner, smart cities exhibit the potential to support many applications using VLC technology, namely, transportation, public safety, government applications, etc.

The above communication architecture in the smart city offers the scope for enabling smart transportation support as well. Listed below are a few such applications of smart transportation systems:

- *Greent trafficm management*: Primarily, most accidents occur at the road-intersection units due to heavy-load carrying trucks. Consequently, to circumvent this drawback and to enhance the safety of the people travelling on roads, necessary management steps are of utmost importance. In this regard, after receiving the data from optical sensors, the subgateway is meant to adjust the timing of the traffic signal (red or green color) in order to allow for the heavy-load carrying trucks or other vehicles to easily pass through the road intersections. Moreover, the gateway can also display a warning message indicating the high traffic prone areas, thereby deviating the vehicles to follow other alternative routes.
- *Parking management*: The subgateway makes a note of the occupancy rate of the parking and transfers to the nearest gateways through the FSO link. In this manner, the vehicles are indicated with the nearest vacant parking area.
- *Accident management*: Upon receiving the occurrence of accident on a specific roadway, the gateway broadcasts this information to other gateways present in the city, thereby displaying an emergency message requesting ambulance support as well as a warning message to other vehicles in order to deviate their ways.
- *Emergency case management*: In case of emergency situations where there is a requirement for an ambulance or help to prevent fire accidents, the gateways should take immense measures to broadcast the location of the ambulance. It should also see to it that, if the ambulance got stuck in the traffic, where the traffic light displays a red color, then appropriate adjustments should be made to display a green color in order to allow it to pass to the accident prone area.
- *Vehicle tracking*: The major goal of enabling intelligence to the vehicles is to ensure that road accidents can be prevented. In this regard, it is vital to transmit information regarding the distance between two moving vehicles, assessing the speed of the vehicles, etc. This is generally performed by means of installation of light-based optical sensors.

From the above discussion, it is evident that research is progressing rapidly to meet the demands of the users as well as efficient technological advancements were evolving to render communication aid throughout the world. To date, several research communities, industries, etc., endeavored compelling efforts to enable a city to emerge as a smart city. Much relevant work pertaining to this discussion can be found in [279], where the authors present an approach that can lead to the emanation of a smart city with the aid of an energy efficient public illumination system. The noteworthy feature of this energy efficient public illumination system is that it can offer pervasive communication throughout the city. Furthermore, this public lighting system is capable to render dimming support as well. For instance, based upon the needs of the applications, i.e., upon sensing the surrounding environment, the intensity of the lights is automatically adjusted. Thus, this aspect is very helpful in the perspective of energy savings. Additionally, a case study depicting the scenario of road illumination is also the focus of this contribution. The simulation results of this work signifies that such smart illumination system imparts high data rate communication.

A practical approach of integrating VLC and power line communication (PLC) technology along with various applications of VLC were discussed in [41]. The integration of VLC technology with PLC emerges as an attractive solution for the new smart city applications, and this approach makes VLC more practical, efficient and cost effective. The smart grid which is based on PLC can provide the broadband data access and power for the VLC node (bulb) to ensure both wireless data connectivity and illumination. Utilizing the advantage of infinite bandwidth of VLC, one key concern is to develop a merging solution which provides both power and data from the same power line to the VLC node. It is also worthy in mentioning that the aforementioned solution has been recommended by the industry, and there is an increasing interest in finding such fascinating solutions to promote VLC adequately.

One of the important components of a smart city is smart tourism. Particularly, it plays a vital role in contributing toward economic growth throughout the world. More specifically, smart tourism makes use of the ICT in order to formulate several tools and approaches for attracting tourists. A suitable work that is closely associated with smart tourism is reported in [401], where visible light communicating devices can be employed as data access points to promote tourism support in a smart city. In this work, the authors have employed street lighting units as auxiliary points to render basic information aid to the tourists who are new to the city. The information pertaining to the place, direction, and nearby tourist places are of paramount important to the tourists visiting a new area. Smart city desires almost everything originating right from the infrastructure to the services to be smart, accurate and intelligent. In recent times, research efforts focused to enable a city to evolve as a smart city and several such applications in the areas of healthcare, vehicular applications, smart homes, industries, airport, multinational companies, etc., are shown in Fig. 5.32. Consequently, the thirst for enabling seamless connectivity and to assure reliable services to the users sparked an enormous interest to exploit significant technical aspects relevant to VLC. Taking into consideration the earlier literature, this section outlines in detail the amalgamation of public lighting systems with the infrastructural units present in the smart city. Furthermore, this section provides a comprehensive discussion on the exploitation of VLC for traffic management.

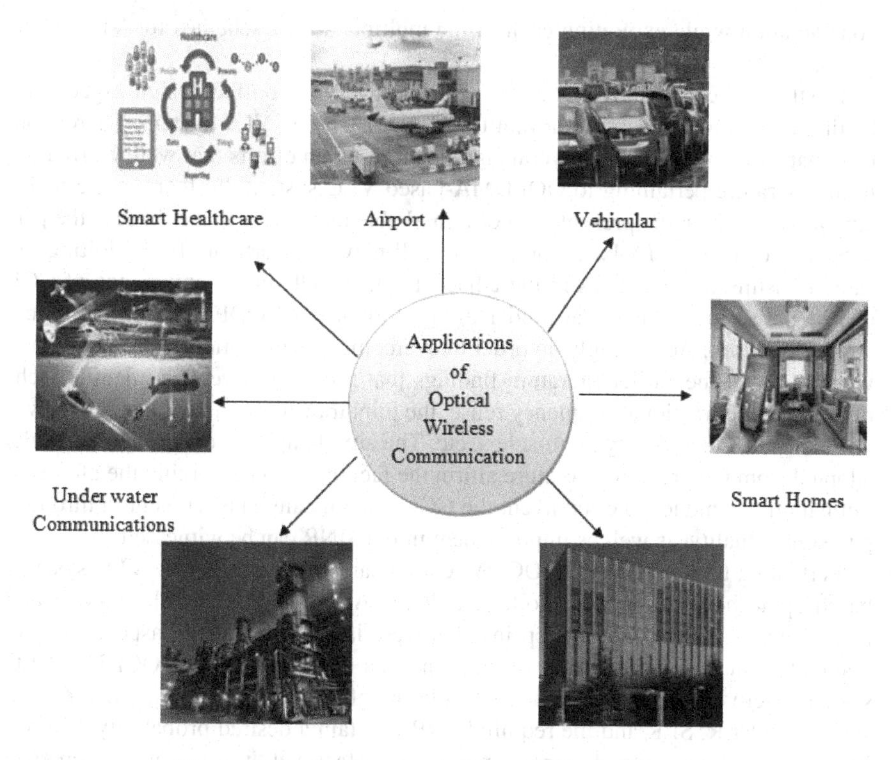

Figure 5.32: Applications of VLC for a smart city.

5.5 CONCLUSION

This chapter presents a thorough review on different orthogonal multiple access techniques which are compatible with IM/DD systems, namely, OOFDMA, OCDMA, OTDMA and OSDMA, along with a comprehensive analysis of the emerging concepts of non-orthogonal multiple access technique like NOMA and its integration with VLC systems. This chapter mainly focuses on the system architectures of the VLC systems exploiting the aforementioned multiple access techniques (i.e., both orthogonal and non-orthogonal), application scenarios, etc. A critical review on the current state-of-the-art research aspects that include both design issues and implementation of the aforesaid multiple access techniques for the realization of cost-effective VLC systems are the major emphasis of this chapter. Moreover, this chapter makes it easier for the reader to identify the underlying performance trade-offs, and thus paves the platform for a scope for future research work by taking into consideration the challenging problems associated with the implementation of each multiple access scheme in the optical domain. Even though many of the techniques and approaches pertaining to these multiple access schemes have been borrowed from the RF domain, they cannot be implemented in a straightforward manner for VLC. Consequently, this chapter also details the design aspects that are mandatory to take into consideration while exploiting each of the multiple access schemes for VLC applications.

Firstly, this chapter gives a comprehensive overview on the design aspects depicting a downlink scenario relevant to OOFDMA-based VLC system. Thereupon, this chapter also focuses on several significant research efforts that were carried out in the literature pertaining to OOFDMA-based VLC system. Furthermore, a major emphasis is laid on the presentation of remarkable techniques that enhances the performance of OOFDMA-based optical attocell network. Particularly, exploiting the same transmission resources in the adjacent cells results in the emergence of CCI, one of the serious issues that hinder the performance of OOFDMA-based optical attocell network. Accordingly, in order to overcome this limitation, this chapter provides in detail the earlier literature findings that proposed several mitigation techniques like the fractional frequency reuse, the joint transmission approach, handover mechanisms, game theory approaches, etc. The simulation and experimental results adopted from the previous literature affirm the fact that upon enforcing the aforesaid mitigation techniques to circumvent the CCI, a significant enhancement in throughput, signal quality, as well as improvement in the SINR can be witnessed.

Pertaining to the scenario of OCDMA, we characterized OCDMA-VLC systems based upon the codes they exploit, i.e., OCDMA-VLC systems with unipolar and bipolar codes. Furthermore, deep insights were laid on the design aspects of several codes, and the performance of both unipolar and bipolar-based OCDMA VLC systems were analyzed and compared in terms of the number of users that can be supported, BER, SER and the required SNR to attain a desired probability of error. Most of the earlier research works emphasize the fact that the correlation properties of the codes play a vital role in determining the capacity of VLC-CDMA system. Even though extensive research efforts were carried out in several areas, the performance of VLC-CDMA systems is limited by the design issues of signature codes. Therefore, the design aspects of signature codes remain as an open challenge problem in CDMA-based VLC systems.

Pertaining to the VLC-CDMA systems that are employing unipolar codes, combating the effects of multiple access interference in such systems poses as a challenge since the cross-correlation properties of unipolar codes are not ideal. On the other hand, VLC-CDMA systems based upon bipolar codes exhibited remarkable performance only at the expense of increased complexity in the design aspects, as it is obligatory for such kinds of codes (bipolar) to satisfy the requirements of IM and DD. Eventually, exploring more suitable signature codes that are capable enough to ameliorate the user capacity, i.e., supporting as many users as possible, allowing for the realization of less complex systems and exhibiting the ability to subdue the multiple access interference will be the ultimate future research topics.

Dealing with the scenario of OTDMA and OSDMA-based VLC systems, this chapter presents an elaborate literature of the pros and cons associated with these two orthogonal multiple access schemes. Furthermore, a detailed system model was presented which highlights the comparisons of performance of OTDMA-based and OSDMA-based VLC systems in terms of achievable spectral efficiency. Many of the simulated result analyses taken from the previous literature suggest the superiority of OSDMA multiple access scheme when compared to OTDMA system. Even though angle diversity-based OSDMA transmitter exhibits remarkable performance in terms of achievement of higher average spectral efficiency, it is complex to implement. Therefore, future research should be directed toward implementation of less complex mechanisms to generate multiple directional beams for serving multiple users simultaneously. Even though the multiple access schemes exhibit their remarkable advantages in rendering services to multiple mobile terminals, they allow the users to share either the time, frequency or space. Therefore, in this regard, with the intent to further enhance the throughput and to serve more users simultaneously by allowing them to share the same time,and frequency resources, NOMA has been exploited for VLC. This chapter provides an elaborate review of the exploitation of this non-orthogonal multiple access scheme for VLC-based applications.

Starting from the basic underlying principle associated with NOMA, this chapter gives a thorough description of the applicability of NOMA for VLC systems. Furthermore, different power allocation mechanisms and exploitation of NOMA for MIMO-VLC system are also the major emphasis of this chapter. This chapter also highlights the research efforts strived to mitigate the inter cell interference, as well as gives briefs about the interface of NOMA with other orthogonal multiple access schemes along with comparison of the performance analysis of the joint usage of NOMA and OMA techniques in VLC systems. Additionally, this chapter presents comprehensive details about the current state-to-the-art research efforts of the usage of NOMA for VLC systems focusing on several areas, namely, proposal of optimal power allocation algorithms, tuning the opto-electronic elements, exploiting different user grouping algorithms, addressing the non-linearity aspects of LEDs, combined exploitation of RF and VLC networks, etc. to improve the user fairness and spectral efficiency of the networks. Moreover, the future directions which are associated with the implementation of NOMA for VLC systems are also the focus of this chapter. From the previous literature, it is concluded that there exist meaningful opportunities for the applicability of NOMA in the context for VLC systems. Eventually, an exhaustive design of optical NOMA-based VLC systems can be regarded to appreciably meet the capacity demands for 5G and beyond 5G networks.

VLC offers a great potential in the formulation of a smart city. It guarantees high density connectivity mostly in the scenarios where there is a necessity to transmit sensitive data between multiple connected devices present in the city. Furthermore, it assures for low latency services, as well as avoids the kind of interruption that might arise during congestion time. Furthermore, this chapter presents a public intelligent lighting system that enables seamless connectivity in a smart city. Additionally, with the advent of vehicle safety applications to enhance the safety of road transportation, smart and energy-efficient public illumination system can be employed to monitor several events such as accident management, park management, vehicle tracking, etc. Finally, this chapter presents an in-depth, up-to-date review of several multiple access schemes like OOFDMA, OTDMA, OCDMA, OSDMA and NOMA-based VLC systems. The result analysis of several signified research contributions articulates the fact that distinguished technological advancements are being carried to render high data rate communication for the new smart city applications.

6 INTEGRATION OF VLC WITH PLC

6.1 INTRODUCTION

The emergence of wireless mobile devices created an alarming increase in the demand for data. In addition, this demand is further accelerated with the emanation of the internet of things (IoT). Consequently, this huge thirst for data will assuredly stipulate the future generation communication systems to ensure rapid connectivity at all times all over the world. Eventually, this huge pursuit for wireless internet access connectivity impoverishes the prevailing radio frequency spectrum. This stems out as the most sensitive aspect that needs to be paid a significant amount of attention. Subsequently, in order to cope with this tremendous increase in the demand for data, new wireless communication technologies, sophisticated modulation, and multiple access techniques, robust network architectures have originated. Among such wireless communication technologies, VLC has proven to be a potential candidate among many of its counterparts for 5G networks and beyond 5G. Predominantly, the stunning advantages offered by VLC are two-fold: Firstly, it is wide, has license-free bandwidth, is eco-friendly and has low-cost deployment, where the simple, energy-efficient and sustainable opto-electronic illuminating devices are sufficient enough to ensure broadband communication, thereby enabling it to evolve as a green communication technology. The second advantage of VLC that attracted the research community is its assurance of high speed, high capacity, secured means of transmission and wide-scale availability with reduced interference.

Nonetheless, in order to facilitate the fruitful realization of VLC system to impart communication as well as to play the role of an access point, it necessitates for VLC has to rely on a ubiquitous network. Taking into account that a VLC system by itself cannot be utilized as a source of information, it is very important for VLC to be interfaced with a backbone network in order to circumvent the phenomenon of turning out into an information isolated island. Substantially, this aspect resulted in the emanation of a diverse number of approaches mainly with the intent to serve as a strong communication backbone aid for VLC systems. A few such backbone approaches are Ethernet, optical fiber, power line communications (PLC), etc. The underlying functionality of any backbone network is to offer the flexibility to be easily coupled with VLC systems as well as to deliver high data rate communications. On the other hand, the PLC technology is evolving as a promising medium to impart secured data communication in which the existing power lines can be employed for the dissipation of both power and data. From several earlier findings, it is ascertained that PLC seems to be the most appealing solution to fix the problem of VLC, where the prevailing power lines which are inherently present in the indoor networks, not only function to supply power to the illumination unit but also dispense information to it.

In recent times, narrowband PLC (NBPLC) has been widely employed in several applications of smart homes, smart cities and smart grids. On the other hand, the wide-scale applications of VLC for both indoors and outdoors are numerous. Thus, both PLC and VLC, even though being two independent technologies where the former is a wired, while the latter is a pure wireless technology, can be seen as excellent candidates to be deployed for several indoor applications. For instance, PLC exploits the existing power-line cables, while VLC makes use of the cost-effective illumination fixtures. Substantially, both these technologies are economical technologies, where PLC can be regarded as the effective logical backbone for VLC systems. Thus, the most effective and smart way to avail the advantages offered by both these technologies is to simply amalgamate them in a manner suitable for several applications. Therefore, the research communities, academicians and industries strived to enable the penetration of the combined PLC-VLC systems into future generation wireless communication systems. From a list of earlier findings, over the past decade, a significant improvement in the emergence of PLC as an alternative to the Ethernet for rendering data communication aid has been observed. This emanation of PLC as a complement candidate to Ethernet is due to the advent of the robust multicarrier modulation format like OFDM. Moreover, the potential of PLC over Ethernet is obvious since, in case of PLC, the existing power cables are sufficient enough to render the dual purpose of powering up the devices as well as offering data connectivity. Much similar to the integration of Wi-Fi with Ethernet, PLC can be interfaced with VLC to facilitate wireless connectivity. This integration of PLC with VLC seems to be as the most efficient and cost-effective approach to ensure broadband access support to the indoor users. The beauty of the integrated PLC and VLC system is that PLC acts as a backbone network since it is an indispensable and integral network element for an indoor VLC system.

The widespread availability of the power lines played a significant role to enable PLC to become increasingly popular as an economical and feasible candidate for guaranteeing last mile access for indoor communications in homes, offices, buildings, smart grids, etc. In several studies, it is revealed that PLC systems can be employed as proficient backhaul network to VLC systems by enabling the connectivity of these systems to the existing power-line cables. Consequently, in this manner, existing power lines or power cables in PLC systems not only supply power, but also the necessary information signal to the VLC systems. In this regard, the amalgamation of both the technologies, i.e., PLC and VLC, will definitely bestow the most economical and enticing solution to guarantee high data rate transfer of information. Even though the stunning benefits of PLC, which includes its ubiquity and the availability of the required infrastructure, enable it to be deployed as a backbone network for VLC, the nature of the power lines, much particularly, the frequency selective nature of the power lines imposes a challenge for PLC to furnish high data rate communication. Nonetheless, with the onset of efficient multicarrier modulation techniques, the research community was successful in converting the frequency-selective nature of the PLC channel into flat-fading channel. Initially, power lines were designed to facilitate power transfer only for low frequencies. However, upon applying sophisticated and advanced modulation and coding techniques, the challenging aspects imposed by PLC were overcome including the high background and impulsive noise, the time-varying and topology-dependent channel responses, as well as interference which emanates from the connected loads to the power lines. Eventually, this breakthrough enabled the scientific and research communities to utilize PLC for data communication support as well. The interface of VLC with the backbone network like PLC, where PLC can deliver data to the VLC transmitter turns out to be an appealing architecture for providing seamless services to the indoor applications. Moreover, this is further fueled by the successful standardization of high data rate PLC through the release of IEEE 1901 and ITU-T G.9960/61, and it is apparent that VLC is standardized through the release of the IEEE standard i.e., IEEE 802.11bb. Thus, an integrated PLC-VLC system will assuredly offer high-speed wireless services to the endusers.

Listed below are several benefits of the integrated PLC-VLC systems:

1. Affirmation of high-speed transfer of data: The integrated system, i.e., PLC-VLC system, offers significantly higher data rates than the Wi-Fi systems over a large distribution area. By exploiting the existing infrastructure and flexible multiple access techniques, seamless connectivity can be assured to the users which are located within the vicinity of the installed lighting fixtures.
2. Enhanced security: When compared to traditional Wi-Fi systems, the signal in the integrated PLC-VLC system is less vulnerable to the effects of interference, jamming and eavesdropping. This is because, upon coupling PLC with VLC, inherent security is ensured due to the fact that the light signal cannot penetrate through walls and other objects thus preventing the intruder to steal the data.
3. Cost-effective nature: The implementation of PLC-VLC system is very simple and cost-effective because by using the already available power-line infrastructure, the LEDs can be powered up as well data communication aid can be facilitated. Thus, this joint exploitation of PLC-VLC systems overcomes the problems which usually exist in RF-based technologies.

The major intent of this chapter is to provide a comprehensive outline about the performance analysis of the integrated PLC-VLC systems. The performance of any communication system ultimately depends on the channel environment in which it operates; therefore, in this regard, in addition to VLC channel modeling, it is very important to have a clear understanding about PLC channel modeling. Thus, this chapter is directed toward an elaborate presentation of PLC channel modeling along with the characterization of different noises that usually prevail in the power lines. Furthermore, this chapter gives an in-depth analysis of the up-to-date research aspects carried out in this emerging area, i.e., PLC-VLC integrated systems, and discusses the applications of PLC-VLC systems in day-to-day life.

6.2 BASIC SYSTEM MODEL

The first interface of PLC with VLC technology can be witnessed in [273], where this renowned work illustrates the magnificent nature of the developed integrated system which exhibits its caliber to avail the already installed power lines to render support for data communication as well. Power lines are omnipresent, every electronic appliance or device generally requires power to properly function. In case of indoor room environments like homes, offices, etc., most of the data equipments as well as the electric appliances will be connected to the previously installed power cables and outlets, thereby avoiding the necessity to rely on additional tangeled cables for enabling data communications. Based on these grounds, the authors in [273] propose to employ the already installed power lines for facilitating VLC. Generally, VLC exploits simple, sustainable opto-electronic devices like LEDs. These LED luminaires requires power supply for their lighting purposes. Thus, these power lines can be deployed to enable data communication between the WLEDs and other fixed networks present in the room environment. Eventually, this set-up will eradicate the wiring problem which usually arises in WLED-based VLC systems.

The schematic representation of the first integrated PLC-VLC system model is represented in Fig. 6.1. The system model as depicted in the figure is very simple to implement due to its easy installation facilities and wiring. In this framework, it was assumed that the power-line modem was already plugged into and a power-line network was also built. Typically, the signal waveform in the power line looks like

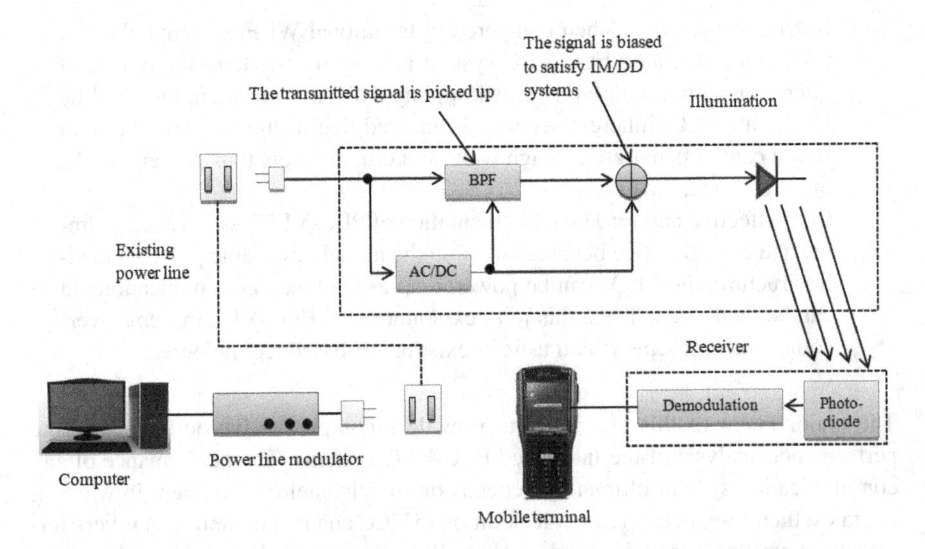

Figure 6.1: Intergrated PLC-VLC system.

the waveform as shown in Fig. 6.2. Much similar to intensity modulation, the signal which is intended to be transmitted is added to the cyclic waveform of the alternating current (AC). It is to be noted that the power supply originates from the already

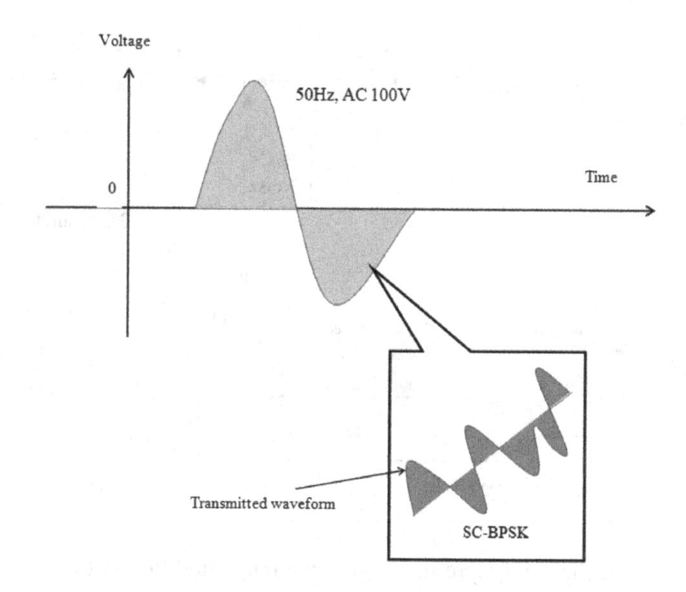

Figure 6.2: Transmitted waveform in VLC+PLC system.

installed power line that is present in the ceiling. After being rectified, the powerline is firstly given to the band pass filter (BPF) followed by the bias circuit. Then on the other side, it is fed to the 100 V AC waveform, which is the transmitted signal. This signal is an AC signal and taking into account the non-coherent characteristics of LED, where it can pass only a direct current signal; therefore, it is vital to enforce the biasing phenomena. In this manner, the power of the LED is varied in accordance with the transmitted signal. In general, the transmitted waveform's frequency is short enough to ensure that the variations in the transmitted signal remain unnoticed by the human eye. Thereupon, the light signal is intercepted by the photodiode that is present at the receiving end. Finally, the signal is demodulated to yield the desired output.

6.3 PLC CHANNEL MODELING

Figure 6.3 depicts the integrated PLC-VLC system which clearly illustrates that the signal which is transmitted from the PLC transmitter to the VLC receiver is drastically affected by the noise inherently present in both PLC and VLC channels. The PLC channel environment is highly frequency selective and time variant. It is very burdensome to model the frequency selective fading channel environment with time variance nature. The lack of flat-fading nature, huge chances for the occurrence of different kinds of interference and noise issues dominate, this eventually enablely, the PLC channel to emanate very harsh, thereby impeding the scope for the possibility to enable data communication. Therefore, in order to jointly exploit the advantages of both PLC and VLC, it a big prerequisite to model the harsh PLC channel

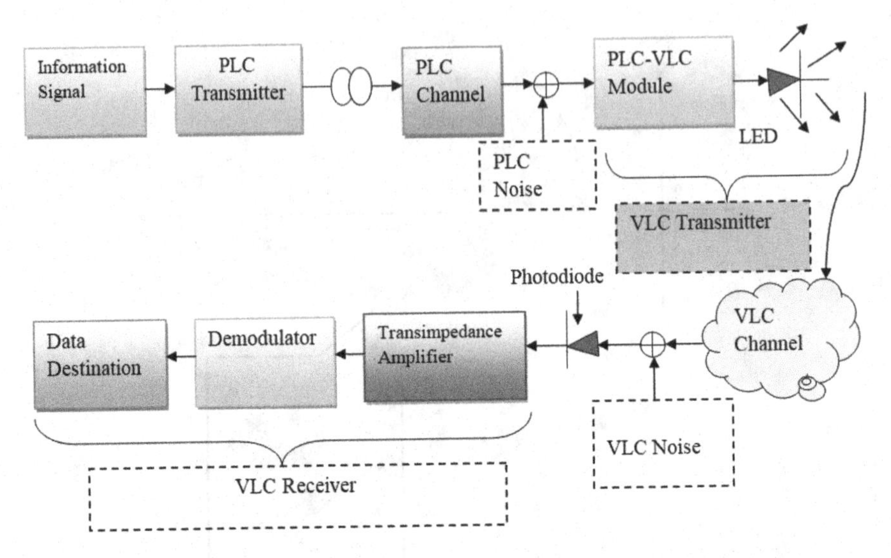

Figure 6.3: Pictorial representation of the integrated PLC-VLC system.

environment. Within this context, it is very important to perceive the impact of PLC channel on the performance of VLC channel and vice versa. On top of this, it is also vital to have immense knowledge on the behavior of the cascaded PLC-VLC channel modeling in order to render high data rate communication.

The time-variant and frequency-selective nature of the PLC channel environment enables the PLC channel to be interpreted as a random, time-varying channel with a frequency-dependent SNR over a certain bandwidth. There exist two different approaches for PLC channel modeling, i.e., the top-down approach and the bottom-up approach. In the former approach, the PLC channel is severely affected due to the multipath effects. This effect is a result of the emanation of several branches and mismatches of impedance that prevails within the PLC channel. Whereas, in the latter approach, the PLC channel can be regarded as a black box model, where an analysis can be generated from the yielded response. Conventionally, in order to fetch the transfer function of the PLC channel, this sort of approach utilizes the transmission line theory. In wireless communication, we are interested in modeling the multipath channel effect due to the multipath propagation environment; consequently, the PLC channel modeling pertaining to the scenario of multipath effects is the major emphasis of this chapter. The impulse response of the echo-based PLC channel with N different paths can be described by the following expression

$$\underset{PLC}{h}(t) = \sum_{l=0}^{N-1} \alpha_l \delta(t-\tau_l) \tag{6.1}$$

From the mathematical expression, as represented by equation (6.1), the channel coefficients specified by α_l, τ_l denote the attenuation and the paths echo delays respectively. Later on, Zimmermann and Dostert [584] proposed an echo-channel model with low-pass behavior which usually observed in practical networks. This model generally represents the combination of multipath propagation, frequency and length-dependence attenuation present in the power-line networks. Consequently, the channel model as described by the mathematical expression which is elucidated in (6.1) is characterized by means of a weighting factor β_l, an attenuation factor which is illustrated by $e^{-(\gamma_0+\gamma_1 f^{\alpha})d_l}$, and a delay factor which is characterized by $e^{-\frac{j2\pi f d_l}{v_P}}$ respectively. By grouping all these factors, the response of the PLC channel in the frequency domain is given by

$$\underset{PLC}{H(f)} = \sum_{l=1}^{N} \beta_l e^{-(\gamma_0+\gamma_1 f^{\alpha})d_l} e^{-\frac{j2\pi f d_l}{v_P}} \tag{6.2}$$

From the mathematical expression as exemplified by equation (6.2), the parameters γ_0, γ_1 and α signify the attenuation parameters which can be determined from the magnitude of the frequency response of the complex channel's transfer function. Based upon the information available from the impulse response, the path parameters β_l and d_l are determined and the speed of the light in the cable is denoted by v_p.

6.3.1 POWER-LINE NOISE

Extensive analysis can be reported from the literature where power line noise was evaluated. In general, the noise in the power line emerges due to the connectivity of the electric appliances to the power lines. Therefore, pertaining to the scenario of PLC, the statistical behavior of this man-made induced noise (i.e., the rapid association of different electric appliances to the powerline) is absolutely distinctive from that of the AWGN. Based upon the earlier experimental validations and measurements, the noise in the power line can be categorized into three classes: stationary continuous noise, cyclic stationary continuous noise and cyclic impulse noise which is synchronous to the mains. The first category of noise, i.e., the stationary continuous noise is not variant with respect to time, i.e., it is time-invariant. Additionally, the power of this noise is almost constant over time. The second class of noise as the name itself implies that this noise is repeatedly varying and also it is cyclically synchronous to the mains frequency, i.e., 50 or 60 Hz. The third category of noise comprises of impulses that occur at repetitive intervals much higher than the frequency of the mains, i.e., between 50 and 200 KHz. Generally, these impulses are generated by means of switch mode power supplies (SMPS).

The power-line noise that is characterized by means of the combination of the aforementioned categories of noise can be assumed to be cyclic-stationary additive Gaussian noise comprising of zero mean. Thus, for this narrow-band power-line noise, the variance of the noise waveform $n(t)$ which is dependent on both time and frequency is manifested as

$$\sigma^2(t,f) = \sigma^2(t)\,a(f) \tag{6.3}$$

where,

$$\sigma^2(t) = \sum_{i=1}^{N} A_i(t)\, sin(2\pi f_i t + \theta_i) \tag{6.4}$$

and

$$a(f) = \frac{e^{-af}}{\int_{f_0}^{f_0+W} e^{-af} df} \tag{6.5}$$

From (6.3), the instantaneous power of the noise is indicated by $\sigma^2(t)$, while, $a(f)$ interprets the power spectral density (PSD) of the noise which is normalized by means of total noise power corresponding to the range f_0 to f_0+W. Furthermore, from the mathematical expression as specified by equation (6.4), the total number of carrier waves is denoted by N, while the amplitudes, frequencies and phases are proclaimed by A_i, f_i and θ_i, respectively.

6.3.2 VLC CHANNEL MODELING

As discussed in Chapter 2 about VLC channel modeling, VLC channel can be described as a space between the installed LED and the photodiode. The illumination that is received by any point of the receiver plane comprises of both LOS component from the installed LED luminaires and reflections emanating from the walls, floor, ceiling and several other objects present within the indoor room environment. Since the LOS component is the direct path from the transmitting LED and the receiving photodiode, the channel response of the LOS contributions can be represented by means of Dirac pulses, while the reflections which can also be specified as the diffuse portion can be signified by means of an integrating sphere model. Thus, the generalized expression for the VLC channel transfer function comprising of both LOS components and the diffusion portion can be mathematically expressed as

$$\underset{VLC}{H}(f) = \sum_{l=1}^{N} \eta_{LOS,l} e^{-j2\pi\Delta\tau_{LOS,l}} + \eta_{DIFF} \frac{e^{-j2\pi f \Delta\tau_{DIFF}}}{1+\frac{jf}{f_0}} \tag{6.6}$$

From (6.6), the channel transfer function corresponding to VLC is denoted as $\underset{VLC}{H}(f)$. The parameters η_{LOS} and η_{DIFF} indicate the channel gains relevant to the LOS and diffuse signals, respectively. Meanwhile, the corresponding signal delays are specified by $\Delta\tau_{LOS,l}$ and $\Delta\tau_{DIFF}$, respectively. The frequency component f_0 in Hz enumerates the 3-dB cut-off frequency of the purely diffuse channel. Further, from equation (6.6), the LOS gain corresponding to the lth LED luminary can be formulated as

$$\eta_{LOS,l} = \frac{A_R(m+1)}{2\pi D^2} cos^m(\phi_l)\, cos(\psi_l) \tag{6.7}$$

From the LOS gain as expressed in equation (6.7), the parameter A_R in m^2 denotes the affective area of the receiver. The distance between the transmitter LED and the receiver photodiode is specified by D, and m represents the Lambert's Mode number which determines the LED's semi-angle at half-power illumination. Meanwhile, ϕ

and ψ determine the irradiance and incidence angles, respectively. The diffuse signal gain also depends on the effective area of the receiver as well as the properties of the room and is given by the following expression

$$\eta_{DIFF} = \frac{A_R}{A_{ROOM}} \frac{\rho}{1-\rho} \tag{6.8}$$

From equation (6.8), the area of the room is denoted by A_{ROOM} and the average reflectivity is described by ρ.

6.3.3 ANALYSIS OF CASCADED PLC-VLC CHANNEL

In cascaded PLC-VLC channel environment, the transfer functions as well as the noise components are superimposed. Nonetheless, from the practical measurement campaign carried out in [371], in terms of real-time implementations, the direct convolution of the channel is not feasible while dealing with practical scenarios. Additionally, in most of the realizations of cascaded PLC-VLC systems, it is assumed that PLC noise does not cross through the VLC domain. However, such kind of assumptions are proved to be wrong in [371]. According to [371], the LED supply that connects the PLC channel with VLC channel has a non-negligible effect on the overall cascaded system. It can even act as a transfer function as well [372]. Consequently, it is appropriate to introduce the factor $\delta(f)$ while computing the cascaded PLC-VLC channel transfer function. Particularly, the factor $\delta(f)$ determines the correlation between the input and output of the PLC-VLC system module. Predominantly, this factor $\delta(f)$ can clip, attenuate or amplify the noise signal present in the PLC channel. Thus, the cascaded PLC-VLC channel can be mathematically expressed as [371]

$$\underset{PLC-VLC}{H(f)} = \left(\delta(f) \times \underset{PLC}{H(f)}\right) \times \underset{VLC}{H(f)} \tag{6.9}$$

From (6.9), $\underset{PLC-VLC}{H(f)}$ denotes the overall cascaded PLC-VLC channel transfer function. While, $\underset{PLC}{H(f)}$ and $\underset{VLC}{H(f)}$ represent the individual transfer functions of PLC and VLC systems. Further, the mathematical expression as represented by (6.9) can be solved to attain

$$\underset{PLC-VLC}{H(f)} = \delta(f) \times \underset{PLC,VLC}{H(f)} \tag{6.10}$$

From (6.10), $\underset{PLC-VLC}{H(f)}$ stipulates the channel gain in the frequency domain. Predominantly, the convolution of the PLC and VLC channel gains in the time-domain results in multiplication of their channel gains in the frequency domain. Subsequently, by taking into consideration the channel gains of both PLC and VLC, i.e., $\underset{PLC}{H(f)}$, $\underset{VLC}{H(f)}$ as expressed by the equations (6.2) and (6.6), the resultant channel gain $\underset{PLC,VLC}{H(f)}$ can be formulated as follows:

$$\underset{PLC,VLC}{H(f)} = \sum_{l=1}^{N} \left(\beta_l e^{-(\gamma_0+\gamma_1 f^{\alpha})d_l} e^{-\frac{j2\pi f d_l}{v_p}} \right) \times \left(\eta_{LOS,l} e^{-j2\pi\Delta\tau_{LOS,l}} + \eta_{DIFF} \frac{e^{-j2\pi f \Delta\tau_{DIFF}}}{1+\frac{jf}{f_0}} \right) \quad (6.11)$$

Further, (6.11) can be solved and rearranged as follows

$$\underset{PLC,VLC}{H(f)} = \sum_{l=1}^{N} \left(\beta_l e^{-(\gamma_0+\gamma_1 f^{\alpha})d_l} e^{-\frac{j2\pi f d_l}{v_p}} \eta_{LOS,l} e^{-j2\pi\Delta\tau_{LOS,l}} \right) + \sum_{l=1}^{N} \left(\beta_l e^{-(\gamma_0+\gamma_1 f^{\alpha})d_l} e^{-\frac{j2\pi f d_l}{v_p}} \eta_{DIFF} \frac{e^{-j2\pi f \Delta\tau_{DIFF}}}{1+\frac{jf}{f_0}} \right) \quad (6.12)$$

Upon solving for the exponential terms, equation (6.12) can be reduced to

$$\underset{PLC,VLC}{H(f)} = \sum_{l=1}^{N} \beta_l e^{-(\gamma_0+\gamma_1 f^{\alpha})d_l} \left[\eta_{LOS,l} e^{-j2\pi f\left(\Delta\tau_{LOS,l}+\frac{d_l}{v_p}\right)} + \eta_{DIFF} \frac{e^{-j2\pi f\left(\Delta\tau_{DIFF}+\frac{d_l}{v_p}\right)}}{1+j\frac{f}{f_0}} \right] \quad (6.13)$$

Further upon exploiting the fact that $\frac{d_l}{v_p} >> \delta_{LOS}$ and $\frac{d_l}{v_p} >> \delta_{DIFF}$, equation (6.13) can be further solved to attain

$$\underset{PLC,VLC}{H(f)} = \sum_{l=1}^{N} \beta_l e^{-(\gamma_0+\gamma_1 f^{\alpha})d_l} e^{-\frac{j2\pi f d_l}{v_p}} \left[\eta_{LOS,l} + \frac{\eta_{DIFF}}{1+j\frac{f}{f_0}} \right] \quad (6.14)$$

Thereupon, in order to yield the final expression for the channel gain corresponding to the cascaded PLC-VLC system, equation (6.14) can be substituted into (6.10); therefore, the resultant PLC-VLC channel gain is given by [371]

$$\underset{PLC-VLC}{H(f)} = \delta(f) \times \sum_{l=1}^{N} \beta_l e^{-(\gamma_0+\gamma_1 f^{\alpha})d_l} e^{-\frac{j2\pi f d_l}{v_p}} \left[\eta_{LOS,l} + \frac{\eta_{DIFF}}{1+j\frac{f}{f_0}} \right] \quad (6.15)$$

Thus, from equation (6.15), it can be surmised that the cascaded PLC-VLC channel varies significantly with respect to the frequency. Furthermore, there are several parameters which are associated with PLC and VLC channels that influence the cascaded channel gain. Particularly, the PLC-VLC channel frequency response is affected by the number of PLC taps, reflections, i.e., NLOS components and the LED supply parameter designated by δ. In order to illustrate the effect of varying the number of PLC taps, along with reflections and the LED supply parameter, extensive simulations were carried out which emphasize that the frequency response of the PLC-VLC channel varies throughout the entire frequency band and is characterized by means of notches as well. Furthermore, it is clear that at some frequencies, the VLC channel even attenuates the PLC channel's frequency response. Upon increasing the number of PLC taps from 5 to 12, a deterioration in the frequency response of the cascaded PLC-VLC channel can be observed. Thus, from this analysis, it can be inferred that the cascaded PLC-VLC channel is affected by both PLC and VLC parameters [371].

6.3.4 STATE-OF-THE-ART RESEARCH EFFORTS ASSOCIATED WITH PLC-VLC CHANNEL MODELING

The work in [368] models and estimates the cascaded PLC-VLC channel by employing the non-parametric Welch windowed method. The size of the window that was employed in this method was a function of both time and frequency. Furthermore, the authors present an overview of both PLC and VLC channel models. The cascaded channel is statistically analyzed by means of non-parametric windowed Welch method. Experimental results were carried out in order to determine the frequency response of the cascaded PLC-VLC channel environment by taking into consideration both PLC and VLC parameters which include the number of PLC taps, LOS and NLOS components of the VLC channel, etc. Moreover, the cascaded PLC-VLC channel is affected by the propagation distance and the size of the room as well. Since the integrated PLC-VLC system is time-varying due to the varying impedance of the indoor power-line networks, it is vital to yield the random channel generator for the design of communication algorithm. In this regard, the research efforts in [446] propose a random channel generator for the cascaded PLC-VLC channel. The work in [371] reports a channel frequency measurement of a hybrid cascaded PLC-VLC system. From the set of measurements that were carried out in the laboratory environment, it can be surmised that both the power-line activities and the LED supply significantly influence the channel response of the hybrid PLC-VLC system. The authors in this work give a detailed analysis of overall characterization, model and spectral analysis of the hybrid PLC-VLC channels. The simulation results affirm that the effect of LED supply on both PLC and VLC channel is analyzed. The channel frequency response of the hybrid PLC-VLC system is determined by taking into consideration both PLC and VLC mutual noise scenarios. The narrow-band PLC frequencies result in observable non-linearity with that of the VLC channel, while, at the broad-band frequencies, a linear activity is attained.

6.4 PERFORMANCE OF OFDM-BASED PLC-VLC SYSTEM

A variety of distinguished features offered by PLC technology such as its mature coded modulation technique [544, 578], channel estimation approach [141, 142] and its noise cancellation method [312, 313, 543], enabled PLC to turn out as the most pertinent solution for its combination with VLC with the major incitement to serve as a backbone for VLC network. The power line not only supplies sufficient amount of power to the LEDs of the VLC system, but also contemporaneously acts as a signal source for the VLC system. Thus, this sort of integration of the two communication methods results in the emergence of a cost-effective communication system that exhibits the potential to realize a new generation high data rate transmission system. As discussed earlier, the first integration of VLC and PLC as reported in the literature dates back to 2003 [273], where the authors exploit a relatively low rate data transfer. Inspired by this integration, PLC exploiting other modulation formats like FSK [432], OOK [365, 410], PPM [62] and so on have emerged. Later on, PLC adopted the radical multicarrier modulation format like OFDM due to a myriad reasons, like its ability to counteract the harsh frequency-selective time-varying PLC channel effects, potential to deliver high data rate communication as well as to combat the effects of ISI, etc. Few such significant works that employ this high data rate transmission scheme like OFDM can be reported from [29, 272]. Furthermore, the research contributions in [143, 454] analyzed the performance of broadcasting integrated PLC-VLC systems and signified that the broadband PLC-VLC systems are essential to furnish the high data rate communication. Figure 6.4 elucidates the schematic representation of the hybrid PLC and VLC system employing OFDM. The schematic representation clearly elucidates that at the power-line modulator, the

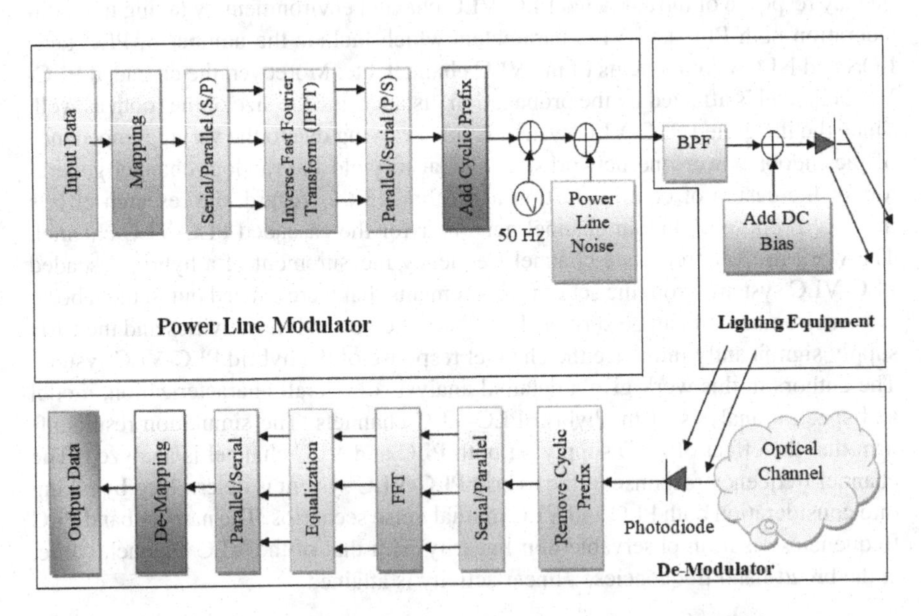

Figure 6.4: Integrated PLC-VLC system exploiting OFDM.

input signal is modulated by employing the methodology of the OFDM technique. As delineated from the figure, at the input side, the incoming stream of data is modulated by using QPSK, and then these sets of complex symbols are transmitted in parallel with the help of serial to parallel converter. Thereupon, these frequency domain symbols are converted into time-domain symbols by means of IFFT and then these time-domain samples are converted into a serial signal with the help of a parallel to serial converter (P/S). Further, in order to combat the effects of ISI, a suitable amount of cyclic prefix or guard interval is added followed by addition of the attained signal to the cyclic waveform of the alternating current. In this manner, the generated signal from the power-modulator block is transmitted to the LED lighting unit by means of a power line. At the LED lighting unit, the OFDM modulated signal is picked up by the band pass filter (BPF). In case of VLC, IM and DD are the most viable modulation format, and the data is modulated by intensity modulating the LED at a rapid pace in a manner that the switching speeds of the LEDs is imperceptible by the human eye. Accordingly, in order for the transmitted signal to comply with the requirements of IM and DD, it is vital for it to be both a real and non-negative valued signal. Therefore, in order to satisfy a positive-valued signal transmission, some amount of DC bias is added prior to intensity modulation via the LED. Thus, at the LED lighting terminal, the power of each LED is varied based upon the waveform of the modulated OFDM signal.

At the receiving terminal, this light signal after propagation through the VLC channel environment, is intercepted by means of a photodiode, where this light signal is converted into a suitable electrical signal in order to facilitate the reception of the transmitted data. Furthermore, inverse operations are performed at the receiving end, where the cyclic prefix is removed, followed by conversion of this serial signal into parallel signal. Thereupon, the received time-domain signal is converted into frequency domain signal by means of FFT and then the appropriate de-mapping scheme is implemented in order to fetch the final output. Furthermore, the performance of this integrated system was evaluated on narrowband OFDM powerline. It is evident from the simulation results as emphasized in [272], the narrowband power-line frequency in the range of 10 to 450 KHz is divided into 32 carriers, while the cyclic duration of the alternating current is 50 Hz.

When amalgamating new technologies with VLC, it is an important prerequisite to maintain the fundamental functionalities of VLC. Assurance of simultaneous illumination and communication is the main purpose of VLC. Therefore, pertaining to illumination, which is the foremost task of VLC, it is important to take into consideration flickering and dimming of the illuminating sources. Thus, when VLC is interfaced with PLC, it is appropriate to discuss the effect of flickering of the lighting equipment. As shown in Fig. 6.4, the lighting equipment emits the light that is varied in accordance to the OFDM signal including the noise that is present in the power line. Consequently, it is essential to know the effect of flicker in such integrated PLC and VLC system. Flicker refers to the extent of the fluctuations in the levels of brightness that can be perceived by the human eye. The limit value of flicker that is caused by the electrical equipment is postulated by IEC1000-3-3, which is particularly an international standard that specifies the values of the fluctuations in the maximum voltage and flicker for equipment that has a rated input current per phase of not more than 16 A. According to this standard, the maximum relative voltage should not be more than 4%, i.e., 0.04. Accordingly, in order to verify whether the integration of PLC with VLC affects the functionality of the lighting equipment, simulations were carried out illustrating the relationship between the average SNR on the narrowband power line which is indicated by SNR_{pl} versus the maximum changes in the relative voltage. The simulation results of this work emphasize the fact that, with the increase in the average SNR on the power line, a rapid decrease in the maximum relative voltage can be observed. The primary reason behind this fact is that flicker of the LED lighting is dependent on the noise on the power line. Also, even though the average SNR on the power line is 0 dB, the maximum relative voltage does not exceed 0.04. Eventually, this analysis confirms the fact that the influence of PLC with VLC does not affect the functionality of the lighting system of VLC. Hence, the integrated PLC and VLC system can be envisaged to render seamless transfer of data for several indoor applications. Furthermore, in order to illustrate the performance of the integrated narrowband-OFDM-PLC and VLC system, extensive simulations were carried out in order to highlight the BER and SNR_{pl} (SNR_{pl} corresponds the SNR on the power lines) performance against each SNR_{vl}. SNR_{vl} corresponds to the SNR on the visible light wireless, where the simulation results confirm that upon increasing the SNR_{vl}, the BER performance of the entire integrated narrowband-OFDM-PLC+VLC system corresponds to the BER performance at the power line. For instance, when the number of subcarriers chosen are 2 and 4, then the BER performance of the entire integrated system is dependent on the BER performance at power line when the SNR_{vl} is 34 and 40 dB, respectively.

6.5 ON THE PERFORMANCE OF DWT-BASED PLC-VLC SYSTEM

The contribution in [46] proposes a discrete wavelet transform (DWT)-based OFDM-PLC-VLC system in order to tune the physical parameters in accordance with the channel characteristics of a hybrid PLC-VLC system. Both the theoretical and the simulation results confirm that the BER performance of DWT-OFDM-based PLC-VLC system is superior when compared to the OFDM-based PLC-VLC system. The schematic representation of a DWT-OFDM-based PLC-VLC system is delineated in Fig. 6.5. The orthogonal multicarrier modulation technique like DWT significantly enhances the reliability of the input data for the PLC-VLC transceiver. When compared with the traditional Fourier transform-based OFDM system, the wavelet transform excellently circumvents the effects of ICI and ISI, thereby optimizing the overall performance of the data-driven indoor communication system. As depicted in the DWT-OFDM-PLC-VLC transceiver, at the transmitter input side, the incoming huge stream of serial data is fed to a serial/parallel converter in order to facilitate parallel transmission. Thereupon, a suitable M-ary constellation technique is employed in order to map these data sets into symbols before getting applied as input to the inverse DWT (IDWT) block.

The noteworthy feature of the wavelet transform is that much similar to the traditional FFT-based OFDM system, the data is characterized in the frequency domain where the corresponding symbols overlap in the time-domain due to various scale resolutions in the DWT. Thus, due to the fact that wavelets comprise of both frequency and location (i.e., time), it avoids the necessity of cyclic prefix in time which is mandatory in case of a traditional OFDM system. This stems out as an

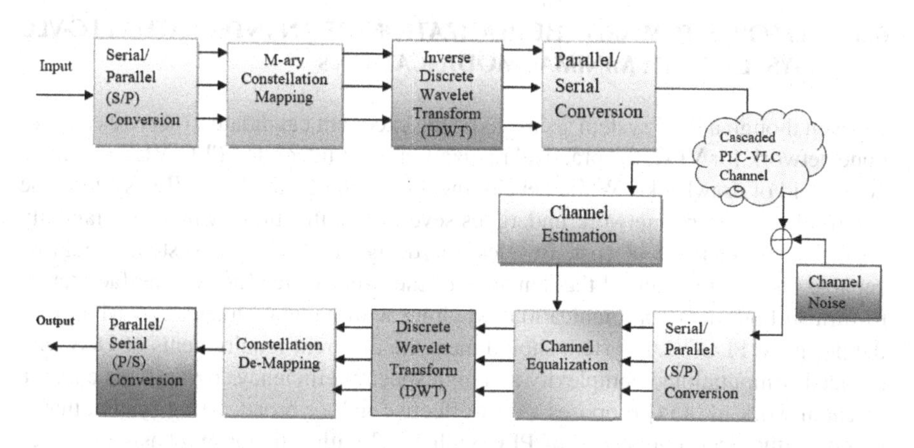

Figure 6.5: Schematic representation of DWT-OFDM-based PLC-VLC system.

important advantage, since the precious resources like the bandwidth and the additional requirement in the total transmitted power are saved. Moreover, the user capacity is enhanced where the bandwidth is effectively exploited for data transmission and redundancy is avoided. Consequently, the transmitted signal is passed through the cascaded PLC-VLC channel where both PLC and VLC noise is added to it. The serialized transmitted signal at the receiving end is transmitted in parallel with the aid of a serial to parallel converter. Furthermore, in order to counteract the effect of channel impairments on the transmitted signal, appropriate equalization techniques can be enforced. The channel state information can be estimated by means of channel estimation algorithms like the LS, MMSE, interpolation techniques, etc.

Thereupon to the received equalized signal, DWT is enforced by exploiting analysis filter banks. Finally, the demapping technique is utilized to ensure reliable recovery of data. The performance of the DWT-OFDM-based PLC-VLC system is evaluated by means of simulation analysis. Furthermore, its performance in terms of BER and $\frac{E_b}{N_0}$ is compared with the traditional FFT-based OFDM system over a cascaded PLC-VLC channel, where the simulated result affirms that the DWT-OFDM system outperforms DFT-based OFDM system.

6.6 RELATED WORK ON THE COMBINATION OF PLC WITH VLC

This section highlights several significant research efforts that have been carried out in the literature to determine the expediency of the joint amalgamation of PLC technology with VLC. In order to determine the significance of the integrated PLC-VLC system, this section categorizes the research advancements carried out in the major aspects which include state-of-the-art aspects relevant to the exploitation of multiple access schemes to render multiuser services, realization of the integrated system with minimal complexity, etc.

6.6.1 EFFORTS TOWARD THE REALIZATION OF AN INTEGRATED PLC-VLC SYSTEM WITH MINIMAL MODIFICATIONS

Even though a PLC system seems to be the excellent candidate to serve as a backbone network for VLC, the practical realization of an integrated PLC-VLC system is not devoid of drawbacks. While developing a fully integrated PLC-VLC system, the layout of the original network undergoes several modifications, which substantially results in the increase of cost. Besides, the designed PLC-VLC system should be capable enough to meet all the demands of the users which include the facilitation to furnish both data communication aid along with navigation services. Moreover, the designed PLC-VLC system should meet the different requirements of QoS with reduced computational complexity and high spectral efficiency. Toward this end, the research work in [454] proposes a cost-effective indoor broadcasting system that is based on the deep integration of PLC with VLC with a major emphasis to reduce the modification of the layout of the original network as well as to ensure a reduced network protocol complexity. From this work, a two-lamp network demo can be witnessed where the performance of the proposed system was evaluated in the laboratory. Furthermore, it can be reported that the authors in this work develop an efficient indoor broadband broadcasting PLC-VLC system where the signal in the power line is amplified and forwarded to the LED without following any decoding phenomena. All LEDs belonging to one group transmit the same signal. In this manner, the developed broadcasting network can be regarded as a homogeneous network which is ably characterized as a single frequency network (SFN).

Consequently, this sort of homogeneous network overcomes the necessity to rely on complicated networking switches for all the devices wandering between different LED lamps. Furthermore, the developed integrated PLC-VLC system prototype comprising of two LEDs with an aggregate payload data rate over 48 Mbps within 8 MHz bandwidth was validated experimentally. The experimental results infer that the proposed broadcasting system inherits the benefits of both technologies, i.e., PLC and VLC. Thus, the proposed indoor broadcasting PLC-VLC system turns out to be an appealing solution which can be implemented in hospitals, shopping complexes, stadiums, auditoriums, etc. Most of the research works deploy an integrated PLC-VLC system with basic structure as delineated in Fig. 6.6. From this integrated PLC-VLC system, it is clear that the PLC modem is an important component which plays an active role to couple the signal either from the Ethernet or data source to the power line. Later on, this signal is transformed from the data source to the modulation scheme and matches with the characteristics of the PLC system in order to facilitate for PLC transmission. Meanwhile, at the LED luminary, it is vital to exploit another PLC modem along with DF module in order to facilitate the remodulation of the transmission signal in a manner to reconcile with the requirements of VLC system. From the traditional PLC-VLC network, it is a pre-requisite to enforce two necessary modifications, out of which the first modification is to deploy a PLC modem from the data source to the power line. While the second modification is to install an additional PLC modem as well as the DF module at the LED luminary which poses as a great modification to the existing layout of the power line. Thus, in order

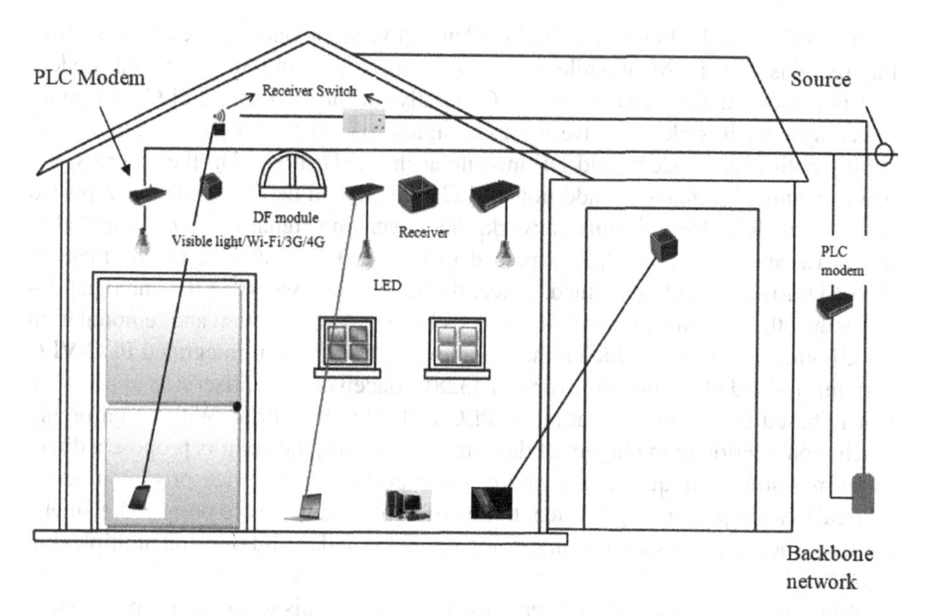

Figure 6.6: Integrated PLC-VLC system.

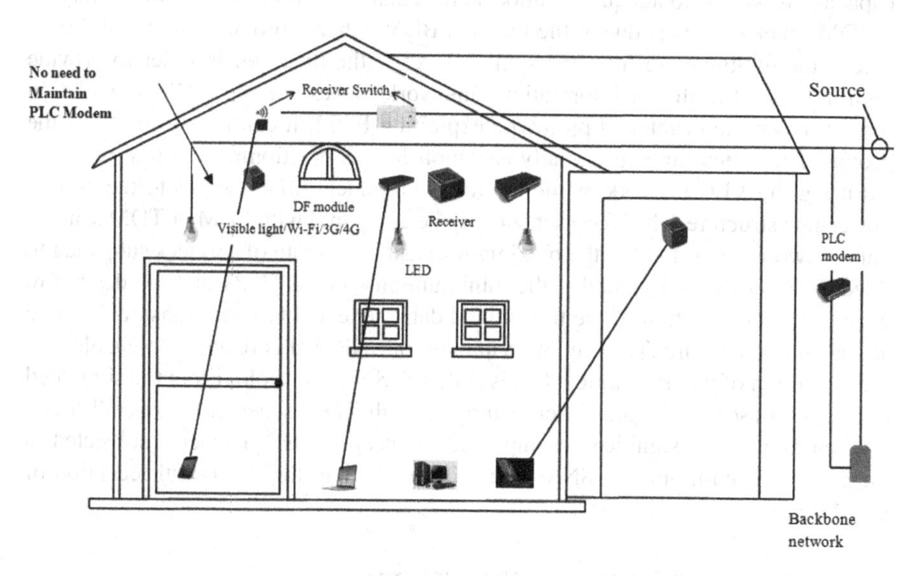

Figure 6.7: Integrated PLC-VLC system with minimal modifications.

to overcome these modifications, it is essential to develop integrated PLC-VLC systems with minimal or no modifications. Toward this end, the integrated PLC-VLC system without incurring any modification to the original network is shown in Fig. 6.7.

As outlined in the figure, the PLC modem between the data source and the power line remains as it is. Meanwhile, the major goal is to abandon the second modem that is present at the LED luminary. Generally, in the traditional PLC-VLC integrated system, in order to drive the PLC signal toward the LED, it is required to deploy both PLC modem, and DF module at the LED mode. On the contrary, the new structure eliminates the additional PLC modem, and the DF modem is replaced with AF module. Here, in this network, the power-line signal is simply coupled to the LEDs and the AF module is placed in the LED lamps; thus, in this manner, the LED-driven signal is attained. Since, the AF module works in the analogue domain, the other intermediate signalling steps like the demodulation and remodulation are eliminated, thereby reducing the complexity of the overall integrated PLC-VLC system. Toward this end, the work in [328] conceives a multiservice application that is based upon the integration of PLC and VLC networks. Without enforcing much modification to the layout of the original network, the authors propose a direct retransmission technique. Furthermore, for enabling multiservice provision, more particularly the positioning service, three different schemes were proposed, namely, the frequency domain scheme, time domain scheme and the bit division multiplexing (BDM) scheme.

While dealing with multiple services, it is a prerequisite to ensure positioning support as well. Furthermore, in order to attain an enhancement in multiuser channel capacity, as well as to adequately allocate the channel resources, it is vital to rely on a BDM scheme. This is due to the fact that BDM scheme provides the flexibility to extend the multiplexing from the symbol level to the bit level. In order to provide high data rate transfer of information, this work makes use of OFDM modulation format. From the developed prototype exploiting FDM, it can be affirmed that the proposed schemes can substantially accomplish both positioning and multiservices aid using the VLC network, while not imposing much modifications to the current power-line structure. The BDM scheme can be compared with FDM or TDM scheme only in two aspects. Firstly, the maximum transmission rate of services subjected to fixed SNR levels and, secondly, the minimum amount of SNRs that are desired to enable perfect reception of the transmitted data subjected to fixed transmission rate are taken into account. The minimum amount of SNR that is required to enable perfect reception of the transmitted data is called as SNR threshold. From the simulated results, it is ascertained that when compared with TDM scheme, the overall transmission rate can be significantly improved by means of BDM scheme subjected to the differentiated minimum SNRs necessitated to yield the successful reception of BDM scheme.

6.6.2 ASSURANCE OF MULTIUSER SUPPORT

In recent times, a PLC network seems to be the best choice for its integration with VLC systems, since it can utilize the existing lighting infrastructure of the already available illumination fixtures. Added to this, the PLC network can facilitate high data rate communication that is sufficient enough for indoor data links [83]. The combined characteristics of the integrated PLC and VLC systems can be reported from a list of earlier findings [254, 323, 324, 367]. Its potential to render multiple access support is also confirmed and reported in several works. The research contribution in [326] proposes a multicarrier integrated PLC-VLC system that is proficient enough to impart indoor high-speed downlink access support with the aid of a symbiotic relationship between the PLC and VLC technologies. The noteworthy feature of this contribution is that the proposed downlink PLC-VLC system alleviates the problem of high PAPR by incorporating the SO-OFDM mechanism at the VLC transceiver. Furthermore, different subcarrier allocation schemes were proposed in order to fully exploit the frequency selectivity of the integrated PLC-VLC channels as well as to enable the multiuser and multitransmitter diversity. The other important aspect of this research work is that the widely employed multiple access schemes like the OFDMA and TDMA were compared for the proposed integrated PLC-VLC system. Corresponding to the aforementioned multiple access schemes, several polynomial-time subcarrier allocation algorithms were proposed. Pertaining to the multiuser scenario, OFDMA-based integrated PLC-VLC system outperforms TDMA-based PLC-VLC system only at the expense of increase in the computational complexity.

In spite of VLC exhibiting superior performance when compared to RF-based wireless communication in terms of assurance of unlicensed, unregulated spectrum and spatial reuse, similar to RF-based wireless communications systems, it does not completely avoid the interference problem. Even though the light signal does not penetrate through the opaque objects like the walls, etc., the coexistence of multiple LED luminaires within a close proximity inside the room environment with the primary motive to furnish uniform levels of illumination stems out as the major source of interference. The primary reason for the emergence of this interference is the overlapping of the emitted light from all the neighboring LED luminaires. As a result of this overlapping, the overall performance of the VLC system degrades severely. Thus, in this regard, it is very important to maintain coordination among different LED luminaires with the aid of backbone networks. The coordinated VLC system seems to be the best choice to substantially improve the signal-to-interference-plus-noise ratio (SINR) at the user by up to 30 dB when compared to an uncoordinated VLC system [323]. PLC turns out to be the most pragmatic approach as a backbone network, since it can leverage on the existing infrastructure. Besides, the modern broadband PLC systems deliver data rates in the order of up to several Mbps or Gbps sufficient enough for indoor applications and a few such PLC-Wi-Fi systems can be demonstrated in [83, 559].

The integration of VLC and PLC dated back to the mid-2000s [29, 273]. Pertaining to the coordinated role of PLC in a hybrid PLC-VLC system, the PLC modem is connected to the outside access network to play a dominant role to serve as a data source for the VLC transceiver. Additionally, in recent times, from a list of earlier findings [35, 322, 561], it can be shown that PLC acts as a central coordinator among multiple LED luminaires. Taking into consideration the harsh channel environments of PLC, in order to mitigate the frequency-selective nature of the PLC channel, PLC systems employed OFDM. However, the major limitation that is associated with OFDM to render high data rate communication is that the time-domain signal is prone to high PAPR. As seen in the earlier chapters, the signal distortion emanates due to the advent of high PAPR in OFDM system. To alleviate this issue, several techniques for PAPR reduction have emerged. However, while adopting any PAPR reduction techniques for optical domain, it is vital to take into consideration the real and non-negativity constraints of the time-domain signal. Toward this end, SO-OFDM seems to be the better option to reduce the amount of PAPR in optical OFDM systems. As discussed in Chapter 3, SO-OFDM facilitates the transmission of subcarriers over the subsets of LEDs belonging to a luminaire, where spatial summing is incorporated to yield the entire OFDM signal. The integrated hybrid PLC-VLC system which exploits SO-OFDM technique [326] and which intends to facilitate downlink optical wireless access support for indoor applications will be discussed below.

The schematic representation of the integrated PLC-VLC system exploiting SO-OFDM in order to ensure the dual functionality of overcoming the harsh frequency-selective PLC channel environment and high PAPR of the time-domain signal is delineated in Fig. 6.8. Also, the block diagram depicts a downlink scenario dedicated to one user. As elucidated in the figure, the OFDM signal transmission in both the PLC and VLC links should fulfill the Hermitian Symmetry criteria in the frequency domain. Firstly, at the PLC transmitting terminal, the input data stream is mapped by exploiting a complex constellation like M-QAM and then followed by transmission of these sets of symbols in parallel with the aid of serial to parallel converter. Similar to traditional optical OFDM system, in order to attain a real valued signal, the input to the IFFT transform is constrained to satisfy Hermitian Symmetry criteria. Then these sets of real time-domain symbols are transmitted as one composite signal by making use of a P/S converter. Finally, before getting added to the cyclic waveform of alternating current, a suitable amount of cyclic prefix is appended. Thereupon, this signal is propagated through the PLC channel where some amount of PLC noise is added. Thus, the main functionality of the PLC modem is to broadcast this OFDM modulated signal to every LED luminaire comprising of independent information-carrying subcarriers denoted by N_p.

SO-OFDM methodology is applied at the VLC-hop where the total number of available subcarriers at VLC is signified by N_v. At the kth LED luminaire, the process of remodulation takes place where N_k of the received PLC data symbols are remodulated onto the N_k subcarriers out of the available N_v VLC subcarriers. The remaining unused subcarriers, i.e., $N_v - N_k$ are set to zero. Note that the aforementioned downlink system is intended to render services to N_U users located in the same indoor room environment comprising of N_L luminaires. With the coordination among these N_L LED luminaires, the N_U users are served. From this figure, it can be affirmed that the PLC plays an active role to serve as a backbone network that not only supplies data to the multiple VLC-equipped LED luminaires, but also maintains coordination among them. Furthermore, these LED luminaires function as the full-duplex relay which significantly processes this received PLC signal and finally forward it to the indoor users through VLC link. The principal reason behind

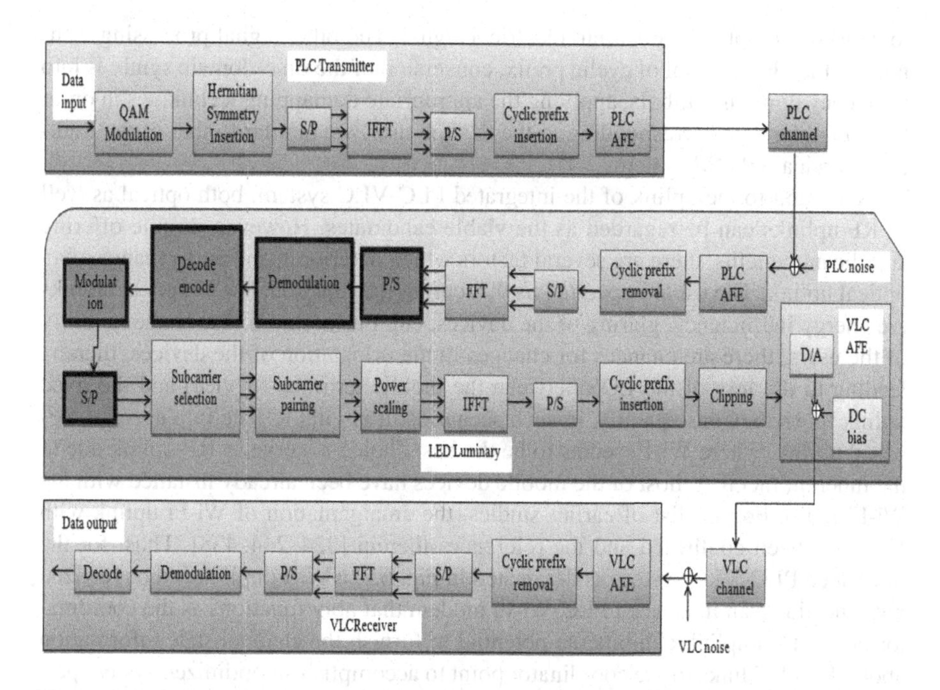

Figure 6.8: Schematic representation of integrated PLC-VLC system exploiting SO-OFDM emphasizing a downlink scenario between one LED luminaire and a single user. The bold blocks are not present when the LED luminaire operates in the amplify-and-forward mode.

applying SO-OFDM methodology across different multiple luminaires is that each LED luminaire emits only a subset of data symbols from the received PLC signal; this in turn plays a major role to limit the amount of PAPR of the OFDM signal at each LED luminaire.

Inherently, the subsets of subcarriers at different LED luminaires are disjoint following the condition:

$$\sum_{k=1}^{N} N_k = N_p \tag{6.16}$$

Furthermore, in order to fully exploit the frequency diversity of both PLC and VLC channel, a subcarrier pairing approach is implemented across each LED luminaire to accurately match the incoming subcarriers with the outgoing subcarriers. Amplify-and-forward (AF) and decode-and-forward (DF) are the widely employed operating modes for the LED luminaire relays. Generally, in case of the AF mode operation, the LED luminaire demodulates the PLC signal followed by scaling of the selected subcarrier signals and then applies the subcarrier pairing phenomenon to remodulate this signal. Even the DF-mode LED luminaire also decodes the received signal. Furthermore, at the VLC receiver terminal of the user, a photodiode is inherently present to convert the optical signal into electrical signal. The other signal processing techniques, like the removal of cyclic prefix, conversion of the time-domain symbols into frequency domain symbols, applying the appropriate demapping techniques in order to extract the data corresponding to the desired user, are similar to that of the traditional optical OFDM system.

Pertaining to the uplink of the integrated PLC-VLC system, both optical as well as RF uplinks can be regarded as the viable candidates. However, despite offering significant benefits, there are several factors which deteriorate the performance of an optical uplink. The usual impediments that curtail the functionality of optical uplinks are energy inefficiency, glaring of the devices, etc. Furthermore, due to the mobility of the users, there are chances for changes in the orientation of the devices, thereby leading to the loss of the LOS between the mobile terminal (device) and the fixed uplink receiver. Consequently, in such scenarios, it is vital to take into account RF-based uplink, where Wi-Fi seems to be the best choice to serve as RF-uplink due to the inherent fact that most of the mobile devices have been already installed with the Wi-Fi radio. From a list of earlier studies, the amalgamation of Wi-Fi uplink with VLC has been confirmed and the references therein [224, 244, 438]. Thus, for the integrated PLC-VLC system as delineated in Fig. 6.9, a Wi-Fi uplink can be realized with the aid of an interfaced PLC-Wi-Fi modem that ably functions as the coordinator point. This uplink exhibits the potential to furnish the channel state information about the VLC links to the coordinator point to accomplish an optimized system performance.

Several works illustrate the performance analysis of cascaded PLC-VLC systems. The research contribution in [367] demonstrates a cascaded PLC-VLC channel, in which the PLC channel is employed to act as the backbone for the VLC channel. This cascaded system is employed to ensure a full link transmission. In a cascaded PLC-VLC system, a PLC modem plays a vital role to deliver data to the VLC system. In this work, QPSK combined with OFDM was employed as a suitable modulation format for the PLC link. Whereas, CSK modulation format was deployed for the VLC link to transmit the information. Furthermore, behavior of the cascaded channel was analyzed and the attained variance was also verified by means of simulation. For different multiple combinations of the variances of the two channels, the BER performance of the cascaded system was evaluated.

The aforementioned works on integrated PLC-VLC systems, as discussed earlier [254, 323, 326, 367], exploit orthogonal multiple access schemes for the integrated PLC-VLC systems to render services to multiple users. Contrarily, the recent research contribution in [310], for the first time, exploits the non-orthogonal multiple access scheme, i.e., NOMA to ensure efficient utilization of the resources. The significance of NOMA is that it allows the users to easily and effectively utilize both time and frequency resources, thereby leading to an enhancement in the spectral efficiency. The authors in [310] introduce the concept of NOMA in an integrated PLC-VLC system as well as analyzes the sum rate. Furthermore, with the primary motive to accomplish fairness among the users, a proportionality fair (PF) scheduling strategy was proposed. In order to provide seamless indoor coverage, the integrated

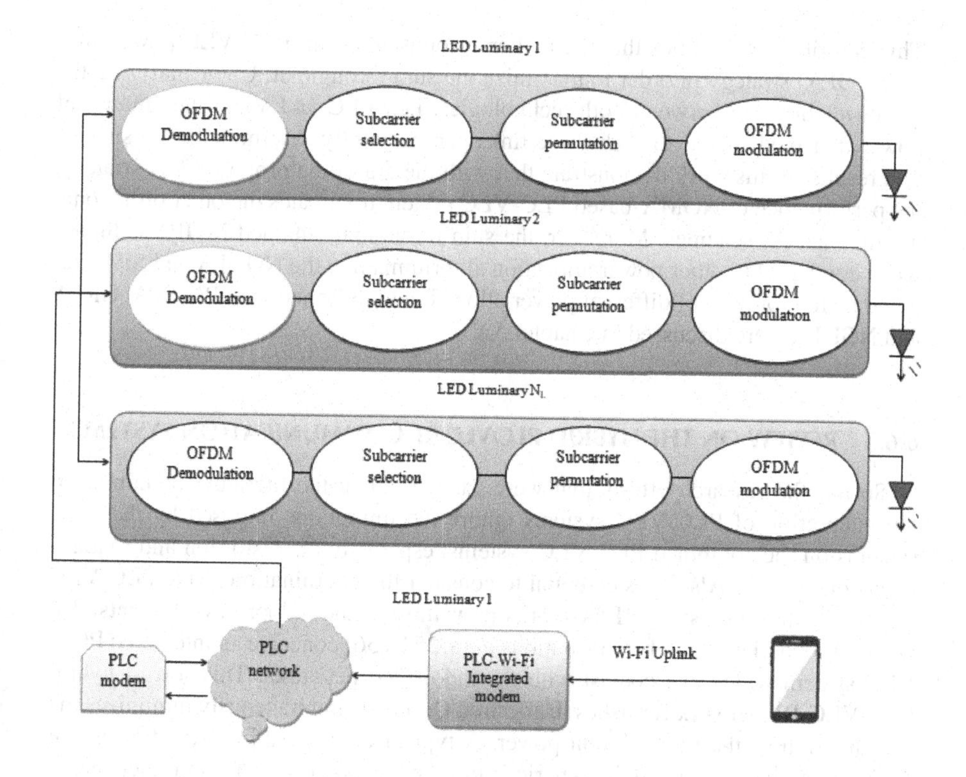

Figure 6.9: Interpretation of uplink scenario in a integrated PLC-VLC system.

PLC-VLC system with four LEDs was considered, and to maximize the sum rate while considering the fairness among the users, an optimal power allocation algorithm was taken into account to determine the optimal solution by transforming the non-convexity of the original formulated model into a convex programming. The simulation results of this work portray the superiority of the optimal power allocation algorithm (OPA) in a NOMA-based integrated PLC-VLC system. These results depict the performance of different power allocation algorithms in a NOMA-based PLC-VLC system in terms of achieved sum rate with respect to the increase in the number of users. The proposed OPA algorithm of NOMA in PLC-VLC system is compared with the other power allocation algorithms like the GRPA, FPA with power allocation factors, i.e., $\alpha = 0.3$ and 0.4, respectively. Among these three power allocation algorithms, the proposed OPA algorithm exhibits a superior performance in terms of achievable sum rate. Moreover, with the increase in the number of users, the sum rate of the aforementioned power allocation algorithms increases highlighting the superiority of NOMA. This is valid because NOMA bestows the flexibility to efficiently exploit both time and frequency resources.

A much similar work that is related to the exploitation of the robust multiple access scheme like NOMA for an integrated PLC-VLC system is reported in [167].

This contribution signifies that the authors proposed a joint PLC-VLC power allocation (JPA) strategy in order to maximize the sum throughput. Upon analyzing the power-relationship between both technologies, i.e., PLC and VLC, the amount of power that is allocated to both these links can be jointly determined. The simulation results of this work demonstrate that with the aid of the proposed JPA strategy, the performance of NOMA-based PLC-VLC system dominates the other orthogonal multiple access schemes. Moreover, the sum throughput attained by JPA is higher when compared to other power allocation algorithms like the NGDPA and SPA (detailed methodologies of different power allocation algorithms like FPA, SPA, GRPA and NGDPA were discussed in Chapter 5.).

6.6.3 REVIEW ON THE HYBRID PLC/VLC/RF COMMUNICATION SYSTEMS

So far, the research efforts that were carried out in the literature pertaining to the integration of PLC-VLC systems ignore the limitations imposed by the VLC systems on the combined PLC-VLC systems, especially the distortion and requirements of lighting. Also, it is essential to consider the coordination of the PLC-VLC systems with the existing RF networks prevailing in the indoor environments. Toward this end, the research contributions in [8,252,256] conceive an integrated PLC-VLC system which can coexist with the indoor RF networks. This sort of hybrid PLC-VLC-RF network furnishes the desired QoS and simultaneously minimizes the consumption of the transmission power. A typical cascaded PLC/VLC/RF system depicting an indoor downlink scenario is outlined in Fig. 6.10. From this schematic representation, it is evident that the source is transmitting data to a destination terminal by means of two parallel communication links. The RF communication channel forms the first communication link between the source and the destination; while the second communication link is determined as the cascaded link comprising of a PLC channel originating from the PLC source to the VLC transmitter, as well as the VLC channel emanating from the VLC transmitter to the destination. It is assumed that the destination node consists of multiple signal reception interfaces that facilitate the establishment of concurrent connection to both wireless transmitters, thereby expediting the destination to amalgamate the available resources.

The work in [256] proposes the combination of hybrid PLC/VLC system with RF system in order to reap the benefits resulting from both systems. Specifically, this cascaded PLC/VC/RF system operates in indoor room environment. In order to correlate the transmissions of RF network with that of the PLC/VLC system, this work evaluates the power allocation in two links, namely, the cascaded PLC/VLC link and the RF communication link. In this work, the total transmission power is constrained by means of a maximum allowable power budget. Furthermore, in order to maximize the total achievable data rate at the destination terminal, the authors developed an algorithm that optimally allocates power equally among both cascaded PLC/VLC link and RF communication link. In order to quantify the attained data rate gains, this work compares the performance of the proposed system with the conventional RF communication systems for the same amount of transmission power. A hybrid

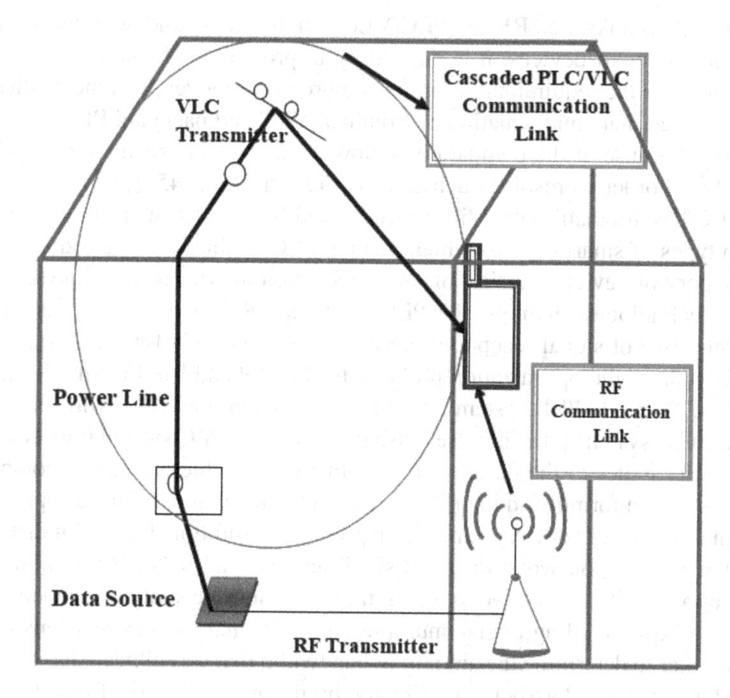

Figure 6.10: Intergrated PLC-VLC-RF communication system.

PLC/VLC/RF system outperforms the RF only communication system. The other major contribution which extends this work has been reported in [252] where the authors carried out investigations on the power allocation problem of a hybrid system composed of a cascaded PLC-VLC link which works in parallel to the RF wireless link. The transmission power minimization problem was formulated and converted into a convex form in order to facilitate the effective solving capability. The simulated result implies that the transmission power of the hybrid PLC-VLC-RF system is 5 times less than that of the single transmitter RF system. The difference between the works portrayed in [256] and [252] is that Kashef et al. [256] considered the rate maximization problem to be constrained by the maximum allowable transmission power and the decodability constraint. Moreover, this rate maximization problem was solved for the scenarios pertaining to non-frequency selective PLC and VLC channel environment. Whereas, the work in [252] takes into account the problem of minimization of transmission power under a general QoS constraint. Additionally, more generalized PLC and VLC channel models were considered in order to ensure a significantly improved performance of the hybrid PLC/VLC/RF system.

In the hybrid PLC/VLC/RF systems as discussed in the aforementioned works [252, 255, 256], a high data rate communication was possible by facilitating the mobile nodes to avail the services simultaneously from both RF and cascaded PLC/VLC communication links. Primarily, the mobile nodes were in close association with both communication links, i.e., RF and PLC/VLC. On the other hand, with the substantial growth in the smart devices, it is mandatory to provide moderate data rates which in turn increase the requirement of more resources. Moreover, in some studies it has been perceived that amalgamation of fronthaul VLC and backhaul PLC seems to be ineffective because of the comparatively low data rate reinforcement of PLC. Consequently, in order to resolve such aspects, Yeduri et al. [457] proposed a hybrid PLC/VLC/RF fronthaul with a fiber based wired backhaul system. In general, there are two types of smart devices which include PLC-enabled and PLC-disabled. The first category of devices consists of signal reception interfaces for all three communication technologies, namely, the PLC, VLC and RF. Whereas, the second category comprises of signal reception interfaces dedicated only for VLC and RF. This work formulates the optimization problem for the AP and bandwidth allocation for the hybrid PLC/VLC/RF systems. Furthermore, when compared with the conventional cellular systems, the impulse noise present in the PLC system imposes serious non-linearity issues to the optimization problem, thus a hierarchical decomposition method has been formulated in order to convert the resultant non-linear optimization problem into a set of convex optimization problems. Additionally, a joint distribution algorithm based upon worst device reshuffling and load balancing techniques has been proposed with the motive to maximize the achievable sum rate capacity. Moreover, upon exploiting Lagrangian multiplier method, analytical expressions were derived in order to determine the amount of bandwidth that was allocated to each smart device for a given AP association. Finally, by means of numerical results, the performance of the proposed system was evaluated. From the result analysis, it can be deduced that the proposed hybrid PLC/VLC/RF system significantly improves the sum rate capacity when compared to the traditional systems as reported in literature.

The other prime design parameter which needs to be considered while designing a wireless network is the energy efficiency. It is an indispensible factor that needs to be paid attention because the designed wireless communication network should consume less power in order to minimize the emission of greenhouse gases. To date, the majority of the works concentrated on the concept of energy efficiency in the context of VLC/RF systems [257] and, for the first time, the problem of maximization of energy efficiency was addressed in [8] in the context of hybrid PLC/VLC/RF indoor communication systems by means of power and backhaul flow optimization. In this work, the problem of energy efficiency maximization by means of power and backhaul flow optimization has been formulated as a non-convex problem, and then by means of Dinkelbach's approach, this non-convex problem has been transformed into a convex one. The simulated results of this work emphasize that the energy efficiency gain of the hybrid system is superior when compared to the RF only system. Furthermore, in order to quantify the achievable energy gain of the hybrid PLC/VLC/RF systems, the proposed energy efficient algorithm was compared with the benchmark scheme like equal power allocation (EPA) algorithm.

Based upon the extensive review carried out in the aforementioned research contributions as portrayed in [29, 182, 252, 253, 272, 273, 321, 329, 454, 455] specifically related to the domain of integrated PLC-VLC systems utilize the deterministic channel modeling in order to analyze the performance of the PLC-VLC systems. Only a few works can be reported on the study of cooperative relay-assisted PLC-VLC systems that analyze various statistical parameters. In practice, there are several parameters which affect the overall performance of the integrated PLC-VLC systems. These parameters which include the distribution of users at the VLC end, the length and the number of branches of the power-line cable expanding from the base network hub to the user terminals, the number of appliances that were attached to the power cables at the PLC terminal, etc., are not deterministic in nature. These parameters are random in nature; therefore, a more suitable manner to evaluate the performance of these types of systems is to model them by exploiting the statistical channel modeling approaches. Even though the research work in [182] evaluated the performance of the integrated PLC-VLC systems by exploiting statistical PLC and VLC channel models, as well as illustrated a significant improvement in the outage performance of the overall system at the cost of decrement in the end-to-distance, the authors could not formulate the end-to-end statistics of the SNR of the overall system as well as failed to interpret the end-end-end error performance of the integrated PLC-VLC system.

Thus, by taking into consideration these aspects, the authors in [239] carried out investigations as well as statistically evaluated a mixed cooperative PLC-VLC system dedicated for indoor broadcasting applications. Particularly, a mixed cooperative DF-relay-assisted PLC-VLC system has been proposed. This system model exploits the stochastic PLC and VLC channel models that take into account the randomness of various parameters present in both links. With the aid of the DF-link, the indoor VLC system was connected to the PLC backhaul network. Furthermore, closed-form expressions for the CDF and probability density function (PDF) of the end-to-end SNR were derived. Additionally, closed-form expressions for the system outage probability and the average BER were derived. By means of simulation results, the derived mathematical expressions were validated and the BER performance of the integrated DF-relay-based PLC-VLC system was investigated under various impulse noise scenarios and indoor parameters. From the simulated results, when compared with high impulsive noise scenarios, it is evident that the integrated system attains an improvement in the BER performance for the weak impulsive noise scenarios. Furthermore, from the result analysis, it can be deduced that the PLC system seems to be the appropriate backhaul solution for the VLC link, thereby facilitating the integrated system to be envisaged as an effective indoor broadcasting communication system.

6.7 APPLICATIONS OF INTEGRATED PLC-VLC SYSTEMS

6.7.1 EXPLOITATION OF PLC-VLC SYSTEMS FOR HEALTHCARE

In recent times, with the emanation of pandemic diseases, like the outbreak of the deadliest novel coronavirus in China in 2020, it is crucial for the people to pay immense attention to their health. As a powerful revolutionary to the traditional healthcare services, E-HEALTH has emerged to give a better experience for patients as well as to facilitate higher diagnostic efficiency. E-HEALTH includes a variety

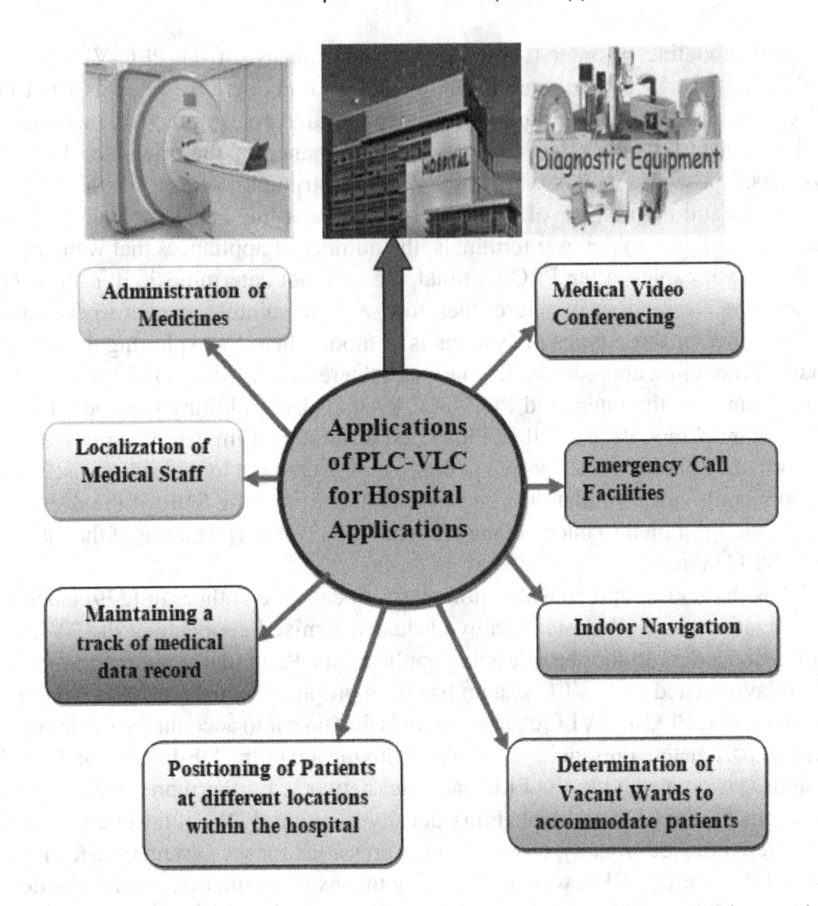

Figure 6.11: Depiction of different localization and communication activities carried out within hospitals [143].

of services, namely, telemedicine, electronic medical records, health management knowledge, etc. Consequently, these sorts of services impose the necessity for reliable communication systems with huge capacities dedicated to exchange and distribute large amounts of data within the hospital. On the other hand, while dealing with the safety of the patients and precision of the medical instruments, it is mandatory for the electromagnetic interference to be as minimal as possible. Eventually, such stringent requirements will definitely lead to abandonment of several RF-based wireless communication systems within the hospitals. For any communication based technology to be deployed within hospitals, it should be able to facilitate certain localization and communication oriented services as delineated in Fig. 6.11. The type of communication technology that is to be deployed predominantly within hospitals should be electromagnetic interference-free and should ensure several services like managing both medicine and assets, ability to locate both patients and medical staff, proficiency to facilitate indoor navigation, emergency treatment, etc. Therefore, this

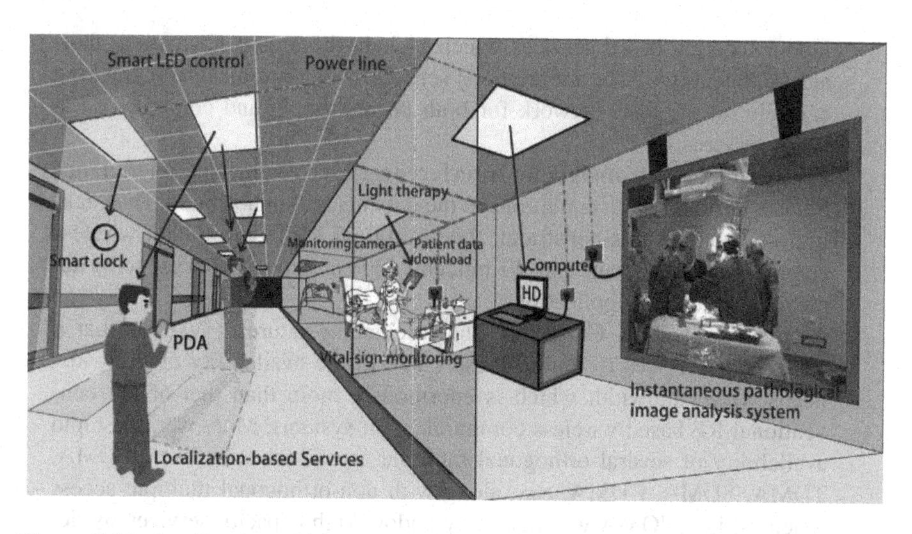

Figure 6.12: Application of integrated PLC-VLC system in the field of healthcare in hospitals [143].

implies that enabling localization aid for hospital applications is a paramount aspect to be taken into account while deploying any wireless communication technology. Thus, to fulfill all the aforesaid tasks within the hospitals, VLC-PLC systems seem to be the best choice. This is due to the fact that conventional RF based indoor localization techniques are abandoned not only due to their imprecise nature, but also their radiation sensitivity which prohibited them in some hospitals. Thus, positioning aid can be easily facilitated by using simple, sustainable illuminating sources like LEDs, where the location information of the target can be obtained upon receiving the unique ID from the corresponding LED lamp. A list of several applications fulfilled by the integrated PLC-VLC system is depicted in Fig. 6.12. More closely associated research contribution that works for the first time toward developing an integrated PLC-VLC system exclusively for hospital applications can be portrayed in [143]. The developed PLC-VLC system exploits the high-dimensional multicarrier modulation format like OFDM along with a precompensation technique in order to overcome the attenuation at higher frequencies. Furthermore, the real-time implementations of the developed integrated OFDM-based PLC-VLC system comprising of a data rate of 48 Mbps within a bandwidth of 8 MHz within the laboratory can be witnessed. This developed integrated PLC-VLC system exhibits its potential to render several services dedicated for hospital applications. Listed below are a few applications of the integrated PLC-VLC system meant for hospital applications:

- **Localization Support:** The developed system can serve as an excellent candidate to enable indoor positioning support due to the exceptional directionality of the light signals. By using an ample amount of VLC indoor localization techniques, this system renders the support to serve several indoor location based services which include the indoor navigation, first-aid treatment, etc. The monitors or sensors which are meant for tracking can rely on the PLC network for both power supply and communication based services.

- **Freedom of Electromagnetic Interference:** Electromagnetic interference induces serious health conditions to the patients within hospitals. Thus, taking into account this constraint, the traditional RF-based 3G/4G wireless communication systems and Wi-Fi technology are not appropriate to be deployed for hospital applications.
- **Assurance of High Capacity:** The fascinating nature of VLC is that it can render services to multiple users due to the availability of the wide, license-free bandwidth which is enormously more than that of the conventional RF based wireless communication systems. Moreover, the rapid availability of several orthogonal multiple access schemes like OFDMA, TDMA, SDMA, CDMA, etc., along with non-orthogonal multiple access schemes, i.e., NOMA will definitely endow high capacity services by deploying several techniques more suitable for both localization and communication based services. Besides, the PLC is capable to render high data rate communication in the range of several orders of Gbps. Thus, the cascaded PLC-VLC systems can work well in hospitals furnishing services to a large number of users with high capacities in several hospital applications which include the medical video conferences, dispensing of patient's data, etc.
- **Cost-effective Nature:** The deployment of hybrid PLC-VLC systems within hospitals seems to incur less cost when compared with other hybrid systems. The underlying principle behind this easy deployment with less cost is that LED-based VLC exploits the ubiquitous power line for power backup as well as the data provider. Thus, installation costs can be drastically reduced, since the same power line to the LED renders support to power up the LED as well as acts as a signal source to it.
- **Ensuring Secured Services:** Since the light signals cannot penetrate through walls and other opaque objects, the intruder cannot hijack the patient's data. Moreover, in order to attain an enhancement in security, simple encryption algorithms can be imposed directly to the hybrid PLC-VLC systems.
- **Medication:** Some of the impressive medical applications of the integrated PLC-VLC systems include light therapy for the skin. For instance, the LED light has been exploited for the treatment of skin which is ably known as light therapy. As shown in Fig. 6.13, the LED therapy device transmits light waves deep into the skin to trigger a natural intracellular reaction. In this manner, many skin disorders which include acne and several other infections can be treated. In the integrated PLC-VLC system, the LED which is meant to be exploited for light therapy, can be interfaced with the overall communication system and thus can be intelligently controlled.

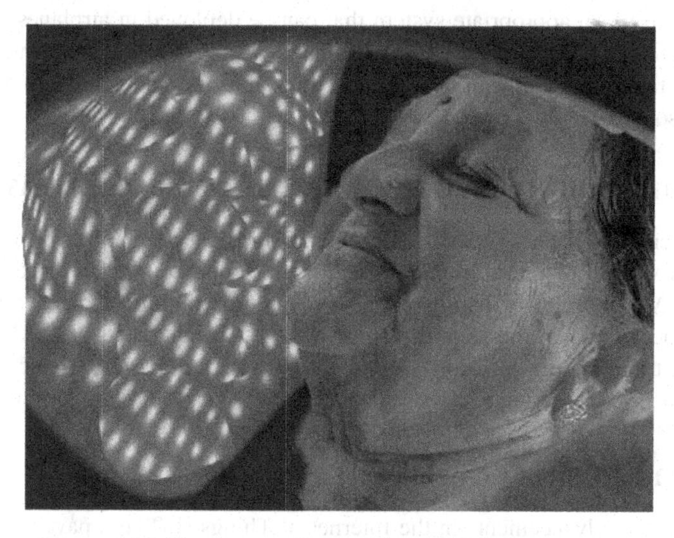

Figure 6.13: Treatment of skin using LED light therapy.

6.7.2 DEPLOYMENT OF PLC-VLC SYSTEMS IN AIRPLANES

The exploitation of VLC technology in airplanes is demonstrated in Fig. 6.14. Particularly, RF based communication technology cannot be utilized in airplanes due to the emanation of severe electromagnetic interference. In such scenarios, the most appealing solution is to deploy LEDs due to features like their security and ability to facilitate seamless data transmission in order to provide in-flight entertainment, other emergency services, etc. Therefore, in this regard, an integrated PLC-VLC system

Figure 6.14: Application of integrated PLC-VLC system in airplanes.

turns out to be an appropriate system that can be deployed in airplanes rather than Wi-Fi technology. This is due to the fact that power lines are meant to supply power to the LED lighting lamps in the airplane, as well as the same power line can act as a signal source.

6.7.3 UTILIZATION OF PLC-VLC SYSTEMS IN EMERGENCY AREAS

In several emergency applications which include defence, power plants, military, etc., it is vital to maintain immense security in order to protect the data from the intruder. Moreover, it is also essential to overcome the effect of electromagnetic interference. Thus, in this regard, deployment of the integrated PLC-VLC systems turns out to be the best possible way to ensure a secure as well as electromagnetic interference-free communication.

6.8 CONCLUSION

The recent advancements in the Internet of Things (IoT) era paved the platform for the deployment of a multitude number of smart devices. Furthermore, replenishing high data rate services to these enormous numbers of smart devices relying exclusively on the RF based communication systems is inconceivable. Bestowing high data rate communication in a reliable manner emerges as a challenging task for RF based wireless communication owing to its spectrum crisis. Besides, the other complementary technologies like PLC, VLC, etc., alone cannot offer effective solutions to furnish high data rates. Thus, in this regard, the integration of different communication technologies turns out to be the most appealing solution to meet the demands of the users. Toward this end, the integration of PLC with VLC seems to be attractive for future generation communication systems. This chapter presents an in-depth analysis of up-to-date research aspects of the integrated PLC-VLC systems. Particularly, PLC offers the potential to serve as a backbone network for an indoor VLC system much similar to the role Ethernet plays toward a Wi-Fi system. In the integrated PLC-VLC systems, the role of PLC is not only to act as a supplier of power to the LEDs, but also as a signal source to it. From most of the earlier and current research findings, it can be deduced that combined PLC-VLC systems provide a secured means of communication with low deployment costs.

The noteworthy feature of integrated PLC-VLC systems is that they can be regarded as green communication systems which are very power efficient due to their low consumption of power when compared with other conventional technologies. Moreover, from a list of significant research findings, it is clear that the combined PLC-VLC systems will definitely accomplish high data rate access for indoor applications, thereby exhibiting their potential to formulate next generation wireless networks. Therefore, taking into consideration the significant research efforts carried out in the literature, this chapter presents the up-to-date research aspects carried out in the broad areas of cascaded PLC-VLC channel modeling, performance evaluation of the integrated PLC-VLC-RF systems and exploitation of orthogonal as well as non-orthogonal multiple access schemes to render multiuser support.

Furthermore, this chapter evaluates and compares the performance of both OFDM and DWT based multicarrier integrated PLC-VLC systems. It is noteworthy that upon integrating VLC with PLC, the basic illumination functionalities of VLC are not disturbed. Additionally, even while developing an integrated PLC-VLC systems, it is vital to ensure that the original layout of the network does not undergo several modifications. In this regard, this chapter elaborates on research efforts that strived hard to design an integrated PLC-VLC system with minimal modifications.

Even though VLC offers several significant benefits, the interference problem is inherently present, which generally emerges due to overlapping of emitted light between the LED luminaires. Thus, there is an urge to address the interference problem while dealing with downlink and uplink scenarios in an indoor room environment. Additionally, it is also vital to alleviate high PAPR which usually emanates while exploiting multicarrier modulation format like OFDM. Toward this end, this chapter provides a thorough representation of a downlink and uplink integrated PLC/VLC system rendering services to multiple users located in an indoor room environment. In order to overcome high PAPR, this system makes use of the SO-OFDM methodology. Thus, the integrated SO-OFDM-based PLC-VLC system exhibits its potential to render high data rate communication to multiple users with reduced system complexity along with good power efficiency. In addition, several applications of the integrated PLC-VLC systems in hospitals, airplanes, etc. is outlined. Finally, we believe that this chapter draws significant insights into several research efforts carried out in the literature pertaining to the integrated PLC-VLC systems. We hope that the provided information spurs further interest in the research communities and industries to formulate several enticing solutions and techniques for the amalgamation of different communication technologies with the major emphasis to furnish high data rate communications.

7 VLC FOR VEHICULAR COMMUNICATIONS

7.1 BRIEF OVERVIEW

With the explosive number of automobile industries accessing the traffic infrastructure, the number of road casualties is also increasing at an alarming rate. This enables one to draw the conclusion that road accidents are the main cause of death. Additionally, many accidents are witnessed by young people, exclusively aged between 15 and 22 years. Hence, it can inferred that road accidents represent a leading cause of mortality rate. Consequently, this sad situation emphasizes the necessity to address the road fatality rates. Thus, based upon these grounds, it is predicted by the World Health Organization (WHO) that handling these road fatality rates turns out to be an indispensable task. Eventually, the government regulation authorities, automobile industries and scientific communities have joined hands and carried out adequate efforts for the purpose of enhancing the road and vehicular safety. Therefore, the automobile industry is rapidly advancing by improving the safety and comfort of vehicles through several measures like adding intelligence to vehicles or exploiting sensors or instrumentation systems in order for the vehicles to sense the environmental behavior and adapt accordingly. Consequently, these tremendous efforts enforced by the above agencies have led to a cutting-edge technology in vehicular safety. By making use of extensive active safety systems, earlier research efforts strived to help people survive road accidents, but, in the current day scenario, research efforts are directed to help people avoid road accidents. In this context, the invulnerability and the efficiency of the transportation sector can be extensively elevated by exploitation of wireless communication based technologies to facilitate the seamless exchange of real-time data between vehicles and traffic infrastructure.

Thus, much of the distinguished works in the literature as suggested in [360] affirms that by bringing together vehicle to vehicle (V2V) and vehicle to infrastructure (V2I) communication, 81% of all the vehicle crashes could be prevented at a larger scale. The most suitable manner to prevent road crashes and associated victims is to create vehicular awareness. Thus, it is vital to deploy novel communication based technologies for reinforcing the safety of people travelling on roads. Nonetheless, promoting the intervehicle communication is not that straightforward due to contradictory channel characteristics as well as high quality of service (QoS). Accelerated by the recent amelioration of technology, particularly the rapid excellence in the fields of mobile computing, remote sensing and wireless communication are nowadays enabling the intelligent transportation system (ITS) to leap forward. Thus, ITS exploits the current state-of-the art cooperative communication based technologies with the primary motive to substantially reduce the vulnerability of road accidents as well as to improve the efficiency of transportation system. Thereby, toward this end, the emissions of carbon dioxide (i.e., CO_2) can be reduced drastically. In recent times, vehicles are already embedded with sophisticated on-board control devices. The new key elements which need to be added to these modernized vehicles is to add new wireless communication oriented methodologies, computing techniques and sensing capabilities. Thus, in ITS, wireless communication based technologies are mandated to allow for V2V and V2I communication, which exemplifies as the basis for the communication-based active safety applications. In particular, ITS generally adds value to the transportation system by furnishing access to real-time traffic relevant information. The primary point to enhance the safety of the vehicles is

Figure 7.1: Pictorial representation of vehicular communication, where the real time data is exchanged between vehicles in a city.

to ensure that cooperation between vehicles or with the road infrastructure is well maintained in heavy density traffic on highways or in suburbs configurations. Hence, it is more vital to communicate several information regarding the state of the vehicle which includes vehicle braking, speed, acceleration, engine failure, etc. Moreover, it is also necessary to gather the traffic related information like the current state of trafficlights, traffic jams, accidents, etc. A typical scenario of V2V communication in a city is delineated in Fig. 7.1 where the vehicles are seamlessly exchanging traffic-related and vehicle-oriented information. Withal, the important functionality of the ITS is to collect traffic-related information, analyze it effectively and disseminate the collected data for the purpose of enabling the acquaintance among vehicles moving on roads. This accumulated data is very appropriate to accurately manage the transportation system by increasing efficiency, reducing traffic jams and automatically routing the transportation system to different traffic situations. Thus, by enumerating intelligence to the transportation system, an effective monitoring and efficient management of traffic can be attained, which ultimately helps in reducing the congestion as well as in bestowing the most optimized alternative routes depending upon received traffic updates.

7.2 EXPLOITATION OF VLC IN VEHICULAR COMMUNICATIONS

7.2.1 ON THE PERFORMANCE OF RF-BASED DEDICATED SHORT RANGE COMMUNICATION TO BUTTRESS V2V COMMUNICATION

Several wireless communication based technologies were examined to facilitate communication between vehicles and road side units (RSUs). In this regard, RF-based wireless communication is considered as one such communication which can be exploited between vehicles and RSUs effectively in order to gather traffic-related information. The evolution of Wi-Fi, Bluetooth, and radio mobile networks have substantiated the efficiency of such technology. The stringent efforts to ensure vehicular safety opened the gates for the formulation of dedicated short range communications (DSRC) for vehicular safety applications. The enormous fascination in this area is confirmed through the release of IEEE 802.11*p* standard for easing the wireless access in vehicular environments [393]. Furthermore, in 1999, the Federal Communications Commission (FCC) dedicated 75 MHz of bandwidth where the spectral band is ranging from 5.850 to 5.9 GHz. It was estimated by the U.S. Department of Transportation (DOT) that vehicular communication directed by DSRC can potentially reduce 82% of road crashes in the U.S. [262]. The basic prerequisite for the deployment of DSRC is to effectively handle the collision prevention applications. More exclusively, these applications rely on rapid exchange of data among the vehicles and between vehicles and road side infrastructural units. From the earlier records it can be evidenced that there is a huge requirement of high reliability and low latencies for the promotion of communication-based safety applications. DSRC is regulated by the IEEE 802.11*p* standard for wireless access in vehicular environments (WAVE). The authors in [360] identify the vehicle safety applications by analyzing the frequency of occurrence and impact of different categories of accidents. As per the authors in [121], the vehicle safety communication project categorized the utmost representative safety applications and, among them, 8 were contemplated as the top superior ones. Table 7.1, which was adopted from [81], states that primarily the topmost high priority applications enforce severe limits pertaining to the latencies.

It is clear from the table that relatively all of the application scenarios demand latencies below 100 ms other than the curve speed warning application and since pre-crash sensing is regarded as the utmost priority application, so it desired latencies below 20 ms. Furthermore, from the table it can be inferred that the maximum communication range varies between 50 to 300 m. From this data analysis, it is imperative that the chance of encountering an accident with a vehicle located 1000 m away is quite low. Thus, this emphasizes that in order to increase the reliability, it is envisaged to have shorter communication distances. Moreover, with the increase in intervehicle distance, the driver will be provided with an ample amount of time to react to the most dangerous situation. In the scenario of lane changing, it is more likely for an accident to occur only when it comes in the vicinity of other vehicles. Dealing with the message generation rate for all the aforesaid applications as indicated in the table, short messages are generated up to 10 times per second. The inter vehicle distances under different traffic conditions are outlined in the Table 7.2. Upon having a glance at inter vehicle distances under different traffic conditions, it is probable that the majority of the involved distances were below 160 m.

Table 7.1
Safety Applications as high priority aspects by Vehicul Safety Communications Consortium [81].

Safety Applications Regarded as High Priority					
Application	Max Range [m]	Rate [s]	Max. Latency [ms]	Message Length [bit]	Type
Traffic Signal Violation	250	10	100	528	I2V
Curve Speed Warning	200	1	1000	235	I2V
Emergency Electronic Brake Light	300	10	100	288	V2V
Pre-Crash Sensing for Cooperative Collision Mitigation	50	-	20	435	V2V
Cooperative Forward Collision Warning	150	10	100	419	V2V
Left Turn Assistant	300	10	100	904/208	I2V/V2I
Lane Change Warning	150	10	100	288	V2V
Stop Sign Movement Assistant	300	10	100	208/416	V2V/I2V

Table 7.2
Representation of intervehicle distance under different traffic conditions [81].

Conditions	Intervehicle Distance [m]
Traffic Jam	< 35
Roadway in Urban Area	35 – 49
Urban Highway Rush Hour	50 – 66
Urban Highway	67 – 100
Rural Highway	101 – 159
Rural Area	> 160

Diverse technologies have been proposed to enable the vehicles and infrastructure to communicate with each other; a few of which are: Bluetooth, long term evolution (LTE), 3G or a combination of the aforesaid technologies. However, the strong focus is directed toward the exploitation of RF-based DSRC for vehicular communication. DSRC is a medium-range communication which is intended to augment the public safety over the road-side to vehicle and V2V communication channels. IEEE 802.11p and IEEE 1609.x sets of standards specify the DSRC. Notably, the physical layer of IEEE 802.11p standard stems out from the IEEE 802.11a/g Wi-Fi standards and this is predestined to intensify the tolerance levels against multipath fading environments most specifically for the high-speed outdoor-vehicular applications. When compared with IEEE 802.11a/g standards, there is a significant improvement in the physical and medium access control (MAC) layers in IEEE 802.11p standard. Moreover, this is designed to replenish a high level of robustness as well as to reconcile with the fast mobility conditions imposed by the vehicular applications. Going into the details of DSRC, the channel of DSRC is branched into 7 channels comprising of bandwidth 10 MHz with a 5 MHz guard band at the low level. Each channel is devoted to different applications. Further, each channel is subdivided into 52 subchannels encompassing a bandwidth of 156.25 KHz. Moreover, the FCC has designated each channel either as a Service Channel (SCH) or as the Control Channel (CCH). CCH, which is also referred to as center channel, plays a major role to broadcast all safety related messages. With the aim to reduce the latency of high priority messages, the messages are categorized into 4 priority messages. It uses the well-known carrier sense multiple access/collision avoidance (CSMA/CA) as a collision prevention mechanism. DSRC is based on half-duplex communication with data rates ranging from 3 to 27 Mbps, and OFDM is used as the modulation format. To achieve a communication range of up to 1000 meters is aspired by DSRC.

In spite of significant traits offered by DSRC by meeting difficult scenarios encountered in vehicular environment, there are several other challenges which need to be addressed when exploiting DSRC. According to numerous studies, the issues related with DSRC will be presented below:

1. **Channel Congestion and Broadcasting Storm**: In general, channel congestion hinders the performance of communication and turns out to be as a major obstacle to render high data rate communication. Since communication based vehicular applications intend to exchange a large scale of dynamic data and, moreover, channel congestion is determined by the vehicle density, the rate of generation of message and the range of communication, this sort of scenario manifests as a serious issue which needs to be handled [241]. Broadcasting storm phenomena are the important problem which arises due to channel congestion. The research work in [476] details the broadcasting phenomena which emanate in vehicular adhoc networks (VANETS) because such kind of networks mainly rely on broadcasting phenomena to disseminate the data packets to the nodes which are located in certain geographical locations. However, the major problems like contention and collision in transmission among the neighboring nodes are

encountered while broadcasting the data packets. Consequently, the authors in the aforesaid work have proposed several broadcast suppression techniques to combat the broadcasting storm. Therefore, the quality of the channel plays a significant role to determine the QoS in VANETS. The reason behind this is that the channel modifies randomly in time and it becomes difficult to predict.

2. **Failure of CSMA/CA under Heavy Traffic Densities**: The inadequacy of CSMA/CA during high traffic density conditions has been clearly illustrated in several studies. It is also revealed in many studies that the IEEE 802.11*p* standard, which is based on CSMA/CA, encounters serious issues with congestion at high traffic densities [63, 106, 154, 516]. Especially, in these studies, the authors state that undesirable packet collisions as well as long delays between successful packet receptions at high traffic densities were observed. More specifically, from these studies it can be ascertained that packet decoding capability fails even in the case of packets transmitted from a vehicle situated closeby. The work in [460] through simulations confirms that when the node (vehicle) density grows, then behavior of CSMA approaches that of the ALOHA process, a process where the nodes continuously transmit packets without sensing any other transmissions and the same inference is drawn by the authors in [369]. Under the circumstances of high traffic prone areas, more particularly on highways as well as in crowded cities, the reliability of wireless communication is pretty ambiguous where there is a requirement of lowlatencies below 20 ms [64]. The work in [154] affirms the fact that WAVE fails to render reliable transmission of high priority packets under high density traffic scenarios. Investigations of DSRC in a highway points out that even though the reliability conditions are met, it becomes difficult to face the external collisions. Furthermore, the packet delivery ratio is also affected due to the presence of hidden nodes in the highway scenario [550].
3. **Effect of Doppler Spread and Multipath Propagation Environment on DSRC**: In addition to channel congestion, broadcasting storm, failure of CSMA/CA, the Doppler spread evolves as one of the stringent problems affecting DSRC. Particularly, the Doppler spread causes the spreading of the signal, thereby leading to broadening of the spectrum compared to that of the transmitted signal. Additionally, the abrupt variations in the channel lead to subcarrier interference, which deteriorate the performance of the system. Doppler spread is directly proportional to the speed or velocity with which the vehicles move, as well as the separation between two moving vehicles also play a significant role [320]. The presence of Doppler spread degrades the bit error rate (BER) performance and throughput of the system [320]. Besides Doppler spread, multipath propagation environment also poses a threat to DSRC. Due to the presence of several obstacles, the transmitted signal undergoes several reflections resulting in different path lengths. The high dynamic nature of VANETs makes this area prone to

multipath environments where these multipath components widen the Doppler spectrum.

4. **Presence of NLOS Environment**: The NLOS environment imposes a stringent problem for IEEE 802.15.7 standard. This statement is verified from several significant works which conclude that the presence of NLOS paths hinders the communication performance in VANETS. In [251] it is evident that, pertaining to urban areas, the tall buildings which are placed on the crossroads perturb the communication. The communication is halted because of the presence of vegetation along the roads, more specifically along the tight curves [65]. In these scenarios, the communication connectivity is almost lost the moment when the LOS path is obstructed leading to the loss of packets.

Upon viewing the upper mentioned challenges associated with DSRC, they are affected by several number of factors which include: high traffic densities, presence of NLOS environment, channel congestion, broadcasting storm phenomena, and failure of CSMA/CA under high traffic density conditions. As a result, these factors enforce several constraints by reducing the range of communication, and lead to degradation in the reliability of services, intensification in the occurrence of packet collisions and delays. From the aforementioned literature, the analytical and experimental results interpret that DSRC systems are reliable only under ideal conditions. But, in the scenario of real-time implementations, the performance of DSRC systems is degraded due to diverse factors. In certain situations, it cannot withstand NLOS environments where there is a communication breakdown. Therefore, the primary functionality of DSRC to ensure seamless communication between vehicles is failing and its ability to face the majority of the challenges in vehicular communications is rather questionable.

7.2.2 VLC-BASED VEHICULAR COMMUNICATION

Even though several RF based wireless communication technologies like Wi-Fi, Bluetooth, 3G, LTE, DSRC, etc. have been proposed to enhance the safety of vehicles travelling on road. Nevertheless, these technologies are not appropriate due to the stringent synchronization requirements which are involved for the reception of data frames. Additionally, the looming RF spectrum also necessitates to rely on alternative communication based technologies. Furthermore, the rapid penetration of solid-state lighting units is due to the rapid deployment of opto-electronic devices like LEDs. The remarkable advantages imparted by LEDs like their long lighting hours, high energy-efficiency, sustainability, reliability and their longer lifetime have drawn the attention of the car manufacturers to replace the conventional classical halogen lamps with cost-effective, illuminating devices like LEDs. Therefore, significant interest has been shown by the automobile industry to allow the production vehicles to use the LED based lighting units, and this is delineated in Fig. 7.2.

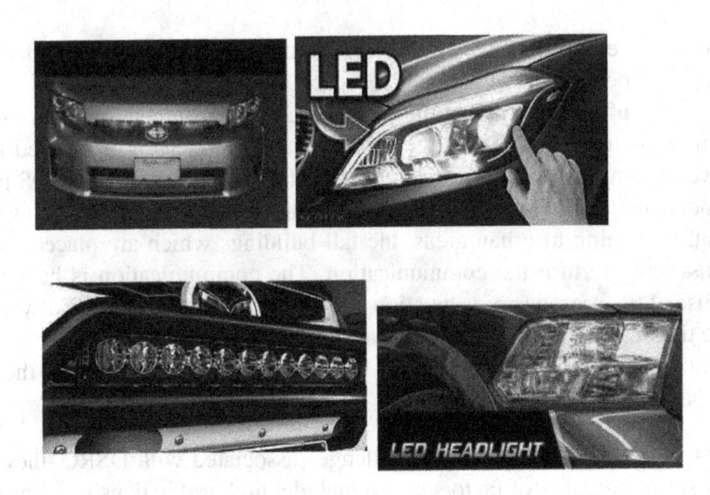

Figure 7.2: Interface of LED-based illuminating devices in vehicles.

Furthermore, the efficiency of the LEDs sparked tremendous interest among the city government authorities to allow their replacement in the traffic lighting systems and street lighting units as well. This new sort of traffic lighting system offers tremendous advantages compared to the traditional traffic lighting systems in terms of their easy maintenance, lower cost, flexibility in operation, smoother and cooler operation, assurance of longer operational lifetime, eco-friendly nature, and better visibility which is aesthetically pleasing to the human eye. Considering these stupendous LED advantages, which are imparted by LEDs, the government authorities in several cities are progressively striving to replace classical existing traffic lights with LED based traffic light systems. Moreover, in the research article as carried out in [153], the size of the LEDs in LED-based traffic light systems is around 200 to 300 millimeter. Moreover, such kind of LED-based traffic light system comprises a huge number of LEDs; approximately 100 to 200 LEDs can be fitted to provide the seamless wireless data transfer, besides the signalling operation which they usually perform. Such kind of LED-based traffic system goes well with the requirements of the traffic regulation standards as well. With the increasing popularity of LED-based illumination appliances, as they can be realized in full color large LED displays, it is anticipated to use such LEDs in road illumination. Much earlier works in the literature [269] proposed a road illumination system based on LEDs in Japan and substantially confirmed that vehicular communications can be carried out by such kind of LED-based road illumination system. The work in [280] illustrates the necessity of employing the energy-efficient illumination system to accomplish the task of fulfilling seamless communication throughout the city, thereby evolving the city as a smart city. Subsequently, by considering the latest trends in the solid-state lighting industry, LEDs have been amalgamated with the traffic signs as well. The integration of LEDs in the traffic infrastructure is shown in Fig. 7.3.

Figure 7.3: Interface of LEDs with traffic infrastructure.

The evolution of LED-based lighting systems, as they are being interfaced with both vehicular and road side infrastructure along with the exploitation of VLC-based technology, will definitely enable the ITS to gather huge amounts of data from an extensive area, thereby facilitating the distribution of widespread communication support as well. Many research works in the literature promulgate the communication between the road side infrastructural units like LED-based traffic lighting system and the vehicles moving on roads. The authors in [370] demonstrated the design and implementation of intelligent transportation system which is making use of VLC technology to allow the heavy load carrying trucks and other vehicles moving on roads to communicate with each other. Additionally, since the maximum number of road accidents are reported due to the heavy-load carrying trucks at the road intersections, it becomes utmost difficult for such trucks to suddenly decelerate, allow for lane changes under complex traffic situations, etc. A typical scenario of vehicles moving around along road intersections is depicted in Fig. 7.4. Consequently, the authors in this work clearly exemplifies that adjusting the duration of the traffic signals, the entire precision of heavy-load carrying trucks can successfully pass through the intersections without sudden deceleration or emergency braking. This sort of communication can be established by fitting the VLC-based transceivers in both heavy-load carrying trucks, cars and in traffic light systems. The moving vehicles on roads communicate with the traffic lights by transferring traffic related updates like the speed with which the vehicles are moving, the distance between two moving vehicles, and the highest priority message like how far is the vehicle located to the intersection.

Theoretically, this sort of communication can be made possible by taking into consideration a bunch of parameters. Let the time duration of red, yellow and green lights be represented as t_r, t_y and t_g, respectively. Additionally, the total number of heavy-load carrying trucks is assumed as n and the length of entire precession of the truck to be as l. The spacing between each truck can be taken as d_1, while d_2

Figure 7.4: Typical scenario of vehicles travelling on roads at intersections.

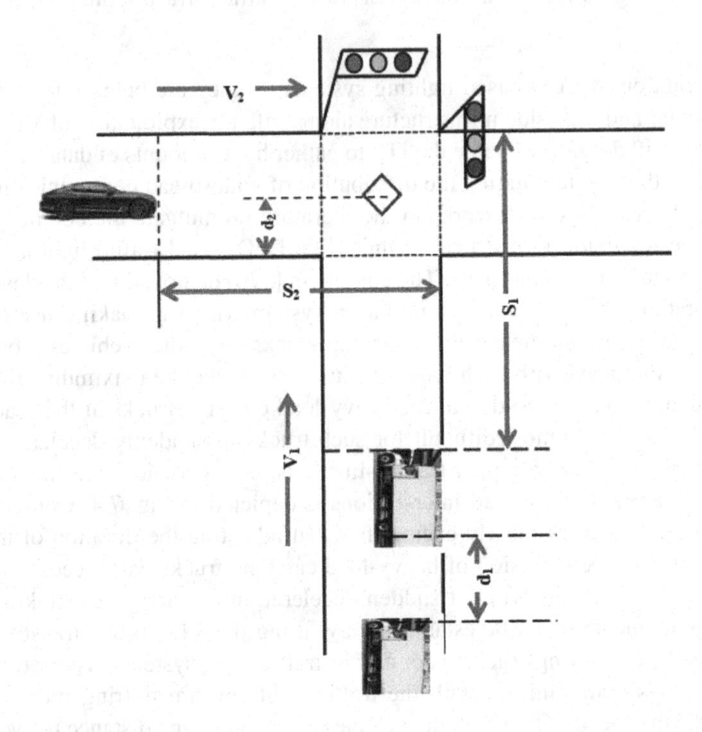

Figure 7.5: Pictorial representation of moving vehicles approaching road intersections.

signifies the distance between the stop point to the central of the intersection point and s_1 specifies the distance between the first load carrying truck and the traffic light. In this manner, by fixing all the parameters, the time of the first truck to arrive at the central point of the intersection can be determined as [370]:

$$t_1 = \frac{s_1 - 2d_2}{v_1} \tag{7.1}$$

from (7.1), the parameter v_1 implies the velocity of the trucks. Consequently, the time for the remaining trucks to pass through the intersection can be calculated as follows

$$\underset{trucks}{t_{rem}} = \frac{(n-1)(d_1 + l)}{v_1} \tag{7.2}$$

Upon calculation of time for the first truck and the remaining trucks to reach the intersection, the traffic light should perform a signalling operation in such a manner that it should extend the duration of the green signal and reduce the duration of the red signal. The duration of the traffic light signals can be adjusted as follows: Let there two scenarios:

- Case 1: **Current signal of traffic light is GREEN when the trucks approached the intersection**
  Upon establishment of VLC communication between the first truck and the traffic light, the traffic light shows a green signal for the truck and a red signal for the car. Therefore, upon extending the duration of the green signal, the entire precession of the trucks can pass through the intersection. It can be assumed that the remaining time duration for the green signal is δt_1. Hence, if there exists a condition that if $\delta t_1 < t_1$, then this implies that the traffic light should change to a red signal upon the arrival of the first truck at the road intersection point. Thus, upon receiving the VLC communication messages, in order to provide the flexibility for the entire trucks to pass through the intersection, the time duration of the green signal should be changed as:

  $$t_g = t_1 + \underset{trucks}{t_{rem}} \tag{7.3}$$

  The other way round when $\delta t_1 > t_1$ entails that the remaining time for the green signal (denoted as δt_3) when the first truck arrived at the road intersection point can be calculated as $\delta t_3 = \delta t_1 - t_1$. Further, it is apparent that the entire precession of trucks cannot pass through the intersection when $\delta t_3 < \underset{trucks}{t_{rem}}$; thus, this situation implies that the traffic light should definitely change the overall duration of the green light as (7.3). However, if $\delta t_3 > \underset{trucks}{t_{rem}}$, then the entire precession of trucks can pass through the road intersection points without the requirement for a change in the status of traffic light signals.
- Case 2: **Current signal of traffic light is RED when the trucks arrived at the road intersection**
  In this scenario, it can be assumed that the current traffic light is red upon the arrival of the truck at the road intersection point and the remaining time

for the red signal as δt_r. If there is a chance for $\delta t_r < t_1$, then the traffic light will be displaying green when the first truck arrives at the road intersection point. Thus, by extending the time duration of the green signal, the entire trucks can pass through the intersection without any deceleration. Also when $\delta t_r > t_1$, it is advisable for the red light switch to yellow the moment the first truck signal arrives at the intersection. Thereupon, when the first truck arrives at the road intersection, the remaining time of the green light needs to be calculated and it should be verified whether it is greater than $t_{\underset{trucks}{rem}}$; if it is greater, then the entire precession of the trucks can pass through the intersection; otherwise, it is required to increase the duration of the green light signal.

From the aforementioned analysis, it is evident that upon enforcing intelligence to the traffic lights, the heavy-load carrying trucks at the road intersection points can be easily allowed to pass through without causing any accidents, thereby enhancing road safety. The work in [370] introduces a VLC communication system between the moving trucks at the road intersections and the traffic lights. From the analysis and numerical results of this work, it is ascertained that upon appropriately adjusting the traffic light signals, the heavy-load carrying vehicles can be easily passed through the intersection points without any emergency braking or deceleration. Much similar work in [281] portrays that a VLC broadcast system that takes into consideration LED-based traffic lights is demonstrated. Additionally, with the intent to minimize the effects of noise, the authors have employed a direct sequence spread spectrum (DSSS) and sequence inverse keying (SIK).

VLC technology can be employed to render seamless communication among the moving vehicles on a highway, and this scenario is represented in Fig. 7.6. Evidently,

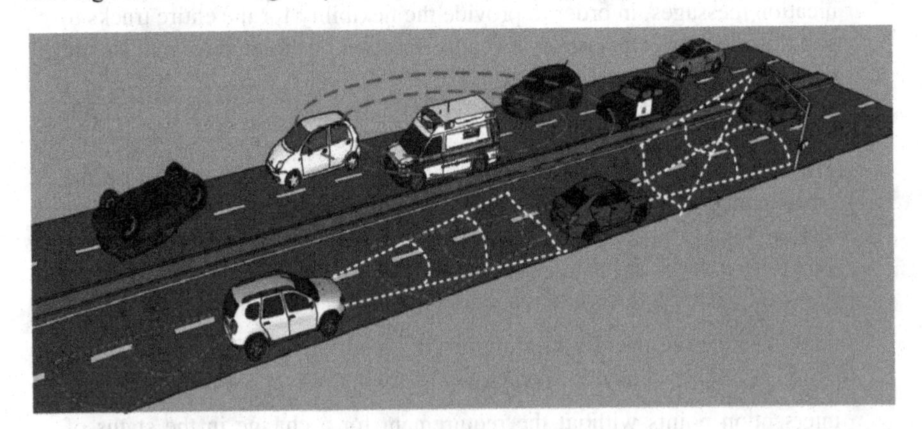

Figure 7.6: Illustration of a highway scenario exploiting VLC technology [81].

illuminating sensors are employed to transmit the data signals among the neighboring cars. Thereupon, by exploiting the head and tail lights of the vehicles, data is propagated among all the vehicles travelling on a highway road. Several important information, which includes the mechanical state of the vehicles, distance between two moving vehicles, highly congested areas, etc., can be communicated in order to enhance the safety and security of the people moving on roads. Both the experimental and simulation results surmise that the proposed system ensures communication distance range over 40 m under three different scenarios including the effect of sunlight. Furthermore, the simulation results of this work also signify that with the amount of received information from the traffic lights, the vehicles gain the potential to implement possible safety correction measures.

7.3 CHALLENGING ASPECTS OF VLC EXPLOITATION IN ENABLING VEHICULAR COMMUNICATIONS

While dealing with the different challenging aspects pertaining to the usage of VLC in enabling vehicular communications, it is vital to have a clear understanding differences between the indoor and the outdoor scenarios, because VLC can be used in both indoor and outdoor environments. In general, the usage of VLC in outdoor conditions enforces different types of challenges, expectations due to the dissimilar channel conditions. Taking into consideration the indoor scenarios, all the indoor applications are desired to accomplish high data rate transfer probably in the range of several Gbps for distances that are usually ranging between 2 to 4 meters. Contrarily, with the outdoor conditions, the scenario is entirely different because, here it is vital to ensure communication distances ranging up to 100 meters; therefore, this implies that the data rate will be significantly reduced when compared with that of the indoor applications. As evident from [81], communication-based vehicular safety applications desire high packet delivery ratio (PDR) and latencies as low as 20 ms, which implies that they must be highly robust to noise and other impediments.

The other major difference between the indoor and outdoor VLC applications arises due to the influence of noise and ambient light interferences. When compared to indoor applications, outdoor applications are mostly affected by the background solar radiation and the interferences occurring due to the other sources of artificial lighting units like street lights made of incandescent lamps, fluorescent lamps and fluorescent lamps geared by electronic ballasts, etc. The outdoor VLC channel is also affected because of the adverse weather conditions which include fog, heavy dust, rain, etc. Thus, the outdoor VLC channel is rather unpredictable when compared with that of the indoor VLC channel. Eventually, a major inference can be drawn that for the scenario of indoor applications, it is vital to enforce sophisticated modulation techniques which facilitate high data rate communication. On the other hand, the major motive of the vehicular communication is to mitigate the effects of such ambient light interferences and to establish a reliable long-range VLC link.

The most vital difference between the indoor and outdoor scenarios is ensuring bidirectional communications. While dealing with indoor applications, it becomes very difficult to establish bidirectional communication links because most indoor devices do not have a lighting function that facilitates data transfer. This eventually makes indoor applications adopt other communication-based technologies to expedite data upload. Even though this sort of approach achieves desirable data rates, the complex system design stems out as the major drawback. Whereas, in relevant to the In outdoor vehicular applications, exploitation of VLC technology is straightforward, since all the road side units like street lights, traffic lighting units, vehicular units, etc. have a lighting function which facilitates the transmission of data. Therefore, in view of the differences between the indoor and outdoor applications, this section illustrates the most challenging aspects which need to be taken into consideration when VLC is exploited to enable a wireless exchange of data between vehicles. In addition to giving a clear-cut discussion of the challenges, this section presents an analysis of comprehensive research efforts to quickly fix the aforesaid challenging aspects.

7.3.1 NOISE EMANATING DUE TO ARTIFICIAL AND AMBIENT LIGHT SOURCES

Unlike indoor applications, the outdoor VLC channel is extremely noisy. Consequently, the noise provoked due to the multitudinal sources of parasitic light agitates the communication in V2V and V2I/I2V applications. Hence, there is an urge to understand the impact of ambient interference in the communication link. From the earlier works, it is evident that the ambient interference emerges due to the prevalence of two different kinds of noise: the artificial light sources and the ambient background solar radiation.

1. **Noise Due to Background Solar Radiation:** Sunlight seems to be the most powerful source of noise which is perturbing the V2V and V2I/I2V communication. Thus, for outdoor VLC applications like vehicular communications, the major challenge arises due to the presence of intense background solar radiation. Both the natural and artificial sources of light produce an appreciable amount of background optical power density or irradiance that undermines the performance of VLC receiver. It is apparent that VLC receiver comprises of a photodetector in order to convert the light intensity signal into an electrical signal. Hence, as a result, photodetectors upon directly getting exposed to the sunlight may easily be saturated, thereby resulting in the degradation in the detection of intensity modulated optical signals. Thus, the strong sunlight, upon being incident on the photoelement, may saturate it thereby making it blind and subsequently obstructing the communication. If there were no saturation problem, there is a chance for the introduction of shot noise in the photodetector due to the induced received solar radiation, and this would be emerging as the dominant source of noise that would limit the performance of the communication system.
2. **Interference Due to Artificial Light Sources:** The other source of interference which impairs the reception of transmitted intensity modulated light is due to the interference of artificial light sources. The majority of the illuminating devices comprises of a large amount of light sources like incandescent lamps with tungsten filaments, halogen and mercury lamps, and fluorescent lamps with different emitting colors, i.e., optical spectra and

fluorescent lamps geared by electronic ballasts. Noise occurring due to the presence of these artificial light sources can be grouped into three categories:

a. Incandescent lamps including halogen lamps, mercury lamps and tungsten filaments.
b. Fluorescent lamps furnished with conventional ballasts.
c. Fluorescent lamps furnished with electronic ballasts.

From the work in [354], it can be inferred that the interfering signal produced by the incandescent lamps is a perfect sinusoid with a frequency of 100 Hz.

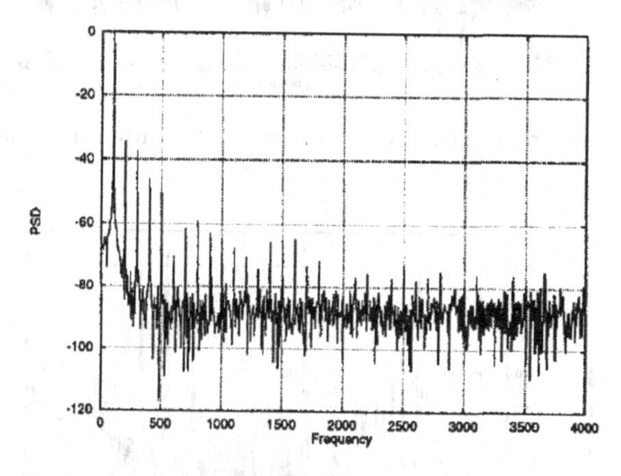

Figure 7.7: Pictorial representation of interference spectrum of an incandescent lamp [354].

The typical interference spectrum of an incandescent lamp is illustrated in Fig. 7.7. As depicted from the figure, in addition to the presence of 100 Hz interfering signal, the harmonics of the signal can extend 2 KHz and upon increasing the frequencies, i.e., for frequencies higher than 800 Hz, all the components are more than 60 dB below the fundamental. Upon proper choice of optical filters, there is a significant reduction in the interference amplitude. Meanwhile, the interference produced by the fluorescent lamp is a distorted sinusoid and the spectrum of fluorescent lamp is much broader than that of the incandescent lamp, because the spectrum of the latter extends up to 20 KHz frequency. The spectrum of fluorescent lamp is shown in Fig. 7.8, and from this plot it can be surmised that for frequencies higher than 5 KHz, the interference power spectral density (PSD) is more than 50 dB below the 100 Hz component. Even the fluorescent lamps geared by electronic ballasts produce band interfering signals in the range of 20 to 40 KHz. The spectrum of interference produced by fluorescent lamp geared by electronic ballasts is depicted in Fig. 7.9. Consequently, it can be stated that the aforementioned light sources in some way or the other lead to degradation in the overall performance of a VLC system. Not only artificial light sources and background solar radiation make the outdoor VLC

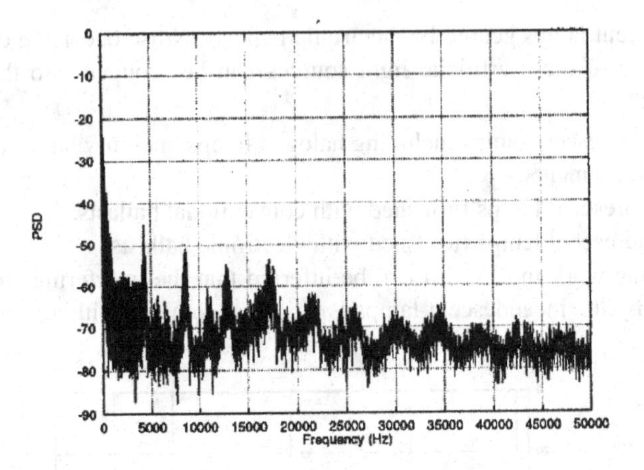

Figure 7.8: Pictorial representation of interference spectrum of a fluorescent lamp [354].

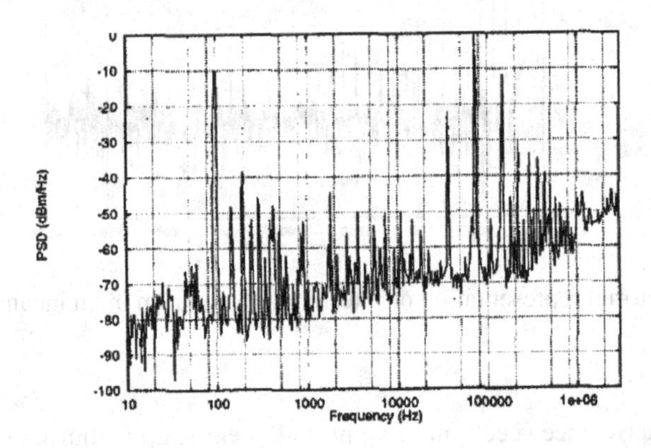

Figure 7.9: Pictorial representation of interference spectrum of a fluorescent lamp geared by electronic ballasts [354].

channel less predictable, but also adverse weather conditions make the VLC channel very unpredictable. Much of the earlier works in the literature illustrate that outdoor vehicular VLC channel is not only affected due to the presence of ambient sunlight but also due to the presence of fog, snow, as well as heavy dust which can obstruct the LOS communication path directly influencing the power of the received signal [222,230]. Additionally, the water particles which are obtained from fog and rain drops contribute to the loss of signal because of absorption, reflection and scattering. Thus, the channel becomes even more uncertain.

Moreover, in addition to the effects of noise which is induced by artificial light sources like the incandescent lamps, fluorescent lamps and the fluorescent lamps geared by the electronic ballasts, natural sources like background solar radiation, long-distance vehicular communications probably in the range of 80 to 100 meters are dependent on the path loss [391, 429]. Therefore, by taking into consideration all the above-mentioned perturbing elements, the signal to noise ratio (SNR) at the receiving end is drastically reduced for such automotive applications. Hence, a conclusion can be drawn that outdoor VLC application is strongly affected by the noise and hence combating its effect for the purpose of facilitating high data rate transfer for long-distance communications turns out to be the most potential task. Extensive research efforts identify a number of perturbing elements that hinder high data rate transfer in outdoor vehicular applications, as well as the tremendous efforts to mitigate them. As discussed earlier, the background solar radiation is the most stringent source of noise that perturbs outdoor vehicular communication in VLC, and hence it turns out to be the most burdensome to counterbalance. In case of a VLC receiver, the receiver's FOV plays a major role to allow the interception of the incident intensity modulated transmitted light signal. Therefore, as already mentioned in the previous chapters, only the signal within the receiver's FOV will be accepted and the rest of the signal which falls outside the receiver's FOV will be discarded. Therefore, in addition to taking the desired signal, there might be chances for the receiver to accept the noise content. Hence, if wider receiver FOV is allowed, then not only the desired signal is captured, but also a large amount of noise content is also accepted. Therefore, one way to counteract the effects of noise is to narrow the receiver's FOV. The authors in [309] experimentally verified that vehicular VLC $\left(V^2LC\right)$ becomes less vulnerable to visible light noise which is occurring from sunlight and other legacy lighting sources by narrowing the receiver's FOV.

The authors in [80] proposed an effective design module of VLC-based receiver which is intended for outdoor vehicular VLC applications. The module as illustrated in the aforementioned work was designed by taking into consideration two important constraints like the limited received power and the interference which is caused by ambient light sources limiting the link performance. The schematic representation of VLC reception module which is suitable for outdoor vehicular communications is shown in Fig. 7.10. As elucidated in figure, the reception unit comprises of an optical collecting system; for facilitating signal reception there is an electronic front stage as well followed by signal processing unit. With the motive to reduce the environmental light coming from the sides, the optical collecting system is mainly composed of a cover as well as a lens to focus the incident light on the photosensitive element of a photodiode. The experimental work confirms the fact that by reducing the receiver's FOV to an angle of $\pm 10^0$, a communication distance of up to 50 meters is achieved while maintaining a BER of 10^{-7}. This turns out to be an effective solution in terms of enhancing SNR, but the major drawback involved is the reduced mobility. Similar such works to reduce the effect of ambient noise as well as to enhance the SNR by using effective optical filters as well as reducing the receiver's FOV can be illustrated in [123, 263]. In addition to usage of effective optical filtering units and by reducing the receiver's effective FOV, optical beamforming also proved to be an effective way to enhance the SNR [267]. Generally, optical beamforming is a technology used for focusing the LED light onto a desired target. Here, rather than the conventional lens, a Fresnel lens is used for SNR enhancement. The experimental results of this

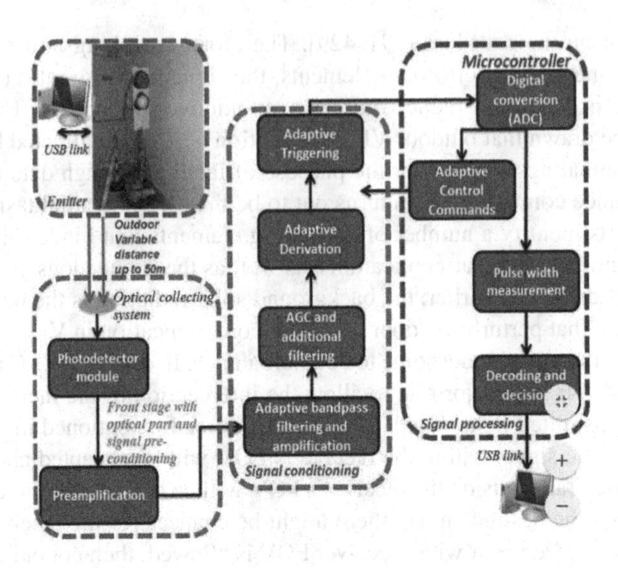

Figure 7.10: Pictorial representation of VLC reception module suitable for outdoor vehicular communications which is comprising of optical filtering unit and other signal processing units [80].

work emphasize that the SNR is improved by 13.4 dB. A much similar work which enhances the SNR under adverse weather conditions like fog is discussed in [268]. Fog is admittedly one of the detrimental atmospheric conditions which really make the optical wireless channel highly unpredictable. A typical scenario of the effect of fog which is affecting outdoor vehicular VLC applications is illustrated in Fig. 7.11. Therefore, to counteract the effects of fog to enhance SNR as well as to facilitate

Figure 7.11: Scenario of V2V communication in the presence of fog [268].

high data transfer between moving vehicles on roads, the authors in this work use a Fresnel lens to effectively focus the incoming light and multiple photodiodes to detect the highest intensity over the NLOS channel conditions. Through experimental results verify that the proposed experimental set-up achieves higher SNRs and a reliable vehicle to vehicle data transfer under heavy fog conditions. Furthermore, in order to enhance the SNR in the presence of daylight conditions, the authors in [287] proposed an analytical daylight noise model which is based on modified Blackbody radiation model to capturing the effect of ambient-light noise and conducted an in-depth analysis highlighting the impact of daylight conditions on the overall system performance. Additionally, the authors proposed a new receiver structure exploiting the selective combining technique in order to combat the effects of background noise. The result analysis in this work infers that the newly proposed receiver based on selective combining technique achieves an improvement in SNR by 5 dB and successful VLC link can be established at any time in the day.

As stated earlier, the influence of ambient light interference on VLC systems stems out as one of the most challenging aspects needing to be addressed. As discussed, both indoor lighting sources like the incandescent lamps, fluorescent lamps or fluorescent lamps geared by electronic ballasts, and outdoor lighting sources like the background solar radiation occupy a wide range of wavelengths which inherently overlap with the transmission wavelength of VLC system, thereby making them prone to in-band interference with other light sources. Consequently, this enforces a constraint on the design of VLC receiver which should have a minimal or the potential to nullify the interference which is emanating due to the presence of ambient lighting sources. Several research efforts have been illustrated to minimize the interference which is arising due to ambient lighting sources. The most vital way to counteract the effects of interference is to design efficient optical filtering techniques or to rely on modulation schemes in the electronic domain, and few such works have already seen earlier. These studies emphasize that VLC systems are intensely sensitive to in-band interference, as they have little tolerance or are not at all capable to reject in-band interference. However, there have been several methods proposed and patented to avoid interference rather than reject it [69, 418]. The underlying concept behind these methods is that interference mitigation can be made possible by carefully selecting the transmission signals with wavelengths which are minimally overlapped with ambient lighting, or by appropriately choosing the transmission signals with minimal interferences to each other. It should be noted that these methods are more suitable for interference avoidance than interference rejection.

Therefore, in order to allow for interference rejection rather than interference avoidance, the usage of the spectrum sensor array at the receiving end is proposed in [92]. Driven by the recent advancements of semi-conductor technologies, spectrum sensors comprising of distinct spectral properties can be integrated into a chip-scale sensor array [91, 282]. Upon appropriate design of weightings corresponding to the individual spectrum sensor, the effective signal to interference noise ratio (SINR) can be maximized. The simulation result analysis in this work illustrates that by making use of the cost-effective spectrum sensor array along with the proposed weighting method, it is possible to achieve robust and adaptive interference rejection. In spite of this significant rejection of interference, it should be noted that the equivalent signal transmission is achieved from a combination of sensors; thus, there are chances for noise amplification also. Additionally, timing synchronization among different sensors is mandatory, thus, making the realization of hardware set-up more complex. Furthermore, in order to prevent the effects of parasitic light on the overall vehicular VLC communications and with the motive to prevent the saturation of the photodiode even under direct exposure to sunlight, the gain of the preamplification stage was calculated. Even though this solution is effective in terms of mitigating the ambient noise, the limited gain significantly reduces the communication distance to a few meters, most probably 14 meters [78]. In addition to the proposal of effective optical filtering techniques to mitigate the effects of ambient noise which is emanating due to the interference arising from ambient light sources, the literature also suggests the usage of direct sequence spread spectrum (DSSS) coding can also provide significant robustness to noise. The work in [314] proposes the usage of DSSS coding to combat the noise and interference which is arising due to the ambient lighting sources. The prototype implementations of this work demonstrate that it is feasible for the realization of a low data rate communication link which is ranging up to 40 meters by using the commercially available LEDs. Much similar work which uses the concept of DSSS coding with sequence inverse keying is depicted in [281], where the authors in this work presented a VLC broadcast system by employing the LED-based traffic light units. Here, the simulation results affirm that data communication range of 40 meters can be obtained. Even the work in [474] implements a DSSS-based VLC transceiver for VLC systems based on field programmable gate array (FPGA) for the purpose of combating the effects of noise and other ambient interference.

The effects of noise can be overcome by exploiting Image Sensors in vehicular VLC systems. Such Image Sensors form the backbone of a typical VLC-based

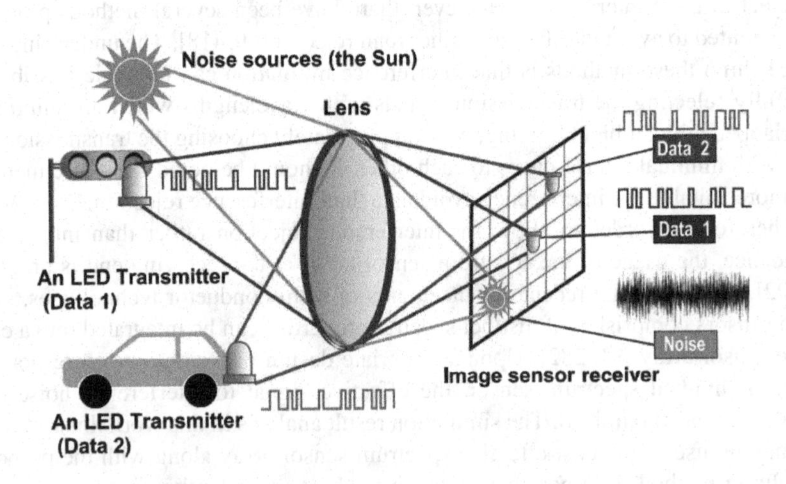

Figure 7.12: Exploiting Image Sensor for enabling spatial separation of multiple sources [187].

vehicular system and have the potential to perform many major safety applications which include the lane detection, pedestrian detection, etc. The striking trait of Image Sensor is that it is capable enough to receive the VLC signals; in particular, it can receive the data signals from the LED traffic lights, LED signage, LED headlights, LED tail lights, etc. In general, an LED-based traffic light system broadcasts the phase and timing relevant information along with additional safety information to the vehicles by blinking its LEDs at a rapid pace. Even the LED tail lights and headlights can also dissipate the internal vehicular information like the latitude, longitude, speed to nearby moving vehicles and roadside stations. Therefore, an Image Sensor is used for the purpose of detection of the position of the objects and the aforesaid transmitted data. Additionally, as depicted in Fig. 7.12, the Image Sensor is adequate enough to spatially separate the multiple objects. Consequently, such kind of ability will enable the VLC receivers to discard the noise which is originating due to sun, street lights as well as other lighting sources and aids in directing the pixels to which the LED light strikes. Thus, high SNRs can be obtained in case of outdoor environments.

7.3.2 ASSURANCE OF LONG DISTANCE COMMUNICATION

In addition to providing robustness to noise as well as ambient interference, enhancing the range of communication in case of outdoor vehicular VLC applications is the most challenging aspect which needs to be given great attention. Very low power levels are achieved at the receiving end in case of long distance communication. Therefore, this implies that the attained SNR is also very low. Eventually, this implies that the SNRs should be enhanced similar to that of the challenge as stated earlier. As evident from the report [121], the Vehicle Safety Communications Consortium (VSCC) which is a group of vehicles like BMW, Ford, Nissan, Toyota, etc. culminated the fact that upon facilitating the ease of exchange of real-time data between moving vehicles on roads by adopting novel and powerful means of wireless communication technologies, the safety and efficiency of the transportation system can be remarkably enhanced. This report interprets some of the fundamental prerequisites of vehicular communication. As per this document, it was enumerated that the communication distance has ranges of up to 1000 meters. This specification led to the formulation of IEEE 802.11p standard for WAVE which can be called DSRC and, as stated in the earlier sections, DSRC uses a dedicated band of 5.9 GHz as well as aims to warrant communication distances of up to 1000 meters. In general, for long distance communications, the effect of interference is also higher and, furthermore, reliability concerns might arise. As illustrated in [154], the performance of IEEE 802.11p standard was analyzed and verified through simulations by taking into consideration several parameters like collision probability, throughput and delay. It can be inferred from this work that even though WAVE can prioritize messages, in case of dense and heavy load scenarios, there is a decrement in throughput and increase in delay for broadcasting the information to the vehicles.

Thus, in order to increase the reliability of services for outdoor vehicular communications, the point of establishing long distance communication in the range of 1000 meters is no longer valid because the probability of arising an accident with a vehicle which is located at a distance of 1000 meters is quite low. This affirms that short distance communication distances must be anticipated for the purpose of increasing the reliability. Moreover, as briefed in Section 8.2.1, VSCC determines eight most essential communication-based vehicular traffic applications, and these are categorized as the most high priority safety applications. The communication distance as specified in these applications clearly specifies that the communication distances below 300 meters are desirable. Additionally, by considering the intervehicle distances in different traffic conditions, it can be inferred that shorter communication distances in the range of 160 meters are sufficient. Consequently, the aforementioned analysis enables to draw a conclusion that communication distance range of 400 meters is sufficient enough to facilitate the high data rate transfer relevant to outdoor vehicular communications. The existing literature much specific to the exploitation of VLC in vehicular applications specifies that VLC renders low error rate communication distances which can range from 1 to 100 meters. Such works are elucidate below.

By exploiting camera-based receiver systems, the achievable communication distance ranges from 110 meters. In general, for the next generation future vehicle to everything (V2X) communication as well as for vehicular automation, VLC which is exploiting LEDs as transmitting modules and camera receivers has been vigorously studied. In the near future, it is predicted that vehicles can be connected to each other anywhere and at anytime by means of optical signals. This is due to the tremendous enhancement in the receiving technology, especially the camera receiver which is exploiting an appropriate complementary metal oxide semiconductor (CMOS) image sensor, more specifically, the optical communication image sensor (OCI). The experimental works confirm that an optical V2V communication system which is exploiting this sort of OCI-camera receiver achieves 10 Mbps optical data transmission between moving real vehicles. The distinguished peculiarity of OCI is that its pixel array comprises of two types of pixels, namely, the image pixels (IPx) which is meant to receive the image scenes, and the communication pixels (CPx) which is mainly used for receiving high-speed optical signals [467]. Therefore, in order to attain higher data rates by exploiting the same OCI-based camera receiver, higher order multicarrier modulation formats like OFDM can be employed [187]. The general schematic representation of optical OFDM system which is using OCI-based receiver is represented in Fig. 7.13. Optical OFDM is capable enough to achieve high data rate communication. As already discussed in the earlier chapters, the optical OFDM signal which is to be transmitted should comply with the requirements of IM/DD systems, where the signal transmission should be assured of its real and positive nature. As depicted, the incoming stream of bits are first encoded by employing FEC coding most probably the convolution coding, and then this encoded bit stream is interleaved followed by transmission in parallel by means of a serial to parallel (S/P) converter. Further, complex mapping techniques like M-ary QAM are generally used to extract the complex symbols followed by assigning them to each subcarrier component resulting in a block of data symbols. Since it is a known fact that for complex mapping of the input data stream using M-QAM, the output of the IFFT will be a complex valued signal. Thus, it entails this stream of data

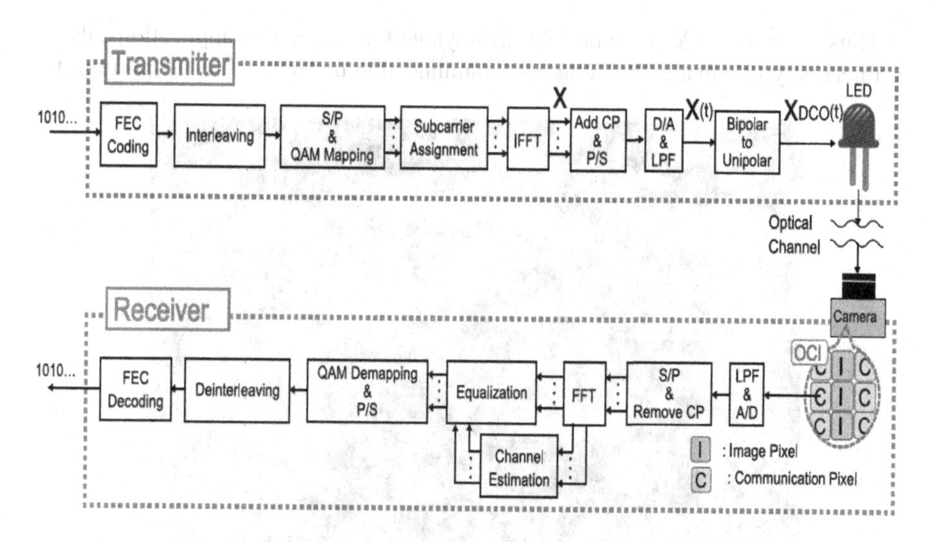

Figure 7.13: Typical system model comprising of OCI-based receiver in optical OFDM compatible for vehicular communication [187].

symbols which are compelled to satisfy the Hermitian Symmetry criteria before getting applied to the IFFT transformation block to generate a real-valued signal. Then to this real valued signal, a suitable amount of cyclic prefix is added in order to overcome the effects of ISI followed by the back conversion into serial stream of data, and then this signal is passed through a digital to analog (D/A) conversion block. Finally, in order to ensure the positive signal transmission, a suitable amount of DC bias is added, and then this signal is intensity modulated through the LED. At the receiving end, this transmitted optical signal is converted into an electrical signal by means of a photodiode in the selected communication pixel of the OCI. Generally, after this step, the rest of the processing steps are similar to that of the conventional optical OFDM receiver. Following the S/P conversion and equalization of the time-domain signal into frequency domain by using FFT transformation module. Channel-equalizing factors are obtained by averaging the training symbols for each subcarrier that are transmitted prior to the data-carrying symbols. The equalized symbols are demodulated and deinterleaved. Finally, the received bits are decoded by the Viterbi soft-decision algorithm. Thus by employing this sort of methodology in the experimental demonstration, the achievable data rates are enhanced from 10 to 55 Mbps while maintaining BER of $< 10^{-5}$. However, the communication distance ranges up to 1.5 meters. The exploitation of CMOS-based image sensor is evident from [542] where the experimental results demonstrate that this sort of CMOS image sensor has the potential to receive high-speed optical signals, and its effectiveness is verified in V2V field trials. The high-speed signal reception capability of the camera receiver along with the image sensor facilitated to achieve data rates of 10 Mbps at a communication distance of 20 meters.

Thus, in order for VLC to be fully deployable for automotive applications, there is a necessity to enhance the assured communication distances. For the purpose to

Figure 7.14: Demonstration of relay-based vehicular communication using VLC.

enhance the safety of the vehicles on roads and to reduce road fatality rates as well as to reduce the emission of green-house gases, the ITS moved a step further to amalgamate the cooperative traffic control and management center which helps in gathering the information pertaining to traffic followed by analyzing the data and then redistributing it. More specifically, for the purpose of enhancing the communication distance, one approach is to incorporate cooperative-based vehicular communication. Generally, in case of VANET, the traffic infrastructural units represent the fixed gateways, while the moving vehicles serve as mobile nodes. It is proven experimentally that cooperative systems generally have the advantage of enabling short to medium communication between the gateways (i.e., RSUs) and the mobile nodes (vehicles), thus preventing mutual interferences. The stunningly noteworthy feature is that in this sort of network, there is a flexibility to transfer messages from one mobile node to the other and, in this manner, the messages are forwarded to the nodes which are not within the vicinity of communication area. This sort of transmission of information is termed multi hop transmission Figure 7.14 illustrates that the communication distance can be enhanced by employing multi hop transmissions. Particularly, the high priority event-driven messages can be easily communicated using multi hop transmissions. The mobile node, i.e., the vehicle which is within the range of the communication area or service area of the gateway (traffic light) receives the information signal and then transfers it to the mobile node which is outside the range of the service area. This sort of incorporation of multi hop transmission has been verified in [79], where the result analysis confirms that the low BER in the range of 10^{-7} has been obtained for the communication distance range of 50 meters.

For the implementation of revolutionary ITS protocols, VLC seems to be a potential candidate. In the scenario of cooperative ITS, interconnection of vehicles is mandatory in order to facilitate for the emergence of many critical automatic and assisted driving applications, car platooning, queue circumvention, etc. Essentially, in case of cooperative vehicular communications, it is an indispensable aspect for the vehicles to relay several critical information pertaining to the safety to the incoming vehicular units or infrastructural units in a reliable manner at a rapid speed. Within this context, the authors in [363] propose and validate a novel infrastructure-to-vehicle-to-vehicle (I2V2V) VLC system for ITS. To enable seamless relaying of safety critical information among all neighboring vehicles, a digital active Decode-and-relay (ADR) stage seems to be the most appealing solution. By embedding ADR module into the traffic infrastructure and vehicles, it is possible to decode and relay the information received from a regular LED traffic light toward the incoming vehicular units. The work in [363] validates the ADR VLC system along with the presentation of thorough statistical analysis of the distribution of packet error rate (PER). The result analysis emphasizes that the proposed system was successful in the transmission of packets for distances up to 50 meters. Furthermore, the simulation results show that the proposed system is highly reliable in terms of attainment of ultra-low, sub-ms latencies for distances up to 30 meters. Furthermore, for distances up to 50 meters, the granted latencies was below 10 ms. When compared with the traditional RF-based wireless technologies, like the Wi-Fi and LTE, the attained latencies of this work are far shorter than the ones achieved by the aforementioned RF-based technologies. Moreover, the proposed ADR-based I2V2V VLC prototype is reconciling with the IEEE 802.15.7 standard.

7.3.3 DEVELOPMENT OF HYBRID RF AND VLC-BASED WIRELESS COMMUNICATION NETWORKS FOR VEHICULAR APPLICATIONS

The global demand for mobile data has been witnessed in the past few years, and this is evident when CISCO published a report stating that the global mobile data traffic is continuously increasing at an alarming rate and it is expected to continually increase. Consequently, in this regard, it is vital to exploit Heterogeneous Networks (HetNets) which is an excellent pragmatic approach to meet this exponential increase in demand for data [225]. In case of indoor applications, for the purpose of ensuring high data rate transfer, it is vital to employ HetNets. Generally, the co-existence of different wireless communication-based technologies aids in facilitating the high data transfer by exploiting the advantages of each wireless technology which forms the HetNet. The works in [414] illustrate that by deploying RF/VLC-based wireless communication technologies in an indoor room environment, enhanced coverage has been obtained. Further, it is apparent that VLC technology is highly referred to as electromagnetic interference-free technology, and it is also anticipated that it can easily go well with RF-based wireless communication-based technologies. A HetNet which is formed by allowing the co-existence of VLC and RF imparts enthralling advantages such as additional bandwidth in order to meet the rapid increase in thirst for Internet related services as well as bandwidth density benefits to alleviate the congestion problems where there is a requirement for high data rate video streaming, a problem which is rapidly encountered in RF-based wireless communications. This could be better understood with the help of a real time scenario, where the provision of non-interfering VLC channels aids in overcoming the Wi-Fi congestion problem which usually emanates in highly crowded public places like road stations, hospitals, railway stations, schools, hotels, multinational companies, etc.

In general, most of the mobile devices which are rapidly employed in an indoor area do not have a specific lighting function to allow for the data uploading. Hence, in this regard, it is foreseen that, it is vital to deploy a second means of wireless communication-based technology for the purpose to render high data rate communication. Thus, multi-Gbps VLC-based wireless communication-based technologies can be used for data receiving, and RF-based wireless communication facilitates the ease of data uploading. Meanwhile, in case of RF-restricted or prohibited areas, VLC-based wireless communication can be used in combination with infrared-based communication. Moreover, VLC transmitter illumination is confined to a small area; therefore, standalone VLC networks can be interfaced with RF-based networks to ensure ubiquitous data coverage [381]. Thus, the combination of such hybrid HetNets is popularly known to enhance the high data rate information transfer, throughput, reliability of services, etc. These aforesaid benefits of HetNets can be considered to facilitate their usage in vehicular communications for the purpose of enhancing the reliability of services in vehicular networks. Despite VLC offering several significant advantages in case of vehicular communications, the main and biggest drawback associated with it is its strict or limited LOS coverage. Thus, the combination of VLC with IEEE 802.15.11*p* standard, i.e., the DSRC which works in the dedicated 5.9 GHz spectrum range will definitely ensure high link reliability, one of the most indispensable aspects in vehicular safety applications. As stated in the previous section, DSRC can be regarded as a mature technology which guarantees long-distance vehicular communications. Contrarily, VLC is highly capable in case of huge traffic densities because of its wide geographical distribution; however, it cannot impart comparable distances like those of DSRC.

Moving a step further, autonomous vehicular technology is gaining a significant momentum which is triggered by Google's driverless car, and in recent times, autonomous vehicular technology is evolving as a burning research topic [72]. Autonomous vehicles have the capability to support autonomous acceleration, steering and braking. It is believed that vehicle platooning applications are a part of autonomous driving. Still, while the advancements in these smart vehicular technologies are rapidly progressing, cooperative adaptive cruise control (CACC) comes into existence where autonomous vehicles cruise themselves by allowing for the interchange of each other's data. Generally, vehicle platooning is a technique that allows for the organization of CACC enabled vehicles into groups of closely following vehicles which can be termed as a platoon. It is ascertained that platoon-based applications tend to improve the safety of the transportation sector, since a faster response to certain events is obtained. Platoons comprise of a platoon driver controlling the platoon, and a set of platoon followers that follow the platoon leader by adjusting their speed. The pictorial representation of vehicular platooning is shown in Fig. 7.15. Particularly, the stability of a platoon is determined when the platoon followers are

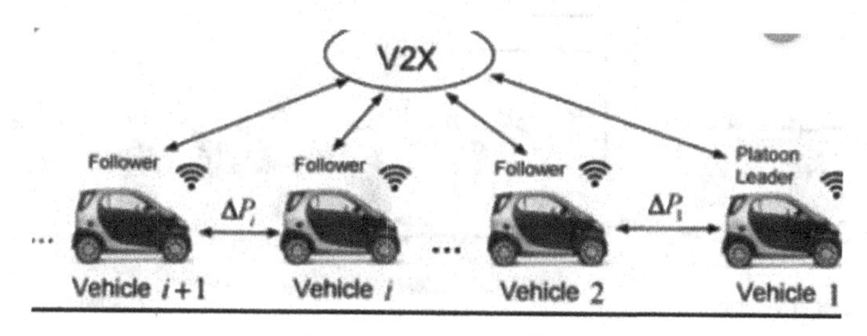

Figure 7.15: Platooning vehicular applications [482].

able to exploit the CACC for adjusting their speed and distance to the leader in terms of variation over time. It is shown in the literature that platooning-based vehicular applications generally adopt the dominant RF-based vehicular technology, which is nothing but the DSRC. However, there are certain drawbacks associated with DSRC: first, DSRC suffers from the limited RF spectrum because the continuous intensification of wireless mobile devices is creating a huge pressure on the existing RF spectrum. A second major drawback of DSRC is that the exploitation of omnidirectional antennas makes DSRC susceptible to all sorts of adversaries within the transmission range. Third, congestion on the DSRC channel can lead to packet collision which seriously degrades the stability of the platoon and undermines the safety of the platoon. Hence, VLC can be regarded as a promising complementary technology having the potential to address the issues which generally arise when DSRC is exploited for platooning vehicular applications. This can be illustrated in [10], where this work exemplifies the application of VLC for exchange of information between the platoon members. A complete VLC system model which was based on ON-OFF keying modulation format has been developed taking into consideration the effects of background noise, incidence angle and receiver electrical bandwidth. Precise BER calculations were made by varying intervehicle distances. Finally, the impact of exploitation of VLC on the performance of platooning controlled under longitudinal and lateral control was evaluated by developing a Simulink model. This study demonstrates the prominence of VLC and its feasibility for application in platooning control even in the presence of optical noise at significant levels, and the developed system model achieved a BER of 10^{-6} which is equivalent to SNR of around 14.6 dB and imparted the potential to achieve LOS communication distance range of up to 7 meters and with up to 40 degrees of road curvature. Thus, there should be a complementary cooperation between these two wireless communication-based technologies like VLC and RF, so that these both technologies can work together and strive to support a diverse number of requirements of the vehicular applications.

Much similar work regarding the usage of VLC for enabling vehicular platooning applications can be witnessed in [11]. This work signifies that VLC technology has been exploited for a platoon of autonomous vehicles. A cost-effective, low-latency and simple outdoor VLC prototype has been presented that allows for easy installation as vehicular tail-lighting system. Fig. 7.16 clearly depicts the usage of VLC as

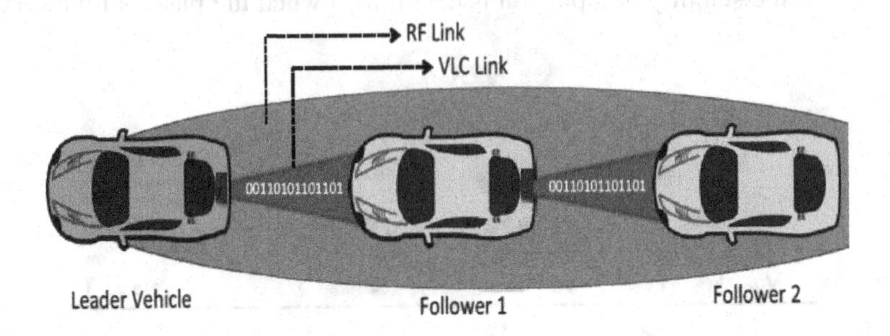

Figure 7.16: Exploitation of VLC for intervehicle platoon communications [11].

vehicular tail-lighting units for the purpose of exchange of information between the platoon members, and this in turn plays a vital role to reduce the channel congestion. The authors in [11] develop the architecture of a VLC system by paying immense attention to the resilience of VLC links to ambient noise and the amount of communication range that can be ensured. A Simulink model has been developed for the purpose of illustrating the benefit of VLC for platooning control. As stated in the earlier challenges where the authors have followed similar approaches for enhancing the range of communication as well as mitigating the effects of ambient noise by taking into consideration the usage of optical filtering at the receiving end as well as by narrowing the receiver's FOV. From the experimental results, it can be analyzed that the developed system has the potential to provide 30 meters of intervehicle communication with 36 ms of latency under sunny day conditions. These aforementioned works convey the prominence of exploiting VLC along with DSRC for improving the performance of platooning applications.

On the other hand, DSRC can be used in certain cases where VLC technology imposes certain drawbacks for vehicular platooning applications. Mainly, using VLC, the messages to the nearby vehicles which are not within the range of LOS, are conveyed through a multi hop manner. This process of transmission of information to the neighboring vehicles introduces a certain amount of delay especially during the transmission of event-driven or high priority messages; therefore, multi hop transmissions definitely result in end-to-end delay. This kind of problem can be overcome by using the combined usage of DSRC along with VLC for reducing the end-to-end delay and enhancing the reliability of the communication link. It would be more reliable to deploy both radio and VLC interfaces to forward high priority event-driven messages from the platoon leader to the nearby vehicles which are also termed as platoon followers. Thus, by employing such kind of approach reduces the delay incurred while delivering messages under RF jamming attacks [232]. In this approach, the platoon leader vehicle communicates the message which includes its identification module, acceleration, speed, position, etc., periodically by using both radio and

VLC interface channels. Under such circumstances, even if one of the neighboring vehicles fails to receive the message disseminated by the leading vehicle sent through radio interface, there is a chance that it will assuredly receive it via VLC interface. In this manner, by employing both RF and VLC interfaces, there is a scope to reduce the end-to-end delay of message delivery from the leader vehicle to the neighboring vehicles which are not within the LOS.

Even though IEEE 802.11*p* standard has a significant potential to be employed for vehicle to vehicle communication, there is the possibility for the emergence of certain concerns regarding its usage for the control of platoons, as it is prone to packet losses in highly dense scenarios due to the prevalence of high congestion. The work in [435] delineates that VLC can be used as a backup for IEEE 802.11*p* for the purpose of augmenting the reliability, scalability and safety of the platooning applications. Thus, it is vital to deploy VLC as a backup/offloading communication technology to control and enhance the safety of platooning applications. The work in [168] elucidates the combination of DSRC with other alternative technologies like IR to address the drawbacks associated with DSRC for platooning applications. The work in [53] signifies the viability of exploiting VLC-based wireless communication technology in vehicular networks for cellular network offloading. The authors in this work firstly characterize the performance of VLC under dynamic outdoor conditions by taking into consideration the major advantages and drawbacks exhibited by VLC. Secondly, in this work, several investigations were carried out to determine the level of its connectivity in an urban environment. Furthermore, this work analyzes the feasibility of VLC to deliver data for delay tolerant applications without making use of cellular network. It can be inferred from the simulation results that by taking into consideration a realistic urban environment, even the limited number of road side units (i.e., traffic lights) which are equipped with VLC creates the flexibility to offload more than 90% of the cellular resources. In [311], an ITS based on the hybrid combination of visible light and radio communications has been developed for positioning applications. The combined advantages of both these technologies, which include the directionality of VLC and the extended coverage of radio communications, can be used to enhance the services to the vehicle drivers on roads.

Vehicles which are located in the platoon generally communicate with each other through the IEEE 802.11*p* standard. In spite of its assurance of long communication range, the IEEE 802.11*p* standard is vulnerable to several security attacks imposed from adversaries. In this regard, VLC exhibits its potential to overcome such vulnerabilities by making use of the directivity and impermeability of light. This is valid because the light signals do not penetrate through several objects like walls, partitions, etc. Thus, the data can be protected from the intruder hijacking it. However, this technology is also not devoid of drawbacks. The primary reason is that when VLC is exploited in a vehicle platoon to render wireless communication support among the vehicles, there is a chance for the degradation of the stability of the platoon due to VLC being sensitive to several environmental effects. Based on these grounds, one best way to overcome the security vulnerabilities in vehicular platooning applications is to utilize the advantages of both IEEE 802.11*p* standard and VLC.

As a result, a hybrid IEEE 802.11*p*/VLC-based security protocol can be designed for vehicular platooning applications.

The research contribution in [483] proposes a hybrid security protocol, namely SP-VLC protocol, exclusively dedicated for guaranteeing the stability of the platoon, acquiring the platoon maneuvers under channel overhearing, jamming, data packet injection and platoon maneuver attacks. A platoon maneuver attack is defined by taking into consideration several scenarios depicting the possibility for a malicious actor to transmit a fake maneuver packet. Thus, by considering the susceptibility of data within the vehicular platoon to several security attacks, this work encompasses several mechanisms for the establishment of secret key, authentication of messages, identification of jamming and transmission of data over both IEEE 802.11*p* and VLC channels. The authors present simulation platforms by exploiting the realistic vehicle mobility model and realistic VLC, IEEE 802.11*p* channel models. The extensive simulations carried out in this work illustrate the performance of SP-VLC protocol under various possible security attacks. From the simulation results, it is evident that the proposed SP-VLC protocol attains less than that of 0.1% difference in the speed of the platoon members and accomplishes any kind of maneuvers without any kind of interference from attackers.

With the rapid increase in the number of vehicles and mobile terminals, VANET is gaining a significant amount of attention in recent times. The distinguished nature of VANET is that it facilitates the exchange of the status of traffic among vehicles in order to enhance the safety of the people travelling on roads and to save millions of lives every year. Additionally, VANETs are capable to offload data flows from the cellular network. Particularly, the communication links in VANETs are established with the motive to exchange information between vehicles. In this context, the distribution characteristics of the link such as the network topology has a pronounced impact on the performance of the network; also it is vital to take into consideration the effects of connectivity, delay, power consumption and interference. Thus, topology control (TC) is one important aspect that facilitates a reliable link establishment in ad-hoc networks. From the earlier studies, it is proclaimed that conventional VANETs exploit radio frequencies in order to allow for vehicles to communicate with each other. The major limitation that is associated with RF-based VANET is that, upon increasing the density of the vehicles, there is a dramatic increase in the amount of interference which eventually results in the reduction of the efficiency of the network. Additionally, the limited RF bandwidth curtails the performance of the network. Therefore, in order to alleviate this limited and overcrowded RF bandwidth, VLC is emerging lately as a complement candidate.

The potential of VLC in the area of vehicular communications is rapidly flourishing due to its stunning benefits, namely, wide licensed-free spectrum, high power efficiency, high spatial reuse ratio and low interference. However, the major drawback associated with VLC is that the limited range of communication offered by VLC impedes a reliable vehicle-to-vehicle communication. Along these lines, by exploiting the combined advantages of both RF and VLC, a hybrid VLC/RF structure turns out to be the most pragmatic approach that allows for the efficient and reliable realization of VANET. On these grounds, it is vital to study the control mechanisms for the hybrid VLC/RF-VANET in order to effectively exploit the advantages of both VLC and RF links by circumventing their drawbacks. Different vehicles within the VANET communicate with each other without the management of any center infrastructure in VANET. Moreover, different vehicles might have different distinctive preferences. In this regard, it is of utmost importance to develop a distributed and non-cooperative TC scheme for hybrid VLC/RF VANET in order to expedite the formulation of VLC and RF links among various vehicles to fulfill some of the properties which include the power consumption, interference control, connectivity and delay requirements. The TC mechanisms that were exclusively designed for RF-based ad-hoc networks cannot be implemented in a straightforward manner for hybrid VLC/RF networks due to the peculiar features offered by both VLC and RF links.

Based on these grounds, the research contribution in [103] models the TC problem for the hybrid VLC/RF VANET as a potential game. By exploiting the different features offered by both VLC and RF links, a distributed TC algorithm is proposed to handle the TC game. Based upon the vehicle's local knowledge about the topology of the VANET, both the RF and VLC transmitting power of each vehicle is locally adapted and can converge to a Nash Equilibrium state. The simulation results of this work demonstrate the fact that when compared with conventional RF-based VANET, the proposed TC algorithm for the hybrid VLC/RF-VANET significantly reduces the power consumption and interference among the links without sacrificing the delay-limited connectivity. Thus, this challenge can be concluded by stating that the combination of wireless communication-based technologies is required to enhance the reliability of services for vehicular communications.

7.3.4 ENSURING HIGH DATA RATE VEHICULAR COMMUNICATIONS BY USING COMPLEX MULTICARRIER MODULATION FORMATS

In contrast to the indoor applications, huge amount of data rates in the range of several tens of Gbps can be ensured by exploiting robust multicarrier modulation formats like optical OFDM and its variants or other multilevel codes. However, while dealing with the outdoor scenarios, it is quite difficult to achieve high data rate transfer due to dissimilar channel conditions. However, from the literature, it can be evidenced that several research efforts strived to enhance a high data rate V2V or V2I communication by exploiting VLC as wireless communication-based technology. The earlier works show that upon using photodiodes as receiving components in a VLC-based vehicular system, significantly better communication range has been attained at the cost of reduction in data rate. However, by employing Image Sensors-based receivers in VLC-based vehicular systems, the data rate can be enhanced. The authors in [429] proposed a new tracking method by exploiting both Image Sensor and photodiode for enabling visible light road to vehicle communication. In this sort of communication system, Image Sensor with a wide FOV is meant to receive the tracking signal, and the photodiode with narrow FOV is used to receive the communication signal. The reason behind employing high FOV Image Sensor is to ensure a wide service range, and when compared to the conventional methods, a large amount of communication signal is received. Furthermore, the photodiode with narrow FOV helps to overcome the noise which is induced by the presence of ambient sunlight and other artificial light sources. The numerical analysis of this proposed method delineates that a data rate of 10 Mbps can be obtained when the communication service range was 48 meters. However, as the range of communication increases to 90 meters, the data rate deteriorates to 4.8 Kbps. Thus, the literature articulates that higher data rate visible light vehicular communication can be accomplished by making use of camera-based VLC receivers.

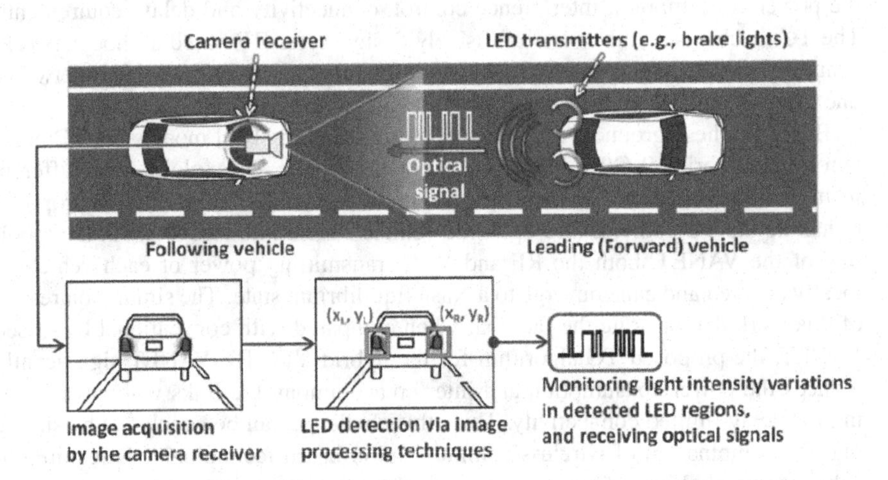

Figure 7.17: Pictorial representation of vehicle to vehicle communication where the moving vehicle is fitted with camera receiver [467].

The typical scenario of optical vehicle to vehicle communication is illustrated in Figure 7.17, where the LED transmitters are fitted into the leading vehicle (LV), and the following vehicle (FV) has the camera receiver. Particularly, the vehicle LED light sources such as the tail lights, brake lights and head lights can be exploited as the LED transmitters. In general, the functionality of the LV is to collect the relevant data regarding the speed, etc., and to transmit this data to the FV by means of optical signals. The camera receiver that is located on the FV captures the images as well as exhibits its potential to locate the LED regions by means of image processing techniques. In this manner, based upon the detected LED regions, the receiver system monitors the variations in the intensity of light and receives the optical signals. An important aspect of the camera receiver is that it is not only used for the detection of LEDs, but also for rendering safety and comfort applications which include lane detection and pedestrian detection.

Few such distinguished works that strived to enhance the data rates can be found in the references [187, 466, 467, 542]. The work in [466] reports that the authors have employed a LED-based transmitter and a camera receiver for automotive applications. Typically, the automotive applications require data rates in the orders of Mbps, and they must be capable to quickly detect the LEDs from an image. This task could be made possible by incorporating certain improvements to the Image Sensor mounted on a camera receiver. This work signifies that a novel OWC system that is much suitable for automotive applications has been designed by taking into consideration the OCI that has been developed by means of CMOS technology. The speciality of this type of receiver is discussed earlier where OCI comprises of CPx to promptly respond to the variations in optical intensity. Additionally, a new quick LED detection technique based upon 1-bit flag image was formulated. It is interesting to note that the communication pixels, image pixels and the relevant circuits which include the 1-bit flag image output circuits were then interfaced into the OCI. The experimental demonstrations of this work illustrate the potential of the proposed system to render data rate of 20 Mb/s/pixel without the LED detection and imparts a data rate of 15 Mb/s/pixel with a 16.6 ms real-time LED detection. Similarly the work [467] proposes a V2V communication system which exploits LED transmitter and a camera receiver that is based on the special CMOS Image Sensor, i.e., OCI. The OCI has the CPx for responding to the variations of light intensity as well as flag image in which only high-intensity light sources such as the LEDs have evolved. In this work, various experiments were carried out under real driving and outdoor lighting conditions. From these experimental field trials, various vehicle internal data which include the speed and image data (320 × 240 color) were transmitted successfully. Besides, 13.0 fps image data reception was accomplished while driving outside.

The work in [542] presents the V2I and V2V communications using VLC technology. Very similar to the earlier mentioned works as stated in [466] and [467], this work also makes use of an Image Sensor for enabling V2I and V2V communications. This work illustrates two different scenarios, where the first scenario demonstrates the realization of a V2I-based VLC system that exploits an LED array as the transmitting unit and a high-frame rate CMOS Image Sensor as the receiver. Furthermore, with the aid of field trials, the developed system, i.e., V2I-VLC, was capable of accomplishing a real-time transmission of an audio signal ensuring a data rate of 32 Kbps. Meanwhile, the V2V-based VLC system exploits the OCI with the primary motive to receive high-speed optical signals. In order to evaluate the performance of the developed V2V-VLC system, upon conducting field trials, the system illustrates its potential to render data rate of 10 Mbps. Besides, the experimental results demonstrate that the camera receiver is proficient enough to enable perfect reception of a high-speed signal along with the concurrent acquisition of several vehicle internal data such as the speed of the vehicle, the vehicle's ID, etc. Similarly, the work in [187] also describes the adequacy of a V2V-based VLC system that exploits the OCI-based camera receiver. In order to effectuate a data transmission of 54 Mbps that is quite often regarded as the maximum data rate that can be accomplished by IEEE 802.11*p* standard for enabling vehicle to everything (V2X) communication, this work develops a more progressive version of OCI-based automotive VLC system. The distinguished aspect that needs to be taken into consideration is that the developed system considers the robust multicarrier modulation format like OFDM to furnish a higher data rate transmission of information. The experimental results of this work exemplify that upon introduction of OFDM in the proposed V2V-based

VLC system, a five-fold increase in data rate can be substantiated. The experimental BER analysis of this work shows that the proposed system exhibits its caliber to achieve 45 Mbps without the presence of any bit errors, while delivering a data rate of 55 Mbps at a BER of 10^{-5}, respectively.

Furthermore, other distinguished work as delineated in [486] specifies that the authors evaluated the performance of a V2V-VLC system by making use of realistic automotive light sources. Specifically, stunning results can be evidenced from this work where the V2V-VLC system takes into account a measured headlamp beam pattern model. Additionally, this work considers the impact of the road reflected light on the performance of the overall system. The result analysis emphasizes that depending upon the location of the photodiode in the car, significantly higher data rates in the orders of 50 Mbps which cover a distance of up to 70 m can be ensured. Thus, by taking a glance of these published articles, it can be deduced that as the technology advances, it is vital to ensure higher data rates for long-distance applications as well. Thus, on these grounds, future research should be directed toward exploiting and analyzing the characteristics of the indoor modulation formats that can be employed for outdoor applications as well. In this manner, a high data rate communication can be ensured for long distances.

7.3.5 AUGMENTING MOBILITY OF V2V COMMUNICATIONS

As discussed earlier, in order to combat the effects of noise in vehicular communications to attain an enhancement in the SNR, several methods have been proposed in the literature, such as narrowing the receiver's FOV with the intent to combat the background noise which seems to be the simplest technique to combat the noise in vehicular communications. This solution is most appealing to yield an enhancement in the SNR; however, the stringent requirement of LOS between the VLC transmitter and receiver will impose a major problem. This is because upon narrowing the receiver's FOV, the reception angle becomes narrow which in turn curtails the mobility of the vehicles. Consequently, this implies that in order for VLC to be exploited for vehicular communication, it should adhere to the requirements of mobility of the vehicles. Along these lines, several research efforts worked to augment the mobility of the vehicles when VLC is exploited as a wireless communication medium between them. By means of experimental validations, the research contribution in [447] affirms the establishment of VLC connection between two moving vehicles. This work introduces vehicle-to-vehicle messaging services by exploiting off-the-shelf LEDs and photodiodes. The light which is emanated from the brake lamps of the vehicle is employed to disseminate messages dedicated for emergency hand braking in a manner to warn the vehicle following the main/leading vehicle so that it can take the necessary precautionary measures to circumvent accidents. The authors in this work designed a prototype for V2V communication which exhibits low complexity and high reliability.

The experimental validations of this work infer that the designed prototype is capable to detect hard brakes from a distance of 20 m. Additionally, in order to reduce the occurrence of accidents and to enhance the safety of the drivers travelling on roads, the designed prototype can also furnish early warning to the drivers who are maintaining speeds up to 80 Km/h. Thus, in this manner, the designed V2V prototype imparts warning messages related to emergency braking, forward collision and control loss. Eventually, this information is very vital to the drivers to help them slow down the speed of their vehicles in a manner to avert mishaps. Nonetheless, these kinds of experimental prototypes pertaining to V2V technology are implemented by relatively aligning the VLC transmitter and receiver that are installed within the vehicles. However, pertaining to the real-time scenarios, it is always not feasible for the transmitter and receiver belonging to the corresponding leading and following vehicles to be perfectly aligned. For instance, in certain scenarios, the vehicles also communicate with the road side units, like traffic lights and so on. The traffic lights installed at heights between 2.5 and 5 m above the road pose a critical issue due to the fact that such installations at those heights will substantially affect the overall performance of the I2V and V2V communication systems due to the limited service area offered.

Therefore, in order to tackle this issue, the best possible solution was to deploy a tracking mechanism to enable the road-to-vehicle communication. The related work can be reported in [428] and [431]. The proposed tracking method comprises of Image Sensor with a wide FOV and a photodiode with a narrow FOV. The Image Sensor is meant to receive the tracking signal, whereas the photodiode is dedicated to receive the communication signal. At the same FOV, when the photodiode is compared with the Image Sensor, it is observed that photodiode is severely affected by the background noise including sunlight, road light and noise originating from other artificial light sources which are received along with the desired signal and simultaneously processed by the photodiode. Contrarily, the Image Sensor segments the received noise into pixels in order to counteract its effect. Thus, this promulgates that it is necessary to narrow the FOV of the photodiode in order to reduce the effects of background noise than that of the Image Sensor. However, in doing so, the range of communication becomes shorter. Thus, in order to alleviate this drawback, i.e., with the motive to enhance the range of communication, the work in [428] introduces a tracking mechanism that jointly exploits both photodiode and Image Sensor receiver. Since the FOV of the Image Sensor is wider, the communication range can be enhanced. Particularly, when compared with the conventional methods, the service range, where the communication signal can be received is enhanced. On the other hand, upon narrowing the FOV of the photodiode, a considerable amount of reduction in the effect of background noise can be witnessed. Thus, this facilitates the possibility to deliver high-speed communication.

The proposed tracking mechanism in the aforementioned work, i.e., [428], was verified by means of both numerical analysis and experimental validations. From the numerical analysis, it is evident that for a data rate of 10 Mbps, the common service range of the proposed method was 48 m. Thus, when compared to vehicle information and communication system (VICS) infrared beacon system, the proposed tracking mechanism was able to impart 137 times more bits reception. Additionally, from the experimental results, tracking and receiving data over 90 m away from the LED traffic light was demonstrated, while the achievable data rate was only 4.8 Kbps. In the same manner, [431] also worked in the same direction to enhance the range of communication. The authors in this work proposed a tracking mechanism for facilitating road-to-vehicle communication. The authors rely on imaging optics in order to receive information over long distance, as well as two cameras were employed to interpret the change in position of both transmitter and receiver. Predominantly, the major advantage associated with camera is perceiving light from a long distance. Furthermore, a vibrational correction technique was also incorporated to the receiver for the sake of minimizing the vibrational problems which generally occur in dynamic conditions. Thus, these aforesaid research contributions employ the joint combination of photodiode and Image Sensors/camera-based receivers to exploit the advantages of both, and the proposed tracking mechanism is efficient in terms of mobility, communication range, and robustness to noise. Despite offering these remarkable benefits, the wide scale acceptability of this method is quite low due to high complexity in its implementation.

A comparatively complementary solution which extends the region of signal reception was demonstrated for an infrastructure-to-car communication system that employs VLC as a means of wireless communication technology [240]. In contrast to the aforementioned works in [428] and [431], this work replaces the camera system by means of light sensor array that is utilized to evaluate the transmitter-receiver received power. Based upon the changes in the received signal power, the receiver sensor's elevation angle is adjusted in order to compensate for the alignment angle variation. In this set-up, a VLC transmitter is placed with its axis parallel to the bottom surface and a photo receiver is mounted on a rotatable board that is equipped with two ambient light sensors. The main functionality of the photo receiver is to track the transmitter by simply rotating the receiver board comprising of two ambient light sensors. Upon comparing the measured optical powers of the two sensors, the receiver decides the direction of large power and keeps track of that particular direction by changing the elevation angle. Furthermore, an optical lens is exploited as well in order to increase the amount of received power. The proposed method was experimentally validated, and the result analysis implies that the signal reception region increases substantially. In addition to the aforementioned works, the other way to enhance the range of communication is to exploit more photodiodes that are aligned for different angles of reception. Thereupon, the signal processing unit should exhibit its potential to evaluate the signals pertaining to each photodetector which will in turn enable it to decide which signal should be employed for reliable message reconstruction. Additionally, sensors can also be placed on each side of the vehicle [13]. If at least one of the sensors is properly aligned with the transmitter, then the incoming data pertaining to the desired vehicle can be successfully received. Moreover, it is vital for this kind of solution to consider the optimal configuration analysis for both receivers and emitters. The optimal configuration parameters play a major role in determining the performance of the vehicular links.

Specifically, the important parameters which need to be taken into account in order to determine the optimal configuration are the mounting height and the mounting angle of the road side unit's transmitter, i.e., the traffic light and the mounting angles corresponding to the on-vehicle receiver. Therefore, while determining the optimal configuration for the road side units and for the on-vehicle receiver, a compromise occurs between mobility versus the range of communication (i.e., communication distance). Based upon [443], the optimal configuration implies the longest available communication distances. In several studies in the literature it is stated that in order to enhance the communication distance, it is vital to employ the optimized irradiation pattern of the LED and to optimally place the LED within the LED traffic light. Additionally, when the VLC transmitter within the vehicular unit is intended to serve long-distance communication, it is appropriate for the LED to have a narrow angle emission pattern. However, even though the range of communication is increased, the mobility is decreased. Therefore, toward this end, an effective solution to ensure an enhanced mobility without compromising the range of communication is to increase the number of LEDs present inside the road side unit, i.e., the traffic light. Thus, narrow emission angle LEDs can facilitate long range communication, while the wide angle emission LEDs enable short range wide angle transmissions; therefore, the mobility enhances. There are many solutions proposed in the literature in an effort to enhance the mobility pertaining to VLC-based vehicular communications. The work in [127] proposes an improved V2V and V2I model by taking into consideration the position and posture of the vehicles. With the intent to achieve an enhanced diversity gain as well as to overcome the interruption between moving vehicles, this work proposes a cooperative diversity scheme. A typical scenario depicting the VLC-based cooperative diversity scheme is depicted in Fig. 7.18. Therefore,

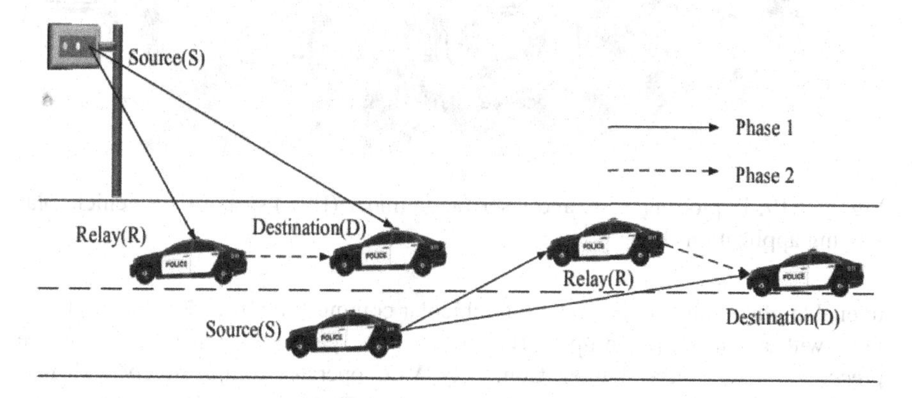

Figure 7.18: VLC-based cooperative diversity [127].

to enhance the performance of VLC in VANETs, a cooperative diversity scheme can be employed. As evident from the figure, there are three types of nodes present, i.e., the source node, relay node and the destination node, respectively. Particularly, the road side unit (i.e., traffic light), either the vehicles that transmit optical signals or the vehicles that request data downloading are categorized as the source nodes. The cooperative diversity scheme comprises of two phases denoted by Phase 1 and Phase 2. Firstly, in Phase 1, the source node broadcasts the optical signals which contain data pertaining to the request data, and so on. The broadcasted optical signals are received by both the destination nodes as well as other vehicles. While in the second phase, the relay accomplishes the task of forwarding the optical signals to the destination node. Finally, at the destination node the received signal that is attained from the source and the relay is combined by means of maximum ratio combining (MRC) technique. Furthermore, from the simulation results, it can affirmed that the cooperative diversity scheme significantly decreases the BER as well as circumvents the VLC interruption that usually emanates due to the motion of the vehicles.

The work in [9] takes into account the laser range finder (LRF) to serve as a backup for the VLC link in case of autonomous driving applications. When compared to VLC technology, the LRF can significantly offer more coverage area and improved viewing angle. Consequently, the benefits offered by both VLC and LRF can be fully exploited, where the VLC technology renders support to exchange information among the members of the platoon, while LRF is dedicated to estimate the intervehicle distances. By assessing the different metrics which include the LRF confidence value, vehicle angular orientation and VLC link latency, a hand-over algorithm has been proposed to manage the switching process for the scenario of occurrence of any failure. Figure 7.19 illustrates the combination of both technologies

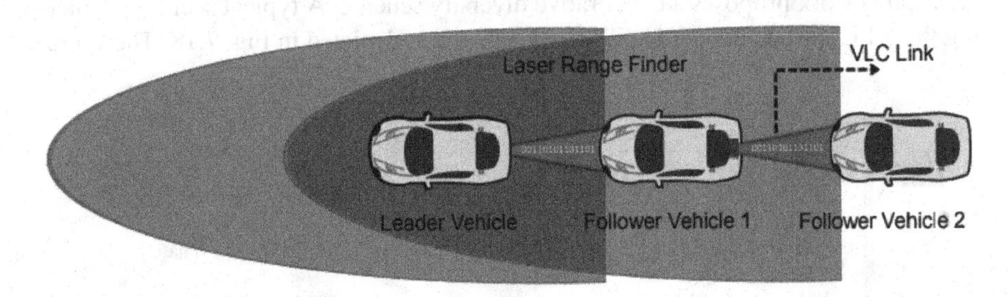

Figure 7.19: Exploiting VLC and laser range finder (LRF) systems for vehicle platooning applications [9].

to enable platooning applications for vehicular communications. LRF constitutes of FOV with a sensing range up to 100 m, which is quite sufficient for tracking the preceding vehicle in a platoon. Contrarily, VLC operates comparatively with narrow FOV in order to filter out the ambient noise. Therefore, the work as portrayed in [9] combines the joint advantages of both the technologies for platooning applications, where the vehicle members within the platoon can seamless exchange the status regarding the vehicles, GPS positions, etc. The important benefit by integrating VLC with LRF is that in some particular situations, especially due to mobile conditions where the VLC technology fails to render communication support, due to the exploitation of LRF, the vehicles in the platoon are capable to main intervehicle distances, thereby facilitating the seamless exchange of communication. Similarly, the authors in [415] solve the problem of mobility in vehicular communications by formulating a LOS link with the aid of a relay vehicle in order to enable the establishment of communication link.

By taking into consideration the estimated transmitter-receiver angles, an optimized solution pertaining to the mobility problem can be formulated and substantially this paved a platform for the emergence of optimized receiver designs. Specifically for the V2V communications, in several studies it is illustrated that the transceivers are relatively aligned to the vertical axis which implies that the receiver should comprise of a wide FOV for the horizontal axis in order to facilitate the data reception capability from the entire width of the road; as for the vertical axis, it is desired for the receiver to have a narrow FOV in order to counteract the effects of parasitic light, thereby enabling an enhanced range of communication. Conversely, for I2V communications, for the VLC transmitters, i.e., traffic lights, located at several meters above the road, it is mandatory for the FOV of the receiver to be narrow for the vertical axis and wide enough for the horizontal axis. Consequently, this sort of optimal design for the receiver will result in appreciable amount of enhancement of mobility while retaining robustness to noise. Additionally, the stringent requirement of LOS path which gradually hinders the mobility can be substantially solved by employing multi hop communications [77]. Generally, in case of multi hop communications, the highest priority messages will be easily delivered among the vehicles. In case of indoor applications, mobility can be intensified by exploiting angle diversity receivers which usually comprise of multiple narrow FOVs of different FOV photodiodes [84]. However, pertaining to outdoor communications, a similar such concept has been proposed in [82], where an adaptive FOV receiver can be exploited to analyze the environmental conditions by making use of several sensors. Consequently, based upon the sensed data, the receiver should be able to adjust its FOV. In general, this adjustment of FOV can be made possible by means of simple mechanics or by placing a transparent LCD display in front of the photodiode. However, in spite of the availability of significant research efforts to enhance the mobility pertaining to vehicular communications, future research should further strive to determine the optimal receiver positioning in order to enhance mobility.

7.3.6 ENABLING VISIBLE LIGHT POSITIONING FOR VEHICULAR APPLICATIONS

In order to enhance the well-being of the people travelling on roads, it is of utmost importance to maintain the information pertaining to the position of the vehicle as well as its surroundings. Global positioning system (GPS) is the widely deployed tool that determines the location of the moving vehicle. Despite enabling accurate tracking of the moving vehicle, GPS fails to render its services especially under multipath propagation environments and link blockage scenarios. Thus, GPS cannot be deployed inside tunnels or underground areas or in urban valleys. Consequently, in addition to GPS, several other sophisticated positioning sensors, namely, radar or light detection and ranging (LIDAR) technologies, were adopted by modern vehicles in order to gather the information from the surrounding areas. The accuracy of these vehicles comes at the expense of increased cost to install them. Moreover, in the automotive field, radar sensors are widely employed mainly to ensure comfort as well as safety to moving drivers on roads. The rapid increase in the number of automotive radar sensors operating close to each other at the same time will result in reception of multiple signals from the other radar sensors. Thus, this sort of scenario will definitely lead to the emergence of interference which in turn leads to emanation of several problems like ghost targets aided with the significant drop in the SNR. Figure 7.20 clearly depicts the automotive interference scenario with one target and one interferer. The ghost targets are very difficult to handle because they create an illusion of real targets and confuse the receiver in isolating the correct target. Interference is defined as the coherent superimposition of two or more waves at one particular point in space. Consequently, several changes in the received signal's magnitude and phase can be observed due to this coherent addition of several signals. The received signal's amplitude can be either constructively added or decreased, i.e., the interference can be either in a form of constructive interference or destructive interference. As a result, the strength of the received signal fluctuates, thereby deteriorating the SNR and the overall system performance. Explicitly, even in the scenario of automotive applications, interference is present. The main source of interference in automotive radar applications is the disturbances caused by the presence of other radio devices like the vehicular radar, surveillance radar and other fixed links. These disturbances get intercepted at the receiving antenna of the corresponding radar, thereby leading to failure of the reception of the exact signal. Thus, these detrimental effects induced from other radar units into the vicinity of the victim radar receiver turn down into mutual interference of the radar. In the same manner, mutual interferences can appear in LIDAR as well [264, 265].

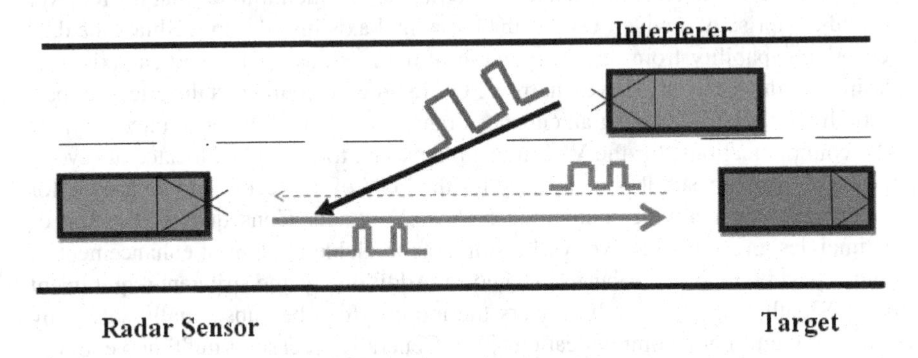

Figure 7.20: Simple interference scenario when vehicles were installed with radar sensors.

Taking into consideration the above scenarios, the major task of VLC technology in addition to ensuring of contemporaneous illumination and communication, is to ensure a cost-effective, computationally efficient and high accuracy positioning aid for the automotive applications. The efficiency of exploitation of VLC technology to guarantee positioning support has already been confirmed and verified for indoor applications. A high number of publications can be reported in the literature and few such renowned work can be found in [542]. Broadly, localization or positioning is one of the alluring applications of VLC technology. Even though several global navigation satellite systems, like the GPS, Compass, etc., are widely deployed navigation devices for cars, smart phones, etc., their performance is limited in indoor areas where the signals from satellites fail to reach the indoors. Thus, this will result in utter failure of these devices to detect the positions of the objects as well as users present in the indoor environment. Contrarily, the LEDs installed on top of the ceiling create a provision of the formulation of a small scale optical attocell network. Thus, this sort of isolated LED link exhibits its position by rendering the positioning information from the LED.

In an attempt to accurately detect the position of the target object, the strength of the received signal of the photodiode is fed as input to the sensor. The other methods are employed to detect the position of an object by using photodiode is through the time of arrival, time difference of arrival, phase of arrival and phase difference of arrival. However in case of white phosphorescent LEDs, the limited modulation bandwidth of the LEDs limits the localization accuracy. This is further fueled by the exploitation of photodiode at the receiving end. This is due to the fact that photodiode, upon reception of the incoming light, detects its intensity and fails to detect the angle of arrival (AOA) of the incoming light signal. Thus, it is vital to rely on Image Sensor which has the potential to detect the AOA of the incoming light signal along with its intensity. The specific advantage of Image Sensor is its capability to spatially separate sources because of the availability of a massive number of pixels. In addition to this feature, one more stunning feature of Image Sensor is that it has the ability to receive and process multiple transmitting sources. Figure 7.21 represents that the data transmitted from two different LEDs can be captured concurrently. As

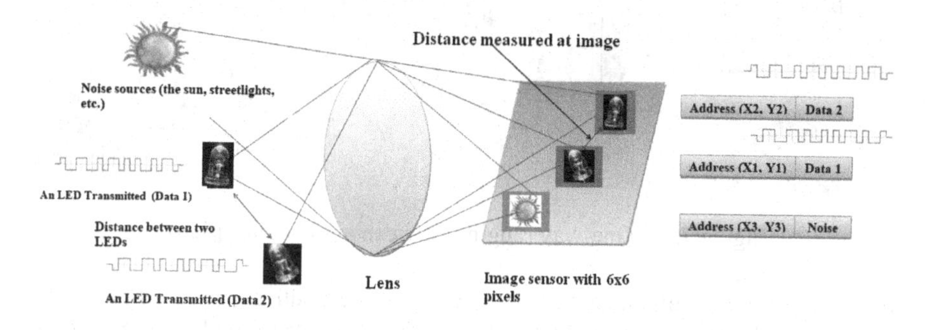

Figure 7.21: Demonstration of VLC positioning using Image Sensor.

delineated from the figure, the output of the Image Sensor forms a digital electronic representation of the captured scene and the relative position of the two LEDs is projected onto the Image Sensor. Thus, the relative positions of the two LEDs along with the distance between them are available at the receiver, then the positioning of the receiver can be performed by means of triangulation. Typically, positioning operation is very flexible to be implemented by means of VLC technology due to the following reasons:

1. Without any necessity for preinstalled data, LEDs possess the ability to transmit identification numbers (IDs) and other relevant data.
2. The fast switching speeds of the LEDs enable them to easily recognize the objects without any requirement of additional complex image processing techniques. It is also possible to recognize the objects in case of fast receiver moving scenarios, i.e., vehicle driving cases.
3. The omnipresence of LEDs fetches sufficient amount of data to enable localization support.

Extensive research efforts strived to enable visible light positioning support in the automotive applications as well. The research efforts in [423] clearly depict that the authors proposed a cost-effective, computationally efficient and high accuracy positioning technique which exploits the existing automobile illumination lighting units like the headlights and the taillights of the car. In case of GPS, in order to determine the time difference of arrival (TDOA) between two signals that are received from two different satellites with known positions, pseudorandom sequences are generally employed. On the other hand, the scenario with visible light positioning is quite different, in order to identify the relative position, typically simple high rate repetitive ON-OFF-Keying pattern can be used for measuring the phase difference of arrival [560]. The concept of one-way positioning is clearly elucidated in

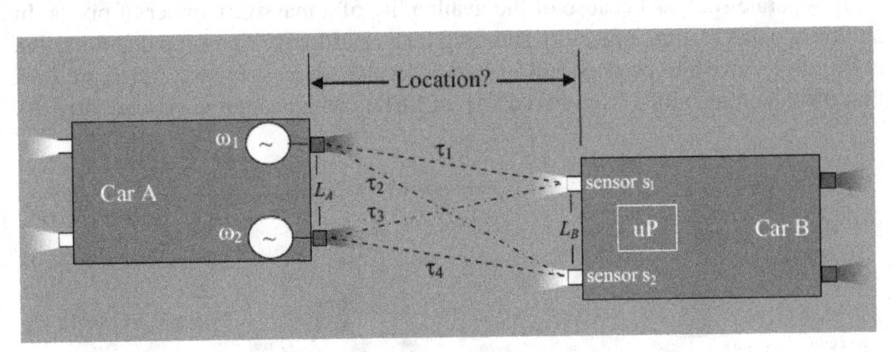

Figure 7.22: One-way ranging between automobiles [423].

Fig. 7.22, where the illuminating devices transmit the positioning signals processed by the photodiode with the intent to calculate the relative position between the cars. As delineated in the figure, each taillight in car A is amplitude modulated with a different frequency tone. Meanwhile, car B's headlight comprises of a photodetector that receives the signal from the taillights as well as processes the tone data in order to determine the relative position between the two vehicles. From these received tone signals, the phase difference of arrival can be determined which can later be translated to find out the distance differences between the two vehicles.

Similar to the research work in [423], taking into consideration the growing demand for vehicle positioning as well as with the motive to create environment awareness, the research contribution in [57] introduced a simple, cost-effective intervehicle distance measurement scheme with the aid of headlamps and taillights of the vehicles suitable for platooning applications. In case of platooning configurations, the leading vehicle is closely followed by means of a following vehicle. With the ease to adjust its trajectory, the following vehicle must be able to calculate its distance to the leading vehicle. This phenomenon is implemented by means of a modified phase-shift range finder. The underlying principle behind the implementation of the phase-shift range finder in the event of estimation of distances between moving vehicles is that it emits a periodic signal at frequency f toward the target. Upon receiving this periodic signal, the target reflects back this signal with a phase-shift of ψ which is proportional to the distance of the system or target designated by D. On measuring the phase shift ψ, the distance D can be determined as follows [57]:

$$D = \frac{c}{2f}\frac{\psi}{2\pi} \tag{7.4}$$

From (7.4), c represents the velocity of light which is given as 3×10^8 m/s, while f represents the modulation frequency, D is the distance and ψ specifies the phase shift. It is to be noted that the choice of the modulation frequency is very crucial because the non-ambiguity range which is denoted by D_{nambi} is given by [57]

$$D_{nambi} = \frac{c}{2f} \tag{7.5}$$

While the distance measurement resolution δD_{min} is defined as [57]

$$\delta D_{min} = \frac{c}{2f}\frac{\delta \psi_{min}}{2\pi} \tag{7.6}$$

From (7.6), the phase measurement resolution is designated by $\delta \psi_{min}$. Thus, from the mathematical expression as emphasized by (7.6), both the distance measurement and the phase measurement resolutions are dependent on the modulation frequency.

Typically, radio or infrared monochromatic laser sources which are coherent in nature incorporate the phase-shift distance measurement methodology. It is because, in such systems, the reflected beam is very strong such that it can be easily detected by the system. Nonetheless, the headlamps of the following vehicle in platooning applications emits a white light which is usually polychromatic and non-coherent in nature. Consequently, this results in the loss of a major portion of the optical power after the emitted signal gets reflected from the target. Therefore, to countervail this loss in optical power, the leading vehicle rather than reflecting the signal should be able to receive, process and re-emit it. Eventually, in order to implement this phenomenon, it is necessary to rely on the concept called Distance Estimation via Asynchronous Phase Shift (DEVAPS), which was first implemented for radio sensors [520]. DEVAPS facilitates restoration of the energy of the signal in such a manner that the echo received by the following vehicle can be substantially detected and processed.

Thereupon, it is followed by comparison of this processed signal with the emitted signal in order to determine the phase-shift which ultimately helps in computing the distance between two vehicles. In this manner, by using the concept of DEVAPS, vehicular positioning in case of platooning applications is implemented.

The proposed system in [57] exploits the methodology of phase-shift range finder, where the proposed system makes use of the non-coherent visible light emitted by the headlamps and the taillights of the vehicles to determine the distance between moving vehicles. This work presents a Simulink model of the proposed system over a path loss AWGN channel environment. From the simulation results, it can be surmised that the proposed system achieves a resolution of 10 cm and exhibits its ability to measure the intervehicle distances of up to 25 m, while 30 cm resolution was attained for distances up to 30 m at a refresh rate of 267 Hz. However, higher measurement ranges of up to 50 m can also be accomplished but with the emanation of error over 1 m. Furthermore, the analysis of this work affirms that unlike other technologies, the proposed system fails to furnish both longitudinal and lateral intervehicles distances. The work in [54] extends the concept of visible light range finder to enable relative positioning aid of two consecutive vehicles. By making use of the vehicles, LED-based headlamps and taillights, this work presents a relatively simple positioning system that is based on phase-shift measurement. Furthermore, the proposed system is validated by conducting a simulation study over a realistic environment. The performances of both longitudinal and lateral range findings were analyzed. The simulation result of this work signifies that at a refresh rate of 2 KHz, the proposed system achieved a resolution of 12.4 cm for a longitudinal measurement over 18 m and a resolution of 22.6 m for a lateral measurement of 3.5 m.

Further, the work in [55] extends the work as portrayed in [57] and [54] by demonstrating a proof of concept of the visible light range finder for the first time. Earlier works in [57] and [54] verify the concept of visible light range finder through simulations. On the contrary, this work presents a prototype, where the light source of the following and leading vehicles in a platoon uses a commercial off-the-shelf (CTOS) white headlamp, photodiode and analog signal processing cards were built and tested in order to validate the range finding capabilities of the proposed system. The developed prototype was proficient enough to determine the distance of up to 25 m as well as ensure a resolution of 24 cm at a distance of 10 m with a refresh rate of 506 Hz. The research efforts in [56] for the first time develops a VLC range finder where the system is competent to enable simultaneous vehicle-to-vehicle communication and range finding operations by utilizing the headlamps and taillights of the vehicles. The following and leading vehicles in a platoon can simultaneously transfer information by means of exchange of the clock-signal that is contained in the Manchester-encoded signals, thereby facilitating the estimation of their interdistance by means of phase shift measurement. The developed system is verified theoretically and validated by means of Simulink. Furthermore, both range finding and communication functions were evaluated experimentally. From the result analysis, it is ascertained that the proposed system achieves the distance measurement by up to 25 m while maintaining a resolution of around 24 cm at 10 m. It is also fascinating to note that the proposed link delivers a data rate of 500 Kbps while maintaining a BER below 10^{-6} up to 30 m.

Taking into consideration the tremendous applications offered by indoor localization in the fields of public safety and several indoor applications, several techniques have been proposed in the literature for indoor location sensing, out of which the most widely used positioning technique was triangulation, fingerprinting, scene analysis and proximity. In order to implement the triangulation technique, it is a primary prerequisite to determine the angle or distance between a reference point and a mobile terminal. Consequently, in order to accomplish this task, it is vital to rely on several other methods like AOA, time of arrival (TOA), TDOA and the intensity of the received signal strength (RSS). As of now, in the earlier discussion, it is evident that visible light positioning can be accomplished by using visible light range finding techniques. In addition to visible light range finder, the TDOA technique also renders visible light positioning support for indoor and outdoor applications. Relevant work can be evidenced from [43], where instead of one photosensitive element, the receiver comprises of two photosensitive elements in order to detect the light intensity. The authors in this work demonstrate visible light positioning with the aid of traffic light and two photodiodes that are mounted in front of a vehicle. In this sort of scenario, the light signal that is emitted from the traffic light illustrating the positioning information is received by the two photodiodes that are mounted in front of the vehicle. Based upon the TDOA of the light signal to the two photodiodes, the position of the vehicle can be estimated. Furthermore, visible light positioning can be performed in vehicular applications based upon the intensity of RSS. However, the accuracy ensured by this method can be sustained only for several meters. On these grounds, several techniques were proposed in the literature to combine VLC technology with them in order to furnish accurate visible light positioning support for vehicular communications.

The work in [464] proposes a bidirectional Laser Radar Visible Light Bidirectional Communication Boomerang System (LRVLB-ComBo) for enabling real-time two-way V2V communication. Even though the proposed system is very complex to implement, it enables positioning reliability. This work makes use of the combination of laser radar with visible light boomering systems to enable highly reliable intervehicle ranging and the feasibility of bidirectional V2V communication. Furthermore, this work makes use of the time hopping spread spectrum (THSS) technique to provide additional data like the size of the vehicle, several other information like the braking, acceleration as well facilitates for the exchange of change of direction of the vehicle. Furthermore, the work in [430] presents a simple and cost-effective ranging technique to determine the intervehicle distance. This method makes use of the geometric similarity between an object on the rear surface of the vehicle and its corresponding image on the image sensor. As reference objects, a license plate and distance between taillamps was employed. By making use of taillamps, night-time measurements were performed and from the obtained data points, an error ratio less than 5% was obtained. Based upon the license plate, a ranging method for daytime was studied.

In addition to the aforementioned techniques, Image Sensors also exhibit their potential in determining the AOA and, consequently, research on enabling ranging and positioning operations by employing Image Sensors in the area of vehicular communications is rapidly progressing. A much recent literature in [477] reports that the authors proposed an intelligent transport system that is meant for calculating the distance between the vehicles with the aid of Image Sensor. From this work, it is evident that the proposed positioning algorithm employs two Image Sensors and it relies only on LED for determining the distance between the two moving vehicles. Particularly, this article focuses on camera-based VLC to estimate the distance between two moving vehicles. Predominantly, in order to capture the LED of the target vehicle, a camera is mounted on the estimating vehicle. The distance between the two vehicles is identified by employing the picture that has been captured by the camera. As delineated in Fig. 7.23, the two cameras which are mounted on the estimating vehicle capture the pictures of the LEDs of the target vehicle. Thereupon, by making use of these captured images, the process of data extraction takes place in order to determine the coordinates of the LED of the image. Consequently, by making use of the coordinates of the captured image, the distance between two vehicles can be determined. Even till today, extensive research efforts were carried out in order to determine the intervehicle distances. Within this context, intervehicle ranging has attracted an immense interest in the research community. The research contribution in [540] proposes a simple and cost-effective ranging system much suitable for VLC-based vehicular communication. The proposed system comprises of a visible light LED acting as transmitter and an array of avalanche photodiode (APD) as receiver. These components were fitted into each vehicle. From this work, it can be analyzed that the proposed system was able to determine the round trip distance between the rear vehicle and the front vehicle. Furthermore, by exploiting LS and

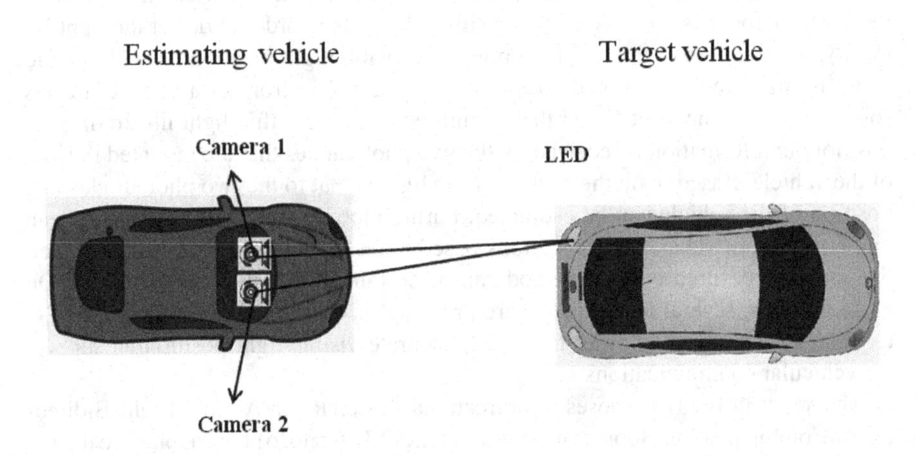

Figure 7.23: Vehicular communication using Image Sensors mounted on the vehicle.

Levenberg-Marquardt algorithms, the position of the front vehicle was estimated. Both simulation and experimental results of this work confirm that the proposed system functions efficiently and effectively, where it can determine the absolute distance up to 14 m with an error less than 5 cm, and for distances up to 19 m, the error was less than 20 cm.

In contrast to the existing methods for visible light positioning, where there is a necessity for the requirement of spatially separated co-operating nodes possessing strict synchronization requirements, the research work in [453] proposes a novel visible light positioning technique which exploits a single VLC receiver that is able to measure the AOA on the receiving vehicle. The sophisticated nature of this method is that it does not enforce any constraints on the VLC subsystem. Additionally, it does not require any cooperation with the transmitting vehicle or prior knowledge of any channel parameters or other node locations. The only requirement is that the transmitting vehicle must disseminate the real-time speed and heading information via the VLC link. In order to determine the $2D$-positioning information of the transmitting vehicle relative to the receiver via triangulation, the visible light positioning method exploits the speed, heading data and two consecutive AOA samples. Under realistic road and VLC channel conditions, the simulation results affirm that the proposed method exhibits cm-level positioning accuracy > 50 Hz.

The work in [145] employs a new positioning algorithm and a modified version of Kalman filter in order to present a VLC-based vehicle-to-vehicle tracking system. For the purpose of transmitting the positioning signals to other vehicles, this work makes use of the LED headlight and taillights of the vehicles. In order to receive these signals, two CMOS dashboard cameras were installed on each vehicle. These cameras are generally meant to capture the images of the LEDs of the targetted vehicle. Thereupon, the instantaneous position of the target vehicle is determined by making use of the geometric relationship between the two cameras and the images captured by them. This position tracking of the target vehicle is possible only if the LED of the target vehicle is within the view frame of the two cameras. The attained discrete positioning is not devoid of errors. Predominantly, the types of errors which usually prevail are the systematic errors that occur due to the CMOS rolling shutter artifact and the weak spatial separability of the sensor. In addition to these errors, even some random errors also occur. Accordingly, the work in [145] addresses the systematic errors by proposing a new positioning algorithm with two compensation mechanisms. Furthermore, in order to filter out the random errors and to accomplish a smooth and an accurate tracking result with the purpose to determine the position of the vehicle, this work makes use of a modified Kalman filter. By means of simulations, the effectiveness of the proposed tracking system is verified. Furthermore, the simulation results show that the aforementioned methodology significantly improves the accuracy of the attained positioning result.

7.4 PERFORMANCE OF CAR-TO-CAR VLC

This section illustrates the performance of car-to-car communication employing VLC technology [317, 318]. Similar to the conventional RF-based wireless communication systems like the DSRC which facilitates support for enabling car-to-car communication, the other way to enable car-to-car communication is through the usage of visible light emitted by the LEDs to support concurrent illumination and communication support. When compared to traditional RF-based DSRC technologies, car-to-car communication employing VLC technology is proficient enough to guarantee the distinctive benefits listed below:

1. Cost-effective nature and low complexity involved in the installation of LED lamps in vehicles, street lamps, traffic lights, etc.
2. Rapid availability of excellent high precession VLC-based positioning techniques play a crucial role in determining the location of the vehicles [319].
3. High level scalability and low level interference offered by VLC technology when compared with RF-based wireless communication technologies especially at peak rush hours.
4. Stunning ability of VLC when employed in vehicular communications to facilitate the vehicles to receive signals from neighboring vehicles, thereby overcoming the signal congestion as well as interference levels when compared with RF-based technologies. Moreover, vehicular communication exploiting RF-based wireless communication undergoes abominable packet collisions, higher latencies, and the rate of reception of the packet is very low.

The important aspects which highlight the fundamental differences between both RF-based wireless communication technologies, i.e., DSRC and VLC technologies, can be interpreted from Table 7.3. The pictorial representation of a car-to-car communication by using their head lamps is clearly illustrated in Fig. 7.24. It is to be

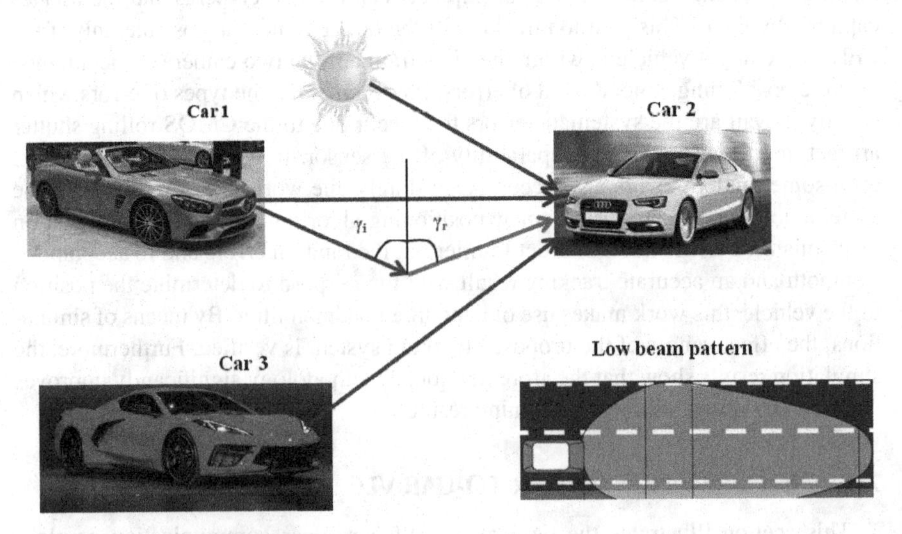

Figure 7.24: Car-to-car communication using visible light communication.

Table 7.3
Differences between VLC- and RF-based wireless communication technologies for vehicular communication.

Mode of Communication Technology	VLC	RF (Dedicated Short Range Communication)
Communication Range	Direct Line of Sight (LOS): Point to Point	Broadcasting of information from a single point to multiple points
Range of Carrier Frequency	400-780 THz	5.85-5.925 GHz
Licensing of Spectrum	Unlicensed	Licensed
Coverage Area offered	Low and Narrow	Long and Wide
Mobility	Low-Medium Mobility offered. (However, with the advent of several sophisticated techniques like exploitation of Image Sensors, multi-hop communications, etc., mobility can be enhanced.)	High amount of Mobility is offered

noted that the projected beam pattern is unique, where the received light comprises of both LOS and NLOS components. The NLOS components emanate due to the reflections originating from the road surface. Particularly, the reflected rays arise due to surface materials of the road, the adverse weather conditions which include fog, rain, snow, etc. Furthermore, these induced reflections will definitely pave the platform for the emergence of multipath induced ISI, thereby impeding high data rate communication. Moreover, the dominant noise sources in vehicular communications that hinder the performance of communication link are the ambient noise which originates due to the sunlight especially at the daytime and the noise stemming out at night-time from other light sources like the moving vehicles, road side infrastructural units, street lights, etc. Since the communication distance between two cars is made possible through the aid of their head-lamps, it is vital to model the vehicle's headlamp. In order for the vehicles to accomplish the desired levels of illumination while not imposing any glaring effect to the other vehicles present on the road, the Economic Commission of Europe [485] and Federal Motor USA Vehicle Safety Standards [458] regulations mandated that the vehicular units, which include the headlamps, reflective devices and other related equipment, adhere to the requirements imposed in [524].

Consequently, modeling the vehicle's headlamp is not that straightforward and hence we cannot exploit the widely used LED Lambertian model. Generally, the combination of both high and low beam headlamps of the vehicle are employed to establish enhanced safety along with aesthetically appealing lighting conditions for the people (vehicle drivers) during both day time and night time, as well as in all adverse weather conditions. Specifically, the combination of both high-beam and low-beam headlamps of the vehicle render ample amount of safety aid to the drivers on roads. Particularly, the high-beam headlamps are utilized to ensure long-distance visibility with no oncoming cars, while the low-beam headlamps with asymmetrical light pattern are meant to furnish both maximum forward and lateral illuminations and at the same time accomplishes the crucial task of minimizing the glare toward oncoming vehicles and other users travelling on roads. Primarily, the Lambertian model, which comprises of symmetrical profile, has been widely deployed for indoor VLC LED modeling, and hence, it is not suitable to be modeled for a vehicle's headlamp. Therefore, in order to attain an enhanced reliability of services, a market-weighted headlamp beam pattern can be exploited as a part of the VLC channel model for the vehicular applications. Principally, this headlamp was developed by the Transportation Research Institute of the University of Michigan [434]. Figures 7.25 and 7.26

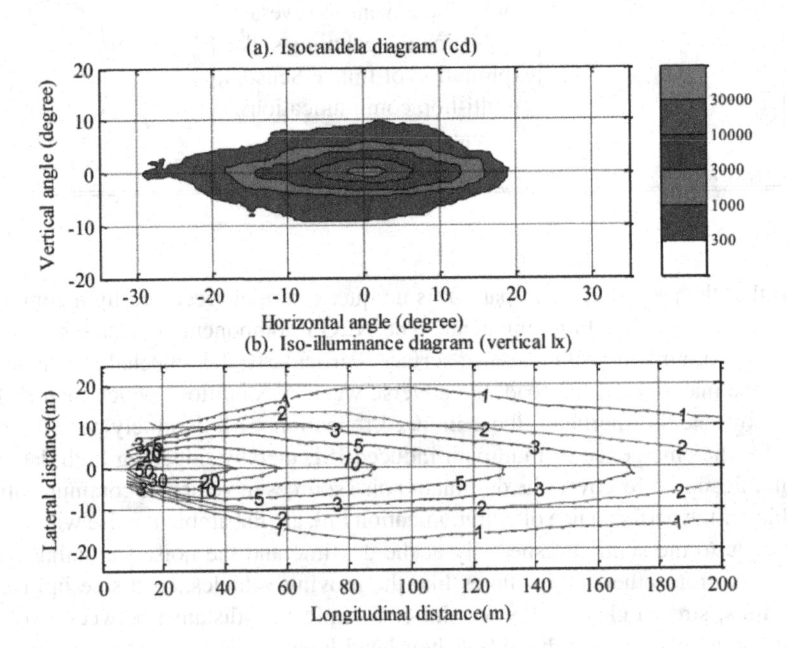

Figure 7.25: Pictorial Interpretation of Isocandela and Isoilluminance of the road surface from a pair of high-beam headlamps [317].

delineate the isocandela and isoilluminance diagrams corresponding to the road surface originating from a pair of high-beam and low-beam headlamps, respectively. As evident from the figure, pertaining to the high-beam headlamps, a narrow flat beam

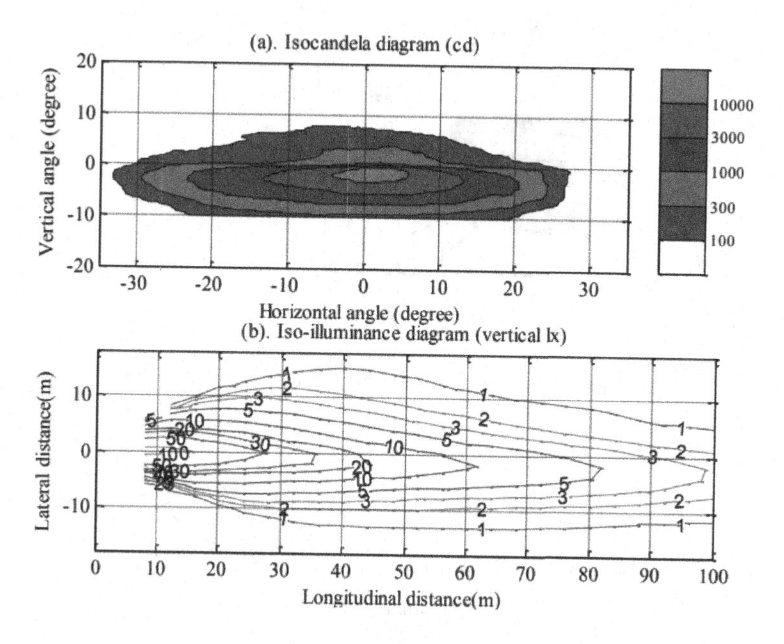

Figure 7.26: Depiction of Isocandela and Isoilluminance of the road surface from a pair of low-beam headlamps [317].

is projected in the horizontal direction of a few degrees to the left, thereby exhibiting symmetrical illumination pattern on the surface of the road. Besides, asymmetrical illumination pattern is offered by the low-beam headlamps in order to guarantee tolerable levels of forward and lateral illuminations while regulating the induced glare by reducing the amount of light being directed toward the eyes of the other vehicle drivers. Therefore, the illumination which is denoted by E on the surface of the road can be expressed as [315]

$$E = \frac{d\phi}{dS} = \frac{d\phi}{d\omega}\frac{d\omega}{dS} = I(\zeta,\xi)\frac{d\omega}{dS} = I(\zeta,\xi)\frac{\cos\gamma}{d^2} \tag{7.7}$$

From (7.7), the luminous flux in lumens is denoted by $d\phi$, S represents the area of the surface of road in m^2, while the solid angle in steradian is signified by ω. The luminous intensity in candela is denoted by $I(\zeta,\xi)$, where ζ and ξ specify the horizontal and the vertical angles, respectively. Furthermore, the distance between the light source and the small area dS is exemplified by d, while the angle between the normal of the road surface and the incident direction is represented by γ. The entire process portraying the calculation of illumination is interpreted in Fig. 7.27. As discussed, in car-to-car-based VLC system, the received optical power comprises both LOS and NLOS components. The reflection pattern from the surface of the road which is nothing but the NLOS path is assumed to have a Lambertian profile as delineated in Fig. 7.28. Upon assuming $m = 1$, the reflected radiant intensity $R(\phi)$ is

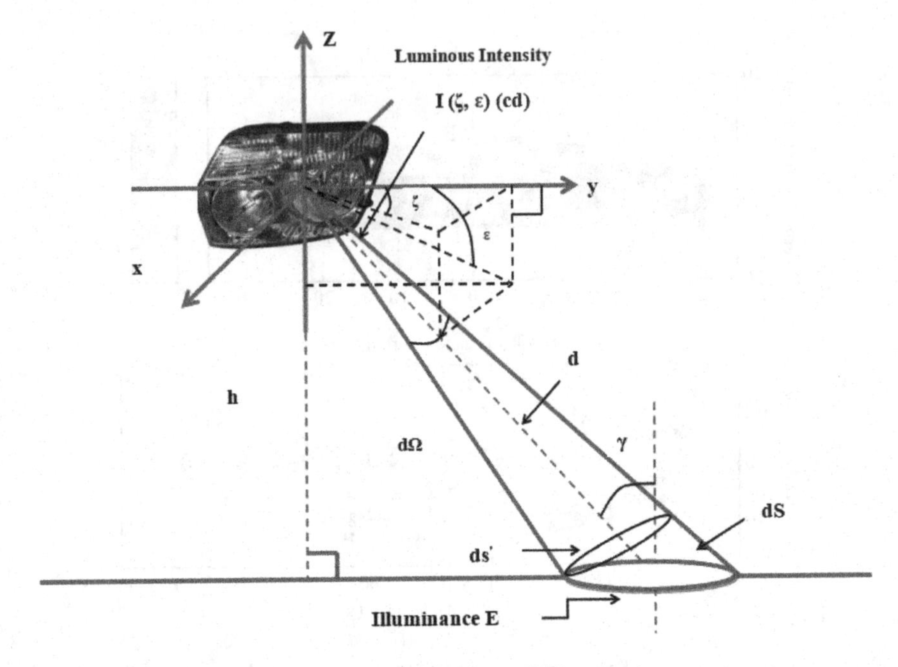

Figure 7.27: Determination of Illumination.

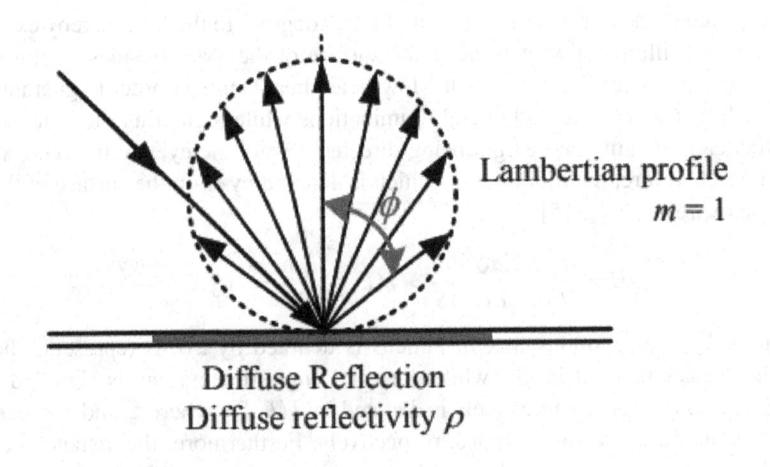

Figure 7.28: Road surface reflection [317].

expressed as [248, 317]

$$R(\phi) = \frac{\rho cos(\phi)}{\pi} \tag{7.8}$$

From the mathematical expression as shown by equation (7.8), the diffuse reflectivity is given by ρ, and in general, this varies with different pavement materials, while the polar angle of the scattered light is given by ϕ. A typical configuration of a VLC-based car-to-car communication system along with a low-beam pattern is illustrated in the Fig. 7.29. To have a clear-cut illustration, this figure conveys the rays emanating from the right headlamp. In general, the photodetector of the receiver vehicle receives rays of light from both the right and left headlamps. As emphasized in [450], both the left and right headlamps exhibit the same output light distribution, and hence, the mathematical analysis is confined only to the right side headlamp (RH). Thus, both the LOS and NLOS paths from a single transmitter are calculated below. Therefore, by making use of (7.7), the vertical illuminance at a particular

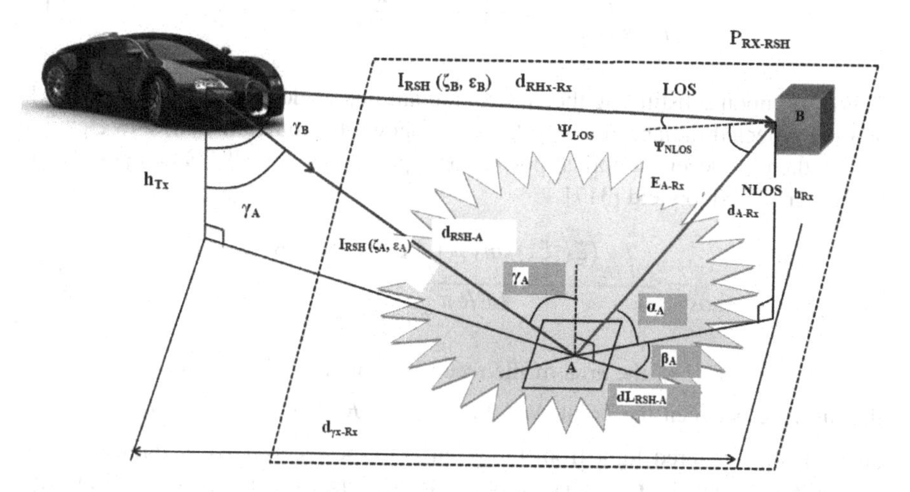

Figure 7.29: Configuration of VLC-based car-to-car communication system.

location A within the area dS can be interpreted as

$$E_{RH-A} = \frac{I_{RH-A}(\zeta_A, \xi_A)\, sin\gamma_A}{d^2_{RH-A}} \tag{7.9}$$

It is apparent from Fig. 7.29 that the parameter $I_{RH}(\zeta_A, \xi_A)$ specifies the luminous intensity of the right-side headlamp from the direction (ζ_A, ξ_A), the angle between the normal of the road surface corresponding to the point A and the incident direction is given by γ_A. The path length from the right-side headlamp to the point A is signified by d_{RH-A}. The vertical radiant flux which is denoted by dP_{RH-A} at the point A can be put up as [317]

$$dP_{RH-A} = \frac{E_{RH-A}.dS}{LER} \tag{7.10}$$

Upon substituting the vertical illuminance as shown by equation (7.9) into (7.10), the following expression is obtained

$$dP_{RH-A} = \frac{I_{RH-A}(\zeta_A, \xi_A) \sin\gamma_A}{LER \, d^2_{RH-A}} dS \tag{7.11}$$

From the mathematical expressions (7.10) and (7.11), LER is the luminous efficacy of radiation of a high-power phosphor-coated WLED which is given by 250.3 lumen/watt [216]. Eventually, the received optical power from a single reflected path at the receiver placed at the position B is formulated as [317]

$$dP_{Rx-RH-NLOS} = \frac{dP_{RH-A} R(\phi_A) A_{Rx} \cos(\psi_{NLOS})}{d^2_{A-Rx}} \tag{7.12}$$

Therefore, upon substituting the vertical radiant flux as depicted by equation (7.11) and the reflected radiant intensity $R(\phi_A)$ as shown by equation (7.8) into equation (7.12), then the received optical power corresponding to the NLOS component can be further solved to yield [317]

$$dP_{Rx-RH-NLOS} = \frac{I_{RH-A}(\zeta_A, \xi_A) \sin(\gamma_A) A_{RX} \rho \cos(\phi_A) \cos(\psi_{NLOS})}{LER \, \pi \, d^2_{RH-A} \, d^2_{A-Rx}} dS \tag{7.13}$$

From equation (7.13), the area of the receiver photodiode is specified by A_{Rx}, while, the distance between the point A to the receiver Rx is denoted by d_{A-Rx}, the polar angle of the scattered light from the point A to the receiver Rx is represented by ϕ_A and the parameter ψ_{NLOS} corresponds to the angle of incidence of NLOS link as viewed from the photodiode. Subsequently, for the right-side headlamp, the total received optical power corresponding to the NLOS components emanating from all the reflected paths can be expressed as [317]

$$P_{Rx-RH-NLOS} = \begin{cases} \iint_S dP_{Rx-RH-NLOS} \, dS & 0 \leq \psi_{NLOS} \leq \psi \\ 0 & \psi_{NLOS} > \psi \end{cases} \tag{7.14}$$

From (7.14), the half-angle of the photodiode's FOV can be expressed as ψ, while the total area of the road surface that has been illuminated is given by S. Furthermore, the total amount of received power corresponding to the LOS link can be manifested as [317]

$$P_{Rx-RH-LOS} = \begin{cases} \frac{I_{RH}(\zeta_B, \xi_B)}{LER \, d^2_{RH-Rx}} A_{Rx} \cos(\psi_{LOS}) & 0 \leq \psi_{LOS} \leq \psi \\ 0 & \psi_{LOS} > \psi \end{cases} \tag{7.15}$$

From (7.15), the parameters $I_{RH}(\zeta_B, \xi_B)$ and ψ_{LOS} correspond to the luminous intensity for the angle pair (ζ_B, ξ_B) and the incident angle of the LOS, respectively.

Finally, the total received optical power relevant to the right-side headlamp is expressed as the sum of optical power corresponding to both LOS and NLOS components, respectively. Accordingly, the total received optical power can be represented as

$$P_{Rx-RH} = P_{Rx-RH-NLOS} + P_{Rx-RH-LOS} \quad (7.16)$$

Thus, the total received power analogous to both right-side and left-side headlamp (LH) can be expressed as

$$P_{Rx} = P_{Rx-RH} + P_{Rx-LH} \quad (7.17)$$

Apart from the varying lateral position, the received power pertaining to the left-side headlamp has the same mathematical analysis as that of the right-side headlamp. Thus, this analysis clearly depicts the calculation of overall received optical power. However, in order to analyze the overall performance of the C2C-based VLC system in terms of BER, it is vital to have the analysis of noise power which will be seen in the subsequent subsection.

7.4.1 ANALYSIS OF NOISE IN VLC-BASED CAR-TO-CAR COMMUNICATION SYSTEM

As discussed earlier, the two major sources of ambient-light noise that hinder the performance of outdoor vehicular communication systems are the noise emanating due to the background solar radiation and the noise originating from the artificial light sources which include street lights, dash-boards, sign-boards, advertising panels and the noise occurring from other vehicular head lights and taillights [126]. In general, the background solar radiation comprises of both direct and scattered radiation. Particularly, the direct solar radiation emanates as the most stringent noise source which is relying on several factors such as weather conditions, time of the day, i.e., whether morning, noon or evening. Also, the intensity of the solar radiation also varies in accordance with the month, and even the position of the sun differs during the day and throughout the year as well [151]. On the other hand, modeling of the scattering solar radiation is quite complex and is dependent on the surrounding environment. Based upon the earlier findings as stated in [560], the measured electrical power spectrum of the solar radiation relatively approaches as the direct current (DC), which can be subsequently removed by means of AC coupling. Nonetheless, the major source of noise that prevails during the daytime for VLC-based car-to-car communication system is the shot-noise induced by the solar radiation.

Furthermore, the interference originated due to the artificial light sources is mostly concentrated at the low-frequency region of the communication spectrum. Moreover, when compared to the solar radiation, the intensity of the artificial light sources is quite small. Consequently, an inference can be drawn that the contribution of the artificial light sources is quite negligible especially during the daytime, where the most dominant source of noise is the solar radiation. Therefore, along with the noise occurring due to the ambient light, the other important noise which needs to be taken into account is the thermal noise associated with the receiver circuitry. Subsequently, the shot noise that is induced by the solar radiation and the thermal noise can be modeled as AWGN. Therefore, the total noise variance can be represented as

$$\sigma^2_{Total} = \sigma^2_{Shot} + \sigma^2_{Thermal} \tag{7.18}$$

However, the shot-noise and the thermal-noise variances can be expressed as [277]

$$\sigma^2_{Shot} = 2eR \underset{Rx-Signal}{P} B_s + 2eI_{bg}I_2B_s \tag{7.19}$$

$$\sigma^2_{Thermal} = \frac{8\pi kT_k}{G}\eta \underset{Rx}{A} I_2B_s^2 + \frac{16\pi^2 kT_k\Gamma}{g_m}\eta^2 \underset{Rx}{A^2} I_3B_s^3 \tag{7.20}$$

From the mathematical expression as represented by equation (7.19), the electronic charge in $1.602\times10^{-19}C$ is given by e, the average responsivity of the Photodetector is specified by R, while the average received optical power pertaining to the desired signal is indicated by $\underset{Rx-Signal}{P}$, whereas the parameters B_s, I_{bg} and I_2 signify the bandwidth of the system, the received background noise current and the noise bandwidth factor pertaining to the background noise [275]. Meanwhile, the parameters described by equation (7.20) include the Boltzmann's constant which is indicated by k, the absolute temperature T_k, open-loop voltage gain G, fixed capacitance of the photodetector per unit area η, the field-effect transistor (FET) channel noise factor Γ, the transconductance of the FET g_m and the noise bandwidth factor I_3 is given by 0.0868 [287].

7.4.2 PERFORMANCE OF VLC-BASED CAR-TO-CAR COMMUNICATION SYSTEM

Based upon the mathematical analysis carried in the previous subsections in order to determine the desired signal optical power and the noise power, this subsection is directed toward evaluating the performance of VLC-based car-to-car communication system. Upon exploiting the widely employed and the simplest modulation format like OOK-based IM and DD, with a AWGN channel environment, the electrical SNR at the receiving terminal is determined as

$$SNR = \frac{R\underset{Rx}{P^2}}{\sigma^2_{Total}} \tag{7.21}$$

From the electrical SNR as represented by equation (7.21), the responsivity of the photodiode is specified by R which in general changes with respect to the wavelength.

In general, the probability of error, i.e., BER for OOK-based communication system, is expressed in terms of Q-function as follows

$$P_e = Q\left(\sqrt{SNR}\right) = Q\left(\frac{R\left(\underset{Rx-RH}{P} + \underset{Rx-LH}{P}\right)}{\sigma_{Total}}\right) \tag{7.22}$$

Particularly, the Q-function gives the tail probability of the standard Gaussian distribution which is expressed as

$$Q(z) = \frac{1}{\sqrt{2\pi}} \int_{z}^{\infty} e^{-\frac{y^2}{2}} dy \tag{7.23}$$

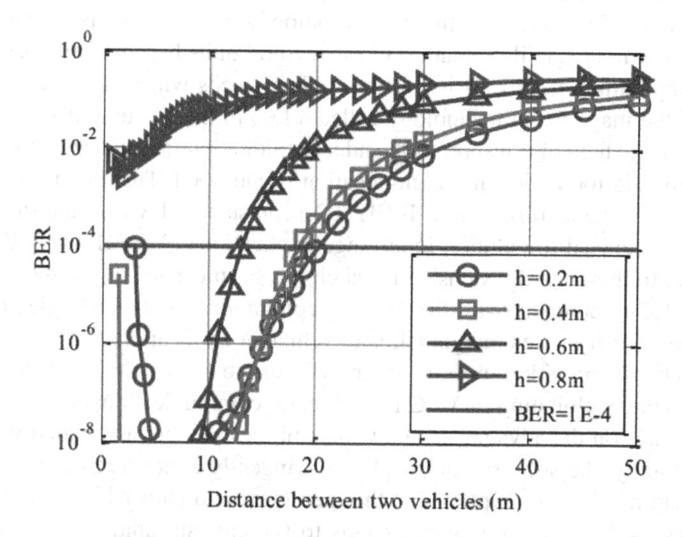

Figure 7.30: BER Performance of VLC-based car-to-car communication system with respect to varying distances between two cars for different ranges of height of the photodiode (h) from the ground [317].

Figure 7.30 elucidates the BER performance of VLC-based car-to-car communication system against the varying distances between two cars for different ranges of height of the photodiode. The figure shows that at a particular range of probability of error, for lower values of h, the amount of coverage area offered is high. For instance, at a BER of 10^{-4} and height h of 0.2 m, the range of communication path is approximately 20 m. Furthermore, the range of communication path worsens at higher orders of h, i.e., $h = 0.8$ m respectively.

7.5 CONCLUSION

In order to minimize the number of road accident victims, as well as to drastically reduce the road fatality rates, the ITS aspired to propose several modes of vehicular communications, i.e., V2I and V2V communications. This is further triggered by the development of DSRC for vehicular safety applications and the publication of the IEEE 802.11p standard for facilitating short to medium range intervehicular communication. Consequently, ITS struggled to reconcile several aspects like humans, roads and vehicles through the sate-of-the-art communication and information technology. The major aim of ITS is to bring up effective solutions for efficient traffic flow, frame up solutions for road transportation problems, etc. Over many years, wireless communication was dominated by RF technologies. With time, the rapid advancements in the solid state lighting industry paved a way for the reinforcement of LEDs which offers multiple advantages.

Eventually, this has drawn a significant amount of enthusiasm in the automotive industry to replace the halogen lighting units with the LED illumination infrastructural units. Consequently, the transportation field, traffic lights, street lights and several other road side illumination infrastructural units became LED oriented. The noteworthy nature of LED is its fast switching speeds which resulted in the emergence of the magnificent technology called VLC. In this context, the omnipresence of LEDs in the field of transportation and its exploitation for rendering wireless communication aid for vehicular communication sound good. This fact is further intensified with the standardization of IEEE 802.11*bb* standard, where the PHY I of this standard is devoted to outdoor, long-range applications like V2I and V2V communications. In this regard, extensive research has been carried out in the literature to enable VLC to compete with the existing communication technologies in order to exhibit its caliber in the emerging area of vehicular applications.

This chapter provides an in-depth analysis of the advancements done so far pertaining to the exploitation of VLC for vehicular communications. Firstly, this chapter elaborates on the advantages and vulnerabilities that are associated with DSRC. Predominantly, the success rate of ITS is stringently dependent on its penetration. Inadequate penetration implies that the inefficiency to gather huge amount of data facilitates its distribution as well as fails to accomplish rapid connectivity among moving vehicles. In this regard, if we take into consideration the RF-based solutions in an effort to enable sufficient amount of penetration, installation of RF-based infrastructural units at every intersection, street and in every vehicle is very cumbersome as well as incurs a huge cost. On the contrary, the arresting benefit of VLC over RF is its cost-effectiveness and, moreover, the inherent fact that this technology is already interfaced with the existing traffic infrastructural units, street illumination systems, transportation infrastructural units, etc., thereby enabling VLC to evolve as a universal technology.

On the other hand, DSRC offers a large coverage range, but the drawback associated with it is that under high dense traffic conditions, the amount of interference increases. Thus, by exploiting the advantages of both VLC and RF, a hybrid VLC/RF network suitable for vehicular applications can be developed, and the literature efforts that have been made so far pertaining to the hybrid VLC/RF-VANETs are briefed in this chapter. Even though VLC has gained significant popularity in recent times in the emerging area of vehicular communication, its application in this area is not devoid of drawbacks. Consequently, in order to make the implementation of VLC more flexible for vehicular applications, this chapter presents the contemporary quick fixes for each challenge that hinders the exploitation of VLC technology to furnish its services to enable V2V and V2I communication. In contrast to indoor VLC channel, outdoor VLC channel is subjected to several impairments due to the presence of other artificial light sources as well as the ambient light source like solar radiation. Hence, in this regard, extensive research efforts were carried out in the literature to date suggest that several techniques and suitable modifications in the receiver circuitry were performed to reject this parasitic light. Furthermore, a scenario is given depicting the intelligent traffic light system which adjusts the lights of the traffic signals in order to enhance the safety of the people travelling on roads especially at the road intersections. In addition to the ability to combat the effects of noise emanating due to other parasitic light sources and ambient light sources, VLC should exhibit the potential to enable range measuring and positioning functionality. In the field of visible light positioning for outdoor vehicular applications, enormous research efforts were carried out such as the exploitation of Image Sensors, proposal of suitable AOA range finding algorithms, etc. to determine the distance between two vehicles.

Accordingly, this chapter is directed toward elaborating the current state-of-the-art techniques that strived to enable visible light positioning aid for vehicular communications. Moreover, in addition to positioning aid, VLC should be competent enough to disseminate information among the moving vehicles. In this regard, it is imperative to rely on multi shop transmissions along with the exploitation of decoding and relaying techniques to enable the transmission of information among vehicles in a cooperative manner. Toward this direction, this chapter concise several multi hop transmission and relaying techniques carried out in vehicular platooning applications. Finally, this chapter presents several up-to-date research efforts that were carried out in the literature in this emerging area of VLC-based vehicular communications. The future research pertaining to road-to-vehicle and vehicle-to-vehicle communication should propose techniques to attain an enhancement in the range of communication, combat the effects of noise, and maintain a high data rate communication. We anticipate that the provided information spurs further interest in the research communities as well as automobile industries to formulate several enticing solutions and techniques for the amalgamation of VLC technology with vehicular communications in order to furnish an enhancement in the safety of the people travelling on roads as well as to substantially decrease road fatality rates.

8 RESEARCH CHALLENGES ASSOCIATED WITH VLC

8.1 EXPLOITATION OF ORGANIC LIGHT EMITTING DIODES FOR VLC APPLICATIONS

While this chapter highlights the challenging aspects associated with VLC, this section presents a comprehensive overview about the significant research efforts carried out in the literature relevant to the exploitation of organic light emitting diodes (OLEDs) and superluminescent diode (SLD) for VLC applications. This section is directed toward presentation of several remarkable research efforts that were carried out to render high data rate transfer of information from a OLED-based VLC system. From a list of earlier research findings, it can be surmised that the stunning noteworthy features of OLEDs accelerated the interest in research communities allowing OLEDs to emerge as appropriate candidates for rendering concurrent 'illumination' and 'data communication' as well. The distinguished characteristic features of OLEDs which drew the attention of research era are: durability, wide viewing angle, mechanical flexibility, low weight, ease of utilization in high-definition television displays due to their high contrast ratio, high-end gadgets, a faster response less than 0.01 ms, etc. This is further triggered by the rapid availability of low-cost solution-based processing methods as well as their power efficient nature which play a dominant role in facilitating easy penetration into the solid state lighting (SSL) industry. Similar to the inorganic LEDs discussed throughout this book, OLED-based VLC systems are also proficient enough to support 'illumination' and 'communication'.

Predominantly, the major drawback associated with the implementation of the well-known inorganic opto-electronic devices like WLEDs and RGB LEDs is that they make the VLC system encounter a significant limitation in terms of scalability. The primary reason behind this limitation is that basically the inorganic LEDs are realized by exploiting thermal epitaxial evaporation methods where the metal alloys like gallium nitride (GaN) results in brittle crystals that eventually fails them to offer large display areas. Consequently, the most feasible solution is to rely on small molecule polymer-based OLEDs in situations where inorganic LEDs fail to find their applicability in the realization of large panel displays and screens. Notably, OLEDs sparked a great deal of public attention since they are called flat and flexible panel visible light sources exclusively meant for displays and lighting. Importantly, when compared with liquid crystal displays (LCDs), OLEDs are thinner as well as their renowned properties like their self-emissive and luminous nature enable them to glow significantly without the requirement of any additional back-lighting. Intuitively, these remarkable characteristic features result in low production costs, less power consumption and less consumption of area when compared with LCDs.

Upon proceeding further, the first breakthrough was reported in early 1987 by Tang and Van Slyke of Kodak when they for the first time promulgated the efficient and low-voltage OLEDs that were competently called as small molecule OLEDs (SMOLEDs) emerged by means of p-n heterostructure devices using thin films of vapor deposited organic materials [471]. The second breakthrough was achieved in 1990 with the invention of electroluminescence in polymer LEDs [75]. OLEDs comprise many of the properties of the traditional inorganic LEDs with a major difference in manufacturing; in contrast to inorganic LEDs, OLEDs consist of thin films of organic material (i.e., light emitting in nature) sandwiched between two electrodes which are acting as anode and cathode, respectively. It even comprises of a charge transport layer fabricated using polymer poly (3,4-ethylenedioxythiophene)-poly(styrenesulfonate) (PEDOT:PSS) [285]. A few of the myriad varieties of manufacturing processes which incur low production costs are solution processing-based, spray coating, doctor blading, spin coating and ink-jet printing. The structure of a typical OLED is delineated in Fig. 8.1. Predominantly, the structure of OLED en-

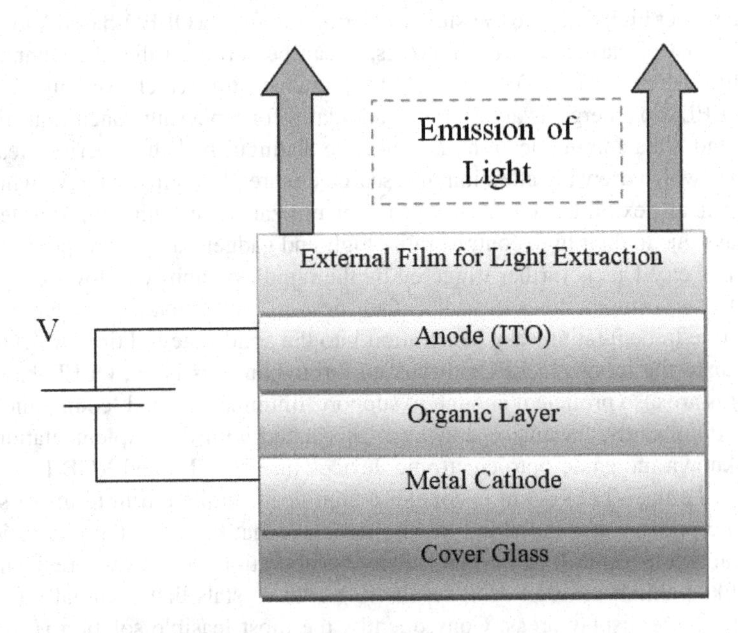

Figure 8.1: Pictorial representation of The structure of OLED.

compasses a substrate which can be a glass, metal or plastic. As evident from the figure, this is the layer to which all other layers which include the metal cathodes and anodes, organic photoactive layers and an external light extraction film were deposited. As already discussed, the organic layers are generally sandwiched between the transparent anode, i.e., indium tin oxide (ITO) and a metal cathode which is usually aluminum or silver. Principally, these OLEDs can be either SMOLEDs or they can be in the form of long-chain polymers i.e., polymer LEDs. It is also interesting to note that upon Exploiting flexible substrates like polyethylene terephthalate results in the emission of OLED panels or displays that can be either curved or rolled. Applications of OLEDs include mobile phone displays, televisions, etc.

The high illumination levels usually obtained by exploiting a single panel enable OLEDs to exhibit large photoactive areas. These ultimate features will allow for their exploitation for the realization of VLC links. However, in spite of its potential to offer a large photoactive area, the major limitation associated with OLEDs is that their limited available raw bandwidth. Furthermore, when compared with inorganic LEDs, the mobility of the charge carrier is less. The underlying reason behind this drawbacks is that OLEDs exhibit a capacitance like behavior and demonstrate a low-pass filter transfer function possessing a cut-off frequency which is given by [207]:

$$f_c = \frac{1}{2\pi RC} \tag{8.1}$$

From the mathematical expression of cut-off frequency as elucidated in (8.1), the parameters R and C specify the series resistance and capacitance, respectively, where, the capacitance of the OLEDs is given by

$$C = \frac{A\varepsilon_0\varepsilon_r}{d} \tag{8.2}$$

From (8.2), the OLED photoactive area and the thickness are represented by A in (m^2) and d in (m), respectively. Meanwhile, the permittivity of free space and the relative dielectric constant of the organic layer are signified by ε_0 and ε_r. Furthermore, it can be surmised from equations (8.1) and (8.2) that there is an inverse relationship between the plate area and the frequency. Broadly, applications where large photoactive areas are desirable articulate to limit the offered bandwidth by OLEDs. Toward this end, a comprehensive research work was carried out to address the limited bandwidth of the OLEDs.

Accordingly, in order to ameliorate the frequency response of the OLEDs, the proposal of numerous equalization techniques can be witnessed from the earlier research findings. The work in [203] investigates the performance of OLED-based VLC system by imposing different equalization techniques. With the intent to enhance the higher data rates, this work investigates the performance of two separate VLC links, where the first link exploits the silicon LED and an organic photodetector (OPD), while the second link comprises of OLED and a silicon photodetector. Both of these VLC links exploit OOK modulation technique. This work reports that in order to combat the effects of ISI, this work utilizes three different equalizers, namely, RC high pass equalizer, a fractionally spaced zero-forcing equalizer (FSZFE) and an ANN. Additionally, with the intent to enhance the performance of the digital equalizers, a predistortion scheme can be employed. The experimental results of this work emphasize that with a raw bandwidth of 30 and 93 KHz, both systems deliver a bit rate of 750 and 550 Kbps, respectively. Furthermore, with the major rationale to enhance the data rates, several works in the literature report the feasibility of DMT in VLC systems. Several significant works in the literature demonstrate that DMT achieves a superior performance when compared with unequalized time-domain modulation techniques like OOK in VLC systems. Using this fact, the work in [204] for the first time experimentally validates the performance of OLED-based VLC systems exploiting DMT modulation. The DMT format utilized in this work illustrates the spectral efficient nature, where QAM modulation scheme comprising of orthogonally spaced frequencies was employed. The experimental results of this work illustrate that the proposed OLED-based VLC link exploiting 32-QAM in conjugation with 1024 subcarriers achieved a data rate of 1.4 Mbps, drastically higher than that of the research work as portrayed in [203].

Furthermore, with the advent of time, rapid progress to furnish high data rate transfer is evidenced from the literature. The research effort in [206] demonstrated through experimental analysis an OLED-based VLC system was able to deliver 2.7 Mbps data rate. For the first time, this work employs an online ANN equalizer to combat the impact of baseline wander, as well as bandwidth limitation which usually arises in OLED-based VLC system due to the low charge carrier mobility. Due to these drawbacks, ISI emanates thereby leading to the deterioration of the system performance. Based on these grounds, the ANN equalization technique plays a major role to mitigate all such effects which hinder the assurance of high data rate communication. In order to determine how well the proposed equalization techniques work in the perspective of real world scenario, this work implements the performance of this equalization technique using TI TMS320C6713 digital signal processing (DSP) board. By exploiting online and offline filtering techniques and making use of 4-PPM modulation format, the reported data rates are 2.65 and 2.7 Mbps, respectively. Furthermore, much higher data rates can be reported from the research contribution as delineated in [212]. This work develops a polymer LED-based VLC link which makes use of ON-OFF keying modulation format and real time equalization format on a field programmable gate array. The experimental results of this work infer that at a BER of 4.6×10^{-3}, the achievable transmission speed was 10 Mbps, thereby imparting its dominance when compared with the former works.

The work in [197] reports the doubling of data rate from 10 Mbps which was achieved in [212] to 20 Mbps with the aid of exploiting multilayer perception (MLP) ANN classifier as the equalization technique due to its exceptional MSE convergence and BER performance. An experimental demonstration of VLC link exploiting 350 KHz organic polymer LED and ON-OFF keying modulation format can be depicted. In this work, the developed LVC link was able to deliver data rate of 19 Mbps while maintaining a BER of 10^{-6} and 20 Mbps at the forward error correction limit. When compared to the previously published works, the data rate from this work corresponds to a 55-fold increase. The work in [198] elucidates much higher data rates in order of ~ 55 Mbps can be achieved from a VLC link that exploits polymer LEDs and a commercial silicon photodetector. Firstly, the authors in this work make use of transmitters operating at RGB wavelengths in a manner to realize of wavelength division multiplexing (WDM), and ANN was used as the equalization technique.

In addition to the organic LEDs at the transmitting end, research progressed to employ organic-based photodiodes at the receiving end, due to the fact that organic photodiodes exhibit a better responsivity when compared to silicon-based photodiodes in the visible spectrum. Exploitation of organic photodetectors can be reported in a few such publications. The work in [211] manifests the realization of a VLC link that exploits organic photodiode at the receiving end. In order to mitigate the effects of ISI, an ANN was employed as the equalization technique. Experimental validations were performed to illustrate the performance of the organic photodiode-based VLC link that makes use of the digital modulation formats like NRZ-ON-OFF-keying and 4-PPM. Without enforcing any equalization technique, the VLC link that employs ON-OFF-keying modulation delivers a data rate of 200 Kbps, while 4-PPM-based VLC link imparts a data rate of 300 Kbps, respectively. Upon enforcing ANN, a significant improvement in data rate can be witnessed, where the ON-OFF-keying-based VLC link was able to achieve a data rate of 2.8 Mbps, while the VLC link employing 4-PPM modulation format delivers 3.75 Mbps data rate. Thus, these were the highest data rates that have been recorded for the first time upon making use of the organic photodiode.

The work in [205] for the first time demonstrates through experimental illustrations the realization of a VLC link that exploits organic opto-electronic devices at both transmitting and receiving ends. This work makes use of the ON-OFF-keying modulation format and ANN as equalization format to circumvent the effects of ISI. The developed VLC link delivers a data rate of 1.15 Mbps. Typically, organic devices are band-limited in hundreds of kilohertz bandwidth. Consequently, to alleviate this limited bandwidth offered by organic devices, the work in [99] proposes an all-organic VLC system that exploits higher order multicarrier modulation format like OFDM. Furthermore, this work takes into account the bit and power loading algorithms to effectively mitigate the frequency selectivity of the channel effects as well as to enhance the data rate. From the simulation results, it can be interpreted that the proposed system attains a data rate of 1.9 Mbps. Much higher data rates can be witnessed from the work as portrayed in [101] where the authors develops an experimental set-up for a VLC system exploiting blue OLED at the transmitting end and avalanche photodiode at the receiving end. This work experimentally characterizes the static and dynamic communication properties by varying the properties of the input signals. This work evaluates the performance of optical OFDM system in conjugation with offset QAM and compares its performance with DCO-OFDM system. With the motive to enhance the communication performance, this work employs adaptive bit and power loading algorithms along with the joint exploitation of equalization algorithm to combat the frequency selectivity of the channel as well as increase the data rate. Furthermore, in order to guarantee high data rate communication while maintaining a desired BER, a joint linear MMSE (LMMSE) and DFE techniques were implemented at the receiving end to countervail the ISI and ICI which generally emanate due to the limited modulation bandwidth of OLEDs as well as the non-linearity issues of OLEDs. From the simulation results, it can be deduced that the OLED-based VLC link was able to accomplish a data rate of 51.6 Mbps while fulfilling the BER requirement below FEC limit of 3.8×10^{-3}.

Apart from data rate oriented-based applications, the research work in [298] focuses on the non-linearity aspects of OLEDs. Despite OLEDs offering several advantages over the inorganic LEDs, the non-linearity of OLEDs may drastically deteriorate the performance of OLED-based VLC system. This is an indispensable aspect which needs to be prioritized especially when dealing with high PAPR prone OFDM. On these grounds, the work in [298] experimentally analyzes the non-linearity of OLEDs by making use of Volterra series as well as applies the Volterra-based nonlinear equalizer for enabling the demodulation process. Furthermore, this work reports that the proposed system could accomplish an end-to-end transmission distance of 3 meters, the longest ever reported distance that is achieved when compared with the existing OLED-based VLC literature. By making use of a time domain equalization (TDE) technique, the proposed system was capable to accomplish a BER of 2.9×10^{-3} and due to the extremely low modulation bandwidth of OLEDs, the system could achieve only 80 Kbps of data rate.

The applicability of OLEDs in VLC applications can be found in the recent published literature as well and the relevant references herewith [76, 89, 100, 134, 364, 506]. Remarkable characteristic features like their high levels of brightness, high quality color rending capability and low production costs enabled the rapid deployment of OLEDs for home and enterprise environments. As a consequence, they are chosen as attractive candidates for the implementation of ultra-low-cost visible light optical links in free-space and guided wave communications. Despite offering several significant advantages, the major drawback associated with the applicability of OLEDs in VLC systems is their lower data rates. Consequently, to address this drawback, several research efforts strived to enhance the high data rate communication, and this is clear from the earlier discussion. In addition to exploitation of several equalization techniques, the work in [134] focuses on scaling down the sizes of the organic devices. This work infers that upon reducing the size of the device, a significant improvement in the offered bandwidth can be witnessed. This work presents a detailed simulation model which illustrates that upon scaling down the dimensions of the device from 9 to 0.12 mm^2, a significant improvement in bandwidth from 8 to 50 MHz can be evidenced.

The work in [76] analyzes the performance of a VLC link that employs polymer LEDs. By exploiting the commercial light-emitting polymers, blue and green emitting PLEDs were prepared followed up by the characterization of the emission of the device in terms of spectrum and power. Furthermore, this work determines the bandwidth of the devices over a period of continuous driving, i.e., $\sim$ 4 hours. The experimental analysis shows that there is a decrement in bandwidth from an initial value of 750 KHz to a steady state of approximately 250 KHz for blue polymer LEDs. Whereas, in case of green emitting devices, stabilization in the bandwidth from $\sim$ 1.5 to $\sim$ 850 KHz can be observed. Besides, the steady state bandwidth of the blue polymer LEDs was increased by a factor of 3 upon exploiting a first order RC filter. Furthermore, upon making use of RC filter and ON-OFF-keying modulation format, the developed system exhibits its potential to allow for a data transmission in the speed of 1 Mbps.

The work in [89] outlines a survey on the emergence of OLED-based VLC systems along with the challenges associated with OLEDs, and it provides significant research efforts to mitigate them. Upon incorporation of fluorescence dyes with metal-organic framework (MOF), two new types of OLEDs for VLC systems have been developed. The authors in [506] present spectrally flat white OLED that is based on red-yellow-green-blue dye-loaded metal organic framework and linearly polarized warm-yellow metal organic framework-OLED. From the earlier discussions, it is apparent that ANN-based equalization technique exhibits its prominence to alleviate the limited bandwidth offered by OLEDs. Accordingly, the work in [364] investigates the effects of different learning algorithms in an ANN equalizer for a feed-forward multilayer perceptron configuration in VLC systems exploiting low bandwidth OLED at the transmitting end. Furthermore, this work evaluates the per-

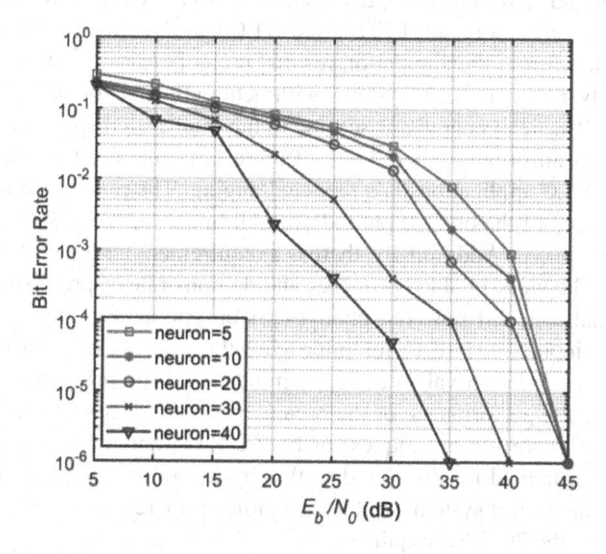

Figure 8.2: BER performance of ANN equalizer-based OLED-VLC system operating over an increasing range of neurons [364].

formance of different training algorithms like conjugate-gradient (CG), conjugate-gradient backpropagation (CGB) and Levenberg-Marquardt back propagation (LM) algorithms over a range of $\{5,10,20,30,40\}$ neurons. Among these training algorithms, LM illustrates a dominant performance in terms of BER. Furthermore, the impact of varying the number of neurons on the BER performance in OLED-based VLC system exploiting ANN equalizer is delineated in Fig. 8.2. As evident, the BER performance with 40 neurons is superior than the other neurons. Finally, from the rest of the simulation result analysis of this work, it can be deduced that the choice of the training algorithm incurs a trade-off between complexity and performance.

The research contribution in [100] for the first time performs experimental validations and models the radiation pattern of a bent OLED panel. The OLED panel is bent in different curvatures to evaluate the characteristics of a flexible OLED panel.

Accordingly, the obtained radiation patterns that describe the angular distribution of the intensity of light were also analyzed. Upon comparison with the Lambertian pattern of the traditional inorganic LEDs, the OLED panel offers a larger semi-angle at half intensity with a larger curvature in the bent direction. This confirms the superiority of OLED when compared with inorganic LEDs. In order to determine the rotational asymmetry of the light emitted from the OLED, this work proposes an improved mixed Gaussian analytic representations. Additionally, by making use of the OLED panel model, this work analyzes the characteristics of OLED-based VLC channel. Since the semiangle of OLED is larger in the bent direction, this can lead to better channel characteristics. Furthermore, through simulations, this work studies the characteristics of channel which includes the root mean square (RMS) delay spread and optical path loss. The numerical results of this work show that both the RMS delay spread as well as the optical path loss of OLED-based VLC system were less than that of the inorganic LED-based VLC system which contains the Lambertian emission. From the simulation results, it can be surmised that OLEDs offer remarkable advantages to VLC systems with better channel conditions and result in reduction of ISI which generally arises due to the multipath dispersion.

Even though an organic photodiode offers a large photoactive area, its implementation for VLC applications is rather challenging. The reason is that the optical-to-electrical speed of the organic photodiode is limited due to the slow charging and discharging processes. Additionally, there is a requirement for external bias to enable its operation. The work in [113] proposes and demonstrates a pre-distortion scheme for PAM signal to a self-powered phenyl-C61-butyric acid methyl ester (PCBM): Poly(3-hexylthiophene) (P3HT) for improving the transmission performance of a VLC system. In order to validate the prominence of the proposed schemes, this work evaluates the performance of the organic photodiode-based VLC system with and without predistortion techniques for both 2-PAM and 4-PAM modulation formats. The experimental results elucidate that by exploiting the proposed predistortion scheme, the overall system exhibits a significant increase in data rate by about 5 times satisfying the 7% FEC requirement.

The recent significant work as portrayed in [90] investigates the optical characteristics of OLEDs embedded in a furnished office room environment. The performance of the flat and half-circular-based OLED-VLC link was analyzed in terms of RMS delay spread and BER. The developed OLED-based VLC link as signified in this work affirms a data rate of 4 Mbps for the transmitter's half-beam angle within the range of $\pm 90°$ and $\pm 53°$ with a BER below the forward error correct limit.

M-CAP-BASED OLED SYSTEM

As discussed earlier, the enormous benefits offered by CAP, especially high spectral efficiency with reduced complexity, focused interest in the research communities in recent times to exploit it as an intriguing alternative to optical OFDM. Accordingly, the research contribution in [88] proposes a flexible OLED as a light source for m-CAP-based VLC system suitable for short range device-to-device indoor applications. The schematic representation of m-CAP-VLC system that

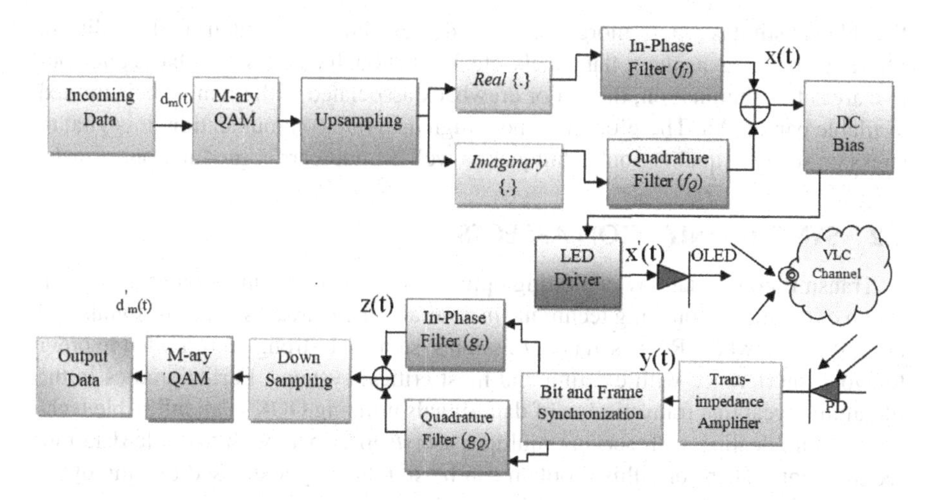

Figure 8.3: Schematic representation of m-CAP-based VLC system exploiting organic light emitting diode (OLED).

exploits a flexible OLED is shown in Fig. 8.3. As delineated, m independent pseudo-random data streams designated by $d_m(t)$ were generated and then mapped into a set of complex symbols by using complex constellation like M-QAM. The real and imaginary components denoted by $R(a_i)$ and $I(b_i)$ were then applied to the SRRC filters. The in-phase and the quadrature components are modulated onto the sine and cosine waves, respectively, to form a Hilbert transform pair. Finally, the outputs of the filters are summed to yield the m-CAP signal $x(t)$ which is used to intensity modulate the OLED. Then, at the receiving end, followed up by the optical detection and amplification, the generated electrical signal which is represented as $y(t)$ is fed to the bit and frame synchronization module. Consequently, the output of this module is applied to the two receive filters, i.e., in-phase and quadrature, which are specified by g_I and g_Q, respectively. These filters are time-reversed SRRC filters. The outputs of the two filters is summed to yield the signal $z(t)$ which is downsampled prior to getting applied to the M-QAM demapper block in order to yield the final estimates of the transmitted data signal $d'_m(t)$.

The work in [88] analyzes the performance of VLC system exploiting a curved as well as flat OLED making use of the high-dimensional spectrally efficient modulation scheme like m-CAP with $m = 2$ at different angles of incidence on the optical receiver. The experimental results of this work elucidate that when compared with flat OLED, the VLC-system exploiting curved OLED attains a better BER performance with the optical receiver moving along a circular path for the viewing angles greater than 40°.

Finally, research is rapidly progressing to enable OLEDs to penetrate the SSL market. Several studies in the literature as seen earlier claim that OLEDs cannot be seen as direct replacement to inorganic-based LEDs; however, they can be termed as feasible substitutes where inorganic-based devices fail to find their applicability in offering large area displays, flat panels, etc. Even though OLEDs offer large photoactive areas for illumination, the major drawback associated with them is their limited available bandwidth. Therefore, to render high data rate communication, it is vital to stress on several equalization techniques as well as multicarrier modulation formats.

8.2 SYNCHRONIZATION ASPECTS

Transmission of data by exploiting optical wireless communication as a medium is one of the most promising technique in several critical areas, such as hospitals and aircraft cabins where RF-based transmission systems are strongly prohibited in order to avoid interference with existing and most critical systems. Earlier studies in the literature reveal that transmission of data signals utilizing OOK is an inflexible technique while dealing with serving multiple users/subscribers with variable data rate requirements. Therefore, this problem can be significantly addressed by employing OFDM to intensity modulate the LED.

For cost-effective realization of VLC systems which are based on white LEDs, IM/DD is the most preferable modulation format. However, without employing optical filtering and equalization, the modulation bandwidth of IM/DD-based VLC systems is usually limited to several MHz. Particularly, it is mandated to have precise synchronization when VLC systems exploit highly spectral-efficient multicarrier modulation schemes like OFDM. In spite of its significance, the vulnerability of OFDM to the synchronization errors, represented by the so-called frequency and timing offsets, results as the most vital principal disadvantages of OFDM. Predominantly, an OWC system which is based on OFDM facilitates for the effective allocation of the bandwidth among competing users. Furthermore, it provides the flexibility for adaptive selection of modulation as well as coding schemes in order to accomplish the data rates in accordance with the traffic demands. Conventionally, the transmission of data signals using OFDM relies on the fact that the subcarrier frequency components must be orthogonal to each other. The orthogonality among the subcarriers can be lost because of the occurrence of frequency and timing offsets. Timing synchronization/frame detection plays a significant role in optical OFDM systems for VLC. Imperfect timing synchronization enables the receiver to capture a mixture of interfering signals at its sampling instant. As a result, these interfering signals lead to the emergence of ISI in the transmitted data streams. Hence, this results in the increase of error rate, thereby degrading the overall performance of the system. So, this entails addressing the aspects of synchronization problems which include analyzing the performance analysis of the overall system in the presence of offsets and estimation of different offsets for reducing the probability of error.

Frequency offset (FO) leads to a reduction of desired signal amplitude level in the output decision variables, besides, the small frequency mismatches between the transmitter and the receiver leads to the loss of orthogonality among the subcarrier components; hence, inter-carrier interference (ICI) emanates. Timing induced offsets results in the rotation of OFDM subcarrier constellation, thereby hindering the recovery of the transmitted signal when high constellation order QAM is implemented. Predominantly, the synchronization process involved in OFDM can be divided into a coarse frequency and timing acquisition and a fine frequency and timing offsets estimation [481]. Generally, coarse frequency and timing synchronization at the receiving end can be achieved by correlating the received and the original synchronization preamble in the frequency and time domain, respectively. Even though these offsets might be small, due to the extremely high synchronization requirement of the system, it is of paramount to estimate and to compensate them. This could be done either by demodulating the OFDM synchronization preamble which is an OFDM symbol used for training purposes, or by inserting the pilot carriers within the OFDM symbols.

Several studies have been reported regarding OFDM synchronization [32,33,217, 316,353,491]. Earlier studies report that synchronization errors are due to three different effects, i.e., carrier error, clock error and sampling timing error. The difference between the local oscillator in the receiver and the carrier frequency of the transmitting signal leads to carrier error. Meanwhile, the difference in the sampling clock in the receiver and the transmitter is usually defined as clock error. Besides, the difference between the optimum sampling time in the receiver and the actual sampling time is called sampling timing error. In the literature, it is reported that carrier error and clock error together can be stated as FO, while sampling timing error is recognized as timing offset. Consequently, due to the high sensitivity of OFDM systems to different synchronization errors, it is worthwhile to investigate the deleterious effects of these offsets when OFDM is adopted for IM/DD systems in VLC.

The synchronization problems encountered in optical OFDM are mainly focused on timing synchronization, i.e., frame detection and frequency synchronization. It is well known that the advantages of OFDM can be utilized as long as the orthogonality among the subcarriers is maintained. The moment, the orthogonality is disrupted by any means, then ICI and intersymbol interference (ISI) emanate. In general, the frequency offset in case of optical OFDM arises due to the occurrence of Doppler shift in a mobile environment. Moreover, the symbol time offset (STO) hinders the detection capability at the receiver side, because due to improper frame misalignments, the current OFDM symbol overlaps with the previous and the next symbols; eventually, this leads to the emanation of ICI and ISI. Therefore, it is vital to estimate these offsets, and then it is required to compensate them. Subsequently, this enforces relying on several synchronization algorithms like Classen, Moose, Training symbol assisted, Minimum difference and Maximum Correlation methods. This augmenting technology like VLC provides the flexibility of creation of a small scale cellular communication network within an indoor room environment by making use of the already utilized LED lighting fixtures. In this scenario, each installed LED lighting fixture will act as an OAP rendering services to several roaming mobile stations which are within the vicinity of LEDs. Therefore, in this scenario it is vital to explore synchronization aspects when multiple access schemes are exploited for IM/DD systems.

With the primary motive to have an adequate knowledge on the detrimental effects of STO and FO on the received OFDM signal, this section presents a proof-of-concept of the mathematical representation illustrating the effects of STO and FO on the received OFDM signal pertaining to DCO-OFDM-based VLC system. Figure 8.4 shows DCO-OFDM system model developed with the inclusion of FO ε and STO δ. Elaborate mathematical analysis is done showing the detrimental effects of FO and STO. Here, at the transmitter side, the incoming bit stream is first mapped with

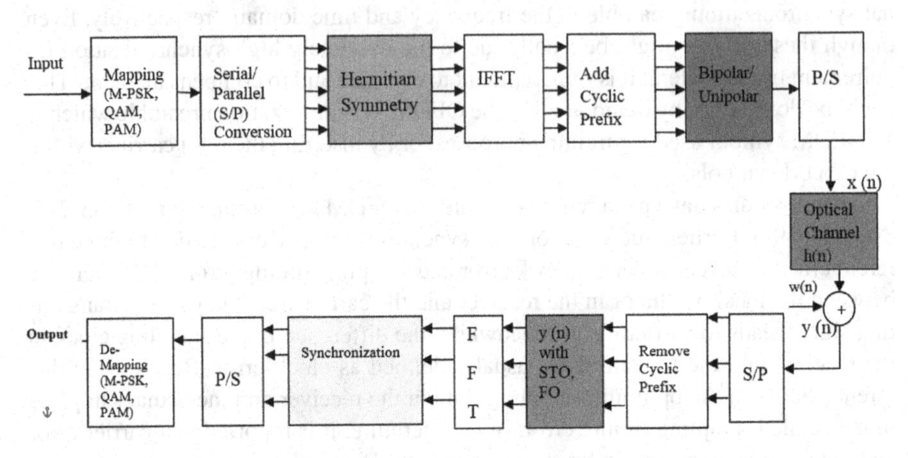

Figure 8.4: Synchronization in optical OFDM for VLC [496].

the help of modulation formats such as M-PSK, M-QAM and M-PAM, respectively, and then this serialized data stream is parallelized, which is denoted as $X(k)$. Since a modified version of OFDM is a prerequisite of optical transmission systems; hence, DMT is incorporated, which allows us to work with real valued signal in spite of the usage of IFFT. So, here this parallelized signal is constrained to satisfy Hermitian Symmetry condition, i.e., if there are N number of subcarriers, then according to Hermitian Symmetry constraint, only $\frac{N}{2}$ are utilized for transmission of data

$$X[N-k] = X^*[k], \quad k = 1,2,\cdots\frac{N}{2} \tag{8.3}$$

The first and the middle subcarriers are assigned zero to avoid the imaginary component in the time-domain signal

$$X[0] = X\left[\frac{N}{2}\right] = 0 \tag{8.4}$$

Here, the signal which is fed to the IFFT block can be treated as a vector with the following representation

$$X = \left[0, X_1, X_2 \cdots X_{\frac{N}{2}-1}, 0, X^*_{\frac{N}{2}-1}, \cdots, X^*_2, X^*_1\right] \tag{8.5}$$

8.2.1 MATHEMATICAL ILLUSTRATION OF THE EFFECTS OF FO

The prevalence of Doppler Shift in both indoor as well as outdoor room environments results in the loss of orthogonality among the subcarrier frequency components; thus, the advantages of OFDM can no longer be warranted. Consequently, this results in a huge amount of degradation in the BER performance of the overall system. Then, under the effect of FO ε, the received signal is derived as follows: It is a widely known fact that,

$$y_l[n] = IFFT[Y_l[q]] \tag{8.6}$$

Where $Y_l[q] = X_l[q]H_l[q] + W_l[q]$, upon substitution of the aforementioned condition in (8.6), the time domain signal when affected with FO of ε can be obtained as

$$y_l[n] = \frac{1}{N}\sum_{q=0}^{N-1} X_l[q]H_l[q]e^{\frac{j2\pi(q+\varepsilon)n}{N}} + w_l[n] \tag{8.7}$$

$w_l[n]$ in (8.7) represents the AWGN which is given as $w_l[n] = \frac{1}{N}\sum_{q=0}^{N-1} W_l[q]e^{\frac{j2\pi qn}{N}}$. For the purpose of attaining a real-valued signal, the frequency domain symbol as represented by $X_l[q]$ is constrained to satisfy Hermitian Symmetry criteria. Accordingly, the time domain signal as represented in (8.7) when affected with FO is derived to fetch the frequency domain signal as [496]

$$\begin{aligned} Y_l[k] = {} & \left(\underset{RC}{X_l[k]} + j\underset{IC}{X_l[k]}\right)H[k]\left[\frac{sin\pi\varepsilon}{Nsin\left(\frac{\pi\varepsilon}{N}\right)}\right]e^{\frac{j\pi\varepsilon(N-1)}{N}} + \\ & \left(\underset{RC}{X_l[k]} - j\underset{IC}{X_l[k]}\right)H[k]\left[\frac{sin\pi(\varepsilon+2k)}{Nsin\pi\left(\frac{\varepsilon+2k}{N}\right)}\right]e^{\frac{-j\pi(2k+\varepsilon)(N-1)}{N}} + \\ & \sum_{\substack{q=1\\q\neq k}}^{\frac{N}{2}-1}\left(\underset{RC}{X_l[q]} + j\underset{IC}{X_l[q]}\right)H[q]\left[\frac{sin\pi(q-k+\varepsilon)}{Nsin\pi\left(\frac{q-k+\varepsilon}{N}\right)}\right]e^{\frac{j\pi(q-k+\varepsilon)(N-1)}{N}} + \\ & \sum_{\substack{q=1\\q\neq k}}^{\frac{N}{2}-1}\left(\underset{RC}{X_l[q]} - j\underset{IC}{X_l[q]}\right)H[q]\left[\frac{sin\pi(q+k+\varepsilon)}{Nsin\pi\left(\frac{q+k+\varepsilon}{N}\right)}\right]e^{\frac{-j\pi(q+k+\varepsilon)(N-1)}{N}} + W_l[k] \end{aligned} \tag{8.8}$$

As seen here in (8.8), the 3rd and 4th terms in the summation represent ICI from other subcarriers and are denoted by $I_l[k]$ which can be expressed as [496]

$$\begin{aligned} I_l[k] = {} & \sum_{\substack{q=1\\q\neq k}}^{\frac{N}{2}-1}\left(\underset{RC}{X_l[q]} + j\underset{IC}{X_l[q]}\right)H[q]\left[\frac{sin\pi(q-k+\varepsilon)}{Nsin\pi\left(\frac{q-k+\varepsilon}{N}\right)}\right]e^{\frac{j\pi(q-k+\varepsilon)(N-1)}{N}} + \\ & \sum_{\substack{q=1\\q\neq k}}^{\frac{N}{2}-1}\left(\underset{RC}{X_l[q]} - j\underset{IC}{X_l[q]}\right)H[q]\left[\frac{sin\pi(q+k+\varepsilon)}{Nsin\pi\left(\frac{q+k+\varepsilon}{N}\right)}\right]e^{\frac{-j\pi(q+k+\varepsilon)(N-1)}{N}} \end{aligned} \tag{8.9}$$

Finally, upon rearranging equation (8.8), the final expression of the frequency domain signal can be obtained as

$$Y_l[k] = \left(\underset{RC}{X_l[k]} + j\underset{IC}{X_l[k]}\right) H[k] \left[\frac{sin\pi\varepsilon}{Nsin\left(\frac{\pi\varepsilon}{N}\right)}\right] e^{\frac{j\pi\varepsilon(N-1)}{N}} + \left(\underset{RC}{X_l[k]} - j\underset{IC}{X_l[k]}\right) H[k] \left[\frac{sin\pi(\varepsilon+2k)}{Nsin\pi\left(\frac{\varepsilon+2k}{N}\right)}\right] e^{\frac{-j\pi(2k+\varepsilon)(N-1)}{N}} + I_l[k] + W_l[k] \tag{8.10}$$

It is clearly evident from equation (8.10) that ICI is occurring from neighboring subcarriers into the kth subcarrier, which further implies that the orthogonality among the subcarriers is destroyed.

8.2.2 MATHEMATICAL DEPICTION OF THE EFFECTS OF STO ON THE PERFORMANCE OF DCO-OFDM-BASED VLC SYSTEM

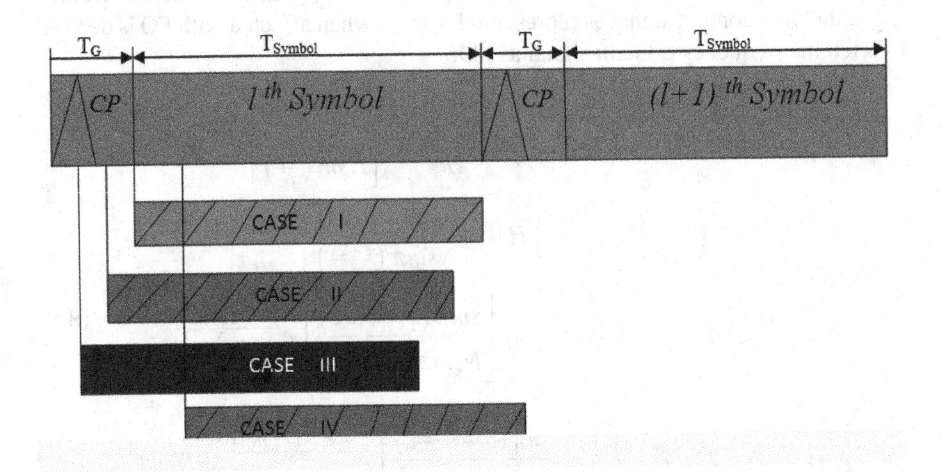

Figure 8.5: STO on the received signal [496].

Perfect timing synchronization is required to ensure a reliable communication. In this regard, it is essential to know the effect of STO on the received signal. Figure 8.5 shows the elaborate view of the effects of STO on the received signal. From the figure, it is clearly evident that there are four scenarios of timing offset where, in the first case, the estimated starting point perfectly matches with the exact timing instance; for the second scenario it is a little earlier; whereas, in the third case, it is too early; and in the last case, it is little later than the exact timing instance [111].

- **CASE I:** In this scenario, there is a perfect timing synchronization because there is a perfect coincidence between the estimated starting point of the optical OFDM symbol and the timing interval, thereby preserving the orthogonality among the subcarriers. Hence, in this scenario there is perfect recovery of the symbol without getting affected by any type of interference.

- **CASE II:** Here, the estimated point is earlier than the timing instant. FFT and IFFT play a vital role at both the transmitting and receiving end of an optical or conventional OFDM system. The essential prerequisite for performing the N-point FFT operation at the receiving end is to ensure that the exact samples of the transmitted signal are obtained within the symbol duration. Consequently, the received signal in DCO-OFDM system under the effect of STO is represented as

$$
\begin{aligned}
Y_l[k] &= \underset{RC}{X_l[K]} e^{\frac{j2\pi k\delta}{N}} + \frac{sin 2\pi k}{N sin\left(\frac{2\pi k}{N}\right)} \underset{Real}{X_l[K]} e^{\frac{-j2\pi k\delta}{N}} e^{\frac{-j2\pi(N-1)k}{N}} \\
&+ \left(\underset{RC}{X_l[K]} + \underset{IC}{X_l[k]}\right) e^{\frac{j2\pi k\delta}{N}} \\
&+ \frac{1}{N} \sum_{\substack{u=1 \\ u\neq k}}^{\frac{N}{2}-1} \left(\underset{RC}{X_l[u]} + j\underset{IC}{X_l[k]}\right) e^{\frac{j2\pi u\delta}{N}} \left[\frac{sin(\pi(u-k))}{N sin\left(\frac{\pi(u-k)}{N}\right)}\right] e^{\frac{j\pi(u-k)(N-1)}{N}} \\
&+ \frac{1}{N} \sum_{\substack{u=1 \\ u\neq k}}^{\frac{N}{2}-1} \left(\underset{RC}{X_l[u]} - j\underset{IC}{X_l[u]}\right) e^{\frac{-j2\pi u\delta}{N}} \left[\frac{sin(\pi(u+k))}{N sin\left(\frac{\pi(u+k)}{N}\right)}\right] e^{\frac{-j\pi(u+k)(N+1)}{N}} + W_l(k)
\end{aligned}
\tag{8.11}
$$

On carefully examining (8.11), the received signal is not only affected by amplitude and phase distortion, but also by ICI from other subcarriers. Hence, the above equation emphasizes that the orthogonality is no longer warranted.

- **CASE III:** In this scenario, as depicted from the figure, the present symbol overlaps with the previous symbol. This signifies that this scenario is prone to ISI leading to the disruption in the orthogonality among the subcarrier components, and, besides, ICI can also occur.
- **CASE IV:** This scenario evidences a timing mismatch, i.e., there is no perfect timing synchronization as the signal within the timing instant consists of a part of the current optical OFDM symbol, $x_l[n]$ and some portion of the next symbol, i.e., $x_{l+1}[n]$. Therefore, the signal received under such scenario can be represented as [111]

$$
y_l[n] = \begin{cases} x_l[n+\delta] & for\ 0 \leq n \leq N-1-\delta \\ x_{l+1}[n+2\delta-N_g] & for\ N-\delta \leq n \leq N-1 \end{cases} \tag{8.12}
$$

Under such scenario which emphasizes the overlap of the present/current frame with the next frame, the frequency domain representation of the signal under the detrimental effects of STO is given by [496]

$$Y_l[k] = \left(\frac{N-\delta}{N}\right)\left(\underset{RC}{X_l}[k] + j\underset{IC}{X_l}[k]\right) e^{\frac{j2\pi\delta k}{N}} + \left(\underset{RC}{X_l}[k] - j\underset{IC}{X_l}[k]\right)\left[\frac{sin\left(\frac{2\pi(N-\delta)k}{N}\right)}{N sin\left(\frac{2\pi k}{N}\right)}\right] e^{\frac{-j2\pi(N-1)k}{N}} +$$

$$\sum_{\substack{u=1\\u\neq k}}^{\frac{N}{2}-1}\left(\underset{RC}{X_l}[u] + j\underset{IC}{X_l}[u]\right) e^{\frac{j2\pi\delta u}{N}}\left[\frac{sin\left(\frac{\pi(u-k)(N-\delta)}{N}\right)}{N sin\left(\frac{\pi(u-k)}{N}\right)}\right] e^{\frac{j\pi(u-k)(N-\delta-1)}{N}} +$$

$$\sum_{\substack{u=1\\u\neq k}}^{\frac{N}{2}-1}\left(\underset{RC}{X_l}[u] - j\underset{IC}{X_l}[u]\right) e^{-\frac{j2\pi\delta u}{N}}\left[\frac{sin\left(\frac{\pi(u+k)(N-\delta)}{N}\right)}{N sin\left(\frac{\pi(u+k)}{N}\right)}\right] e^{-\frac{j\pi(u+k)(N-\delta-1)}{N}} +$$

$$\sum_{n=N-\delta}^{N-1}\left[\frac{2}{N}\sum_{u=1}^{\frac{N}{2}-1}\left(\underset{RC}{X_{l+1}}[u] cos\left(\frac{2\pi(n+2\delta-N_g)u}{N}\right) - \underset{IC}{X_{l+1}}[u] sin\left(\frac{2\pi(n+2\delta-N_g)u}{N}\right)\right)\right] e^{\frac{-j2\pi nk}{N}} + W_l[k] \quad (8.13)$$

equation (8.13) emphasizes that the signal has been severely affected by the presence of STO. The first two terms of (8.13) represent that the signal has been affected by amplitude and phase distortions. Meanwhile, the 3rd and 4th term illustrate the ICI from the neighboring subcarrier components and the 5th term represents the ISI due to the presence of next optical OFDM symbol $\underset{RC}{X_{l+1}}[u]$.

Upon having a glance on the deterioration of the received signal due to the presence of both STO and FO, the below simulation result further confirms the distortion of the received signal constellation. Figures 8.6 and Fig. 8.7 depict the constellation diagrams with and without the effect of FO. From Fig. 8.7, it can be viewed that the constellation is distorted, as it has a circular shape. Thus, this entails to stress on robust synchronization techniques in order to estimate these offsets as well as to compensate them. In order to estimate these offsets, several significant work can be reported from the literature. Previous studies in the literature as stated in [39, 173, 175, 183, 343] illustrate the effect of synchronization error on the modulating signal constellation in optical communication system. The work in [175] gives the theoretical analysis of the effects of synchronization error in OOK and PPM-based optical communication systems. The authors in [39] propose an inverse PPM (IPPM) modulation method to encapsulate communication data in the time domain without adversely affecting the illumination property. Inherently, this method necessitates an effective synchronization aid. Consequently, this work evaluates the effect of clock time shift and timing jitter on the BER performance of the proposed system. Accordingly, the authors derived effective models with minimal requirements of the synchronization system.

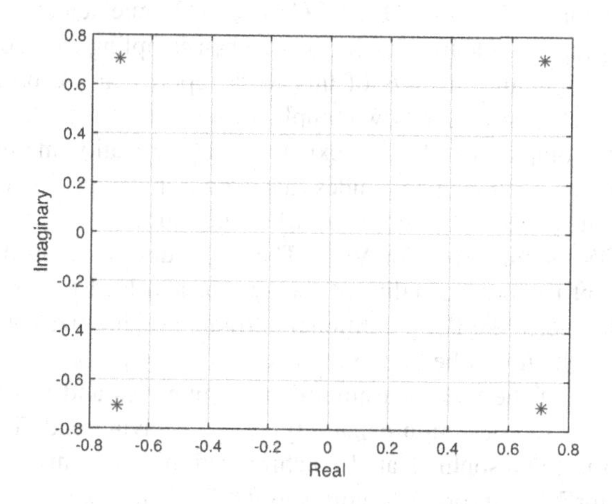

Figure 8.6: Constellation without the effect of FO [496].

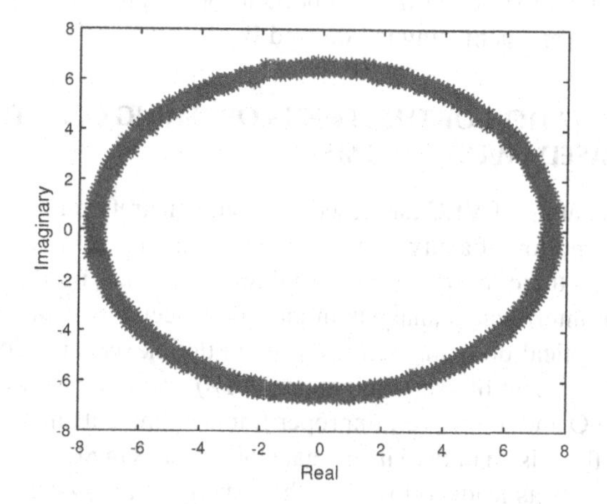

Figure 8.7: Constellation with the effect of FO [496].

The research work in [173] illustrates that the authors have proposed coding framework to address the synchronization aspects of pulse position modulation formats. Further, the authors investigated the effects of clock time shift and jitter on the BER performance of the system. The works in [183,343] depict the impact of synchronization error on optical OFDM system and also present techniques to correct it. Much recent literature pertaining to the investigations of the effects of frequency and timing offsets can be delineated in [242], where the authors proposed a low-complexity maximum likelihood-based timing synchronization method for DCO-OFDM systems. The interesting aspect associated with this research contribution is that the timing synchronization method enables both frame detection as well as sampling clock synchronization for DCO-OFDM-VLC systems. Frame detection can be made possible by exploiting a single training block and then sampling clock offset can also be estimated. The simulation results of this work report that the proposed timing synchronization mechanism offers low complexity as well as demonstrates superior performance when compared with other existing synchronization methods.

Thus, this research challenge concludes that synchronization is one of the indispensable issues which need to be addressed when the robust multicarrier modulation scheme like OFDM is employed for VLC. The slight deviations in the transmitter and receiver disrupt the orthogonality of the system and lead to the emanation of deleterious interferences like ICI. Additionally, frame synchronization is one of the crucial aspects which should be addressed because the improper timing alignment leads to the overlap of the present frame with the previous and next frames. This phenomenon hinders the detection capability at the receiving end. Therefore, suggestion of appropriate and sophisticated synchronization algorithms compatible with IM/DD systems for VLC in order to compensate the signal with the estimated FO and STO stems out to be the major concern. Additionally, while analyzing the high-speed communication systems, it is vital to take into consideration synchronization jitter because, proper clock function is important for the purpose of providing accurate synchronization and achieving the desired BER.

8.2.3 INTERPRETATION OF THE EFFECTS OF TIMING OFFSETS IN DCO-SC-FDMA-BASED IM/DD SYSTEMS

The renowned nature of VLC has eased its rapid penetration into several applications like the proomotion of a city as a smart city, vehicular communications, health care, etc. However, there are major challenging aspects that need to be addressed, among which the interference analysis in an uplink scenario is the most vital one, especially in the optical domain. Generally, in an uplink scenario doppler shift occurs due to the movement of user equipments (UEs) or mobile stations with respect to BS leading to FO and, moreover, improper frame alignments make STO unavoidable. Eventually, there is an urge to investigate these issues in SC-FDMA because the presence of these offsets leads to loss of orthogonality of the system, thereby resulting in overlapping of the current symbol with the previous and the next symbol. This gives rise to a very well-known interference called inter-carrier interference (ICI). Furthermore, ISI also prevails in indoor room environment due to sudden blockage of signals which arise because of shadowing, as well as reflections occur due to the presence of several obstacles like furniture, people, etc. In an uplink scenario the different timing misalignments between the roaming mobiles, i.e., UEs and BS, lead to the most detrimental issue called multi-user interference (MUI) or multiple access interference (MAI). If MUI/MAI prevails in the uplink scenario, then it hinders the detection capability of multiple users, resulting in deterioration of system performance. Tackling these issues in optical domain is not straightforward due to real and unipolar nature of the time-domain signal. Consequently, with this motive, this chapter presents in detail an elaborate discussion on the uplink scenario in SC-FDMA-based VLC system. The basic schematic of the uplink VLC system

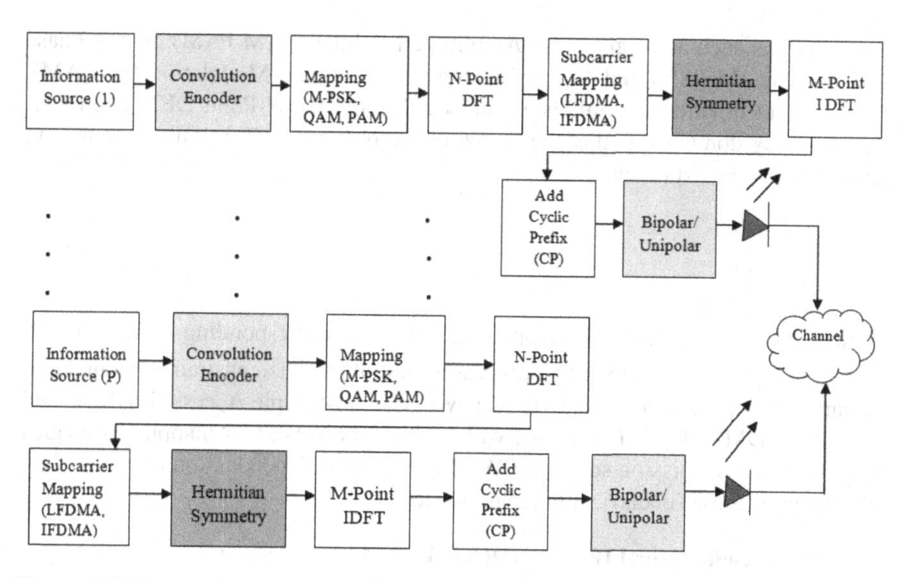

Figure 8.8: Transmitter pertaining to the P subscribers communicating to the base station (BS).

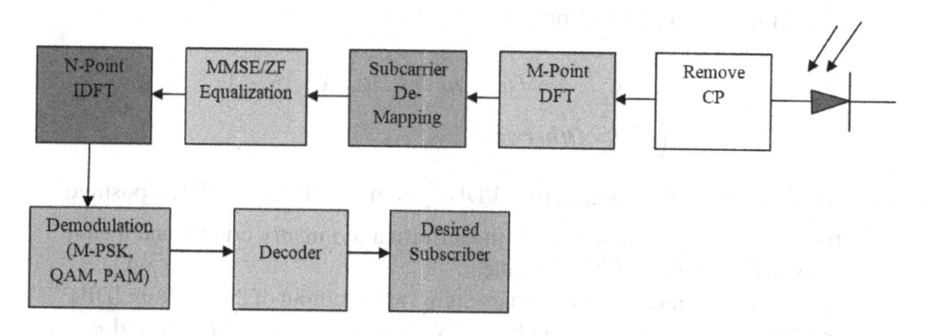

Figure 8.9: Typical schematic representation of base station. Depiction of uplink scenario in DCO-SC-FDMA system for VLC.

employing the phenomena of SC-FDMA is interpreted in Figs. 8.8 and 8.9. From Figure 8.8, it can be evidenced that, P subscribers/users are indulged in carrying on their transactions with the base station (BS) through their relevant independent optical channels. Looking into the details of transmitter and receiver schematic, we proceed with the elaborate discussion by utilizing the rth subscriber as the desired/-corresponding subscriber. In this scenario, we assumed that the total number of subcarriers are M and each subscriber is allocated with a certain set of subcarriers for facilitating uplink transmission.

At the transmitting terminal, the huge sets of data stream is encoded with the help of a convolutional encoder and then mapped by exploiting different orders of constellations such as M-ary Pulse Amplitude Modulation (M-PAM), M-ary Phase Shift Keying (M-PSK) and M-ary Quadrature Amplitude Modulation (M-QAM). The resultant modulated symbols are passed through the N-Point DFT. Therefore, the frequency domain signal at the output of the N-Point DFT pertaining to the rth subscriber can be expressed as

$$X_k^{(r)} = \sum_{n=0}^{N-1} x_n^{(r)} e^{\frac{-j2\pi nk}{N}} \tag{8.14}$$

From (8.14), $x_n^{(r)}$ signifies the mapped data symbols corresponding to the rth subscriber and $X_k^{(r)}$ denotes the DFT processed output. $X_k^{(r)}$ is constrained to undergo subcarrier mapping. Interleaved Frequency Division Multiple Access (IFDMA) and Localized FDMA (LFDMA) are the widely deployed subcarrier mapping strategies in the uplink transmission scenario. PAPR reduction is dependent on the way of arrangement of these subcarriers mapping strategies.

- **DC-Biased Optical IFDMA (DCO-IFDMA)**
  In case of IFDMA, the DFT outputs are distributed and allocated among the subscribers in an equidistant manner as $\frac{M}{N} = Q$, where Q represents the spreading factor.
  The allocation of $X_k^{(r)}$ is represented as [111]

$$\tilde{X}_k^{(r)} = \begin{cases} X_{\frac{k}{Q}}^{(r)}, \; k = Qn_1, \;\; n_1 = 0,1,2,3,\cdots N-1 \\ 0, \;\; Otherwise \end{cases} \tag{8.15}$$

  Since the major prerequisite of IM/DD system is to affirm a real and positive transmission, $\tilde{X}_k^{(r)}$ is enforced with Hermitian Symmetry criteria and then it is loaded to M-Point IDFT module.
  Hence, the discretized time-domain signal at the output of the M-Point IDFT module can be formulated with $m = Nq + n$, $q = 0,1,2,\cdots Q-1$ and $n = 0,1,2,3\cdots N-1$ as

$$\tilde{x}_m^{(r)} = \frac{1}{M} \sum_{\substack{k=0 \\ k \in Z_r}}^{M-1} \tilde{X}_k^{(r)} e^{\frac{j2\pi mk}{M}} \tag{8.16}$$

$$\tilde{x}_m^{(r)} = x_{Nq+n}^{(r)} = \frac{1}{NQ} \sum_{\substack{n_1=0 \\ n_1 \in Z_r}}^{N-1} X_{n_1}^{(r)} e^{\frac{j2\pi(Nq+n)Qn_1}{NQ}} \tag{8.17}$$

  In general, (8.17) infers that the term $X_{n1}^{(r)}$ specifies the mapped data corresponding to the rth subscriber on the n_1th subcarrier. Precisely, $n_1 \in Z_r$, where Z_r represents the total number of subcarriers which are dedicated to the rth user and $\cup_{r=1}^{P} Z_r = \{0,1,2,\cdots N-1\}$. Furthermore, Hermitian Symmetry constraint must be imposed in (8.17). Therefore, the signal $X_{n_1}^{(r)}$ can be treated as a vector to have the following representation

$$X_{n_1}^{(r)} = \left[0, X_1^{(r)}, X_2^{(r)}, \cdots X_{\frac{N}{2}-1}^{(r)}, 0, X_{\frac{N}{2}-1}^{*(r)} \cdots X_2^{*(r)}, X_1^{*(r)}\right] \tag{8.18}$$

From (8.18), it is evident that the first and the middle subcarriers are forced to zero, in order to avoid the presence of any imaginary component at the output of the IDFT/IFFT block.

$$X_0^{(r)} = X_{\frac{N}{2}}^{(r)} = 0 \tag{8.19}$$

Therefore, from the representation of (8.19), it is evident that all of the subcarriers are not utilized for data transmission because only half are meant for data transmission and the remaining half are flipped conjugate versions of the previous ones. This can be put up as

$$X_{(N-n_1)}^{*(r)} = X_{(n_1)}^{(r)},\ n_1 = 1, 2, 3, \cdots \frac{N}{2} \tag{8.20}$$

Thus, accordingly upon imposing the aforementioned constraints, (8.17) can be solved to attain

$$\tilde{x}_m^{(r)} = \frac{1}{Q}\frac{1}{N}\left[X_0^{(r)} + \sum_{\substack{n_1=0 \\ n_1 \in Z_r}}^{\frac{N}{2}-1} X_{n_1}^{(r)} e^{\frac{j2\pi n n_1}{N}} + X_{\frac{N}{2}}^{(r)} e^{\frac{j2\pi n N}{2N}} + \sum_{\substack{n_1=\frac{N}{2}+1 \\ n_1 \in Z_r}}^{N-1} X_{n_1}^{(r)} e^{\frac{j2\pi n n_1}{N}}\right] \tag{8.21}$$

It is apparent that, from the above discussion, the first and the middle subcarrier components can be set to zero in (8.21) and followed by change of variable to attain the following representation as

$$\tilde{x}_m^{(r)} = \frac{1}{Q}\frac{1}{N}\left[\sum_{\substack{n_1=0 \\ n_1 \in Z_r}}^{\frac{N}{2}-1} X_{n_1}^{(r)} e^{\frac{j2\pi n n_1}{N}} + \sum_{\substack{n_1'=1 \\ n_1' \in Z_r}}^{\frac{N}{2}-1} X_{N-n_1'}^{(r)} e^{\frac{-j2\pi n\left(N-n_1'\right)}{N}}\right] \tag{8.22}$$

Upon rearranging (8.22), the following expression can be attained

$$\tilde{x}_m^{(r)} = \frac{1}{Q}\frac{1}{N}\left[\sum_{\substack{n_1=0 \\ n_1 \in Z_r}}^{\frac{N}{2}-1} X_{n_1}^{(r)} e^{\frac{j2\pi n n_1}{N}} + \sum_{\substack{n_1=1 \\ n_1 \in Z_r}}^{\frac{N}{2}-1} X_{n_1}^{*(r)} e^{\frac{-j2\pi n n_1}{N}}\right] \tag{8.23}$$

(8.23), can be solved by using simple trigonometric expressions like $e^{j\theta} = cos\Theta + jsin\Theta$ and $e^{-j\theta} = cos\Theta - jsin\Theta$. Therefore, the following expression is attained

$$\tilde{x}_m^{(r)} = \frac{1}{Q}\frac{1}{N}\sum_{\substack{n_1=1\\ n_1 \in Z_r}}^{\frac{N}{2}-1}\left[\left(X_{n_1}^{(r)} + X_{n_1}^{*(r)}\right) cos\left(\frac{2\pi n n_1}{N}\right) + j\left(X_{n_1}^{(r)} - X_{n_1}^{*(r)}\right) sin\left(\frac{2\pi n n_1}{N}\right)\right] \tag{8.24}$$

Upon further solving (8.24), by using simple inequalities like $X_{Real}(t) = \frac{x(t)+x^*(t)}{2}$ and $X_{Imag} = \frac{x(t)-x^*(t)}{2j}$. Therefore, the resultant expression can be put as follows

$$\tilde{x}_m^{(r)} = x_{Nq+n}^{(r)} = \frac{1}{Q}\left\{\frac{2}{N}\left[\sum_{\substack{n_1=1\\ n_1 \in Z_r}}^{\frac{N}{2}-1}\left(\underset{RC}{X_{n_1}^{(r)}} cos\left(\frac{2\pi n n_1}{N}\right) - \underset{IC}{X_{n_1}^{(r)}} sin\left(\frac{2\pi n n_1}{N}\right)\right)\right]\right\} \tag{8.25}$$

It is clearly evident from (8.25) that the time-domain signal corresponding to rth subscriber, i.e., $\tilde{x}_m^{(r)}$ is Q-times scaled version of x_n. In fact, $\tilde{x_m}$, comprises, the characteristics of a single carrier with low peak to average power ratio (PAPR). It is interesting to note that, The obtained time-domain signal is real in nature complying to the requirements of IM/DD systems for VLC rather than a complex signal which is attained in RF.

- **DCO-LFDMA:**
Here, in case of LFDMA, the time-domain signal format at the output of M-point IFFT corresponding to the desired subscriber r is represented as $\tilde{x}_m^{(r)}$, with $m = Qn + q$ for $q = 0, 1, 2, 3, \cdots Q-1$. Therefore, this can be represented mathematically as

$$\tilde{x}_m^{(r)} = x_{Qn+q}^{(r)} = \frac{1}{M}\sum_{\substack{k=0\\ k \in Z_r}}^{M-1}\tilde{X}_k^{(r)} e^{\frac{j2\pi mk}{M}} \tag{8.26}$$

upon substitution of the aforementioned conditions, (8.26) is obtained as

$$\tilde{x}_m^{(r)} = x_{Qn+q}^{(r)} = \frac{1}{NQ}\sum_{\substack{k=0\\ k \in Z_r}}^{N-1} X_k^{(r)} e^{\frac{j2\pi(Qn+q)k}{NQ}} \tag{8.27}$$

It should be noted that, looking into the schematic as represented in Fig. 8.8, the signal prior to IFFT is confined to satisfy Hermitian Symmetry criteria. Therefore, this implies that $\tilde{X}_k^{(r)}$ must satisfy Hermitian Symmetry

constraint and then should be applied as input of IFFT. Hence, (8.27) can be reduced as:
case 1: $q = 0$

$$\tilde{x}_m^{(r)} = x_{Qn}^{(r)} = \frac{1}{Q}\left[\frac{1}{N}\sum_{\substack{k=1\\k\in Z_r}}^{\frac{N}{2}-1}\left(X_{\substack{k\\RC}}^{(r)} cos\left(\frac{2\pi nk}{N}\right) - X_{\substack{k\\IC}}^{(r)} sin\left(\frac{2\pi nk}{N}\right)\right)\right] \quad (8.28)$$

From (8.28), this expression follows a similar approach as that of the IFDMA. But, when $q \neq 0$, the resultant time-domain signal consists of multiplication with additional weighing factors. This can be elucidated mathematically as follows:
case 2: $q \neq 0$
(8.26), can be further solved by substituting

$$X_k^{(r)} = \sum_{p=0}^{N-1} x_p^{(r)} e^{\frac{-j2\pi pk}{N}} \quad (8.29)$$

The resulting equation can be represented as

$$\tilde{x}_m^{(r)} = x_{Qn+q}^{(r)} = \frac{1}{NQ}\sum_{k=0}^{N-1}\sum_{p=0}^{N-1} x_p^{(r)} e^{\frac{-j2\pi pk}{N}} e^{\frac{j2\pi(Qn+q)}{NQ}} \quad (8.30)$$

It is to be noted that in (8.30), the time domain signal generally corresponds to

$$x_p^{(r)} = \frac{2}{N}\sum_{\substack{k=1\\k\in Z_r}}^{\frac{N}{2}-1}\left(X_{\substack{k\\RC}}^{(r)} cos\left(\frac{2\pi pk}{N}\right) - X_{\substack{k\\IC}}^{(r)} sin\left(\frac{2\pi pk}{N}\right)\right) \quad (8.31)$$

So, upon substituting in (8.31) and then rearranging it reduces to

$$\tilde{x}_m^{(r)} = x_{Qn+q} = \frac{1}{NQ}\sum_{p=0}^{N-1}\left(\frac{2}{N}\sum_{\substack{k=1\\k\in Z_r}}^{\frac{N}{2}-1}\left(X_{\substack{k\\RC}}^{(r)} cos\left(\frac{2\pi pk}{N}\right) - X_{\substack{k\\IC}}^{(r)} sin\left(\frac{2\pi pk}{N}\right)\right)\right) \times \sum_{k=0}^{N-1} e^{j2\pi\left(\frac{n}{N}+\frac{q}{NQ}-\frac{p}{N}\right)k} \quad (8.32)$$

(8.32), can be further solved to attain

$$\tilde{x}_m^{(r)} = x_{Qn+q} = \\ \frac{1}{Q}\sum_{p=0}^{N-1}\left(\frac{2}{N}\sum_{\substack{k=1\\k\in Z_r}}^{\frac{N}{2}-1}\left(X_{\substack{k\\RC}}^{(r)} cos\left(\frac{2\pi pk}{N}\right) - X_{\substack{k\\IC}}^{(r)} sin\left(\frac{2\pi pk}{N}\right)\right)\right) \\ \times \frac{sin\pi\left(\frac{(n-p)Q+q}{Q}\right)}{N sin\pi\left(\frac{(n-p)Q+q}{NQ}\right)} e^{j\pi\left(\left(1-\frac{1}{N}\right)\left(n-p+\frac{q}{Q}\right)\right)} \tag{8.33}$$

From (8.33), it is evident that LFDMA incurs a huge amount of PAPR when compared with IFDMA because the time-domain signal is multiplied by several weighing factors.

Looking into (8.25) and (8.33), a real valued signal is attained in accordance with IM/DD systems for VLC. Furthermore, to mitigate the effects of intersymbol interference (ISI) a certain amount of cyclic prefix or guard band is appended to this time-domain signal. Furthermore, in order to certify a real and unipolar (positive) signal transmission, a certain amount of DC bias is added.
Generally, the added DC bias is equal to the absolute value of the maximum negative amplitude of the bipolar signal. This added DC bias is dependent on the constellation size as well as the LED characteristics. Too much DC bias will lead to power inefficiency, but in the literature it is claimed in several studies that it is desired for illumination purpose in VLC. However, the amount of DC bias is defined as

$$B_{DC} = 10log10\left(k^2+1\right) \;\; dB \tag{8.34}$$

In (8.34), k specifies the clipping factor and is defined as $\frac{B_{DC}}{\sqrt{E\left[|y|^2\right]}}$. This type of addition of DC bias is commonly referred to as fixed DC bias. However, since the amount of DC bias added has a direct relationship with the constellation sizes, it is desirable to add DC bias by taking into consideration the LED dynamic range. Every LED has its own turn-on-voltage (TOV), i.e., the region where the LED starts conducting. Therefore, it is more advisable to add a DC bias which equals the sum of absolute value of the maximum negative amplitude of the signal and the LED TOV. Therefore, the DC bias added signal can be put as

$$\tilde{x}_{\substack{m\\DC}}^{(r)} = \tilde{x}_m^{(r)} + B_{DC} \tag{8.35}$$

Further, if any negative peaks are present, then they are clipped; as a result this leads to induction of clipping noise which is denoted by η_{clip}.

$$\tilde{x}_{\substack{m\\DC}}^{(r)} = \tilde{x}_m^{(r)} + B_{DC} + \eta_{clip} \tag{8.36}$$

Therefore, now this real and positive signal invades the receiver after propagating through the channel. Hence, this can be expressed as

$$y_m^{(r)} = \tilde{x}_{m_{DC}}^{(r)} * h + w_m^{(r)} \tag{8.37}$$

$w_m^{(r)}$ in (8.37) represents the additive white Gaussian noise (AWGN), and h represents the channel impulse response. In general, at the receiving end, inverse operations such as removal of cyclic prefix, and passing the signal through a M-point DFT, followed by minimum mean square error (MMSE) and zero forcing (ZF) equalization. Then this equalized signal is passed through N-point IDFT in order to transform the frequency domain signal into time domain. Further, this signal is mapped by employing the corresponding mapping scheme like M-PSK, M-PAM and M-QAM. Since, at the transmitting side convolutional encoder is employed, at the receiving end the signal is decoded by employing Viterbi decoding algorithm. Finally, the signal is mapped to the desired subscriber r.

8.3 IMPACT OF TIMING ERRORS IN DCO-SC-FDMA FOR VLC

In particular, in order to emphasize the detrimental aspects of FO α and STO β on the received signal in IM/DD-based VLC systems, we employ the methodology as illustrated in [413], where the authors have evaluated the system performance for RF domain. The same timing interval misalignments have been incorporated in this work, and mathematical expressions are derived for optical domain. The possibilities of different timing discrepancies in DCO-SC-FDMA system in accordance with IM/DD systems for VLC are highlighted in Figs. 8.10 and 8.11. Typically, the timing induced errors can be both positive as well as negative. It is to be noted that α_u and β_u denote the uth subscriber's frequency offset and STO, respectively.

It is known that,

$$y_m^{(u)} = IDFT\left[Y_k^{(u)}\right] \tag{8.38}$$

Further, (8.38) can be manifested by contemplating the detrimental aspects of timing errors as follows:

$$\tilde{y}_m^{(u)} = \frac{1}{M}\sum_{k=0}^{M-1}\left[\tilde{X}_k^{(u)} H_k^{(u)} + Z_k^{(u)}\right] e^{\frac{j2\pi(k+\alpha_u)(m+\beta_u)}{M}} \tag{8.39}$$

Therefore, the time-domain output sequence of the uth subscriber $\tilde{y}_m^{(u)}$ with $m = Nq+n$, $q = 0,1,2,\cdots Q-1$ and $n = 0,1,2,3\cdots N-1$ and making use of (4.39) can be rearranged as

$$\tilde{y}_m^{(u)} = \frac{1}{NQ}\sum_{\substack{n_1=0 \\ n_1 \in Z_u}}^{N-1}\left[X_{n_1}^{(u)} e^{\frac{j2\pi(Qn_1+\alpha_u)(m+\beta_u)}{NQ}} H_{n_1}^{(u)}\right] + z_m^{(u)} \tag{8.40}$$

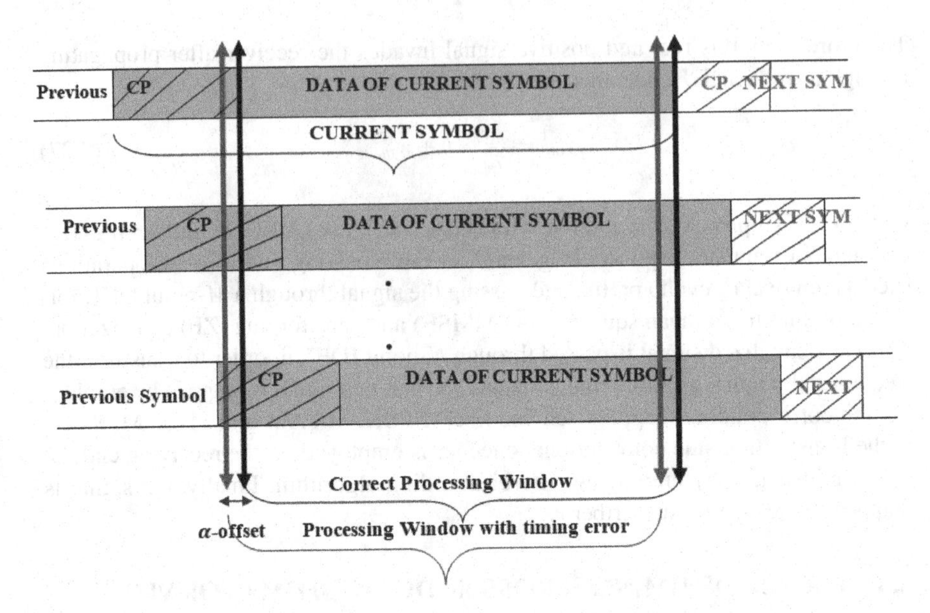

Figure 8.10: Negative timing error in DCO-SC-FDMA system in accordance with IM/DD systems for VLC.

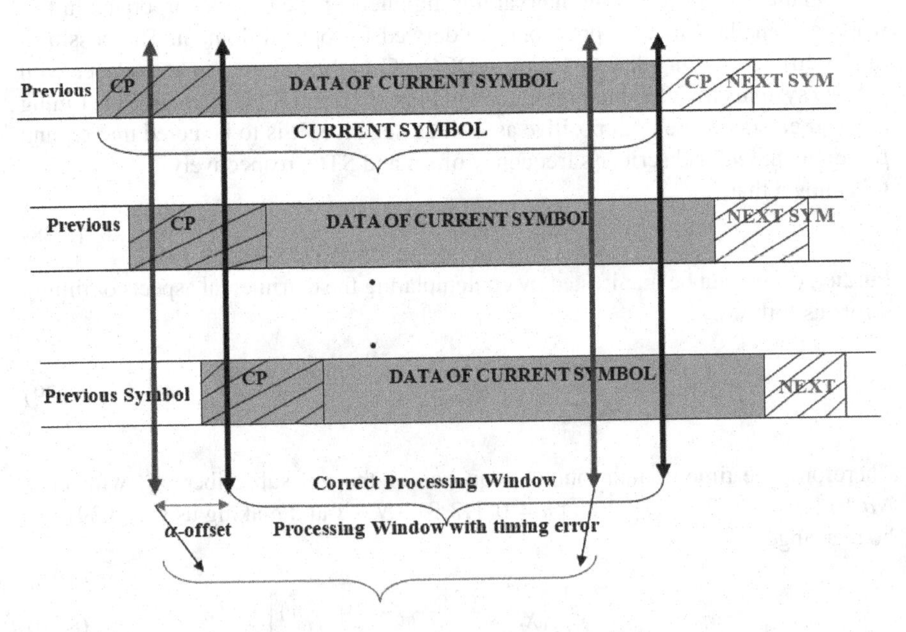

Figure 8.11: Overlap with previous frame.

Further, Hermitian Symmetry constraint must be imposed in (8.40). Accordingly, by incorporating the aforesaid constraints to yield a real signal, (8.40) can be rearranged to obtain the time-domain signal of the uth user

$$\tilde{y}_m^{(u)} = \frac{1}{NQ}\Bigg[\sum_{\substack{n_1=1\\ n_1\in Z_r}}^{\frac{N}{2}-1} X_{n_1}^{(u)} e^{\frac{j2\pi(Qn_1+\alpha_u)(m+\beta_u)}{NQ}} + \sum_{\substack{n_1=\frac{N}{2}+1\\ n_1\in Z_r}}^{N-1} X_{n_1}^{(u)} e^{\frac{j2\pi(Qn_1+\alpha_u)(m+\beta_u)}{NQ}}\Bigg] H_{n_1}^{(u)} + z_m^{(u)} \quad (8.41)$$

Further, by assimilating certain signal processing and performing change in variable (8.41) can be reduced to obtain the corresponding time-domain signal

$$\tilde{y}_m^{(u)} = \frac{1}{NQ}\sum_{\substack{n_1=1\\ n_1\in Z_u}}^{\frac{N}{2}-1}\Bigg[X_{\substack{n_1\\RC}}^{(u)} cos\left(\frac{2\pi(Qn_1+\alpha_u)(m+\beta_u)}{NQ}\right) - X_{\substack{n_1\\IC}}^{(u)} sin\left(\frac{2\pi(Qn_1+\alpha_u)(m+\beta_u)}{NQ}\right)\Bigg] H_{n_1}^{(u)} + B_{DC} + z_m^{(u)} \quad (8.42)$$

In general, $\tilde{y}_m^{(u)}$ represents the time-domain signal with the presence of both timing and frequency discrepancies after passing through the optical channel environment. This is the most pronounced issue which needs to be addressed especially in an uplink environment because both these errors result in loss of orthogonality among the subcarriers. As a result intercarrier interference (ICI) and multiuser interference/multiple access interference (MUI/MAI) emanate due to frequency disparities among the uplink subscribers and, moreover, there might be a chance of induction of discrepancies between the uplink subscribers and the BS.
The major emphasis is laid on depicting the detrimental aspects of these interferences on the received signal. So, we proceed further to the receiving side, where the cyclic prefix and the added DC bias are removed. Further, the received time-domain signal is fed to the FFT transformation block to yield the corresponding frequency domain signal. Consequently, the frequency domain representation of the time-domain signal pertaining to the uth subscriber is signified as $Y_k^{(u)}$. Accordingly, when $\tilde{y}_m^{(u)}$ is passed through M-Point DFT, then

$$Y_k^{(u)} = \sum_{m=0}^{M-1} \tilde{y}_m^{(u)} e^{\frac{-j2\pi mk}{M}} \quad (8.43)$$

Thus, upon substituting the time-domain signal in (8.43), the following frequency domain signal results

$$Y_k^{(u)} = \sum_{m=0}^{M-1} \left[\frac{1}{NQ} \sum_{\substack{n_1=1 \\ n_1 \in Z_r}}^{\frac{N}{2}-1} \left(X_{\substack{n_1 \\ RC}}^{(u)} cos\left(\frac{2\pi (Qn_1 + \alpha_u)(m+\beta_u)}{NQ} \right) - X_{\substack{n_1 \\ IC}}^{(u)} sin\left(\frac{2\pi (Qn_1 + \alpha_u)(m+\beta_u)}{NQ} \right) \right) \right] e^{\frac{-j2\pi mk}{M}} \sum_{l=0}^{L-1} h_l^{(u)} e^{\frac{-j2\pi lk}{L}} + Z_k^{(u)} \quad (8.44)$$

Upon utilizing simple trigonometric inequalities like $cos(\theta) = \frac{e^{j\theta}+e^{-j\theta}}{2}$ and $sin(\theta) = \frac{e^{j\theta}-e^{-j\theta}}{2j}$, (8.44) can be solved to attain the following signal

$$Y_k^{(r)} = \underbrace{\sum_{\substack{n_1=1 \\ n_1 \in Z_r}}^{\frac{N}{2}-1} \left(X_{\substack{n1 \\ RC}}^{(r)} + jX_{\substack{n1 \\ IC}}^{(r)} \right) e^{\frac{j2\pi\beta_r (Qn_1+\alpha_r)}{NQ}} \Upsilon_{n_1 n_1}^{(r)} \sum_{l=0}^{L-1} h_l^{(r)} e^{\frac{-j2\pi lQn_1}{L}} + \sum_{\substack{n_1=1 \\ n_1 \in Z_r}}^{\frac{N}{2}-1} \left(X_{\substack{n1 \\ RC}}^{(r)} - jX_{\substack{n1 \\ IC}}^{(r)} \right) e^{-\frac{j2\pi\beta_r (Qn_1+\alpha_r)}{NQ}} \Upsilon_{n_1 n_1}^{'(r)} \sum_{l=0}^{L-1} h_l^{(r)} e^{\frac{-j2\pi lQn_1}{L}}}_{Desired\ User\ Signal} + ICI + MUI + Z_k^{(r)} \quad (8.45)$$

ICI in (8.45) can be derived as

$$\sum_{\substack{p_1=1 \\ p_1 \in Z_r \\ p_1 \neq n_1}}^{\frac{N}{2}-1} \left(X_{\substack{p1 \\ RC}}^{(r)} + jX_{\substack{p1 \\ IC}}^{(r)} \right) e^{\frac{j2\pi\beta_r (Qp_1+\alpha_r)}{NQ}} \Upsilon_{p_1 n_1}^{(r)} \sum_{l=0}^{L-1} h_l^{(r)} e^{\frac{-j2\pi lQp_1}{L}} + \sum_{\substack{p_1=1 \\ n_1 \in Z_r \\ p_1 \neq n_1}}^{\frac{N}{2}-1} \left(X_{\substack{p1 \\ RC}}^{(r)} - jX_{\substack{p1 \\ IC}}^{(r)} \right) e^{-\frac{j2\pi\beta_r (Qp_1+\alpha_r)}{NQ}} \Upsilon_{p_1 n_1}^{'(r)} \sum_{l=0}^{L-1} h_l^{(r)} e^{\frac{-j2\pi lQp_1}{L}} \quad (8.46)$$

MUI can be expressed as

$$\sum_{\substack{p_1=1 \\ p_1 \in Z_s \\ p_1 \neq n_1}}^{\frac{N}{2}-1} \left(X_{\substack{p1 \\ RC}}^{(s)} + jX_{\substack{p1 \\ IC}}^{(s)} \right) e^{\frac{j2\pi\beta_s (Qp_1+\alpha_s)}{NQ}} \Upsilon_{p_1 n_1}^{(s)} \sum_{l=0}^{L-1} h_l^{(s)} e^{\frac{-j2\pi lQp_1}{L}} + \sum_{\substack{p_1=1 \\ n_1 \in Z_s \\ p_1 \neq n_1}}^{\frac{N}{2}-1} \left(X_{\substack{p1 \\ RC}}^{(s)} - jX_{\substack{p1 \\ IC}}^{(s)} \right) e^{-\frac{j2\pi\beta_s (Qp_1+\alpha_s)}{NQ}} \Upsilon_{p_1 n_1}^{'(s)} \sum_{l=0}^{L-1} h_l^{(s)} e^{\frac{-j2\pi lQp_1}{L}} \quad (8.47)$$

From (8.45), (8.46) and (8.47) the terms $\Upsilon^{(r)}_{n_1n_1}$, $\Upsilon'^{(r)}_{n_1n_1}$, $\Upsilon^{(r)}_{p_1n_1}$, $\Upsilon'^{(r)}_{p_1n_1}$, $\Upsilon^{(s)}_{p_1n_1}$, $\Upsilon'^{(s)}_{p_1n_1}$ can be further solved to obtain

$$\Upsilon^{(r)}_{n_1n_1} = \frac{1}{M}\sum_{m=0}^{M-1} e^{\frac{j2\pi(Qn_1+\alpha_r-Qn_1)m}{M}} = \frac{sin\pi\alpha_r}{Msin\left(\frac{\pi\alpha_r}{M}\right)}e^{j\pi\alpha_r\left(1-\frac{1}{M}\right)} \tag{8.48}$$

$$\Upsilon'^{(r)}_{n_1n_1} = \frac{1}{M}\sum_{m=0}^{M-1} e^{-\frac{j2\pi(Qn_1+\alpha_r+Qn_1)m}{M}} = \frac{sin\pi\left(2Qn_1+\alpha_r\right)}{Msin\pi\left(\frac{2Qn_1+\alpha_r}{M}\right)}e^{-j\pi(2Qn_1+\alpha_r)\left(1-\frac{1}{M}\right)} \tag{8.49}$$

$$\Upsilon^{(r)}_{p_1n_1} = \frac{1}{M}\sum_{m=0}^{M-1} e^{\frac{j2\pi(Qp_1+\alpha_r-Qn_1)m}{M}}$$
$$= \frac{sin\pi\left(Q\left(p_1-n_1\right)+\alpha_r\right)}{Msin\left(\frac{\pi(Q(p_1-n_1)+\alpha_r)}{M}\right)}e^{j\pi(Q(p_1-n_1)+\alpha_r)\left(1-\frac{1}{M}\right)} \tag{8.50}$$

$$\Upsilon'^{(r)}_{p_1n_1} = \frac{1}{M}\sum_{m=0}^{M-1} e^{-\frac{j2\pi(Qp_1+\alpha_r+Qn_1)m}{M}}$$
$$= \frac{sin\pi\left(Q\left(p_1+n_1\right)+\alpha_r\right)}{Msin\left(\frac{\pi(Q(p_1+n_1)+\alpha_r)}{M}\right)}e^{-j\pi(Q(p_1+n_1)+\alpha_r)\left(1-\frac{1}{M}\right)} \tag{8.51}$$

$$\Upsilon^{(s)}_{p_1n_1} = \frac{1}{M}\sum_{m=0}^{M-1} e^{\frac{j2\pi(Qp_1+\alpha_s-Qn_1)m}{M}}$$
$$= \frac{sin\pi\left(Q\left(p_1-n_1\right)+\alpha_s\right)}{Msin\left(\frac{\pi(Q(p_1-n_1)+\alpha_s)}{M}\right)}e^{j\pi(Q(p_1-n_1)+\alpha_s)\left(1-\frac{1}{M}\right)} \tag{8.52}$$

$$\Upsilon'^{(s)}_{p_1n_1} = \frac{1}{M}\sum_{m=0}^{M-1} e^{-\frac{j2\pi(Qp_1+\alpha_s+Qn_1)m}{M}}$$
$$= \frac{sin\pi\left(Q\left(p_1+n_1\right)+\alpha_s\right)}{Msin\left(\frac{\pi(Q(p_1+n_1)+\alpha_s)}{M}\right)}e^{-j\pi(Q(p_1+n_1)+\alpha_s)\left(1-\frac{1}{M}\right)} \tag{8.53}$$

- **Overlap with previous frame:**
  This is the scenario which depicts the overlap of the current frame with the previous frame. Here, we consider the scenario for negative timing error. This scenario evidences that some of the samples of the current symbol of the corresponding frame are lost in the processing window. Therefore, to

highlight this impact, the received signal $Y_k^{(u)}$ can be put as follows:

$$Y_k^{(u)} = \sum_{m=0}^{-\alpha_u - N_{cp}+l-1} \frac{2}{N}\Big[\sum_{\substack{n_1=1 \\ n_1 \in Z_u}}^{\frac{N}{2}-1} X_{\substack{n_1 \\ RC}}^{(u)} cos\left(\frac{2\pi\left(m+N_{cp}+\alpha_u\right)\left(Qn_1+\beta_u\right)}{NQ}\right) -$$
$$X_{\substack{n_1 \\ IC}}^{(u)} sin\left(\frac{2\pi\left(m+N_{cp}+\alpha_u\right)\left(Qn_1+\beta_u\right)}{NQ}\right)\Big] e^{\frac{-j2\pi mk}{M}} \sum_{l=N_{cp}+\alpha_u+1}^{L-1} h_l^{(u)} e^{\frac{-j2\pi lk}{L}} +$$
$$\sum_{m=-\alpha_u-N_{cp}+l}^{M-1} \frac{2}{N}\Big[\sum_{\substack{n_1=1 \\ n_1 \in Z_u}}^{\frac{N}{2}-1} X_{\substack{n_1 \\ RC}}^{(u)} cos\left(\frac{2\pi\left(m+\alpha_u\right)\left(Qn_1+\beta_u\right)}{NQ}\right) -$$
$$X_{\substack{n_1 \\ IC}}^{(u)} sin\left(\frac{2\pi\left(m+\alpha_u\right)\left(Qn_1+\beta_u\right)}{NQ}\right)\Big] e^{\frac{-j2\pi mk}{M}} \sum_{l=0}^{L-1} h_l^{(u)} e^{\frac{-j2\pi lk}{L}} + \sum_{m=0}^{M-1} w_m^{(u)} e^{\frac{-j2\pi mk}{M}} \tag{8.54}$$

Due to overlap with the previous frame, definitely the subcarrier components which are allocated to the desired subscriber r overlap leading to the occurrence of ICI, thereby resulting in loss of orthogonality. This is clearly depicted in Fig. 8.11. Furthermore, in the presence of multiple users, MUI/-MAI occur from other users. Equation (8.54) can be further solved to yield the frequency domain signal of the corresponding user r as follows:

$$Y_k^{(r)} = \underbrace{\left(X_{\substack{n_1 \\ RC}}^{(r)} + jX_{\substack{n_1 \\ IC}}^{(r)}\right) e^{\frac{j2\pi\alpha_r(Qn_1+\beta_r)}{NQ}} \sum_{l=0}^{L-1} h_l^{(r)} e^{\frac{-j2\pi ln_1}{L}} \xi_{n_1 n_1}^{(r)(l)} + \left(X_{\substack{n_1 \\ RC}}^{(r)} - jX_{\substack{n_1 \\ IC}}^{(r)}\right) e^{\frac{-j2\pi\alpha_r(Qn_1+\beta_r)}{NQ}} \sum_{l=0}^{L-1} h_l^{(r)} e^{\frac{-j2\pi ln_1}{L}} \xi_{n_1 n_1}^{\prime(r)(l)}}_{Desired\ Signal} +$$
$$ICI_{\{Current\ Symbol+Previous\ Symbol\}} +$$
$$MUI_{\{Current\ Symbol+Previous\ Symbol\}} + W_k^{(r)} \tag{8.55}$$

From (8.55), the ICI in the current frame occurs due to overlap among the subcarrier components allocated to subscriber r and this is derived as

$$\sum_{\substack{p_1=1 \\ p_1 \neq n_1 \\ p_1 \in Z_r}}^{\frac{N}{2}-1} \left(X_{\substack{p_1 \\ RC}}^{(r)} + jX_{\substack{p_1 \\ IC}}^{(r)} \right) e^{\frac{j2\pi\alpha_r(Qp_1+\beta_r)}{NQ}} \sum_{l=0}^{L-1} h_l^{(r)} e^{\frac{-j2\pi l p_1}{L}} \xi_{p_1 n_1}^{(r)(l)} + \underbrace{\sum_{\substack{p_1=1 \\ p_1 \neq n_1 \\ p_1 \in Z_r}}^{\frac{N}{2}-1} \left(X_{\substack{p_1 \\ RC}}^{(r)} - jX_{\substack{p_1 \\ IC}}^{(r)} \right) e^{\frac{-j2\pi\alpha_r(Qp_1+\beta_r)}{NQ}} \sum_{l=0}^{L-1} h_l^{(r)} e^{\frac{-j2\pi l p_1}{L}} \xi_{p_1 n_1}^{\prime(r)(l)}}_{ICI\ in\ Current\ Symbol} \tag{8.56}$$

The ICI in the previous frame can be solved as

$$\underbrace{\sum_{\substack{p_1=1 \\ p_1 \neq n_1 \\ p_1 \in Z_r}}^{\frac{N}{2}-1} \left(X_{\substack{p_1 \\ RC}}^{(r)(pr)} + jX_{\substack{p_1 \\ IC}}^{(r)(pr)} \right) e^{\frac{j2\pi(\alpha_r+N_{cp})(Qp_1+\beta_r)}{NQ}} \sum_{l=N_{cp}+\alpha_r+1}^{L-1} h_l^{(r)} e^{\frac{-j2\pi l p_1}{L}} \wp_{p_1 n_1}^{(r)(l)} + \sum_{\substack{p_1=1 \\ p_1 \neq n_1 \\ p_1 \in Z_r}}^{\frac{N}{2}-1} \left(X_{\substack{p_1 \\ RC}}^{(r)(pr)} - jX_{\substack{p_1 \\ IC}}^{(r)(pr)} \right) e^{\frac{-j2\pi(\alpha_r+N_{cp})(Qp_1+\beta_r)}{NQ}} \sum_{l=N_{cp}+\alpha_r+1}^{L-1} h_l^{(r)} e^{\frac{-j2\pi l p_1}{L}} \wp_{p_1 n_1}^{\prime(r)(l)}}_{ICI\ emanating\ from\ Previous\ Frame's\ Symbol} \tag{8.57}$$

The MUI/MAI can be characterized as the interference emanating due to interference of subscriber s with subscriber r and this is mathematically represented as:

$$\sum_{\substack{p_1=1 \\ p_1 \neq n_1 \\ p_1 \in Z_s}}^{\frac{N}{2}-1} \left(X_{\substack{p_1 \\ RC}}^{(s)} + jX_{\substack{p_1 \\ IC}}^{(s)} \right) e^{\frac{j2\pi\alpha_s(Qp_1+\beta_s)}{NQ}} \sum_{l=0}^{L-1} h_l^{(s)} e^{\frac{-j2\pi l p_1}{L}} \xi_{p_1 n_1}^{(s)(l)} + \underbrace{\sum_{\substack{p_1=1 \\ p_1 \neq n_1 \\ p_1 \in Z_s}}^{\frac{N}{2}-1} \left(X_{\substack{p_1 \\ RC}}^{(s)} - jX_{\substack{p_1 \\ IC}}^{(s)} \right) e^{\frac{-j2\pi\alpha_s(Qp_1+\beta_s)}{NQ}} \sum_{l=0}^{L-1} h_l^{(s)} e^{\frac{-j2\pi l p_1}{L}} \xi_{p_1 n_1}^{\prime(s)(l)}}_{MUI\ in\ Current\ Symbol} \tag{8.58}$$

Further, the MUI/MAI in the previous frame can be derived as

$$\underbrace{\sum_{\substack{p_1=1\\p_1\neq n_1\\p_1\in Z_s}}^{\frac{N}{2}-1}\left(X_{\underset{RC}{p_1}}{}^{(s)(pr)}+jX_{\underset{IC}{p_1}}{}^{(s)(pr)}\right)e^{\frac{j2\pi(\alpha_s+N_{cp})(Qp_1+\beta_s)}{NQ}}\sum_{l=N_{cp}+\alpha_s+1}^{L-1}h_l^{(s)}e^{\frac{-j2\pi lp_1}{L}}\wp_{p_1n_1}^{(s)(l)}+\sum_{\substack{p_1=1\\p_1\neq n_1\\p_1\in Z_s}}^{\frac{N}{2}-1}\left(X_{\underset{RC}{p_1}}{}^{(s)(pr)}-jX_{\underset{IC}{p_1}}{}^{(s)(pr)}\right)e^{\frac{-j2\pi(\alpha_s+N_{cp})(Qp_1+\beta_s)}{NQ}}\sum_{l=N_{cp}+\alpha_s+1}^{L-1}h_l^{(s)}e^{\frac{-j2\pi lp_1}{L}}\wp_{p_1n_1}^{'(s)(l)}}_{MUI\ emanating\ from\ Previous\ Frame's\ Symbol} \tag{8.59}$$

where, $\xi_{p_1n_1}^{(u)(l)}$, $\xi_{p_1n_1}^{'(u)(l)}$, $\wp_{p_1n_1}^{(u)(l)}$ and $\wp_{p_1n_1}^{'(u)(l)}$ in (8.55) are given by

$$\xi_{p_1n_1}^{(u)(l)}=\frac{1}{M}\sum_{m=-\alpha_u-N_{cp}+l}^{M-1}e^{\frac{j2\pi(Q(p_1-n_1)+\beta_u)m}{M}} \tag{8.60}$$

$$\xi_{p_1n_1}^{'(u)(l)}=\frac{1}{M}\sum_{m=-\alpha_u-N_{cp}+l}^{M-1}e^{\frac{-j2\pi(Q(p_1+n_1)+\beta_u)m}{M}} \tag{8.61}$$

$$\wp_{p_1n_1}^{(u)(l)}=\frac{1}{M}\sum_{m=0}^{-\alpha_u-N_{cp}+l-1}e^{\frac{j2\pi(Q(p_1-n_1)+\beta_u)m}{M}} \tag{8.62}$$

$$\wp_{p_1n_1}^{'(u)(l)}=\frac{1}{M}\sum_{m=0}^{-\alpha_u-N_{cp}+l-1}e^{\frac{-j2\pi(Q(p_1+n_1)+\beta_u)m}{M}} \tag{8.63}$$

Thus, from the aforementioned mathematical analysis, a final conclusion can be drawn that the possibilities of occurrence of different timing misalignments in uplink scenario for DCO-SC-FDMA system enforces the necessity for frame synchronization and leads to the scope of proposal of robust and sophisticated synchronization algorithms which are much compatible with IM systems.

8.4 AMALGAMATION OF OLED-BASED VLC SYSTEMS FOR AUTOMOTIVE APPLICATIONS

Having seen the prominent advantages offered by OLEDs, they can be exploited to realize VLC links which are flexible for outdoor applications as well. On these grounds, the amalgamation of OLED with VLC technology can be employed to enable road infrastructure to vehicle communications. The primary reason behind this fact is that OLEDs offer a large photoactive area and play a vital role where in-organic LEDs fail their applicability. Moreover, in areas where inorganic LEDs hinder high data rate communication due to the presence of natural as well as artificial light, OLEDs can be oriented in such areas of applications. Additionally, the joint exploitation of OLEDs as well as inorganic LEDs (precisely, the VLC link can be realized by using a combination of both organic and inorganic devices) will definitely improve the data rate transfer. So far, several developments in automotive industry can be witnessed where OLED-based illumination infrastructure was already launched for new car models. Several car manufacturing companies demonstrated that OLED exhibits a stunning property to illuminate light uniformly which greatly reduces the necessity to rely on reflectors.

8.5 FLICKERING AND DIMMING ISSUES

Flickering induces serious health hazards to humans when it is noticed. This is more pronounced in cases of slow data rate communications because the changes become perceptible to the human eye. Therefore, design of sophisticated modulation formats to combat the detrimental aspects of flickering turns out to be a sensitive issue. On the other hand, it is desirable to expeditiously dim a light source based on the needs of the application. Therefore, future research should aim toward designing robust dimming control algorithms so that a desired amount of illumination can be acquired in an appropriate manner. Consequently, flicker mitigation and dimming support are two major challenges which need to be handled in the indoor scenarios.

8.6 INFLUENCE OF AMBIENT NOISE ON THE PERFORMANCE OF VLC SYSTEM

When VLC is exploited for outdoor applications like vehicular communications, the noise emanating due to the presence of background solar radiation and the interference arising due to the other sources of artificial lighting units like street lights made of incandescent lamps, fluorescent lamps and fluorescent lamps geared by electronic ballasts perturb the communication. Due to the presence of these perturbing elements, the SNR at the receiving end is drastically reduced in case of automotive applications. Hence, in conclusion, outdoor VLC application is strongly affected by the noise, and hence combating its effect to facilitate high data rate transfer for long-distance communications turns out to be the most promising task. Therefore, it is vital to employ effective optical filters to eradicate the effects of noise prior to reception of the transmitted signal. Furthermore, in case of a VLC receiver, the receiver's FOV plays a major role to allow for the interception of the incident intensity modulated transmitted light signal. Therefore, as already mentioned in the previous chapters, only the signal which is within the receiver's FOV will be accepted and the rest of the signal which falls outside the receiver's FOV will be discarded. Thus, in addition to taking the desired signal, there might be chances for the receiver to accept the noise content also. Hence, if wider receiver FOV is allowed, then, not only desired signal is captured, but also a large amount of noise content is also accepted. Therefore, one way to counteract the effects of noise is to narrow the receiver's FOV. Still, there is a need to emphasize several other techniques to mitigate the effects of noise in case of outdoor VLC applications.

8.7 RESEARCH CHALLENGES ASSOCIATED WITH VLC IN THE EMERGING AREA OF INDOOR POSITIONING

Predominantly, the positioning systems which are exploited to estimate the location/position of the user are a major basis for navigation-oriented services. Global positioning system (GPS) is the widely deployed positioning system that exhibits its potential in aircraft, vehicles and other portable devices with the major emphasis to furnish real-time positioning and navigation aid to the end-users [249]. However, several studies as well as practical implementations affirm that in most of the challenging scenarios which include urban canyons and indoor areas, GPS positioning and navigation fail to render accurate and continuous services due to the fact that the signals transmitted by satellites are mostly degraded and interrupted by clouds, walls, ceilings and several other obstructions [284,580]. However, to fill in the gap, i.e., with the motive to improve the performance of indoor positioning which GPS fails to do in most of the challenging environments, several other indoor positioning systems (IPS) employing indoor wireless signals such as Wi-Fi [42,582], Bluetooth [221,583], ZigBee, radio frequency identification module (RFID), etc., have been proposed. In recent times, LED-based positioning systems have emerged to furnish localization services, and the potential of VLC to deliver positioning services has drawn significant momentum in diverse areas of industry, academia and research communities. However, in spite of its significance, the application of VLC in the field of positioning is not devoid of drawbacks. Primarily, while designing indoor positioning systems, the limited modulation bandwidth of LEDs should not be a big hurdle for the realization of indoor positioning-based VLC systems. Hence, there is a necessity for the future research to expand the limited modulation bandwidth of LEDs [581]. Furthermore, it is vital for the future research to study as well as validate appropriate methodologies while taking into consideration positioning applications based on different multiple access schemes like optical TDMA, FDMA, CDMA, SDMA and NOMA. The major motive should be toward enhancing a reliable means of communication over the shared medium. Additionally, it is required for the TDOA-based indoor positioning VLC systems to be put into practice along with the experimental realizations of such systems. Even though received signal strength-based localization algorithms perform well in VLC systems, there is a necessity for VLC to be interfaced with complementary positioning systems like Wi-Fi, Bluetooth, etc. The primary reason behind this interface of VLC with other positioning systems is that since light signals cannot penetrate through walls, there might be chances for the occurrence of interference when there exist some media like glass, etc., between rooms. Hence, future research should be directed toward addressing these aforesaid aspects while dealing with the applications of VLC in the area of indoor positioning.

References

1. Cree lmh6. [online]. *Available: http://www.cree.com//media/Files/Cree/LED Components and Modules/Modules/ DataSheets/LEDModules LMH6.pdf.*
2. Cree xlamp xp-e high-efficiency white. [online]. *Available: http://www.cree.com// media/Files/Cree/LED Components and Modules/XLamp/DataandBinning/ XLampX-PEHEW.pdf.*
3. Cree xlamp xr-e. [online]. *Available: http://www.cree.com//media/Files/Cree/LED Components and Modules/XLamp/ DataandBinning/XLamp7090XRE.pdf.*
4. A comparison of fso wavelength system designs. *LightPointe White Paper*, 2002.
5. IEEE standard for local and metropolitan area networks–part 15.7: Short-range wireless optical communication using visible light. *IEEE Std 802.15.7-2011*, pages 1–309, Sept 2011.
6. IEEE standard for local and metropolitan area networks–part 15.7: Short-range optical wireless communications. *IEEE Std 802.15.7-2018 (Revision of IEEE Std 802.15.7-2011)*, pages 1–407, April 2019.
7. A. H. Abdolhamid and D. A. Johns. A comparison of cap/qam architectures. In *ISCAS '98. Proceedings of the 1998 IEEE International Symposium on Circuits and Systems (Cat. No.98CH36187)*, volume 4, pages 316–316/3 vol.4, May 1998.
8. S. Aboagye, A. Ibrahim, T. M. N. Ngatched, A. R. Ndjiongue, and O. A. Dobre. Design of energy efficient hybrid vlc/rf/plc communication system for indoor networks. *IEEE Wireless Communications Letters*, 9(2):143–147, 2020.
9. M. Y. Abualhoul, P. Merdrignac, O. Shagdar, and F. Nashashibi. Study and evaluation of laser-based perception and light communication for a platoon of autonomous vehicles. In *2016 IEEE 19th International Conference on Intelligent Transportation Systems (ITSC)*, pages 1798–1804, 2016.
10. Mohammad Y Abualhoul, Mohamed Marouf, Oyunchimeg Shagdar, and Fawzi Nashashibi. Platooning control using visible light communications: A feasibility study. In *16th International IEEE Conference on Intelligent Transportation Systems (ITSC 2013)*, pages 1535–1540. IEEE, 2013.
11. Mohammad Y Abualhoul, Oyunchimeg Shagdar, and Fawzi Nashashibi. Visible light inter-vehicle communication for platooning of autonomous vehicles. In *2016 IEEE Intelligent Vehicles Symposium (IV)*, pages 508–513. IEEE, 2016.
12. H. Abumarshoud, H. Alshaer, and H. Haas. Dynamic multiple access configuration in intelligent lifi attocellular access points. *IEEE Access*, 7:62126–62141, 2019.
13. A. Agarwal and T. D. C. Little. Role of directional wireless communication in vehicular networks. In *2010 IEEE Intelligent Vehicles Symposium*, pages 688–693, 2010.
14. M. S. Ahmed, S. Boussakta, B. S. Sharif, and C. C. Tsimenidis. Ofdm based on low complexity transform to increase multipath resilience and reduce papr. *IEEE Transactions on Signal Processing*, 59(12):5994–6007, Dec 2011.
15. Nasir Ahmed, T˙ Natarajan, and Kamisetty R Rao. Discrete cosine transform. *IEEE Transactions on Computers*, 100(1):90–93, 1974.
16. K. O. Akande, P. A. Haigh, and W. O. Popoola. On the implementation of carrierless amplitude and phase modulation in visible light communication. *IEEE Access*, 6:60532–60546, 2018.

17. K. O. Akande and W. O. Popoola. Impact of timing jitter on the performance of carrier amplitude and phase modulation. In *2016 International Conference for Students on Applied Engineering (ICSAE)*, pages 259–263, Oct 2016.
18. K. O. Akande and W. O. Popoola. Synchronization of carrierless amplitude and phase modulation in visible light communication. In *2017 IEEE International Conference on Communications Workshops (ICC Workshops)*, pages 156–161, May 2017.
19. K. O. Akande and W. O. Popoola. Experimental demonstration of subband index techniques for m -cap in short-range si-pof links. *IEEE Photonics Technology Letters*, 30(24):2155–2158, 2018.
20. K. O. Akande and W. O. Popoola. Generalised spatial carrierless amplitude and phase modulation in visible light communication. In *2018 IEEE International Conference on Communications (ICC)*, pages 1–6, 2018.
21. K. O. Akande and W. O. Popoola. Mimo techniques for carrierless amplitude and phase modulation in visible light communication. *IEEE Communications Letters*, 22(5):974–977, 2018.
22. K. O. Akande and W. O. Popoola. Subband index carrierless amplitude and phase modulation for optical communications. *Journal of Lightwave Technology*, 36(18):4190–4197, 2018.
23. K. O. Akande and W. O. Popoola. Enhanced subband index carrierless amplitude and phase modulation in visible light communications. *Journal of Lightwave Technology*, 37(23):5867–5874, 2019.
24. Kabiru O Akande, Paul Anthony Haigh, and Wasiu O Popoola. Joint equalization and synchronization for carrierless amplitude and phase modulation in visible light communication. In *2017 13th International Wireless Communications and Mobile Computing Conference (IWCMC)*, pages 876–881. IEEE, 2017.
25. Masako Akanegawa, Yuichi Tanaka, and Masao Nakagawa. Basic study on traffic information system using led traffic lights. *IEEE Transactions on Intelligent Transportation Systems*, 2(4):197–203, 2001.
26. S. Al-Ahmadi, O. Maraqa, M. Uysal, and S. M. Sait. Multi-user visible light communications: State-of-the-art and future directions. *IEEE Access*, 6:70555–70571, 2018.
27. F. M. Alsalami, Z. Ahmad, O. Haas, and S. Rajbhandari. Regular-shaped geometry-based stochastic model for vehicle-to-vehicle visible light communication channel. In *2019 IEEE Jordan International Joint Conference on Electrical Engineering and Information Technology (JEEIT)*, pages 297–301, 2019.
28. B. Aly. Performance analysis of adaptive channel estimation for u-ofdm indoor visible light communication. In *2016 33rd National Radio Science Conference (NRSC)*, pages 217–222, Feb 2016.
29. P. Amirshahi and M. Kavehrad. Broadband access over medium and low voltage power-lines and use of white light emitting diodes for indoor communications. In *CCNC 2006. 2006 3rd IEEE Consumer Communications and Networking Conference, 2006.*, volume 2, pages 897–901, 2006.
30. J. An and W. Chung. Single-led multichannel optical transmission with scma for long range health information monitoring. *Journal of Lightwave Technology*, 36(23):5470–5480, Dec 2018.
31. Lin, Yuan-Pei, and See-May Phoong. BER minimized OFDM systems with channel independent precoders. Ber minimized ofdm systems with channel independent precoders. *IEEE Transactions on Signal Processing*, 51(9):2369–2380, Sep 2003.

32. Kang, Keon Woo, Jaemin Ann, and Hwang Soo Lee. Decision-directed maximum-likelihood estimation of ofdm frame synchronisation offset. *Electronics Letters*, 30(25):2153–2154, Dec 1994.
33. ho Hwang, In, Hwang soo Lee, and Keon woo Kang. Frequency and timing period offset estimation technique for ofdm systems. *Electronics Letters*, 34(6):520–521, March 1998.
34. D. N. Anwar, A. Srivastava, and V. A. Bohara. Adaptive channel estimation in vlc for dynamic indoor environment. In *2019 21st International Conference on Transparent Optical Networks (ICTON)*, pages 1–5, 2019.
35. M. A. Arfaoui, A. Ghrayeb, and C. M. Assi. Secrecy performance of multi-user miso vlc broadcast channels with confidential messages. *IEEE Transactions on Wireless Communications*, 17(11):7789–7800, 2018.
36. M. A. Arfaoui, A. Ghrayeb, and C. M. Assi. Secrecy performance of the mimo vlc wiretap channel with randomly located eavesdropper. *IEEE Transactions on Wireless Communications*, 19(1):265–278, 2020.
37. M. A. Arfaoui, H. Zaid, Z. Rezki, A. Ghrayeb, A. Chaaban, and M. Alouini. Artificial noise-based beamforming for the miso vlc wiretap channel. *IEEE Transactions on Communications*, 67(4):2866–2879, 2019.
38. J. Armstrong and B. J. C. Schmidt. Comparison of asymmetrically clipped optical ofdm and dc-biased optical ofdm in awgn. *IEEE Communications Letters*, 12(5):343–345, May 2008.
39. Shlomi Arnon. The effect of clock jitter in visible light communication applications. *Journal of Lightwave Technology*, 30(21):3434–3439, 2012.
40. K. Asadzadeh, A. A. Farid, and S. Hranilovic. Spectrally factorized optical ofdm. In *2011 12th Canadian Workshop on Information Theory*, pages 102–105, May 2011.
41. S. Ayub, S. Kariyawasam, M. Honary, and B. Honary. A practical approach of vlc architecture for smart city. In *2013 Loughborough Antennas Propagation Conference (LAPC)*, pages 106–111, Nov 2013.
42. Paramvir Bahl and Venkata N Padmanabhan. Radar: An in-building rf-based user location and tracking system. In *Proceedings IEEE INFOCOM 2000. Conference on Computer Communications. Nineteenth Annual Joint Conference of the IEEE Computer and Communications Societies (Cat. No. 00CH37064)*, volume 2, pages 775–784. IEEE, 2000.
43. B. Bai, G. Chen, Z. Xu, and Y. Fan. Visible light positioning based on led traffic light and photodiode. In *2011 IEEE Vehicular Technology Conference (VTC Fall)*, pages 1–5, 2011.
44. Bo Bai, Zhengyuan Xu, and Yangyu Fan. Joint led dimming and high capacity visible light communication by overlapping ppm. In *The 19th Annual Wireless and Optical Communications Conference (WOCC 2010)*, pages 1–5, May 2010.
45. Jurong Bai, Yong Li, Yang Yi, Wei Cheng, and Huimin Du. Papr reduction based on tone reservation scheme for dco-ofdm indoor visible light communications. *Opt. Express*, 25(20):24630–24638, Oct 2017.
46. S. Baig, H. Muhammad Asif, T. Umer, S. Mumtaz, M. Shafiq, and J. Choi. High data rate discrete wavelet transform-based plc-vlc design for 5g communication systems. *IEEE Access*, 6:52490–52499, 2018.
47. Howard Baldwin. Wireless bandwidth: Are we running out of room? *ComputerWorld*, pages 1–4, 2012.

48. K. Bandara, P. Niroopan, and Y. Chung. PAPR reduced ofdm visible light communication using exponential nonlinear companding. In *2013 IEEE International Conference on Microwaves, Communications, Antennas and Electronic Systems (COMCAS 2013)*, pages 1–5, Oct 2013.
49. K. Bandara, S. Tiwari, and Y. H. Chung. Nonlinear companding transform for reducing peak-to-average power ratio of ofdm signals. *IEEE Transactions on Broadcasting*, 50(3):342–346, Sep. 2004.
50. X. Bao, X. Zhu, T. Song, and Y. Ou. Protocol design and capacity analysis in hybrid network of visible light communication and ofdma systems. *IEEE Transactions on Vehicular Technology*, 63(4):1770–1778, 2014.
51. N Bardsley, S Bland, L Pattison, M Pattison, K Stober, F Welsh, and M Yamada. Solid-state lighting research and development multi-year program plan. *US Department of Energy*, 2014.
52. J. R. Barry, J. M. Kahn, W. J. Krause, E. A. Lee, and D. G. Messerschmitt. Simulation of multipath impulse response for indoor wireless optical channels. *IEEE Journal on Selected Areas in Communications*, 11(3):367–379, Apr 1993.
53. Alessandro Bazzi, Barbara M Masini, Alberto Zanella, and Alex Calisti. Visible light communications in vehicular networks for cellular offloading. In *2015 IEEE International Conference on Communication Workshop (ICCW)*, pages 1416–1421. IEEE, 2015.
54. B. Béchadergue, L. Chassagne, and H. Guan. A visible light-based system for automotive relative positioning. In *2017 IEEE Sensors*, pages 1–3, 2017.
55. B. Béchadergue, L. Chassagne, and H. Guan. Vehicle-to-vehicle visible light phase-shift rangefinder based on the automotive lighting. *IEEE Sensors Journal*, 18(13):5334–5342, 2018.
56. B. Béchadergue, L. Chassagne, and H. Guan. Simultaneous visible light communication and distance measurement based on the automotive lighting. *IEEE Transactions on Intelligent Vehicles*, 4(4):532–547, 2019.
57. B. Béchadergue, L. Chassagne, and Hongyu Guan. Visible light phase-shift rangefinder for platooning applications. In *2016 IEEE 19th International Conference on Intelligent Transportation Systems (ITSC)*, pages 2462–2468, 2016.
58. Alexander Graham Bell. On selenium and the photophone. *Electrician*, 5:214–220, 1880.
59. Alexander Graham Bell, WG Adams, WH Preece, et al. Discussion on the photophone and the conversion of radiant energy into sound. *Journal of the Society of Telegraph Engineers*, 9(34):375–383, 1880.
60. G. Berardinelli, L. A. M. Ruiz de Temino, S. Frattasi, M. I. Rahman, and P. Mogensen. Ofdma vs. sc-fdma: performance comparison in local area imt-a scenarios. *IEEE Wireless Communications*, 15(5):64–72, October 2008.
61. Samuel M Berman, Daniel S Greenhouse, Ian L Bailey, Robert D Clear, and Thomas W Raasch. Human electroretinogram responses to video displays, fluorescent lighting, and other high frequency sources. *Optometry and Vision Science: Official Publication of the American Academy of Optometry*, 68(8):645–662, 1991.
62. M. Biagi and A. M. Vegni. Enabling high data rate vlc via mimo-leds ppm. In *2013 IEEE Globecom Workshops (GC Wkshps)*, pages 1058–1063, 2013.
63. Katrin Bilstrup, Elisabeth Uhlemann, Erik G Strom, and Urban Bilstrup. Evaluation of the ieee 802.11 p mac method for vehicle-to-vehicle communication. In *2008 IEEE 68th Vehicular Technology Conference*, pages 1–5. IEEE, 2008.

64. Katrin Bilstrup, Elisabeth Uhlemann, ErikG Ström, and Urban Bilstrup. On the ability of the 802.11 p mac method and stdma to support real-time vehicle-to-vehicle communication. *EURASIP Journal on Wireless Communications and Networking*, 2009(1):902414, 2009.
65. Annette Böhm, Kristoffer Lidström, Magnus Jonsson, and Tony Larsson. Evaluating calm m5-based vehicle-to-vehicle communication in various road settings through field trials. In *IEEE Local Computer Network Conference*, pages 613–620. IEEE, 2010.
66. Deva K Borah, Anthony C Boucouvalas, Christopher C Davis, Steve Hranilovic, and Konstantinos Yiannopoulos. A review of communication-oriented optical wireless systems. *EURASIP Journal on Wireless Communications and Networking*, 2012(1):91, 2012.
67. Stephan Bose-O'Reilly, Kathleen M McCarty, Nadine Steckling, and Beate Lettmeier. Mercury exposure and children's health. *Current Problems in Pediatric and Adolescent Health Care*, 40(8):186–215, 2010.
68. W. Boubakri, W. Abdallah, and N. Boudriga. A light-based communication architecture for smart city applications. In *2015 17th International Conference on Transparent Optical Networks (ICTON)*, pages 1–6, July 2015.
69. AC Boucouvalas. Indoor ambient light noise and its effect on wireless optical links. *IEE Proceedings-Optoelectronics*, 143(6):334–338, 1996.
70. AC Boucouvalas, Periklis Chatzimisios, Zabih Ghassemlooy, Murat Uysal, and Konstantinos Yiannopoulos. Standards for indoor optical wireless communications. *IEEE Communications Magazine*, 53(3):24–31, 2015.
71. R. Bouziane, P. A. Milder, R. J. Koutsoyannis, Y. Benlachtar, J. C. Hoe, M. Glick, and R. I. Killey. Dependence of optical ofdm transceiver asic complexity on fft size. In *OFC/NFOEC*, pages 1–3, March 2012.
72. J Boyd. Japan's plan to speed self-driving cars. *IEEE Spectrum*, 2015.
73. Ronald N Bracewell. The fast hartley transform. *Proceedings of the IEEE*, 72(8):1010–1018, 1984.
74. Vladimir Britanak, Patrick C Yip, and Kamisetty Ramamohan Rao. *Discrete cosine and sine transforms: general properties, fast algorithms and integer approximations*. Elsevier, 2010.
75. JH Burroughes, DDC Bradley, AR Brown, RN Marks, K Mackay, RH Friend, PL Burns, and AB Holmes. Light-emitting diodes based on conjugated polymers. *nature*, 347(6293):539–541, 1990.
76. A. Burton, A. Minotto, P. A. Haigh, Z. Ghassemlooy, H. L. Minh, F. Cacialli, and I. Darwazeh. Optoelectronic modelling, circuit design and modulation for polymer-light emitting diodes for visible light communication systems. In *2019 26th International Conference on Telecommunications (ICT)*, pages 55–59, 2019.
77. A. Cailean, B. Cagneau, L. Chassagne, S. Topsu, Y. Alayli, and M. Dimian. Visible light communications cooperative architecture for the intelligent transportation system. In *2013 IEEE 20th Symposium on Communications and Vehicular Technology in the Benelux (SCVT)*, pages 1–5, 2013.
78. Alin Cailean, Barthélemy Cagneau, Luc Chassagne, Suat Topsu, Yasser Alayli, and Jean-Marc Blosseville. Visible light communications: Application to cooperation between vehicles and road infrastructures. In *2012 IEEE Intelligent Vehicles Symposium*, pages 1055–1059. IEEE, 2012.
79. Alin-M Cailean, Barthélemy Cagneau, Luc Chassagne, Suat Topsu, Yasser Alayli, and Mihai Dimian. Visible light communications cooperative architecture for the

intelligent transportation system. In *2013 IEEE 20th Symposium on Communications and Vehicular Technology in the Benelux (SCVT)*, pages 1–5. IEEE, 2013.

80. Alin-Mihai Căilean, Barthélemy Cagneau, Luc Chassagne, Mihai Dimian, and Valentin Popa. Novel receiver sensor for visible light communications in automotive applications. *IEEE Sensors Journal*, 15(8):4632–4639, 2015.
81. Alin-Mihai Cailean, Barthélemy Cagneau, Luc Chassagne, Valentin Popa, and Mihai Dimian. A survey on the usage of dsrc and vlc in communication-based vehicle safety applications. In *2014 IEEE 21st Symposium on Communications and Vehicular Technology in the Benelux (SCVT)*, pages 69–74. IEEE, 2014.
82. Alin-Mihai Căilean and Mihai Dimian. Toward environmental-adaptive visible light communications receivers for automotive applications: A review. *IEEE Sensors Journal*, 16(9):2803–2811, 2016.
83. C. Cano, A. Pittolo, D. Malone, L. Lampe, A. M. Tonello, and A. G. Dabak. State of the art in power line communications: From the applications to the medium. *IEEE Journal on Selected Areas in Communications*, 34(7):1935–1952, 2016.
84. J. B. Carruthers and J. M. Kahn. Angle diversity for nondirected wireless infrared communication. *IEEE Transactions on Communications*, 48(6):960–969, June 2000.
85. J. B. Carruthers and J. M. Kahn. Modeling of nondirected wireless infrared channels. *IEEE Transactions on Communications*, 45(10):1260–1268, Oct 1997.
86. Jeffrey B Carruthers, Sarah M Carroll, and Prasanna Kannan. Propagation modelling for indoor optical wireless communications using fast multi-receiver channel estimation. *IEE Proceedings-Optoelectronics*, 150(5):473–481, 2003.
87. Jeffrey B Carruthers and SM Carroll. Statistical impulse response models for indoor optical wireless channels. *International Journal of Communication Systems*, 18(3):267–284, 2005.
88. Z. N. Chaleshtori, A. Burton, Z. Ghassemlooy, and S. Zvanovec. A flexible oled based vlc link with m-cap modulation. In *2019 15th International Conference on Telecommunications (ConTEL)*, pages 1–6, 2019.
89. Z. N. Chaleshtori, P. Chvojka, S. Zvanovec, Z. Ghassemlooy, and P. A. Haigh. A survey on recent advances in organic visible light communications. In *2018 11th International Symposium on Communication Systems, Networks Digital Signal Processing (CSNDSP)*, pages 1–6, 2018.
90. Zahra Nazari Chaleshtori, Stanislav Zvanovec, Zabih Ghassemlooy, Hossien B. Eldeeb, and Murat Uysal. "A Flexible OLED VLC System for an Office Environment." In *2020 12th International Symposium on Communication Systems, Networks and Digital Signal Processing (CSNDSP)*, pp. 1–5. IEEE, 2020.
91. Cheng-Chun Chang and Heung-No Lee. On the estimation of target spectrum for filter-array based spectrometers. *Optics Express*, 16(2):1056–1061, 2008.
92. Cheng-Chun Chang, Yuan-Jun Su, Umpei Kurokawa, and Byung Il Choi. Interference rejection using filter-based sensor array in vlc systems. *IEEE SENSORS Journal*, 12(5):1025–1032, 2011.
93. Chang He, Lie-liang Yang, Pei Xiao, and M. A. Imran. Ds-cdma assisted visible light communications systems. In *2015 IEEE 20th International Workshop on Computer Aided Modelling and Design of Communication Links and Networks (CAMAD)*, pages 27–32, Sep. 2015.
94. C. Chen, D. A. Basnayaka, and H. Haas. Downlink performance of optical attocell networks. *Journal of Lightwave Technology*, 34(1):137–156, 2016.
95. C. Chen, M. Ijaz, D. Tsonev, and H. Haas. Analysis of downlink transmission in dco-ofdm-based optical attocell networks. In *2014 IEEE Global Communications Conference*, pages 2072–2077, 2014.

96. C. Chen, N. Serafimovski, and H. Haas. Fractional frequency reuse in optical wireless cellular networks. In *2013 IEEE 24th Annual International Symposium on Personal, Indoor, and Mobile Radio Communications (PIMRC)*, pages 3594–3598, 2013.
97. C. Chen, D. Tsonev, and H. Haas. Joint transmission in indoor visible light communication downlink cellular networks. In *2013 IEEE Globecom Workshops (GC Wkshps)*, pages 1127–1132, 2013.
98. C. Chen, W. Zhong, H. Yang, and P. Du. On the performance of mimo-noma-based visible light communication systems. *IEEE Photonics Technology Letters*, 30(4):307–310, Feb 2018.
99. H. Chen, S. Li, B. Huang, Z. Xu, W. Li, G. Dong, and J. Xie. A 1.9mbps ofdm-based all-organic visible light communication system. In *2016 IEEE International Conference on Communication Systems (ICCS)*, pages 1–6, 2016.
100. H. Chen and Z. Xu. Oled panel radiation pattern and its impact on vlc channel characteristics. *IEEE Photonics Journal*, 10(2):1–10, 2018.
101. H. Chen, Z. Xu, Q. Gao, and S. Li. A 51.6 mb/s experimental vlc system using a monochromic organic led. *IEEE Photonics Journal*, 10(2):1–12, 2018.
102. Haoshuo Chen, Henrie PA van den Boom, Eduward Tangdiongga, and Ton Koonen. 30-gb/s bidirectional transparent optical transmission with an mmf access and an indoor optical wireless link. *IEEE Photonics Technology Letters*, 24(7):572–574, 2012.
103. J. Chen and Z. Wang. Topology control in hybrid vlc/rf vehicular ad-hoc network. *IEEE Transactions on Wireless Communications*, 19(3):1965–1976, 2020.
104. L. Chen, B. Krongold, and J. Evans. Successive decoding of anti-periodic ofdm signals in im/dd optical channel. In *2010 IEEE International Conference on Communications*, pages 1–6, May 2010.
105. L. Chen, B. Krongold, and J. Evans. Performance analysis for optical ofdm transmission in short-range im/dd systems. *Journal of Lightwave Technology*, 30(7):974–983, April 2012.
106. Qi Chen, Daniel Jiang, and Luca Delgrossi. IEEE 1609.4 dsrc multi-channel operations and its implications on vehicle safety communications. In *2009 IEEE Vehicular Networking Conference (VNC)*, pages 1–8. IEEE, 2009.
107. S. Chen and C. Chow. Color-shift keying and code-division multiple-access transmission for rgb-led visible light communications using mobile phone camera. *IEEE Photonics Journal*, 6(6):1–6, Dec 2014.
108. Z. Chen, D. A. Basnayaka, and H. Haas. Space division multiple access in optical attocell networks. In *2016 IEEE Wireless Communications and Networking Conference Workshops (WCNCW)*, pages 228–232, April 2016.
109. Z. Chen, D. A. Basnayaka, and H. Haas. Space division multiple access for optical attocell network using angle diversity transmitters. *Journal of Lightwave Technology*, 35(11):2118–2131, June 2017.
110. Z. Chen and H. Haas. Space division multiple access in visible light communications. In *2015 IEEE International Conference on Communications (ICC)*, pages 5115–5119, June 2015.
111. Yong Soo Cho, Jaekwon Kim, Won Young Yang, and Chung G Kang. *MIMO-OFDM wireless communications with MATLAB*. John Wiley & Sons, 2010.
112. Su-il Choi. Analysis of vlc channel based on the shapes of white-light led lighting. In *2012 Fourth International Conference on Ubiquitous and Future Networks (ICUFN)*, pages 1–5. IEEE, 2012.

113. C. Chow, H. Wang, C. Chen, H. Zan, C. Yeh, and H. Meng. Pre-distortion scheme to enhance the transmission performance of organic photo-detector (opd) based visible light communication (vlc). *IEEE Access*, 6:7625–7630, 2018.
114. M. I. S. Chowdhury, W. Zhang, and M. Kavehrad. Combined deterministic and modified Monte Carlo method for calculating impulse responses of indoor optical wireless channels. *Journal of Lightwave Technology*, 32(18):3132–3148, Sept 2014.
115. Mostafa Zaman Chowdhury, Md Tanvir Hossan, Amirul Islam, and Yeong Min Jang. A comparative survey of optical wireless technologies: Architectures and applications. *IEEE Access*, 6:9819–9840, 2018.
116. P. Chvojka, K. Werfli, P. A. Haigh, S. Zvanovec, Z. Ghassemlooy, and M. R. Bhatnagar. Multi-band carrier-less amplitude and phase modulation for vlc: An overview. In *2017 First South American Colloquium on Visible Light Communications (SACVLC)*, pages 1–6, 2017.
117. P. Chvojka, K. Werfli, S. Zvanovec, P. A. Haigh, V. H. Vacek, P. Dvorak, P. Pesek, and Z. Ghassemlooy. On the m-cap performance with different pulse shaping filters parameters for visible light communications. *IEEE Photonics Journal*, 9(5):1–12, 2017.
118. P. Chvojka, S. Zvanovec, P. A. Haigh, and Z. Ghassemlooy. Channel characteristics of visible light communications within dynamic indoor environment. *Journal of Lightwave Technology*, 33(9):1719–1725, May 2015.
119. P. Chvojka, S. Zvanovec, K. Werfli, P. A. Haigh, and Z. Ghassemlooy. Variable m-cap for bandlimited visible light communications. In *2017 IEEE International Conference on Communications Workshops (ICC Workshops)*, pages 1–5, 2017.
120. Petr Chvojka, Paul Anthony Haigh, Stanislav Zvanovec, Petr Pesek, and Zabih Ghassemlooy. Evaluation of multi-band carrier-less amplitude and phase modulation performance for vlc under various pulse shaping filter parameters. In *OPTICS*, pages 25–31, 2016.
121. CAMP Vehicle Safety Communications Consortium et al. Vehicle safety communications project: Task 3 final report: identify intelligent vehicle safety applications enabled by dsrc. *National Highway Traffic Safety Administration, US Department of Transportation, Washington DC*, 2005.
122. James W Cooley and John W Tukey. An algorithm for the machine calculation of complex fourier series. *Mathematics of Computation*, 19(90):297–301, 1965.
123. Raffaele Corsini, Riccardo Pelliccia, Giulio Cossu, Amir Masood Khalid, Marco Ghibaudi, Matteo Petracca, Paolo Pagano, and Ernesto Ciaramella. Free space optical communication in the visible bandwidth for v2v safety critical protocols. In *2012 8th International Wireless Communications and Mobile Computing Conference (IWCMC)*, pages 1097–1102. IEEE, 2012.
124. Thomas Cover. Broadcast channels. *IEEE Transactions on Information Theory*, 18(1):2–14, 1972.
125. Kaiyun Cui, Gang Chen, Qunfeng He, and Zhengyuan Xu. Indoor optical wireless communication by ultraviolet and visible light. In *Free-Space Laser Communications IX*, volume 7464, page 74640D. International Society for Optics and Photonics, 2009.
126. Kaiyun Cui, Gang Chen, Zhengyuan Xu, and Richard D Roberts. Traffic light to vehicle visible light communication channel characterization. *Applied Optics*, 51(27):6594–6605, 2012.
127. Z. Cui, P. Yue, and Y. Ji. Study of cooperative diversity scheme based on visible light communication in vanets. In *2016 International Conference on Computer, Information and Telecommunication Systems (CITS)*, pages 1–5, 2016.

128. J. Dang and Z. Zhang. Comparison of optical ofdm-idma and optical ofdma for uplink visible light communications. In *2012 International Conference on Wireless Communications and Signal Processing (WCSP)*, pages 1–6, Oct 2012.
129. J. Dang, Z. Zhang, and L. Wu. A novel receiver for aco-ofdm in visible light communication. *IEEE Communications Letters*, 17(12):2320–2323, 2013.
130. J. Dang, Z. Zhang, and L. Wu. Frequency-domain diversity combining receiver for aco-ofdm system. *IEEE Photonics Journal*, 7(6):1–10, 2015.
131. D. W. Dawoud, F. Héliot, M. A. Imran, and R. Tafazolli. A novel unipolar transmission scheme for visible light communication. *IEEE Transactions on Communications*, pages 1–1, 2019.
132. S. De Lausnay, L. De Strycker, J. Goemaere, B. Nauwelaers, and N. Stevens. A test bench for a vlp system using cdma as multiple access technology. In *2015 17th International Conference on Transparent Optical Networks (ICTON)*, pages 1–4, July 2015.
133. S. De Lausnay, L. De Strycker, J. Goemaere, N. Stevens, and B. Nauwelaers. Optical cdma codes for an indoor localization system using vlc. In *2014 3rd International Workshop in Optical Wireless Communications (IWOW)*, pages 50–54, Sep. 2014.
134. P. de Souza, N. Bamiedakis, K. Yoshida, P. P. Manousiadis, G. A. Turnbull, I. D. W. Samuel, R. V. Penty, and I. H. White. High-bandwidth organic light emitting diodes for ultra-low cost visible light communication links. In *2018 20th International Conference on Transparent Optical Networks (ICTON)*, pages 1–4, 2018.
135. M. Dehghani Soltani, X. Wu, M. Safari, and H. Haas. Access point selection in li-fi cellular networks with arbitrary receiver orientation. In *2016 IEEE 27th Annual International Symposium on Personal, Indoor, and Mobile Radio Communications (PIMRC)*, pages 1–6, Sep. 2016.
136. S. Dimitrov and H. Haas. Information rate of ofdm-based optical wireless communication systems with nonlinear distortion. *Journal of Lightwave Technology*, 31(6):918–929, March 2013.
137. Vappangi, Suseela, and V. V. Mani. Interference analysis and MUI-cancellation in DCO-OFDMA-based IM/DD systems for VLC. *Optics Communications*, 448 (2019):130–146.
138. E. Dinc, O. Ergul, and O. B. Akan. Soft handover in ofdma based visible light communication networks. In *2015 IEEE 82nd Vehicular Technology Conference (VTC2015-Fall)*, pages 1–5, 2015.
139. De-qiang Ding and Xi-zheng Ke. A new indoor vlc channel model based on reflection. *Optoelectronics Letters*, 6(4):295–298, Jul 2010.
140. Jupeng Ding, Kun Wang, and Zhengyuan Xu. Accuracy analysis of different modeling schemes in indoor visible light communications with distributed array sources. In *2014 9th International Symposium on Communication Systems, Networks & Digital Sign (CSNDSP)*, pages 1005–1010. IEEE, 2014.
141. W. Ding, F. Yang, C. Pan, L. Dai, and J. Song. Compressive sensing based channel estimation for ofdm systems under long delay channels. *IEEE Transactions on Broadcasting*, 60(2):313–321, 2014.
142. W. Ding, F. Yang, J. Song, and Z. Niu. Energy-efficient orthogonal frequency division multiplexing scheme based on time–frequency joint channel estimation. *IET Communications*, 8(18):3406–3413, 2014.
143. Wenbo Ding, Fang Yang, Hui Yang, Jintao Wang, Xiaofei Wang, Xun Zhang, and Jian Song. A hybrid power line and visible light communication system for indoor hospital applications. *Computers in Industry*, 68:170–178, 2015.

144. S. D. Dissanayake, K. Panta, and J. Armstrong. A novel technique to simultaneously transmit aco-ofdm and dco-ofdm in im/dd systems. In *2011 IEEE GLOBECOM Workshops (GC Wkshps)*, pages 782–786, Dec 2011.
145. T. Do and M. Yoo. Visible light communication-based vehicle-to-vehicle tracking using cmos camera. *IEEE Access*, 7:7218–7227, 2019.
146. J. G. Doblado, A. C. O. Oria, V. Baena-Lecuyer, P. Lopez, and D. Perez-Calderon. Cubic metric reduction for dco-ofdm visible light communication systems. *Journal of Lightwave Technology*, 33(10):1971–1978, May 2015.
147. R. J. Drost and B. M. Sadler. Constellation design for color-shift keying using billiards algorithms. In *2010 IEEE Globecom Workshops*, pages 980–984, Dec 2010.
148. H. Du, C. Zhang, and Z. Wu. Robust beamforming-aided signal recovery for mimo vlc system with incomplete channel. *IEEE Access*, 7:128162–128170, 2019.
149. Pierre Duhamel and Martin Vetterli. Cyclic convolution of real sequences: Hartley versus fourier and new schemes. In *ICASSP'86. IEEE International Conference on Acoustics, Speech, and Signal Processing*, volume 11, pages 229–232. IEEE, 1986.
150. Pierre Duhamel and Martin Vetterli. Improved Fourier and Hartley transform algorithms: Application to cyclic convolution of real data. *IEEE Transactions on Acoustics, Speech, and Signal Processing*, 35(6):818–824, 1987.
151. MA Dunlap, G Cook, B Marion, C Riordan, and D Renne. Shining on: A primer on solar radiation data. Technical report, National Renewable Energy Lab., Golden, CO (United States), 1992.
152. Matthew J Eckelman, Paul T Anastas, and Julie B Zimmerman. Spatial assessment of net mercury emissions from the use of fluorescent bulbs. *Environmental Science & Technology*, 42(22):8564–8570, 2008.
153. ECN. En 12368: Traffic control equipment - signal heads. ed: European Commitee for Standardization, April, 2006.
154. Stephan Eichler. Performance evaluation of the IEEE 802.11 p wave communication standard. In *2007 IEEE 66th Vehicular Technology Conference*, pages 2199–2203. IEEE, 2007.
155. M. Elamassie, M. Karbalayghareh, F. Miramirkhani, M. Uysal, M. Abdallah, and K. Qaraqe. Resource allocation for downlink ofdma in underwater visible light communications. In *2019 IEEE International Black Sea Conference on Communications and Networking (BlackSeaCom)*, pages 1–6, 2019.
156. H. Elgala and T. D. C. Little. P-ofdm: Spectrally efficient unipolar ofdm. In *OFC 2014*, pages 1–3, March 2014.
157. H. Elgala and T. D. C. Little. See-ofdm: Spectral and energy efficient ofdm for optical im/dd systems. In *2014 IEEE 25th Annual International Symposium on Personal, Indoor, and Mobile Radio Communication (PIMRC)*, pages 851–855, Sep. 2014.
158. H. Elgala, R. Mesleh, and H. Haas. Indoor broadcasting via white leds and ofdm. *IEEE Transactions on Consumer Electronics*, 55(3):1127–1134, August 2009.
159. H. Elgala, R. Mesleh, and H. Haas. An LED model for intensity-modulated optical communication systems. *IEEE Photonics Technology Letters*, 22(11):835–837, June 2010.
160. Hany Elgala and Thomas DC Little. Reverse polarity optical-ofdm (rpo-ofdm): dimming compatible ofdm for gigabit vlc links. *Optics Express*, 21(20):24288–24299, 2013.
161. Hany Elgala, Raed Mesleh, and Harald Haas. Non-linearity effects and predistortion in optical ofdm wireless transmission using leds. *International Journal of Ultra Wideband Communications and Systems*, 1(2):143–150, 2009.

162. Hany Elgala, Raed Mesleh, and Harald Haas. Indoor optical wireless communication: potential and state-of-the-art. *IEEE Communications Magazine*, 49(9):56–62, 2011.
163. Y. S. Eroğlu, Y. Yapici, and I. Güvenc. Effect of random vertical orientation for mobile users in visible light communication. In *2017 51st Asilomar Conference on Signals, Systems, and Computers*, pages 238–242, Oct 2017.
164. Y. S. Eroğlu, Y. Yapıcı, and İ. Güvenç. Impact of random receiver orientation on visible light communications channel. *IEEE Transactions on Communications*, 67(2):1313–1325, Feb 2019.
165. M. Z. Farooqui, P. Saengudomlert, and S. Kaiser. Average transmit power reduction in ofdm-based indoor wireless optical communications using slm. In *International Conference on Electrical Computer Engineering (ICECE 2010)*, pages 602–605, Dec 2010.
166. Ephraim Feig and Shmuel Winograd. Fast algorithms for the discrete cosine transform. *IEEE Transactions on Signal Processing*, 40(9):2174–2193, 1992.
167. S. Feng, T. Bai, and L. Hanzo. Joint power allocation for the multi-user noma-downlink in a power-line-fed vlc network. *IEEE Transactions on Vehicular Technology*, 68(5):5185–5190, 2019.
168. Pedro Fernandes and Urbano Nunes. Platooning with dsrc-based ivc-enabled autonomous vehicles: Adding infrared communications for ivc reliability improvement. In *2012 IEEE Intelligent Vehicles Symposium*, pages 517–522. IEEE, 2012.
169. N. Fernando, Y. Hong, and E. Viterbo. Flip-ofdm for unipolar communication systems. *IEEE Transactions on Communications*, 60(12):3726–3733, December 2012.
170. M. Fuchs, G. Del Galdo, and M. Haardt. A novel tree-based scheduling algorithm for the downlink of multi-user mimo systems with zf beamforming. In *Proceedings. (ICASSP '05). IEEE International Conference on Acoustics, Speech, and Signal Processing, 2005.*, volume 3, pages iii/1121–iii/1124 Vol. 3, March 2005.
171. N. Fujimoto and H. Mochizuki. 614 mbit/s ook-based transmission by the duobinary technique using a single commercially available visible led for high-speed visible light communications. In *2012 38th European Conference and Exhibition on Optical Communications*, pages 1–3, Sept 2012.
172. N. Fujimoto and H. Mochizuki. 477 mbit/s visible light transmission based on ook-nrz modulation using a single commercially available visible led and a practical led driver with a pre-emphasis circuit. In *2013 Optical Fiber Communication Conference and Exposition and the National Fiber Optic Engineers Conference (OFC/NFOEC)*, pages 1–3, March 2013.
173. Yuichiro Fujiwara. Self-synchronizing pulse position modulation with error tolerance. *IEEE Transactions on Information Theory*, 59(9):5352–5362, 2013.
174. Caroline Gabriel. Wireless broadband alliance industry report 2013: Global trends in public wi-fi. *Marvedis Rethink—Wireless Broadband Alliance, Singapore, Tech. Rep*, 2013.
175. Robert M Gagliardi and Sherman Karp. Optical communications. *New York, Wiley-Interscience, 445 p.*, 1976.
176. J. Gao, Yee Hong Leung, and V. Sreeram. Digital filters for carrierless amplitude and phase receivers. In *Proceedings of IEEE Region 10 International Conference on Electrical and Electronic Technology. TENCON 2001 (Cat. No.01CH37239)*, volume 2, pages 575–579, Aug 2001.
177. F. R. Gfeller and U. Bapst. Wireless in-house data communication via diffuse infrared radiation. *Proceedings of the IEEE*, 67(11):1474–1486, Nov 1979.
178. Zabih Ghassemlooy, Luis Nero Alves, Stanislav Zvanovec, and Mohammad-Ali Khalighi. *Visible light communications: theory and applications*. CRC Press, 2017.

179. Zabih Ghassemlooy, Shlomi Arnon, Murat Uysal, Zhengyuan Xu, and Julian Cheng. Emerging optical wireless communications-advances and challenges. *IEEE Journal on Selected Areas in Communications*, 33(9):1738–1749, 2015.
180. Zabih Ghassemlooy, Wasiu Popoola, and Sujan Rajbhandari. *Optical wireless communications: system and channel modelling with Matlab*. CRC Press, 2012.
181. Abdallah S Ghazy, Hossam AI Selmy, and Hossam MH Shalaby. Fair resource allocation schemes for cooperative dynamic free-space optical networks. *Journal of Optical Communications and Networking*, 8(11):822–834, 2016.
182. W. Gheth, K. M. Rabie, B. Adebisi, M. Ijaz, and G. Harris. On the performance of df-based power-line/visible-light communication systems. In *2018 International Conference on Signal Processing and Information Security (ICSPIS)*, pages 1–4, 2018.
183. Birendra Ghimire, Irina Stefan, Hany Elgala, and Harald Haas. Time and frequency synchronisation in optical wireless ofdm networks. In *2011 IEEE 22nd International Symposium on Personal, Indoor and Mobile Radio Communications*, pages 819–823. IEEE, 2011.
184. E. Giacoumidis, S. K. Ibrahim, J. Zhao, J. M. Tang, I. Tomkos, and A. D. Ellis. Experimental demonstration of cost-effective intensity-modulation and direct-detection optical fast-ofdm over 40km smf transmission. In *OFC/NFOEC*, pages 1–3, March 2012.
185. E. Giacoumidis, A. Tsokanos, C. Mouchos, G. Zardas, C. Alves, J. L. Wei, J. M. Tang, C. Gosset, Y. Jaouën, and I. Tomkos. Extensive comparisons of optical fast-ofdm and conventional optical ofdm for local and access networks. *IEEE/OSA Journal of Optical Communications and Networking*, 4(10):724–733, Oct 2012.
186. O. Gonzalez, J. A. Martin-Gonzalez, M. F. Guerra-Medina, F. J. Lopez-Hernandez, and F. A. Delgado-Rajó. Cyclic code-shift extension keying for multi-user optical wireless communications. *Electronics Letters*, 51(11):847–849, 2015.
187. Yuki Goto, Isamu Takai, Takaya Yamazato, Hiraku Okada, Toshiaki Fujii, Shoji Kawahito, Shintaro Arai, Tomohiro Yendo, and Koji Kamakura. A new automotive vlc system using optical communication image sensor. *IEEE Photonics Journal*, 8(3):1–17, 2016.
188. J. Grubor, S. C. J. Lee, K. Langer, T. Koonen, and J. W. Walewski. Wireless high-speed data transmission with phosphorescent white-light leds. In *33rd European Conference and Exhibition of Optical Communication - Post-Deadline Papers (published 2008)*, pages 1–2, Sep. 2007.
189. Jelena Grubor, Sebastian Randel, Klaus-Dieter Langer, and Joachim W Walewski. Broadband information broadcasting using led-based interior lighting. *Journal of Lightwave Technology*, 26(24):3883–3892, 2008.
190. W. Gu, M. Aminikashani, P. Deng, and M. Kavehrad. Impact of multipath reflections on the performance of indoor visible light positioning systems. *Journal of Lightwave Technology*, 34(10):2578–2587, May 2016.
191. X. Guan, Q. Yang, and C. Chan. Joint detection of visible light communication signals under non-orthogonal multiple access. *IEEE Photonics Technology Letters*, 29(4):377–380, Feb 2017.
192. Guangliang Ren, Hui Zhang, and Yilin Chang. A complementary clipping transform technique for the reduction of peak-to-average power ratio of ofdm system. *IEEE Transactions on Consumer Electronics*, 49(4):922–926, Nov 2003.
193. M. F. Guerra-Medina, O. González, B. Rojas-Guillama, J. A. Martín-González, F. Delgado, and J. Rabadán. Ethernet-ocdma system for multi-user visible light communications. *Electronics Letters*, 48(4):227–228, February 2012.

194. M. F. Guerra-Medina, B. Rojas-Guillama, O. Gonzalez, J. A. Martin-Gonzalez, E. Poves, and F. J. Lopez-Hernandez. Experimental optical code-division multiple access system for visible light communications. In *2011 Wireless Telecommunications Symposium (WTS)*, pages 1–6, April 2011.
195. B. Habib and H. Farhat. Channel hardware simulator design and implementation for mimo time-varying 802.15.7 vlc indoor signals. In *2018 IEEE Middle East and North Africa Communications Conference (MENACOMM)*, pages 1–5, 2018.
196. P. A. Haigh, A. Aguado, Z. Ghassemlooy, P. Chvojka, K. Werfli, S. Zvanovec, E. Ertunç, and T. Kanesan. Multi-band carrier-less amplitude and phase modulation for highly bandlimited visible light communications — invited paper. In *2015 International Conference on Wireless Communications Signal Processing (WCSP)*, pages 1–5, 2015.
197. P. A. Haigh, F. Bausi, T. Kanesan, S. T. Le, S. Rajbhandari, Z. Ghassemlooy, I. Papakonstantinou, W. Popoola, A. Burton, H. L. Minh, A. D. Ellis, and F. Cacialli. A 20-mb/s vlc link with a polymer led and a multilayer perceptron equalizer. *IEEE Photonics Technology Letters*, 26(19):1975–1978, Oct 2014.
198. P. A. Haigh, F. Bausi, H. Le Minh, I. Papakonstantinou, W. O. Popoola, A. Burton, and F. Cacialli. Wavelength-multiplexed polymer leds: Towards 55 mb/s organic visible light communications. *IEEE Journal on Selected Areas in Communications*, 33(9):1819–1828, Sept 2015.
199. P. A. Haigh, A. Burton, K. Werfli, H. L. Minh, E. Bentley, P. Chvojka, W. O. Popoola, I. Papakonstantinou, and S. Zvanovec. A multi-cap visible-light communications system with 4.85-b/s/hz spectral efficiency. *IEEE Journal on Selected Areas in Communications*, 33(9):1771–1779, 2015.
200. P. A. Haigh, P. Chvojka, Z. Ghassemlooy, S. Zvanovec, and I. Darwazeh. Non-orthogonal multi-band cap for highly spectrally efficient vlc systems. In *2018 11th International Symposium on Communication Systems, Networks Digital Signal Processing (CSNDSP)*, pages 1–6, 2018.
201. P. A. Haigh, P. Chvojka, A. Minotto, A. Burton, P. Murto, E. Wang, Z. Ghassemlooy, S. Zvanovec, F. Cacialli, and I. Darwazeh. Hybrid super-nyquist cap modulation based vlc with low bandwidth polymer leds. In *2019 IEEE 30th Annual International Symposium on Personal, Indoor and Mobile Radio Communications (PIMRC)*, pages 1–6, 2019.
202. P. A. Haigh, P. Chvojka, S. Zvanovec, Z. Ghassemlooy, S. T. Le, T. Kanesan, E. Giacoumidis, N. J. Doran, I. Papakonstantinou, and I. Darwazeh. Experimental verification of visible light communications based on multi-band cap modulation. In *2015 Optical Fiber Communications Conference and Exhibition (OFC)*, pages 1–3, 2015.
203. P. A. Haigh, Z. Ghassemlooy, H. Le Minh, S. Rajbhandari, F. Arca, S. F. Tedde, O. Hayden, and I. Papakonstantinou. Exploiting equalization techniques for improving data rates in organic optoelectronic devices for visible light communications. *Journal of Lightwave Technology*, 30(19):3081–3088, Oct 2012.
204. P. A. Haigh, Z. Ghassemlooy, and I. Papakonstantinou. 1.4-mb/s white organic led transmission system using discrete multitone modulation. *IEEE Photonics Technology Letters*, 25(6):615–618, March 2013.
205. P. A. Haigh, Z. Ghassemlooy, I. Papakonstantinou, F. Arca, S. F. Tedde, O. Hayden, and E. Leitgeb. A 1-mb/s visible light communications link with low bandwidth organic components. *IEEE Photonics Technology Letters*, 26(13):1295–1298, 2014.

206. P. A. Haigh, Z. Ghassemlooy, I. Papakonstantinou, and H. Le Minh. 2.7 mb/s with a 93-khz white organic light emitting diode and real time ann equalizer. *IEEE Photonics Technology Letters*, 25(17):1687–1690, Sept 2013.
207. P. A. Haigh, Z. Ghassemlooy, S. Rajbhandari, and I. Papakonstantinou. Visible light communications using organic light emitting diodes. *IEEE Communications Magazine*, 51(8):148–154, August 2013.
208. P. A. Haigh, Z. Ghassemlooy, S. Rajbhandari, I. Papakonstantinou, and W. Popoola. Visible light communications: 170 mb/s using an artificial neural network equalizer in a low bandwidth white light configuration. *Journal of Lightwave Technology*, 32(9):1807–1813, May 2014.
209. P. A. Haigh, S. T. Le, S. Zvanovec, Z. Ghassemlooy, P. Luo, T. Xu, P. Chvojka, T. Kanesan, E. Giacoumidis, P. Canyelles-Pericas, H. L. Minh, W. Popoola, S. Rajbhandari, I. Papakonstantinou, and I. Darwazeh. Multi-band carrier-less amplitude and phase modulation for bandlimited visible light communications systems. *IEEE Wireless Communications*, 22(2):46–53, 2015.
210. PA Haigh, Petr Chvojka, Zabih Ghassemlooy, Stanislav Zvanovec, and Izzat Darwazeh. Visible light communications: multi-band super-nyquist cap modulation. *Optics Express*, 27(6):8912–8919, 2019.
211. Paul Haigh, Francesco Arca, Sandro Tedde, Oliver Hayden, Ioannis Papakonstantinou, and Sujan Rajbhandari. Visible light communications: 3.75 mbits/s data rate with a 160 khz bandwidth organic photodetector and artificial neural network equalization [invited]. *OSA Photonics Research*, 1, 07 2013.
212. Paul Anthony Haigh, Francesco Bausi, Zabih Ghassemlooy, Ioannis Papakonstantinou, Hoa Le Minh, Charlotte Fléchon, and Franco Cacialli. Visible light communications: real time 10 mb/s link with a low bandwidth polymer light-emitting diode. *Optics Express*, 22(3):2830–2838, 2014.
213. Paul Anthony Haigh and Izzat Darwazeh. Demonstration of reduced complexity multi-band cap modulation using xia-pulses in visible light communications. In *2018 Optical Fiber Communications Conference and Exposition (OFC)*, pages 1–3. IEEE, 2018.
214. Lajos Hanzo, Harald Haas, Sándor Imre, Dominic O'Brien, Markus Rupp, and Laszlo Gyongyosi. Wireless myths, realities, and futures: from 3g/4g to optical and quantum wireless. *Proceedings of the IEEE*, 100(Special Centennial Issue):1853–1888, 2012.
215. Simon Haykin. *Communication Systems*. John Wiley & Sons, 2008.
216. Guoxing He, Lihong Zheng, and Huafeng Yan. Led white lights with high cri and high luminous efficacy. In *LED and Display Technologies*, volume 7852, page 78520A. International Society for Optics and Photonics, 2010.
217. L. He and F. Yang. Robust timing and frequency synchronization for tds-ofdm over multipath fading channels. In *2010 IEEE International Conference on Communication Systems*, pages 451–455, Nov 2010.
218. Yongqiang Hei, Jiao Liu, Huaxi Gu, Wentao Li, Xiaochuan Xu, and Ray T. Chen. Improved tkm-tr methods for papr reduction of dco-ofdm visible light communications. *Opt. Express*, 25(20):24448–24458, Oct 2017.
219. Yongqiang Hei, Jiao Liu, Wentao Li, Xiaochuan Xu, and Ray T. Chen. Branch and bound methods based tone injection schemes for papr reduction of dco-ofdm visible light communications. *Opt. Express*, 25(2):595–604, Jan 2017.
220. Juha Heiskala and John Terry Ph D. *OFDM wireless LANs: A theoretical and practical guide*. Sams, 2001.

221. AKM Mahtab Hossain and Wee-Seng Soh. A comprehensive study of bluetooth signal parameters for localization. In *2007 IEEE 18th International Symposium on Personal, Indoor and Mobile Radio Communications*, pages 1–5. IEEE, 2007.
222. Farhana Hossain and Zeenat Afroze. Eliminating the effect of fog attenuation on fso link by multiple tx/rx system with travelling wave semiconductor optical amplifier. In *2013 2nd International Conference on Advances in Electrical Engineering (ICAEE)*, pages 267–272. IEEE, 2013.
223. Hsieh S. Hou. The fast Hartley transform algorithm. *IEEE Transactions on Computers*, (2):147–156, 1987.
224. Pengfei Hu, Parth H Pathak, Aveek K Das, Zhicheng Yang, and Prasant Mohapatra. Plifi: Hybrid wifi-vlc networking using power lines. In *Proceedings of the 3rd Workshop on Visible Light Communication Systems*, pages 31–36, 2016.
225. Rose Qingyang Hu and Yi Qian. *Heterogeneous cellular networks*. John Wiley & Sons, 2013.
226. W. Hu. PAPR reduction in dco-ofdm visible light communication systems using optimized odd and even sequences combination. *IEEE Photonics Journal*, 11(1):1–15, Feb 2019.
227. W. Hu and D. Lee. Papr reduction for visible light communication systems without side information. *IEEE Photonics Journal*, 9(3):1–11, June 2017.
228. N. Huang, J. Wang, C. Pan, J. Wang, Y. Pan, and M. Chen. Iterative receiver for flip-ofdm in optical wireless communication. *IEEE Photonics Technology Letters*, 27(16):1729–1732, 2015.
229. N. Huang, J. Wang, J. Wang, C. Pan, H. Wang, and M. Chen. Receiver design for pam-dmt in indoor optical wireless links. *IEEE Photonics Technology Letters*, 27(2):161–164, 2015.
230. Muhammad Ijaz, Zabih Ghassemlooy, Hoa Le Minh, Sujan Rajbhandari, and J Perez. Analysis of fog and smoke attenuation in a free space optical communication link under controlled laboratory conditions. In *2012 International Workshop on Optical Wireless Communications (IWOW)*, pages 1–3. IEEE, 2012.
231. Cisco Visual Networking Index. Global mobile data traffic forecast update, 2015–2020 white paper. *link: http://goo. gl/ylTuVx*, 2016.
232. Susumu Ishihara, Reuben Vincent Rabsatt, and Mario Gerla. Improving reliability of platooning control messages using radio and visible light hybrid communication. In *2015 IEEE Vehicular Networking Conference (VNC)*, pages 96–103. IEEE, 2015.
233. M. S. Islim, D. Tsonev, and H. Haas. A generalized solution to the spectral efficiency loss in unipolar optical ofdm-based systems. In *2015 IEEE International Conference on Communications (ICC)*, pages 5126–5131, June 2015.
234. M. S. Islim, D. Tsonev, and H. Haas. On the superposition modulation for ofdm-based optical wireless communication. In *2015 IEEE Global Conference on Signal and Information Processing (GlobalSIP)*, pages 1022–1026, Dec 2015.
235. M. S. Islim, D. Tsonev, and H. Haas. Spectrally enhanced pam-dmt for im/dd optical wireless communications. In *2015 IEEE 26th Annual International Symposium on Personal, Indoor, and Mobile Radio Communications (PIMRC)*, pages 877–882, Aug 2015.
236. Mohamed Sufyan Islim and Harald Haas. Augmenting the spectral efficiency of enhanced pam-dmt-based optical wireless communications. *Optics Express*, 24(11):11932–11949, 2016.
237. Mohamed Sufyan Islim and Harald Haas. Modulation techniques for li-fi. *ZTE communications*, 14(2):29–40, 2016.

238. Y Jang. Current status of IEEE 802.15. 7r1 owc standardization. In *Proceedings of the International Conference and Exhibition on Visible Light Communications 2015*, 2015.
239. M. Jani, P. Garg, and A. Gupta. Performance analysis of a mixed cooperative plc–vlc system for indoor communication systems. *IEEE Systems Journal*, 14(1):469–476, 2020.
240. J. Jeong, C. G. Lee, I. Moon, M. Kang, S. Shin, and S. Kim. Receiver angle control in an infrastructure-to-car visible light communication link. In *2016 IEEE Region 10 Conference (TENCON)*, pages 1957–1960, 2016.
241. Daniel Jiang, Vikas Taliwal, Andreas Meier, Wieland Holfelder, and Ralf Herrtwich. Design of 5.9 ghz dsrc-based vehicular safety communication. *IEEE Wireless Communications*, 13(5):36–43, 2006.
242. Yufei Jiang, Yunlu Wang, Pan Cao, Majid Safari, John Thompson, and Harald Haas. Robust and low-complexity timing synchronization for dco-ofdm lifi systems. *IEEE Journal on Selected Areas in Communications*, 36(1):53–65, 2018.
243. Steven G Johnson and Matteo Frigo. A modified split-radix fft with fewer arithmetic operations. *IEEE Transactions on Signal Processing*, 55(1):111–119, 2007.
244. A. Jovicic, J. Li, and T. Richardson. Visible light communication: opportunities, challenges and the path to market. *IEEE Communications Magazine*, 51(12):26–32, 2013.
245. Aleksandar Jovicic, Junyi Li, and Tom Richardson. Visible light communication: opportunities, challenges and the path to market. *IEEE Communications Magazine*, 51(12):26–32, 2013.
246. Kahn, Joseph M., William J. Krause, and Jeffrey B. Carruthers. Experimental characterization of non-directed indoor infrared channels. *IEEE Transactions on Communications*, 43.2/3/4 (1995): 1613–1623.
247. Volker Jungnickel, Volker Pohl, Stephan Nonnig, and Clemens Von Helmolt. A physical model of the wireless infrared communication channel. *IEEE Journal on Selected Areas in Communications*, 20(3):631–640, 2002.
248. J. M. Kahn and J. R. Barry. Wireless infrared communications. *Proceedings of the IEEE*, 85(2):265–298, Feb 1997.
249. Elliott Kaplan and Christopher Hegarty. *Understanding GPS: principles and applications*. Artech House, 2005.
250. M. Karbalayghareh, F. Miramirkhani, H. B. Eldeeb, R. C. Kizilirmak, S. M. Sait, and M. Uysal. Channel modelling and performance limits of vehicular visible light communication systems. *IEEE Transactions on Vehicular Technology*, pages 1–1, 2020.
251. Johan Karedal, Fredrik Tufvesson, Taimoor Abbas, Oliver Klemp, Alexander Paier, Laura Bernadó, and Andreas F Molisch. Radio channel measurements at street intersections for vehicle-to-vehicle safety applications. In *2010 IEEE 71st Vehicular Technology Conference*, pages 1–5. IEEE, 2010.
252. M. Kashef, M. Abdallah, and N. Al-Dhahir. Transmit power optimization for a hybrid plc/vlc/rf communication system. *IEEE Transactions on Green Communications and Networking*, 2(1):234–245, 2018.
253. M. Kashef, M. Abdallah, N. Al-Dhahir, and K. Qaraqe. On the impact of plc backhauling in multi-user hybrid vlc/rf communication systems. In *2016 IEEE Global Communications Conference (GLOBECOM)*, pages 1–6, 2016.
254. M. Kashef, M. Abdallah, K. Qaraqe, H. Haas, and M. Uysal. Coordinated interference management for visible light communication systems. *IEEE/OSA Journal of Optical Communications and Networking*, 7(11):1098–1108, 2015.

255. M. Kashef, M. Ismail, M. Abdallah, K. A. Qaraqe, and E. Serpedin. Energy efficient resource allocation for mixed rf/vlc heterogeneous wireless networks. *IEEE Journal on Selected Areas in Communications*, 34(4):883–893, 2016.
256. M. Kashef, A. Torky, M. Abdallah, N. Al-Dhahir, and K. Qaraqe. On the achievable rate of a hybrid plc/vlc/rf communication system. In *2015 IEEE Global Communications Conference (GLOBECOM)*, pages 1–6, 2015.
257. Mohamed Kashef, Muhammad Ismail, Mohamed Abdallah, Khalid A Qaraqe, and Erchin Serpedin. Energy efficient resource allocation for mixed rf/vlc heterogeneous wireless networks. *IEEE Journal on Selected Areas in Communications*, 34(4):883–893, 2016.
258. S Kaur, W Liu, and D Castor. Vlc dimming support IEEE 802.15-09-0641-00-0007.
259. Hemani Kaushal and Georges Kaddoum. Optical communication in space: challenges and mitigation techniques. *IEEE Communications Surveys & Tutorials*, 19(1):57–96, 2016.
260. Mohsen Kavehrad. Sustainable energy-efficient wireless applications using light. *IEEE Communications Magazine*, 48(12):66–73, 2010.
261. H. Kazemi and H. Haas. Downlink cooperation with fractional frequency reuse in dco-ofdma optical attocell networks. In *2016 IEEE International Conference on Communications (ICC)*, pages 1–6, 2016.
262. John B Kenney. Dedicated short-range communications (dsrc) standards in the united states. *Proceedings of the IEEE*, 99(7):1162–1182, 2011.
263. Deok-Rae Kim, Se-Hoon Yang, Hyun-Seung Kim, Yong-Hwan Son, and Sang-Kook Han. Outdoor visible light communication for inter-vehicle communication using controller area network. In *2012 Fourth International Conference on Communications and Electronics (ICCE)*, pages 31–34. IEEE, 2012.
264. G. Kim, J. Eom, and Y. Park. An experiment of mutual interference between automotive lidar scanners. In *2015 12th International Conference on Information Technology - New Generations*, pages 680–685, 2015.
265. G. Kim, J. Eom, and Y. Park. Investigation on the occurrence of mutual interference between pulsed terrestrial lidar scanners. In *2015 IEEE Intelligent Vehicles Symposium (IV)*, pages 437–442, 2015.
266. Jong Kyu Kim and E Fred Schubert. Transcending the replacement paradigm of solid-state lighting. *Optics Express*, 16(26):21835–21842, 2008.
267. Sung-Man Kim and Seong-Min Kim. Performance improvement of visible light communications using optical beamforming. In *2013 Fifth International Conference on Ubiquitous and Future Networks (ICUFN)*, pages 362–365. IEEE, 2013.
268. Yong Hyeon Kim, Willy Anugrah Cahyadi, and Yeon Ho Chung. Experimental demonstration of vlc-based vehicle-to-vehicle communications under fog conditions. *IEEE Photonics Journal*, 7(6):1–9, 2015.
269. Shogo Kitano, Shinichiro Haruyama, and Masao Nakagawa. Led road illumination communications system. In *2003 IEEE 58th Vehicular Technology Conference. VTC 2003-Fall (IEEE Cat. No. 03CH37484)*, volume 5, pages 3346–3350. IEEE, 2003.
270. R. C. Kizilirmak, C. R. Rowell, and M. Uysal. Non-orthogonal multiple access (noma) for indoor visible light communications. In *2015 4th International Workshop on Optical Wireless Communications (IWOW)*, pages 98–101, Sep. 2015.
271. T Komine. Basic study on visible-light communication using light emitting diode illumination. In *Proc. of 8th Int. Symp. on Microwave and Optical Technology, Montreal, Canada*, pages 45–48, 2001.

272. T. Komine, S. Haruyama, and M. Nakagawa. Performance evaluation of narrowband ofdm on integrated system of power line communication and visible light wireless communication. In *2006 1st International Symposium on Wireless Pervasive Computing*, pages 6 pp.–6, 2006.
273. T. Komine and M. Nakagawa. Integrated system of white led visible-light communication and power-line communication. *IEEE Transactions on Consumer Electronics*, 49(1):71–79, 2003.
274. T. Komine and M. Nakagawa. Fundamental analysis for visible-light communication system using led lights. *IEEE Transactions on Consumer Electronics*, 50(1):100–107, Feb 2004.
275. Toshihiko Komine, Jun Hwan Lee, Shinichiro Haruyama, and Masao Nakagawa. Adaptive equalization system for visible light wireless communication utilizing multiple white led lighting equipment. *IEEE Transactions on Wireless Communications*, 8(6):2892–2900, 2009.
276. Toshihiko Komine and Masao Nakagawa. Integrated system of white LED visible-light communication and power-line communication. *IEEE Transactions on Consumer Electronics*, 49(1):71–79, 2003.
277. Toshihiko Komine and Masao Nakagawa. Fundamental analysis for visible-light communication system using led lights. *IEEE Transactions on Consumer Electronics*, 50(1):100–107, 2004.
278. D. Krichene, M. Sliti, W. Abdallah, and N. Boudriga. An aeronautical visible light communication system to enable in-flight connectivity. In *2015 17th International Conference on Transparent Optical Networks (ICTON)*, pages 1–6, July 2015.
279. N. Kumar. Smart and intelligent energy efficient public illumination system with ubiquitous communication for smart city. In *Smart Structures and Systems (ICSSS), 2013 IEEE International Conference on*, pages 152–157, March 2013.
280. Navin Kumar. Smart and intelligent energy efficient public illumination system with ubiquitous communication for smart city. In *International Conference on Smart Structures and Systems-ICSSS'13*, pages 152–157. IEEE, 2013.
281. Navin Kumar, Nuno Lourenço, Domingos Terra, Luis N Alves, and Rui L Aguiar. Visible light communications in intelligent transportation systems. In *2012 IEEE Intelligent Vehicles Symposium*, pages 748–753. IEEE, 2012.
282. Umpei Kurokawa, Byung Il Choi, and Cheng-Chun Chang. Filter-based miniature spectrometers: spectrum reconstruction using adaptive regularization. *IEEE Sensors Journal*, 11(7):1556–1563, 2010.
283. J. K. Kwon. Inverse source coding for dimming in visible light communications using nrz-ook on reliable links. *IEEE Photonics Technology Letters*, 22(19):1455–1457, Oct 2010.
284. Haiyu Lan, Chunyang Yu, Yuan Zhuang, You Li, and Naser El-Sheimy. A novel kalman filter with state constraint approach for the integration of multiple pedestrian navigation systems. *Micromachines*, 6(7):926–952, 2015.
285. Son T. Le, T. Kanesan, F. Bausi, P. A. Haigh, S. Rajbhandari, Z. Ghassemlooy, I. Papakonstantinou, W. O. Popoola, A. Burton, H. Le Minh, F. Cacialli, and A. D. Ellis. 10 mb/s visible light transmission system using a polymer light-emitting diode with orthogonal frequency division multiplexing. *Opt. Lett.*, 39(13):3876–3879, Jul 2014.
286. H. Le Minh, D. O'Brien, G. Faulkner, L. Zeng, K. Lee, D. Jung, and Y. Oh. High-speed visible light communications using multiple-resonant equalization. *IEEE Photonics Technology Letters*, 20(14):1243–1245, July 2008.

287. It Ee Lee, Moh Lim Sim, and Fabian Wai-Lee Kung. Performance enhancement of outdoor visible-light communication system using selective combining receiver. *IET Optoelectronics*, 3(1):30–39, 2009.
288. K. Lee and H. Park. Channel model and modulation schemes for visible light communications. In *2011 IEEE 54th International Midwest Symposium on Circuits and Systems (MWSCAS)*, pages 1–4, Aug 2011.
289. K. Lee, H. Park, and J. R. Barry. Indoor channel characteristics for visible light communications. *IEEE Communications Letters*, 15(2):217–219, February 2011.
290. Kwonhyung Lee, Hyuncheol Park, and John R Barry. Indoor channel characteristics for visible light communications. *IEEE Communications Letters*, 15(2):217–219, 2011.
291. S. C. J. Lee, S. Randel, F. Breyer, and A. M. J. Koonen. Pam-dmt for intensity-modulated and direct-detection optical communication systems. *IEEE Photonics Technology Letters*, 21(23):1749–1751, Dec 2009.
292. B. Li, W. Xu, S. Feng, and Z. Li. Spectral-efficient reconstructed laco-ofdm transmission for dimming compatible visible light communications. *IEEE Photonics Journal*, 11(1):1–14, Feb 2019.
293. C. Li, S. Wang, and K. Chan. Low complexity transmitter architectures for sfbc mimo-ofdm systems. *IEEE Transactions on Communications*, 60(6):1712–1718, June 2012.
294. C. Li, S. Wang, and C. Wang. Novel low-complexity slm schemes for papr reduction in ofdm systems. *IEEE Transactions on Signal Processing*, 58(5):2916–2921, May 2010.
295. H. Li, X. Chen, B. Huang, D. Tang, and H. Chen. High bandwidth visible light communications based on a post-equalization circuit. *IEEE Photonics Technology Letters*, 26(2):119–122, Jan 2014.
296. H. Li, Z. Huang, Y. Xiao, S. Zhan, and Y. Ji. A power and spectrum efficient noma scheme for vlc network based on hierarchical pre-distorted laco-ofdm. *IEEE Access*, 7:48565–48571, 2019.
297. Honglei Li, Xiongbin Chen, Junqing Guo, and Hongda Chen. A 550 mbit/s real-time visible light communication system based on phosphorescent white light led for practical high-speed low-complexity application. *Opt. Express*, 22(22):27203–27213, Nov 2014.
298. X. Li, H. Chen, S. Li, Q. Gao, C. Gong, and Z. Xu. Volterra-based nonlinear equalization for nonlinearity mitigation in organic vlc. In *2017 13th International Wireless Communications and Mobile Computing Conference (IWCMC)*, pages 616–621, 2017.
299. X. Li, Y. J. Kim, and N. Y. Park. A low PAPR WLED communication system using sc-fdma techniques. In *2011 IEEE 73rd Vehicular Technology Conference (VTC Spring)*, pages 1–5, May 2011.
300. Z. Li and C. Zhang. An improved fd-dfe structure for downlink vlc systems based on sc-fdma. *IEEE Communications Letters*, 22(4):736–739, April 2018.
301. J. Lian, Y. Gao, and D. Lian. Variable pulse width unipolar orthogonal frequency division multiplexing for visible light communication systems. *IEEE Access*, 7:31022–31030, 2019.
302. F. Lichtenegger, C. Leiner, C. Sommer, A. P. Weiss, and F. P. Wenzl. Ray-tracing based channel modeling for the simulation of the performance of visible light communication in an indoor environment. In *2019 Second Balkan Junior Conference on Lighting (Balkan Light Junior)*, pages 1–6, 2019.
303. B. Lin, X. Tang, and Z. Ghassemlooy. Optical power domain noma for visible light communications. *IEEE Wireless Communications Letters*, 8(4):1260–1263, Aug 2019.

304. B. Lin, X. Tang, Z. Ghassemlooy, Y. Li, S. Zhang, Y. Wu, and H. Li. An indoor vlc positioning system based on ofdma. In *2016 Asia Communications and Photonics Conference (ACP)*, pages 1–3, 2016.
305. B. Lin, X. Tang, Z. Ghassemlooy, C. Lin, and Y. Li. Experimental demonstration of an indoor vlc positioning system based on ofdma. *IEEE Photonics Journal*, 9(2):1–9, 2017.
306. Bangjiang Lin, Weiping Ye, Xuan Tang, and Zabih Ghassemlooy. Experimental demonstration of bidirectional noma-ofdma visible light communications. *Opt. Express*, 25(4):4348–4355, Feb 2017.
307. S. Lin, J. Wang, J. Wang, J. Wang, and M. Chen. Low-timing-sensitivity waveform design for carrierless amplitude and phase modulation in visible light communications. *IET Optoelectronics*, 9(6):317–324, 2015.
308. X. Ling, J. Wang, Z. Ding, C. Zhao, and X. Gao. Efficient ofdma for lifi downlink. *Journal of Lightwave Technology*, 36(10):1928–1943, 2018.
309. Cen B Liu, Bahareh Sadeghi, and Edward W Knightly. Enabling vehicular visible light communication (v2lc) networks. In *Proceedings of the Eighth ACM International Workshop on Vehicular Inter-networking*, pages 41–50. ACM, 2011.
310. H. Liu, P. Zhu, Y. Chen, and M. Huang. Power allocation for downlink hybrid power line and visible light communication system. *IEEE Access*, 8:24145–24152, 2020.
311. Jiang Liu, Peter Wing Chau Chan, Derrick Wing Kwan Ng, Ernest S Lo, and Shigeru Shimamoto. Hybrid visible light communications in intelligent transportation systems with position based services. In *2012 IEEE Globecom Workshops*, pages 1254–1259. IEEE, 2012.
312. S. Liu, F. Yang, and J. Song. Narrowband interference cancelation based on priori aided compressive sensing for dtmb systems. *IEEE Transactions on Broadcasting*, 61(1):66–74, 2015.
313. S. Liu, F. Yang, and J. Song. An optimal interleaving scheme with maximum time-frequency diversity for plc systems. *IEEE Transactions on Power Delivery*, 31(3):1007–1014, 2016.
314. Nuno Lourenço, Domingos Terra, Navin Kumar, Luis Nero Alves, and Rui L Aguiar. Visible light communication system for outdoor applications. In *2012 8th International Symposium on Communication Systems, Networks & Digital Signal Processing (CSNDSP)*, pages 1–6. IEEE, 2012.
315. N. Lourenço, D. Terra, N. Kumar, L. N. Alves, and R. L. Aguiar. Visible light communication system for outdoor applications. In *2012 8th International Symposium on Communication Systems, Networks & Digital Signal Processing (CSNDSP)*, pages 1–6, 2012.
316. M. Luise and R. Reggiannini. Carrier frequency acquisition and tracking for ofdm systems. *IEEE Transactions on Communications*, 44(11):1590–1598, Nov 1996.
317. P. Luo, Z. Ghassemlooy, H. Le Minh, E. Bentley, A. Burton, and X. Tang. Fundamental analysis of a car to car visible light communication system. In *2014 9th International Symposium on Communication Systems, Networks & Digital Signal Processing (CSNDSP)*, pages 1011–1016, 2014.
318. Pengfei Luo, Zabih Ghassemlooy, Hoa Le Minh, Edward Bentley, Andrew Burton, and Xuan Tang. Performance analysis of a car-to-car visible light communication system. *Applied Optics*, 54(7):1696–1706, 2015.
319. Pengfei Luo, Min Zhang, Xiang Zhang, Guangming Cai, Dahai Han, and Qing Li. An indoor visible light communication positioning system using dual-tone

multi-frequency technique. In *2013 2nd International Workshop on Optical Wireless Communications (IWOW)*, pages 25–29. IEEE, 2013.

320. T Luo, Z Wen, J Li, and H-H Chen. Saturation throughput analysis of wave networks in doppler spread scenarios. *IET Communications*, 4(7):817–825, 2010.
321. H. Ma, L. Lampe, and S. Hranilovic. Integration of indoor visible light and power line communication systems. In *2013 IEEE 17th International Symposium on Power Line Communications and Its Applications*, pages 291–296, 2013.
322. H. Ma, L. Lampe, and S. Hranilovic. Robust mmse linear precoding for visible light communication broadcasting systems. In *2013 IEEE Globecom Workshops (GC Wkshps)*, pages 1081–1086, Dec 2013.
323. H. Ma, L. Lampe, and S. Hranilovic. Coordinated broadcasting for multiuser indoor visible light communication systems. *IEEE Transactions on Communications*, 63(9):3313–3324, Sep. 2015.
324. H. Ma, L. Lampe, and S. Hranilovic. Hybrid visible light and power line communication for indoor multiuser downlink. *IEEE/OSA Journal of Optical Communications and Networking*, 9(8):635–647, 2017.
325. Hao Ma, Lutz Lampe, and Steve Hranilovic. Integration of indoor visible light and power line communication systems. In *2013 IEEE 17th International Symposium on Power Line Communications and Its Applications*, pages 291–296. IEEE, 2013.
326. Hao Ma, Lutz Lampe, and Steve Hranilovic. Hybrid visible light and power line communication for indoor multiuser downlink. *IEEE/OSA Journal of Optical Communications and Networking*, 9(8):635–647, 2017.
327. S. Ma, Y. He, H. Li, S. Lu, F. Zhang, and S. Li. Optimal power allocation for mobile users in non-orthogonal multiple access visible light communication networks. *IEEE Transactions on Communications*, 67(3):2233–2244, March 2019.
328. X. Ma, J. Gao, F. Yang, W. Ding, H. Yang, and J. Song. Integrated power line and visible light communication system compatible with multi-service transmission. *IET Communications*, 11(1):104–111, 2017.
329. Y. H. Ma, P. L. So, and E. Gunawan. Performance analysis of ofdm systems for broadband power line communications under impulsive noise and multipath effects. *IEEE Transactions on Power Delivery*, 20(2):674–682, 2005.
330. Aditi Malik and Preeti Singh. Free space optics: current applications and future challenges. *International Journal of Optics*, 2015, 2015.
331. Y. Mao, H. Luan, W. Liu, R. Yang, M. Jin, X. Jin, and Z. Xu. Experimental investigation of carrierless amplitude-phase transmission for vehicular visible light communication systems. In *2016 IEEE International Conference on Communication Systems (ICCS)*, pages 1–6, 2016.
332. R. M. Marè, C. E. Cugnasca, C. L. Marte, and G. Gentile. Intelligent transport systems and visible light communication applications: An overview. In *2016 IEEE 19th International Conference on Intelligent Transportation Systems (ITSC)*, pages 2101–2106, 2016.
333. G. W. Marsh and J. M. Kahn. Performance evaluation of experimental 50-mb/s diffuse infrared wireless link using on-off keying with decision-feedback equalization. *IEEE Transactions on Communications*, 44(11):1496–1504, Nov 1996.
334. H. Marshoud, D. Dawoud, V. M. Kapinas, G. K. Karagiannidis, S. Muhaidat, and B. Sharif. Mu-mimo precoding for vlc with imperfect csi. In *2015 4th International Workshop on Optical Wireless Communications (IWOW)*, pages 93–97, Sep. 2015.
335. H. Marshoud, V. M. Kapinas, G. K. Karagiannidis, and S. Muhaidat. Non-orthogonal multiple access for visible light communications. *IEEE Photonics Technology Letters*,

28(1):51–54, Jan 2016.

336. H. Marshoud, S. Muhaidat, P. C. Sofotasios, S. Hussain, M. A. Imran, and B. S. Sharif. Optical non-orthogonal multiple access for visible light communication. *IEEE Wireless Communications*, 25(2):82–88, April 2018.
337. H. Marshoud, P. C. Sofotasios, S. Muhaidat, G. K. Karagiannidis, and B. S. Sharif. On the performance of visible light communication systems with non-orthogonal multiple access. *IEEE Transactions on Wireless Communications*, 16(10):6350–6364, Oct 2017.
338. S. A. Martucci. Symmetric convolution and the discrete sine and cosine transforms. *IEEE Transactions on Signal Processing*, 42(5):1038–1051, May 1994.
339. Stephen A Martucci. Convolution-multiplication properties for the entire family of discrete sine and cosine transforms. *Proc. 1992 CISS, Princeton, NJ*, 1992.
340. L. E. M. Matheus, A. B. Vieira, L. F. M. Vieira, M. A. M. Vieira, and O. Gnawali. Visible light communication: Concepts, applications and challenges. *IEEE Communications Surveys Tutorials*, 21(4):3204–3237, 2019.
341. T. K. Matsushima, S. Sasaki, M. Kakuyama, S. Yamasaki, Y. Murata, and Y. Teramachi. A visible-light communication system using optical cdma with inverted mpsc. In *The Sixth International Workshop on Signal Design and Its Applications in Communications*, pages 52–55, Oct 2013.
342. Stephen McCann and Alex Ashley. Official ieee 802.11 working group project timelines. *November2011.[Online]. Available: http://www. ieee802. org/11/Reports/802.11 ˙Ti melines. htm*, 2014.
343. Marcos F Guerra Medina, Oswaldo González, Silvestre Rodríguez, and Inocencio R Martín. Timing synchronization for ofdm-based visible light communication system. In *2016 Wireless Telecommunications Symposium (WTS)*, pages 1–4. IEEE, 2016.
344. Raed Mesleh, Hany Elgala, and Harald Haas. On the performance of different ofdm based optical wireless communication systems. *J. Opt. Commun. Netw.*, 3(8):620–628, Aug 2011.
345. William J Miniscalco and Steven A Lane. Optical space-time division multiple access. *Journal of Lightwave Technology*, 30(11):1771–1785, 2012.
346. F. Miramirkhani, O. Narmanlioglu, M. Uysal, and E. Panayirci. A mobile channel model for vlc and application to adaptive system design. *IEEE Communications Letters*, 21(5):1035–1038, May 2017.
347. F. Miramirkhani and M. Uysal. Channel modeling and characterization for visible light communications. *IEEE Photonics Journal*, 7(6):1–16, Dec 2015.
348. R. Mitra and V. Bhatia. Precoded chebyshev-nlms-based pre-distorter for nonlinear led compensation in noma-vlc. *IEEE Transactions on Communications*, 65(11):4845–4856, Nov 2017.
349. R. Mitra, F. Miramirkhani, V. Bhatia, and M. Uysal. Mixture-kernel based post-distortion in rkhs for time-varying vlc channels. *IEEE Transactions on Vehicular Technology*, 68(2):1564–1577, 2019.
350. E. Monteiro and S. Hranilovic. Constellation design for color-shift keying using interior point methods. In *2012 IEEE Globecom Workshops*, pages 1224–1228, Dec 2012.
351. E. Monteiro and S. Hranilovic. Design and implementation of color-shift keying for visible light communications. *Journal of Lightwave Technology*, 32(10):2053–2060, May 2014.
352. Bai, Bo, et al. The color shift key modulation with non-uniform signaling for visible light communication. *2012 1st IEEE International Conference on Communications in China Workshops (ICCC)*, IEEE, 2012.

353. P. H. Moose. A technique for orthogonal frequency division multiplexing frequency offset correction. *IEEE Transactions on Communications*, 42(10):2908–2914, Oct 1994.
354. Adriano JC Moreira, Rui T Valadas, and AM de Oliveira Duarte. Characterisation and modelling of artificial light interference in optical wireless communication systems. In *Proceedings of 6th International Symposium on Personal, Indoor and Mobile Radio Communications*, volume 1, pages 326–331. IEEE, 1995.
355. M. S. Moreolo. Performance analysis of dht-based optical ofdm using large-size constellations in awgn. *IEEE Communications Letters*, 15(5):572–574, May 2011.
356. M. S. A. Mossaad, S. Hranilovic, and L. Lampe. Visible light communications using ofdm and multiple leds. *IEEE Transactions on Communications*, 63(11):4304–4313, Nov 2015.
357. M. Rea. *The lighting handbook : reference and application*. Tenth edition, 2011.
358. S. H. Muller and J. B. Huber. Ofdm with reduced peak-to-average power ratio by optimum combination of partial transmit sequences. *Electronics Letters*, 33(5):368–369, Feb 1997.
359. S. Muthu and J. Gaines. Red, green and blue led-based white light source: implementation challenges and control design. In *38th IAS Annual Meeting on Conference Record of the Industry Applications Conference, 2003.*, volume 1, pages 515–522, Oct 2003.
360. Wassim G Najm, Jonathan Koopmann, John D Smith, John Brewer, et al. Frequency of target crashes for intellidrive safety systems. Technical report, United States. National Highway Traffic Safety Administration, 2010.
361. X. Nan, P. Wang, L. Guo, L. Huang, and Z. Liu. A novel vlc channel model based on beam steering considering the impact of obstacle. *IEEE Communications Letters*, 23(6):1003–1007, 2019.
362. O. Narmanlioglu, R. C. Kizilirmak, F. Miramirkhani, S. Safaraliev, S. M. Sait, and M. Uysal. Effect of wiring and cabling topologies on the performance of distributed mimo ofdm vlc systems. *IEEE Access*, 7:52743–52754, 2019.
363. T. Nawaz, M. Seminara, S. Caputo, L. Mucchi, F. S. Cataliotti, and J. Catani. IEEE 802.15.7-compliant ultra-low latency relaying vlc system for safety-critical ITS. *IEEE Transactions on Vehicular Technology*, 68(12):12040–12051, 2019.
364. Z. Nazari Chaleshtori, P. A. Haigh, P. Chvojka, S. Zvanovec, and Z. Ghassemlooy. Performance evaluation of various training algorithms for ann equalization in visible light communications with an organic led. In *2019 2nd West Asian Colloquium on Optical Wireless Communications (WACOWC)*, pages 11–15, 2019.
365. A. R. Ndjiongue, H. C. Ferreira, K. Ouahada, and A. J. H. Vinckz. Low-complexity socpbfsk-ook interface between plc and vlc channels for low data rate transmission applications. In *18th IEEE International Symposium on Power Line Communications and Its Applications*, pages 226–231, 2014.
366. A. R. Ndjiongue, T. M. N. Ngatched, and H. C. Ferreira. On the indoor vlc link evaluation based on the rician *k* -factor. *IEEE Communications Letters*, 22(11):2254–2257, 2018.
367. A. R. Ndjiongue, T. Shongwe, H. C. Ferreira, T. M. N. Ngatched, and A. J. H. Vinck. Cascaded plc-vlc channel using ofdm and csk techniques. In *2015 IEEE Global Communications Conference (GLOBECOM)*, pages 1–6, 2015.
368. Alain Richard Ndjiongue, Hendrik Christoffel Ferreira, Jian Song, Fang Yang, and Ling Cheng. Hybrid plc-vlc channel model and spectral estimation using a nonparametric approach. *Transactions on Emerging Telecommunications Technologies*, 28(12):e3224, 2017.

369. Tien V Nguyen, François Baccelli, Kai Zhu, Sundar Subramanian, and Xinzhou Wu. A performance analysis of csma based broadcast protocol in vanets. In *2013 Proceedings IEEE INFOCOM*, pages 2805–2813. IEEE, 2013.
370. Wang Ning, Qiao Yang, Wang Wuqu, Tang Suheng, and Shen Jianhua. Visible light communication based intelligent traffic light system: Designing and implementation. In *2018 Asia Communications and Photonics Conference (ACP)*, pages 1–3. IEEE, 2018.
371. S. M. Nlom, A. R. Ndjiongue, and K. Ouahada. Cascaded plc-vlc channel: An indoor measurements campaign. *IEEE Access*, 6:25230–25239, 2018.
372. S. M. Nlom, A. R. Ndjiongue, K. Ouahada, H. C. Ferreira, A. J. H. Vinck, and T. Shongwe. A simplistic channel model for cascaded plc-vlc systems. In *2017 IEEE International Symposium on Power Line Communications and its Applications (ISPLC)*, pages 1–6, 2017.
373. M. Noshad and M. Brandt-Pearce. Expurgated ppm using symmetric balanced incomplete block designs. *IEEE Communications Letters*, 16(7):968–971, July 2012.
374. M. Noshad and M. Brandt-Pearce. Multilevel pulse-position modulation based on balanced incomplete block designs. In *2012 IEEE Global Communications Conference (GLOBECOM)*, pages 2930–2935, Dec 2012.
375. M. Noshad and M. Brandt-Pearce. High-speed visible light indoor networks based on optical orthogonal codes and combinatorial designs. In *2013 IEEE Global Communications Conference (GLOBECOM)*, pages 2436–2441, Dec 2013.
376. M. Noshad and M. Brandt-Pearce. Application of expurgated ppm to indoor visible light communications part ii: Access networks. *Journal of Lightwave Technology*, 32(5):883–890, March 2014.
377. M. Noshad and M. Brandt-Pearce. Hadamard-coded modulation for visible light communications. *IEEE Transactions on Communications*, 64(3):1167–1175, March 2016.
378. Mohammad Noshad and Maite Brandt-Pearce. Application of expurgated ppm to indoor visible light communications part i: Single-user systems. *Journal of lightwave technology*, 32(5):875–882, 2013.
379. G. Ntogari, T. Kamalakis, J. Walewski, and T. Sphicopoulos. Combining illumination dimming based on pulse-width modulation with visible-light communications based on discrete multitone. *IEEE/OSA Journal of Optical Communications and Networking*, 3(1):56–65, January 2011.
380. G Ntogari, T Kamalakis, JW Walewski, and T Sphicopoulos. Combining illumination dimming based on pulse-width modulation with visible-light communications based on discrete multitone. *Journal of Optical Communications and Networking*, 3(1):56–65, 2011.
381. Mohanad Obeed, Anas M Salhab, Salam A Zummo, and Mohamed-Slim Alouini. Joint optimization of power allocation and load balancing for hybrid vlc/rf networks. *Journal of Optical Communications and Networking*, 10(5):553–562, 2018.
382. Dominic C O'Brien. Visible light communications: challenges and potential. In *IEEE Photonic Society 24th Annual Meeting*, pages 365–366. IEEE, 2011.
383. F. B. Offiong, S. Sinanović, and W. O. Popoola. On PAPR reduction in pilot-assisted optical ofdm communication systems. *IEEE Access*, 5:8916–8929, 2017.
384. F. B. Ogunkoya, W. O. Popoola, A. Shahrabi, and S. Sinanović. Performance evaluation of pilot-assisted PAPR reduction technique in optical ofdm systems. *IEEE Photonics Technology Letters*, 27(10):1088–1091, May 2015.
385. T. Ohtsuki, I. Sasase, and S. Mori. Overlapping multi-pulse pulse position modulation in optical direct detection channel. In *Communications, 1993. ICC '93 Geneva.*

Technical Program, Conference Record, IEEE International Conference, volume 2, pages 1123–1127, May 1993.

386. T. Ohtsuki, I. Sasase, and S. Mori. Differential overlapping pulse position modulation in optical direct-detection channel. In *Communications, 1994. ICC '94, SUPERCOMM/ICC '94, Conference Record, 'Serving Humanity Through Communications.' IEEE International Conference on*, pages 680–684, vol.2, May 1994.
387. T. Ohtsuki, I. Sasase, and S. Mori. Performance analysis of overlapping multi-pulse pulse position modulation (omppm) in noisy photon counting channel. In *Proceedings of 1994 IEEE International Symposium on Information Theory*, page 80, June 1994.
388. T. Ohtsuki, I. Sasase, and S. Mori. Trellis coded overlapping multi-pulse pulse position modulation in optical direct detection channel. In *Communications, 1994. ICC '94, SUPERCOMM/ICC '94, Conference Record, 'Serving Humanity Through Communications.' IEEE International Conference on*, pages 675–679, vol.2, May 1994.
389. Tomoaki Ohtsuki, Iwao Sasase, and Shinsaku Mori. Capacity and cutoff rate of overlapping multi-pulse pulse position modulation (omppm) in optical direct-detection channel: Quantum-limited case. *IEICE Transactions on Fundamentals of Electronics, Communications and Computer Sciences*, 77(8):1298–1308, 1994.
390. Tomoaki Ohtsuki, Iwao Sasase, and Shinsaku Mori. Error performance of overlapping multi-pulse pulse position modulation (omppm) and trellis coded omppm in optical direct-detection channel. *IEICE Transactions on Communications*, 77(9):1133–1143, 1994.
391. Satoshi Okada, Tomohiro Yendo, Takaya Yamazato, Toshiaki Fujii, Masayuki Tanimoto, and Yoshikatsu Kimura. On-vehicle receiver for distant visible light road-to-vehicle communication. In *2009 IEEE Intelligent Vehicles Symposium*, pages 1033–1038. IEEE, 2009.
392. M. I. Olmedo, T. Zuo, J. B. Jensen, Q. Zhong, X. Xu, S. Popov, and I. T. Monroy. Multiband carrierless amplitude phase modulation for high capacity optical data links. *Journal of Lightwave Technology*, 32(4):798–804, 2014.
393. IEEE 802.11 p Working Group et al. IEEE standard for information technology-local and metropolitan area networks-specific requirements-part 11: Wireless LAN medium access control (mac) and physical layer (phy) specifications amendment 6: Wireless access in vehicular environments. *IEEE Std; IEEE: Piscataway, NJ, USA*, 2010.
394. G. Pang, T. Kwan, Chi-Ho Chan, and Hugh Liu. Led traffic light as a communications device. In *Proceedings 199 IEEE/IEEJ/JSAI International Conference on Intelligent Transportation Systems (Cat. No.99TH8383)*, pages 788–793, Oct 1999.
395. Memedi, Agon, and Falko Dressler. Vehicular visible light communications: A survey. In *IEEE Communications Surveys & Tutorials (2020)*.
396. Grantham Pang, Hugh Liu, Chi-Ho Chan, and Thomas Kwan. Vehicle location and navigation systems based on LEDs. *Proc. 5th World Congr. Intelligent Transport Systems*, pages 12–16, 1998.
397. V. K. Papanikolaou, P. D. Diamantoulakis, and G. K. Karagiannidis. User grouping for hybrid vlc/rf networks with noma: A coalitional game approach. *IEEE Access*, 7:103299–103309, 2019.
398. P. H. Pathak, X. Feng, P. Hu, and P. Mohapatra. Visible light communication, networking, and sensing: A survey, potential and challenges. *IEEE Communications Surveys Tutorials*, 17(4):2047–2077, Fourth quarter 2015.
399. Matthew P Peloso and Ilja Gerhardt. Statistical tests of randomness on quantum keys distributed through a free-space channel coupled to daylight noise. *Journal of Lightwave Technology*, 31(23):3794–3805, 2013.

400. Julie Penning, Kelsey Stober, Victor Taylor, and Mary Yamada. Energy savings forecast of solid-state lighting in general illumination applications. Technical report, Navigant Consulting Inc., Washington, DC (United States), 2016.
401. R. Perez-Jimenez, J. Rabadan, J. Rufo, E. Solana, and J. M. Luna-Rivera. Visible light communications technologies for smart tourism destinations. In *2015 IEEE First International Smart Cities Conference (ISC2)*, pages 1–5, Oct 2015.
402. Stanislav Zvanovec Zabih Ghassemlooy Sujan Rajbhandari Petr Chvojka, Stanislav Vitek. Analysis of nonline-of-sight visible light communications. *Optical Engineering*, 56, 2017.
403. V Pohl, V Jungnickel, and C Von Helmolt. Integrating-sphere diffuser for wireless infrared communication. *IEE Proceedings-Optoelectronics*, 147(4):281–285, 2000.
404. W. O. Popoola, Z. Ghassemlooy, and B. G. Stewart. Pilot-assisted PAPR reduction technique for optical ofdm communication systems. *Journal of Lightwave Technology*, 32(7):1374–1382, April 2014.
405. J. G. Proakis. Adaptive equalization for tdma digital mobile radio. *IEEE Transactions on Vehicular Technology*, 40(2):333–341, 1991.
406. Krishna Prasad Pujapanda. Lifi integrated to power-lines for smart illumination cum communication. In *2013 International Conference on Communication Systems and Network Technologies*, pages 875–878. IEEE, 2013.
407. A. A. Purwita, M. Dehghani Soltani, M. Safari, and H. Haas. Impact of terminal orientation on performance in lifi systems. In *2018 IEEE Wireless Communications and Networking Conference (WCNC)*, pages 1–6, April 2018.
408. A. A. Purwita and H. Haas. Iq-wdm for IEEE 802.11bb-based lifi. In *2020 IEEE Wireless Communications and Networking Conference (WCNC)*, pages 1–6, 2020.
409. A. A. Purwita and H. Haas. Studies of flatness of lifi channel for IEEE 802.11bb. In *2020 IEEE Wireless Communications and Networking Conference (WCNC)*, pages 1–6, 2020.
410. Qi Lu, Qingchong Liu, and G. S. Mitchell. Performance analysis for optical wireless communication systems using subcarrier psk intensity modulation through turbulent atmospheric channel. In *IEEE Global Telecommunications Conference, 2004. GLOBECOM '04.*, volume 3, pages 1872–1875, 2004.
411. H. Qian, S. C. Dai, S. Zhao, S. Z. Cai, and H. Zhang. A robust cdma vlc system against front-end nonlinearity. *IEEE Photonics Journal*, 7(5):1–9, Oct 2015.
412. Y. Qiu, S. Chen, H. Chen, and W. Meng. Visible light communications based on cdma technology. *IEEE Wireless Communications*, 25(2):178–185, 2018.
413. K Raghunath and Ananthanarayanan Chockalingam. SIR analysis and interference cancellation in uplink ofdma with large carrier frequency/timing offsets. *IEEE Transactions on Wireless Communications*, 8(5), 2009.
414. Michael B Rahaim, Anna Maria Vegni, and Thomas DC Little. A hybrid radio frequency and broadcast visible light communication system. In *2011 IEEE GLOBECOM Workshops (GC Wkshps)*, pages 792–796. IEEE, 2011.
415. M. M. Rahman, L. Bobadilla, and B. Rapp. Establishing line-of-sight communication via autonomous relay vehicles. In *MILCOM 2016 - 2016 IEEE Military Communications Conference*, pages 642–647, 2016.
416. A Albert Raj and T Latha. *VLSI design*. PHI Learning Pvt. Ltd., 2008.
417. S. Rajagopal, R. D. Roberts, and S. Lim. IEEE 802.15.7 visible light communication: modulation schemes and dimming support. *IEEE Communications Magazine*, 50(3):72–82, March 2012.

418. Sridhar Rajagopal and Farooq Khan. Apparatus and method for interference mitigation and channel selection for visible light communication, August 27 2013. US Patent 8,521,034.
419. S. Rajbhandari, Z. Ghassemlooy, and M. Angelova. Bit error performance of diffuse indoor optical wireless channel pulse position modulation system employing artificial neural networks for channel equalisation. *IET Optoelectronics*, 3(4):169–179, August 2009.
420. Roberto Ramirez-Iniguez, Sevia M Idrus, and Ziran Sun. *Optical wireless communications: IR for wireless connectivity*. Auerbach Publications, 2008.
421. B. Ranjha and M. Kavehrad. Hybrid asymmetrically clipped ofdm-based im/dd optical wireless system. *IEEE/OSA Journal of Optical Communications and Networking*, 6(4):387–396, April 2014.
422. Kamisetty Ramam Rao and Patrick C Yip. *The transform and data compression handbook*. CRC Press, 2000.
423. R. Roberts, P. Gopalakrishnan, and S. Rathi. Visible light positioning: Automotive use case. In *2010 IEEE Vehicular Networking Conference*, pages 309–314, 2010.
424. R. Rodes, M. Wieckowski, T. T. Pham, J. B. Jensen, and I. T. Monroy. Vcsel-based dwdm pon with 4 bit/s/hz spectral efficiency using carrierless amplitude phase modulation. In *2011 37th European Conference and Exhibition on Optical Communication*, pages 1–3, 2011.
425. V. Rodoplu, K. Hocaoğlu, A. Adar, R. O. Çikmazel, and A. Saylam. Characterization of line-of-sight link availability in indoor visible light communication networks based on the behavior of human users. *IEEE Access*, 8:39336–39348, 2020.
426. Patrick S Ryan. Application of the public-trust doctrine and principles of natural resource management to electromagnetic spectrum. *Mich. Telecomm. & Tech. L. Rev.*, 10:285, 2003.
427. Matthew NO Sadiku, Sarhan M Musa, and Sudarshan R Nelatury. Free space optical communications: An overview. *European Scientific Journal*, 12(9), 2016.
428. T. Saito, S. Haruyama, and M. Nakagawa. A new tracking method using image sensor and photo diode for visible light road-to-vehicle communication. In *2008 10th International Conference on Advanced Communication Technology*, volume 1, pages 673–678, 2008.
429. Hara, Toshiki, et al. A new receiving system of visible light communication for ITS. In *2007 IEEE Intelligent Vehicles Symposium*, IEEE, 2007.
430. N. Sasaki, N. Iijima, and D. Uchiyama. Development of ranging method for inter-vehicle distance using visible light communication and image processing. In *2015 15th International Conference on Control, Automation and Systems (ICCAS)*, pages 666–670, 2015.
431. Satoshi Okada, Tomohiro Yendo, Takaya Yamazato, Toshiaki Fujii, Masayuki Tanimoto, and Yoshikatsu Kimura. On-vehicle receiver for distant visible light road-to-vehicle communication. In *2009 IEEE Intelligent Vehicles Symposium*, pages 1033–1038, 2009.
432. T. Schaub. Spread frequency shift keying-a low cost approach to spread spectrum. In *1994 IEEE GLOBECOM. Communications: The Global Bridge*, volume 3, pages 1722–1726, 1994.
433. T. C. W. Schenk, L. Feri, H. Yang, and J. M. G. Linnartz. Optical wireless cdma employing solid state lighting LEDs. In *2009 IEEE/LEOS Summer Topical Meeting*, pages 23–24, July 2009.

434. B Schoettle and MJ Flannagan. A market-weighted description of low-beam and high-beam headlighting patterns in the US. *University of Michigan*, 2011.
435. Michele Segata, Renato Lo Cigno, Hsin-Mu Michael Tsai, and Falko Dressler. On platooning control using IEEE 802.11 p in conjunction with visible light communications. In *2016 12th Annual Conference on Wireless On-demand Network Systems and Services (WONS)*, pages 1–4. IEEE, 2016.
436. Mathini Sellathurai and Simon S Haykin. *Space-time layered information processing for wireless communications*. Wiley Online Library, 2009.
437. F. Shad, T. D. Todd, V. Kezys, and J. Litva. Dynamic slot allocation (dsa) in indoor sdma/tdma using a smart antenna basestation. *IEEE/ACM Transactions on Networking*, 9(1):69–81, Feb 2001.
438. S. Shao, A. Khreishah, M. Ayyash, M. B. Rahaim, H. Elgala, V. Jungnickel, D. Schulz, T. D. C. Little, J. Hilt, and R. Freund. Design and analysis of a visible-light-communication enhanced wifi system. *IEEE/OSA Journal of Optical Communications and Networking*, 7(10):960–973, 2015.
439. Xuancheng Shao and Steven G Johnson. Type-iv dct, dst, and mdct algorithms with reduced numbers of arithmetic operations. *Signal Processing*, 88(6):1313–1326, 2008.
440. Hong Shen, Yanfei Wu, Wei Xu, and Chunming Zhao. Optimal power allocation for downlink two-user non-orthogonal multiple access in visible light communication. *Journal of Communications and Information Networks*, 2(4):57–64, 2017.
441. J. S. Sheu, B. J. Li, and J. K. Lain. Led non-linearity mitigation techniques for optical ofdm-based visible light communications. *IET Optoelectronics*, 11(6):259–264, 2017.
442. J. Shi, Y. Hong, R. Deng, J. He, L. Chen, and G. Chang. Demonstration of real-time software reconfigurable dynamic power-and-subcarrier allocation scheme for ofdm-noma-based multi-user visible light communications. *Journal of Lightwave Technology*, 37(17):4401–4409, Sep. 2019.
443. W. Shieh, W. Lee, S. Tung, B. Jeng, and C. Liu. Analysis of the optimum configuration of roadside units and onboard units in dedicated short-range communication systems. *IEEE Transactions on Intelligent Transportation Systems*, 7(4):565–571, 2006.
444. Da-Shan Shiu and J. M. Kahn. Differential pulse-position modulation for power-efficient optical communication. *IEEE Transactions on Communications*, 47(8):1201–1210, Aug 1999.
445. M. H. Shoreh, A. Fallahpour, and J. A. Salehi. Design concepts and performance analysis of multicarrier cdma for indoor visible light communications. *IEEE/OSA Journal of Optical Communications and Networking*, 7(6):554–562, June 2015.
446. Shuli Gao, Jian Zhang, Jian Song, Hui Yang, and Xi Chu. Random channel generator of the integrated power line communication and visible light communication. In *2017 IEEE International Symposium on Power Line Communications and its Applications (ISPLC)*, pages 1–7, 2017.
447. K. Siddiqi, A. D. Raza, and S. S. Muhammad. Visible light communication for v2v intelligent transport system. In *2016 International Conference on Broadband Communications for Next Generation Networks and Multimedia Applications (CoBCom)*, pages 1–4, 2016.
448. A. B. Siddique and M. Tahir. Joint rate-brightness control using variable rate mppm for led based visible light communication systems. *IEEE Transactions on Wireless Communications*, 12(9):4604–4611, Sep. 2013.
449. R. Singh, T. O'Farrell, and J. P. R. David. An enhanced color shift keying modulation scheme for high-speed wireless visible light communications. *Journal of Lightwave Technology*, 32(14):2582–2592, July 2014.

450. Michael Sivak, Michael J Flannagan, Shinichi Kojima, and Eric C Traube. A market-weighted description of low-beam headlighting patterns in the US. *SAE Transactions*, pages 686–699, 1998.
451. M. D. Soltani, H. Kazemi, M. Safari, and H. Haas. Handover modeling for indoor li-fi cellular networks: The effects of receiver mobility and rotation. In *2017 IEEE Wireless Communications and Networking Conference (WCNC)*, pages 1–6, March 2017.
452. M. D. Soltani, A. A. Purwita, Z. Zeng, H. Haas, and M. Safari. Modeling the random orientation of mobile devices: Measurement, analysis and lifi use case. *IEEE Transactions on Communications*, 67(3):2157–2172, March 2019.
453. B. Soner and S. C. Ergen. Vehicular visible light positioning with a single receiver. In *2019 IEEE 30th Annual International Symposium on Personal, Indoor and Mobile Radio Communications (PIMRC)*, pages 1–6, 2019.
454. J. Song, W. Ding, F. Yang, H. Yang, B. Yu, and H. Zhang. An indoor broadband broadcasting system based on plc and vlc. *IEEE Transactions on Broadcasting*, 61(2):299–308, 2015.
455. J. Song, S. Liu, G. Zhou, B. Yu, W. Ding, F. Yang, H. Zhang, X. Zhang, and A. Amara. A cost-effective approach for ubiquitous broadband access based on hybrid plc-vlc system. In *2016 IEEE International Symposium on Circuits and Systems (ISCAS)*, pages 2815–2818, 2016.
456. Henrik Sorensen, Douglas Jones, C Burrus, and Michael Heideman. On computing the discrete hartley transform. *IEEE Transactions on Acoustics, Speech, and Signal Processing*, 33(5):1231–1238, 1985.
457. Y. Sreenivasa Reddy, M. Panda, A. Dubey, A. Kumar, T. Panigrahi, and K. M. Rabie. Optimisation of indoor hybrid plc/vlc/rf communication systems. *IET Communications*, 14(1):117–126, 2020.
458. Federal Motor Vehicle Safety Standard (FMVSS). Standard no. 108: Lamps, reflective devices and associated equipment, 2003.
459. G Stepniak and J Siuzdak. Experimental investigation of pam, cap and dmt modulations efficiency over a double-step-index polymer optical fiber. *Optical Fiber Technology*, 20(4):369–373, 2014.
460. Sundar Subramanian, Marc Werner, Shihuan Liu, Jubin Jose, Radu Lupoaie, and Xinzhou Wu. Congestion control for vehicular safety: synchronous and asynchronous mac algorithms. In *Proceedings of the Ninth ACM International Workshop on Vehicular Inter-networking, Systems, and Applications*, pages 63–72. ACM, 2012.
461. H. Sugiyama, S. Haruyama, and M. Nakagawa. Brightness control methods for illumination and visible-light communication systems. In *2007 Third International Conference on Wireless and Mobile Communications (ICWMC'07)*, pages 78–78, March 2007.
462. H. Sugiyama and K. Nosu. Mppm: a method for improving the band-utilization efficiency in optical ppm. *Journal of Lightwave Technology*, 7(3):465–472, March 1989.
463. L. Sun, J. Du, and Z. He. Multiband three-dimensional carrierless amplitude phase modulation for short reach optical communications. *Journal of Lightwave Technology*, 34(13):3103–3109, 2016.
464. A. J. Suzuki and K. Mizui. Laser radar and visible light in a bidirectional v2v communication and ranging system. In *2015 IEEE International Conference on Vehicular Electronics and Safety (ICVES)*, pages 19–24, 2015.
465. P. Tabeshmehr, A. Olfat, Z. Ghassemlooy, and S. Zvanovec. Non-orthogonal variable multi-band carrier-less amplitude and phase modulation with reduced subcarriers. In

2019 2nd West Asian Colloquium on Optical Wireless Communications (WACOWC), pages 16–20, 2019.

466. I. Takai, S. Ito, K. Yasutomi, K. Kagawa, M. Andoh, and S. Kawahito. LED and cmos image sensor based optical wireless communication system for automotive applications. *IEEE Photonics Journal*, 5(5):6801418, 2013.
467. Isamu Takai, Tomohisa Harada, Michinori Andoh, Keita Yasutomi, Keiichiro Kagawa, and Shoji Kawahito. Optical vehicle-to-vehicle communication system using led transmitter and camera receiver. *IEEE Photonics Journal*, 6(5):1–14, 2014.
468. Yuichi Tanaka, Shinichiro Haruyama, and Masao Nakagawa. Wireless optical transmissions with white colored led for wireless home links. In *11th IEEE International Symposium on Personal Indoor and Mobile Radio Communications. PIMRC 2000. Proceedings (Cat. No. 00TH8525)*, volume 2, pages 1325–1329. IEEE, 2000.
469. Yuichi Tanaka, Toshihiko Komine, Shinichiro Haruyama, and Masao Nakagawa. Indoor visible communication utilizing plural white leds as lighting. In *12th IEEE International Symposium on Personal, Indoor and Mobile Radio Communications. PIMRC 2001. Proceedings (Cat. No. 01TH8598)*, volume 2, pages F–F. IEEE, 2001.
470. Yuichi Tanaka, Toshihiko Komine, Shinichiro Haruyama, and Masao Nakagawa. Indoor visible light data transmission system utilizing white led lights. *IEICE transactions on communications*, 86(8):2440–2454, 2003.
471. Ching W Tang and Steven A VanSlyke. Organic electroluminescent diodes. *Applied Physics Letters*, 51(12):913–915, 1987.
472. L. Tao, Y. Ji, J. Liu, A. P. Tao Lau, N. Chi, and C. Lu. Advanced modulation formats for short reach optical communication systems. *IEEE Network*, 27(6):6–13, 2013.
473. Anna Tatarczak, Miguel Iglesias Olmedo, Tianjian Zuo, Jose Estaran, Jesper Bevensee Jensen, Xiaogeng Xu, and Idelfonso Tafur Monroy. Enabling 4-lane based 400 g client-side transmission links with multicap modulation. *Advances in Optical Technologies*, 2015, 2015.
474. Domingos Terra, Navin Kumar, Nuno Lourenço, Luis Nero Alves, and Rui L Aguiar. Design, development and performance analysis of dsss-based transceiver for vlc. In *2011 IEEE EUROCON-International Conference on Computer as a Tool*, pages 1–4. IEEE, 2011.
475. I. L. Thng, A. Cantoni, and Yee Hong Leung. Low timing sensitivity receiver structures for cap. *IEEE Transactions on Communications*, 48(3):396–400, March 2000.
476. Ozan K Tonguz, Nawaporn Wisitpongphan, Jayendra S Parikh, Fan Bai, Priyantha Mudalige, and Varsha K Sadekar. On the broadcast storm problem in ad hoc wireless networks. In *2006 3rd International Conference on Broadband Communications, Networks and Systems*, pages 1–11. IEEE, 2006.
477. V. T. B. Tram and M. Yoo. Vehicle-to-vehicle distance estimation using a low-resolution camera based on visible light communications. *IEEE Access*, 6:4521–4527, 2018.
478. Wen-Shing Tsai, Hai-Han Lu, Chung-Yi Li, Ting-Chieh Lu, Chen-Hong Liao, Chien-An Chu, and Peng-Chun Peng. A 20-m/40-gb/s 1550-nm dfb ld-based fso link. *IEEE Photonics Journal*, 7(6):1–7, 2015.
479. D. Tsonev and H. Haas. Avoiding spectral efficiency loss in unipolar ofdm for optical wireless communication. In *2014 IEEE International Conference on Communications (ICC)*, pages 3336–3341, June 2014.
480. Dobroslav Tsonev, Stefan Videv, and Harald Haas. Towards a 100 gb/s visible light wireless access network. *Optics Express*, 23(2):1627–1637, 2015.

481. T. T.Tjhung and B. Caron. SER performance evaluation and optimization of ofdm system with residual frequency and timing offsets from imperfect synchronization. *IEEE Transactions on Broadcasting*, 49(2):170–177, June 2003.
482. Alon Tuchner and Jack Haddad. Vehicle platoon formation using interpolating control: A laboratory experimental analysis. *Transportation Research Part C: Emerging Technologies*, 84:21–47, 2017.
483. S. Ucar, S. C. Ergen, and O. Ozkasap. IEEE 802.11p and visible light hybrid communication based secure autonomous platoon. *IEEE Transactions on Vehicular Technology*, 67(9):8667–8681, 2018.
484. Muhammad Shahin Uddin, Mostafa Zaman Chowdhury, and Yeong Min Jang. Priority-based resource allocation scheme for visible light communication. In *2010 Second International Conference on Ubiquitous and Future Networks (ICUFN)*, pages 247–250. IEEE, 2010.
485. UNECE. Vehicle regulations. 2014.
486. M. Uysal, Z. Ghassemlooy, A. Bekkali, A. Kadri, and H. Menouar. Visible light communication for vehicular networking: Performance study of a v2v system using a measured headlamp beam pattern model. *IEEE Vehicular Technology Magazine*, 10(4):45–53, 2015.
487. M. Uysal, F. Miramirkhani, O. Narmanlioglu, T. Baykas, and E. Panayirci. IEEE 802.15.7r1 reference channel models for visible light communications. *IEEE Communications Magazine*, 55(1):212–217, January 2017.
488. Murat Uysal, Tuncer Baykas, Farshad Miramirkhani, Nikola Serafimovski, and Volker Jungnickel. Tg7r1 channel model document for high rate pd communications. Technical report, 2015.
489. Murat Uysal, Carlo Capsoni, Zabih Ghassemlooy, Anthony Boucouvalas, and Eszter Udvary. *Optical wireless communications: an emerging technology*. Springer, 2016.
490. Murat Uysal, F Miramirkhani, T Baykas, N Serafimovski, and J Jungnickel. Lifi channel models: office, home, manufacturing cell. Technical report, IEEE, 2015.
491. J. J. van de Beek, M. Sandell, and P. O. Borjesson. Ml estimation of time and frequency offset in ofdm systems. *IEEE Transactions on Signal Processing*, 45(7):1800–1805, July 1997.
492. S. Vappangi and V. V. Mani. Performance analysis of fast optical ofdm for vlc. In *2017 20th International Symposium on Wireless Personal Multimedia Communications (WPMC)*, pages 206–211, 2017.
493. S. Vappangi and V. V. Mani. A low papr dst-based optical ofdm (oofdm) for visible light communication. In *2018 21st International Symposium on Wireless Personal Multimedia Communications (WPMC)*, pages 200–205, 2018.
494. S. Vappangi and V. V. Mani. A power efficient dst-based multicarrier and multiple access systems for vlc. In *2018 IEEE International Conference on Advanced Networks and Telecommunications Systems (ANTS)*, pages 1–6, 2018.
495. S. Vappangi and V. V. Mani. Performance analysis of dst-based intensity modulated/direct detection (im/dd) systems for vlc. *IEEE Sensors Journal*, 19(4):1320–1337, 2019.
496. S. Vappangi and V. M. Vakamulla. Synchronization in visible light communication for smart cities. *IEEE Sensors Journal*, 18(5):1877–1886, 2018.
497. Suseela Vappangi and V.V. Mani. Concurrent illumination and communication: A survey on visible light communication. *Physical Communication*, 33:90 – 114, 2019.

498. Suseela Vappangi and Venkata Mani Vakamulla. Channel estimation in aco-ofdm employing different transforms for vlc. *AEU-International Journal of Electronics and Communications*, 84:111–122, 2018.
499. Suseela Vappangi and Venkata Mani Vakamulla. A low PAPR multicarrier and multiple access schemes for vlc. *Optics Communications*, 425:121 – 132, 2018.
500. Martin Vetterli, Henri J Nussbaumer, et al. Simple fft and dct algorithms with reduced number of operations. *Signal processing*, 6(4):267–278, 1984.
501. J. Vucic, C. Kottke, S. Nerreter, K. Habel, A. Buttner, K. Langer, and J. W. Walewski. 125 mbit/s over 5 m wireless distance by use of ook-modulated phosphorescent white leds. In *2009 35th European Conference on Optical Communication*, pages 1–2, Sep. 2009.
502. Jelena Vučić, Christoph Kottke, Kai Habel, and Klaus-Dieter Langer. 803 mbit/s visible light wdm link based on dmt modulation of a single rgb led luminary. In *2011 Optical Fiber Communication Conference and Exposition and the National Fiber Optic Engineers Conference*, pages 1–3. IEEE, 2011.
503. J. Vučić, C. Kottke, S. Nerreter, K. Habel, A. Büttner, K. Langer, and J. W. Walewski. 230 mbit/s via a wireless visible-light link based on ook modulation of phosphorescent white leds. In *2010 Conference on Optical Fiber Communication (OFC/NFOEC), collocated National Fiber Optic Engineers Conference*, pages 1–3, March 2010.
504. J. Wang, Q. Li, J. Zhu, and Y. Wang. Impact of receiver's tilted angle on channel capacity in vlcs. *Electronics Letters*, 53(6):421–423, 2017.
505. J. Wang, J. Wang, B. Zhu, M. Lin, Y. Wu, Y. Wang, and M. Chen. Improvement of ber performance by tilting receiver plane for indoor visible light communications with input-dependent noise. In *2017 IEEE International Conference on Communications (ICC)*, pages 1–6, May 2017.
506. J. Wang, Y. Yu, X. Liu, Y. Zhang, X. Zhou, and Y. Lu. Metal-organic framework based oled for visible light communication. In *2019 18th International Conference on Optical Communications and Networks (ICOCN)*, pages 1–3, 2019.
507. Q. Wang, Z. Wang, and L. Dai. Iterative receiver for hybrid asymmetrically clipped optical ofdm. *Journal of Lightwave Technology*, 32(22):4471–4477, Nov 2014.
508. Qi Wang, Chen Qian, Xuhan Guo, Zhaocheng Wang, David G Cunningham, and Ian H White. Layered aco-ofdm for intensity-modulated direct-detection optical wireless transmission. *Optics Express*, 23(9):12382–12393, 2015.
509. S. Wang and C. Li. A low-complexity papr reduction scheme for sfbc mimo-ofdm systems. *IEEE Signal Processing Letters*, 16(11):941–944, Nov 2009.
510. Y. Wang, L. Tao, X. Huang, J. Shi, and N. Chi. 8-gb/s rgby led-based wdm vlc system employing high-order cap modulation and hybrid post equalizer. *IEEE Photonics Journal*, 7(6):1–7, Dec 2015.
511. Y. Wang, L. Tao, X. Huang, J. Shi, and N. Chi. Enhanced performance of a high-speed wdm cap64 vlc system employing volterra series-based nonlinear equalizer. *IEEE Photonics Journal*, 7(3):1–7, June 2015.
512. Y. Wang, L. Tao, Y. Wang, and N. Chi. High speed wdm vlc system based on multi-band cap64 with weighted pre-equalization and modified cmma based post-equalization. *IEEE Communications Letters*, 18(10):1719–1722, Oct 2014.
513. Y. Wang, X. Wu, and H. Haas. Resource allocation in lifi ofdma systems. In *GLOBECOM 2017 - 2017 IEEE Global Communications Conference*, pages 1–6, 2017.
514. Yiguang Wang, Xingxing Huang, Li Tao, Jianyang Shi, and Nan Chi. 4.5-gb/s rgb-led based wdm visible light communication system employing cap modulation and rls based adaptive equalization. *Optics Express*, 23(10):13626–13633, 2015.

515. Yuanquan Wang, Yiguang Wang, Nan Chi, Jianjun Yu, and Huiliang Shang. Demonstration of 575-mb/s downlink and 225-mb/s uplink bi-directional scm-wdm visible light communication using rgb led and phosphor-based led. *Optics Express*, 21(1):1203–1208, 2013.
516. Zhe Wang and Mahbub Hassan. How much of dsrc is available for non-safety use? In *Proceedings of the fifth ACM international workshop on Vehicular Inter-Networking*, pages 23–29. ACM, 2008.
517. Zhongde Wang and BR Hunt. The discrete W transform. *Applied Mathematics and Computation*, 16(1):19–48, 1985.
518. Zixiong Wang, Wen-De Zhong, Changyuan Yu, Jian Chen, Chin Po Shin Francois, and Wei Chen. Performance of dimming control scheme in visible light communication system. *Optics Express*, 20(17):18861–18868, 2012.
519. J. Wei, Q. Cheng, D. G. Cunningham, R. V. Penty, and I. H. White. 100-gb/s hybrid multiband cap/qam signal transmission over a single wavelength. *Journal of Lightwave Technology*, 33(2):415–423, 2015.
520. C. Wen, R. D. Morris, and W. A. Sethares. Distance estimation using bidirectional communications without synchronous clocking. *IEEE Transactions on Signal Processing*, 55(5):1927–1939, 2007.
521. K. Werfli, P. Chvojka, Z. Ghassemlooy, N. B. Hassan, S. Zvanovec, A. Burton, P. A. Haigh, and M. R. Bhatnagar. Experimental demonstration of high-speed 4 × 4 imaging multi-cap mimo visible light communications. *Journal of Lightwave Technology*, 36(10):1944–1951, 2018.
522. K. Werfli, P. A. Haigh, Z. Ghassemlooy, P. Chvojka, S. Zvanovec, S. Rajbhandari, and Shihe Long. Multi-band carrier-less amplitude and phase modulation with decision feedback equalization for bandlimited vlc systems. In *2015 4th International Workshop on Optical Wireless Communications (IWOW)*, pages 6–10, 2015.
523. K. Werfli, P. A. Haigh, Z. Ghassemlooy, N. B. Hassan, and S. Zvanovec. A new concept of multi-band carrier-less amplitude and phase modulation for bandlimited visible light communications. In *2016 10th International Symposium on Communication Systems, Networks and Digital Signal Processing (CSNDSP)*, pages 1–5, 2016.
524. Wikipedia. Headlamp. 2014.
525. R Wood. Wireless network traffic worldwide: forecasts and analysis 2013–2018. *Research Forecast Report, Analysys Mason Ltd., London*, 2013.
526. D. Wu, Z. Ghassemlooy, H. Le Minh, S. Rajbhandari, M. A. Khalighi, and X. Tang. Optimisation of lambertian order for indoor non-directed optical wireless communication. In *2012 1st IEEE International Conference on Communications in China Workshops (ICCC)*, pages 43–48, Aug 2012.
527. F. Wu, C. Lin, C. Wei, C. Chen, H. Huang, and C. Ho. 1.1-gb/s white-led-based visible light communication employing carrier-less amplitude and phase modulation. *IEEE Photonics Technology Letters*, 24(19):1730–1732, 2012.
528. F. M. Wu, C. T. Lin, C. C. Wei, C. W. Chen, Z. Y. Chen, H. T. Huang, and S. Chi. Performance comparison of ofdm signal and cap signal over high capacity rgb-led-based wdm visible light communication. *IEEE Photonics Journal*, 5(4):7901507, 2013.
529. Fang-Ming Wu, Chun-Ting Lin, Chia-Chien Wei, Cheng-Wei Chen, Zhen-Yu Chen, and Hou-Tzu Huang. 3.22-gb/s wdm visible light communication of a single rgb led employing carrier-less amplitude and phase modulation. In *2013 Optical Fiber Communication Conference and Exposition and the National Fiber Optic Engineers Conference (OFC/NFOEC)*, pages 1–3. IEEE, 2013.

530. Nan Wu and Yeheskel Bar-Ness. A novel power-efficient scheme asymmetrically and symmetrically clipping optical (asco)-ofdm for im/dd optical systems. *EURASIP Journal on Advances in Signal Processing*, 2015(1):3, 2015.
531. D. Wulich and L. Goldfeld. Reduction of peak factor in orthogonal multicarrier modulation by amplitude limiting and coding. *IEEE Transactions on Communications*, 47(1):18–21, Jan 1999.
532. G Wyzecki and WS Stiles. Color science. *Concepts and Methods, Quantitative Data and Formulae*, 2, 1982.
533. N. Xiang, Z. Zhang, J. Dang, and L. Wu. A novel receiver design for pam-dmt in optical wireless communication systems. *IEEE Photonics Technology Letters*, 27(18):1919–1922, 2015.
534. J. Xiao, J. Yu, X. Li, Q. Tang, H. Chen, F. Li, Z. Cao, and L. Chen. Hadamard transform combined with companding transform technique for papr reduction in an optical direct-detection ofdm system. *IEEE/OSA Journal of Optical Communications and Networking*, 4(10):709–714, Oct 2012.
535. Y. Xiao, P. D. Diamantoulakis, Z. Fang, Z. Ma, L. Hao, and G. K. Karagiannidis. Hybrid lightwave/rf cooperative noma networks. *IEEE Transactions on Wireless Communications*, pages 1–1, 2019.
536. Yaoqiang Xiao, Ming Chen, Fan Li, Jin Tang, Yi Liu, and Lin Chen. PAPR reduction based on chaos combined with slm technique in optical ofdm im/dd system. *Optical Fiber Technology*, 21:81 – 86, 2015.
537. Xiaosong Tang and I. L. . Thng. An ns frequency-domain approach for continuous-time design of cap/icom waveform. *IEEE Transactions on Communications*, 52(12):2154–2164, Dec 2004.
538. S. Xie and C. Zhang. Code division multiple access based visible light communication in vehicle adaptive cruise control under emergency situation. In *2013 IEEE International Conference on Information and Automation (ICIA)*, pages 219–224, Aug 2013.
539. J. Xu, W. Xu, H. Zhang, and X. You. Asymmetrically reconstructed optical ofdm for visible light communications. *IEEE Photonics Journal*, 8(1):1–18, 2016.
540. R. Xu and Z. Xu. Led-based vehicular visible light ranging. In *ICC 2019 - 2019 IEEE International Conference on Communications (ICC)*, pages 1–6, 2019.
541. S. P. Yadav and S. C. Bera. Papr analysis of single carrier fdma system for uplink wireless transmission. In *2015 10th International Conference on Information, Communications and Signal Processing (ICICS)*, pages 1–5, Dec 2015.
542. Takaya Yamazato, Isamu Takai, Hiraku Okada, Toshiaki Fujii, Tomohiro Yendo, Shintaro Arai, Michinori Andoh, Tomohisa Harada, Keita Yasutomi, Keiichiro Kagawa, et al. Image-sensor-based visible light communication for automotive applications. *IEEE Communications Magazine*, 52(7):88–97, 2014.
543. F. Yang, J. Gao, and S. Liu. Priori aided compressed sensing-based clipping noise cancellation for aco-ofdm systems. *IEEE Photonics Technology Letters*, 28(19):2082–2085, 2016.
544. F. Yang, K. Yan, Q. Xie, and J. Song. Non-equiprobable apsk constellation labeling design for bicm systems. *IEEE Communications Letters*, 17(6):1276–1279, 2013.
545. Qiu Yang, Chen Hsiao-Hwa, and Meng Wei-Xiao. Channel modeling for visible light communications—a survey. *Wireless Communications and Mobile Computing*, 16(14):2016–2034.
546. R. Yang, X. Jin, M. Jin, and Z. Xu. Experimental investigation of optical ofdma for vehicular visible light communication. In *2017 European Conference on Optical Communication (ECOC)*, pages 1–3, 2017.

547. Z. Yang, M. Jiang, L. Zhang, and H. Tan. Enhanced multiple pulse position modulation aided reverse polarity optical ofdm system with extended dimming control. *IEEE Photonics Journal*, 10(3):1–17, June 2018.
548. Z. Yang, W. Xu, and Y. Li. Fair non-orthogonal multiple access for visible light communication downlinks. *IEEE Wireless Communications Letters*, 6(1):66–69, Feb 2017.
549. ZY Yang, S Yu, LQ Chen, J Zhou, YJ Qiao, and WY Gu. Qam accommodated double-side band fast ofdm based on idct. *Optics Express*, 21(26):32441–32449, 2013.
550. Yuan Yao, Lei Rao, and Xue Liu. Performance and reliability analysis of IEEE 802.11 p safety communication in a highway environment. *IEEE Transactions on Vehicular Technology*, 62(9):4198–4212, 2013.
551. Y. Yapıcı and İ. Güvenç. Noma for vlc downlink transmission with random receiver orientation. *IEEE Transactions on Communications*, 67(8):5558–5573, Aug 2019.
552. Yiling Zheng, Zaichen Zhang, Jian Dang, and Liang Wu. A novel receiver for filp-ofdm in optical wireless communication. In *2015 IEEE 16th International Conference on Communication Technology (ICCT)*, pages 620–625, 2015.
553. L. Yin, W. O. Popoola, X. Wu, and H. Haas. Performance evaluation of non-orthogonal multiple access in visible light communication. *IEEE Transactions on Communications*, 64(12):5162–5175, Dec 2016.
554. L. Yin, X. Wu, and H. Haas. On the performance of non-orthogonal multiple access in visible light communication. In *2015 IEEE 26th Annual International Symposium on Personal, Indoor, and Mobile Radio Communications (PIMRC)*, pages 1354–1359, Aug 2015.
555. L. Yin, X. Wu, and H. Haas. Sdma grouping in coordinated multi-point vlc systems. In *2015 IEEE Summer Topicals Meeting Series (SUM)*, pages 169–170, July 2015.
556. L. Yin, X. Wu, H. Haas, and L. Hanzo. Low-complexity sdma user-grouping for the comp-vlc downlink. In *2015 IEEE Global Communications Conference (GLOBECOM)*, pages 1–6, Dec 2015.
557. K. Ying, H. Qian, R. J. Baxley, and S. Yao. Joint optimization of precoder and equalizer in mimo vlc systems. *IEEE Journal on Selected Areas in Communications*, 33(9):1949–1958, Sep. 2015.
558. Atsuya Yokoi, Jaeseung Son, and Taehan Bae. Csk constellation in all color band combinations. *IEEE 802.15 contribution 15-11-0247-00-0007*, 2011.
559. Larry Yonge, Jose Abad, Kaywan Afkhamie, Lorenzo Guerrieri, Srinivas Katar, Hidayat Lioe, Pascal Pagani, Raffaele Riva, Daniel M Schneider, and Andreas Schwager. An overview of the homeplug av2 technology. *Journal of Electrical and Computer Engineering*, 2013.
560. S. Yu, O. Shih, H. Tsai, N. Wisitpongphan, and R. D. Roberts. Smart automotive lighting for vehicle safety. *IEEE Communications Magazine*, 51(12):50–59, 2013.
561. Z. Yu, R. J. Baxley, and G. T. Zhou. Multi-user miso broadcasting for indoor visible light communication. In *2013 IEEE International Conference on Acoustics, Speech and Signal Processing*, pages 4849–4853, 2013.
562. Z. Yu, R. J. Baxley, and G. T. Zhou. Iterative clipping for papr reduction in visible light ofdm communications. In *2014 IEEE Military Communications Conference*, pages 1681–1686, Oct 2014.
563. R. Zakaria and D. Le Ruyet. A novel filter-bank multicarrier scheme to mitigate the intrinsic interference: Application to mimo systems. *IEEE Transactions on Wireless Communications*, 11(3):1112–1123, March 2012.
564. Z. Zeng, M. D. Soltani, Y. Wang, X. Wu, and H. Haas. Realistic indoor hybrid wifi and ofdma-based lifi networks. *IEEE Transactions on Communications*, pages 1–1, 2020.

565. H. Zhang, Y. Yuan, and W. Xu. Papr reduction for dco-ofdm visible light communications via semidefinite relaxation. *IEEE Photonics Technology Letters*, 26(17):1718–1721, Sep. 2014.
566. Junwen Zhang, Jianjun Yu, Fan Li, Nan Chi, Ze Dong, and Xinying Li. 11× 5× 9.3 gb/s wdm-cap-pon based on optical single-side band multi-level multi-band carrier-less amplitude and phase modulation with direct detection. *Optics express*, 21(16):18842–18848, 2013.
567. Shuailong Zhang, Scott Watson, Jonathan JD McKendry, David Massoubre, Andrew Cogman, Erdan Gu, Robert K Henderson, Anthony E Kelly, and Martin D Dawson. 1.5 gbit/s multi-channel visible light communications using cmos-controlled gan-based leds. *Journal of Lightwave Technology*, 31(8):1211–1216, 2013.
568. T. Zhang, H. Ji, Z. Ghassemlooy, X. Tang, B. Lin, and S. Qiao. Spectrum-efficient triple-layer hybrid optical ofdm for im/dd-based optical wireless communications. *IEEE Access*, 8:10352–10362, 2020.
569. T. Zhang, Y. Zou, J. Sun, and S. Qiao. Design of pam-dmt-based hybrid optical ofdm for visible light communications. *IEEE Wireless Communications Letters*, 8(1):265–268, Feb 2019.
570. X. Zhang, Q. Gao, C. Gong, and Z. Xu. User grouping and power allocation for noma visible light communication multi-cell networks. *IEEE Communications Letters*, 21(4):777–780, April 2017.
571. J. Zhao and A. Ellis. Advantage of optical fast ofdm over ofdm in residual frequency offset compensation. *IEEE Photonics Technology Letters*, 24(24):2284–2287, Dec 2012.
572. Jian Zhao and Andrew D Ellis. Discrete-fourier transform based implementation for optical fast ofdm. In *36th European Conference and Exhibition on Optical Communication*, pages 1–3. IEEE, 2010.
573. Jian Zhao and Andrew D Ellis. A novel optical fast ofdm with reduced channel spacing equal to half of the symbol rate per carrier. In *Optical Fiber Communication (OFC), collocated National Fiber Optic Engineers Conference, 2010 Conference on (OFC/NFOEC)*, pages 1–3. IEEE, 2010.
574. Zhong Zheng, Te Chen, Lu Liu, and Weiwei Hu. Experimental demonstration of femtocell visible light communication system employing code division multiple access. In *2015 Optical Fiber Communications Conference and Exhibition (OFC)*, pages 1–3, March 2015.
575. J. Zhou, Y. Qiao, Z. Cai, and Y. Ji. Asymmetrically clipped optical fast ofdm based on discrete cosine transform for im/dd systems. *Journal of Lightwave Technology*, 33(9):1920–1927, May 2015.
576. J. Zhou, Y. Qiao, T. Zhang, E. Sun, M. Guo, Z. Zhang, X. Tang, and F. Xu. Fofdm based on discrete cosine transform for intensity-modulated and direct-detected systems. *Journal of Lightwave Technology*, 34(16):3717–3725, Aug 2016.
577. Ji Zhou and Yaojun Qiao. Low-papr asymmetrically clipped optical ofdm for intensity-modulation/direct-detection systems. *IEEE Photonics Journal*, 7(3):1–8, 2015.
578. X. Zhou, F. Yang, and J. Song. Novel transmit diversity scheme for tds-ofdm system with frequency-shift m-sequence padding. *IEEE Transactions on Broadcasting*, 58(2):317–324, 2012.
579. Z. Zhou, C. Chen, and M. Kavehrad. Impact analyses of high-order light reflections on indoor optical wireless channel model and calibration. *Journal of Lightwave Technology*, 32(10):2003–2011, May 2014.

580. Yuan Zhuang and Naser El-Sheimy. Tightly-coupled integration of wifi and mems sensors on handheld devices for indoor pedestrian navigation. *IEEE Sensors Journal*, 16(1):224–234, 2015.
581. Yuan Zhuang, Luchi Hua, Longning Qi, Jun Yang, Pan Cao, Yue Cao, Yongpeng Wu, John Thompson, and Harald Haas. A survey of positioning systems using visible led lights. *IEEE Communications Surveys & Tutorials*, 20(3):1963–1988, 2018.
582. Yuan Zhuang, Zainab Syed, You Li, and Naser El-Sheimy. Evaluation of two wifi positioning systems based on autonomous crowdsourcing of handheld devices for indoor navigation. *IEEE Transactions on Mobile Computing*, 15(8):1982–1995, 2015.
583. Yuan Zhuang, Jun Yang, You Li, Longning Qi, and Naser El-Sheimy. Smartphone-based indoor localization with bluetooth low energy beacons. *Sensors*, 16(5):596, 2016.
584. M. Zimmermann and K. Dostert. A multipath model for the powerline channel. *IEEE Transactions on Communications*, 50(4):553–559, 2002.
585. D. Zwillinger. Differential ppm has a higher throughput than ppm for the band-limited and average-power-limited optical channel. *IEEE Transactions on Information Theory*, 34(5):1269–1273, Sep 1988.

Index

Printed in the United States
by Baker & Taylor Publisher Services